PLANETARY SURFACE PROCESSES

Planetary Surface Processes is the first advanced textbook to cover the full range of geologic processes that shape the surfaces of planetary-scale bodies. This comprehensive introduction ranges from microscopic aspects of the soil on airless asteroids to the topography of super-Earth planets.

Using a modern, quantitative approach, this book reconsiders geologic processes outside the traditional terrestrial context. It highlights processes that are contingent upon Earth's unique circumstances and processes that are universal. For example, it shows explicitly that equations predicting the velocity of a river are dependent on gravity; traditional geomorphology textbooks fail to take this into account.

This textbook is a one-stop source of information on planetary surface processes, providing readers with the necessary background to interpret new data from NASA, ESA, and other space missions. Based on a course taught by the author at the University of Arizona for 25 years, it is aimed at advanced students, and is also an invaluable resource for researchers, professional planetary scientists, and space-mission engineers.

H. JAY MELOSH is Distinguished Professor of Earth and Atmospheric Science at Purdue University. His principal research interests are impact cratering, planetary tectonics, and the physics of earthquakes and landslides. He is a science team member of NASA's Deep Impact mission that successfully cratered comet Tempel 1 on July 4, 2005. Professor Melosh was awarded the Barringer Medal of the Meteoritical Society in 1999, the Gilbert prize of the Geological Society of America in 2001, the Hess Medal of the American Geophysical Union in 2008, and was elected to the US National Academy of Sciences in 2003. He has published over 170 technical papers, edited two books and is the author of *Impact Cratering: A Geologic Process* (1989, Oxford University Press). Asteroid #8216 was named "Melosh" in his honor.

T0296396

Cambridge Planetary Science

Series Editors: Fran Bagenal, David Jewitt, Carl Murray, Jim Bell, Ralph Lorenz, Francis Nimmo, Sara Russell

Books in the series

[†] Issued as a paperback

PLANETARY SURFACE PROCESSES

H. JAY MELOSH

CAMBRIDGE
UNIVERSITY PRESS

CAMBRIDGE
UNIVERSITY PRESS

University Printing House, Cambridge CB2 8BS, United Kingdom

Published in the United States of America by Cambridge University Press, New York

Cambridge University Press is part of the University of Cambridge.

It furthers the University's mission by disseminating knowledge in the pursuit of education, learning and research at the highest international levels of excellence.

www.cambridge.org
Information on this title: www.cambridge.org/9780521514187

First published 2011

A catalogue record for this publication is available from the British Library

Library of Congress Cataloguing in Publication data
Melosh, H. J.
Planetary surface processes / H. Jay Melosh.
p.   cm. – (Cambridge planetary science)
Includes bibliographical references and index.
ISBN 978-0-521-51418-7 (hardback)
1. Planets–Surfaces.   2. Geomorphology.   I. Title.   II. Series.
QB603.S95M45 2011
559.9′2–dc23
2011019269

ISBN 978-0-521-51418-7 Hardback

This book is dedicated to the students and colleagues who participated in my class PtyS 554, Planetary Surfaces, and PtyS 594, Planetary Field Geology Practicum, at the Lunar and Planetary Lab of the University of Arizona and at Caltech and Stony Brook before that. In the years stretching from 1976 to 2009 and from the classroom to campfires in unearthly landscapes under star-studded skies, we all learned together.

Contents

vii

Color plates appear between pages 236 and 237

Preface

We are privileged to be living in one of the greatest eras of exploration that humankind has ever undertaken. Our current Age of Space grew out of the dark struggles of World War II when large rockets were developed as agents of mass murder. The subsequent Cold War rivalry between the United States and Soviet Union pushed rocket capabilities to the point that it became possible to send vehicles into Earth orbit and beyond (even though the stated aim was to send missiles carrying nuclear weapons over mere continental distances). The Russians put the first human into Earth orbit. The Apollo missions took American astronauts to the Moon, a target that Russia reached first with its unmanned vehicles: Russia stopped just short of a manned lunar landing. Somehow, amid all this politically motivated grandstanding, a few visionary engineers and scientists accomplished the feat that will be remembered by all future generations: the exploration of our Solar System.

While humans have not yet traveled beyond the Moon, robotic spacecraft with increasingly sophisticated electronic brains and sensory systems have now left the bounds of the Solar System. Spacecraft have visited all of the major planets, with the exception of Pluto (although some now argue that it is not really a "major planet"). Many planets and even asteroids have been flown by, orbited and landed upon by spacecraft. We have yet to bring back samples of any body other than the Moon, comet Wild 2 and asteroid 25143 Itokawa, so there is much more to accomplish, but we are learning about the universe outside our little Earth at a tremendous rate.

When I first started teaching a course in Planetary Sciences in 1977 it was possible to treat each individual planet as a separate entity. Weeks would be spent talking about the Moon and its special attributes. Mars was a Moon-like disappointment after the flybys of Mariners 4, 6, and 7 that all, ironically, imaged nothing but the heavily cratered terrains of the southern highlands. Mariner 10 had just returned our first views of Mercury, which also turned out to be very much like the Moon. As time went on, we learned much more about the planets we had first studied and learned new things about planets that had never before been visited by spacecraft. Mars blossomed into a new world in the wake of the Mariner 9 orbiter, with giant volcanoes, canyons that dwarfed Arizona's Grand Canyon and channels that could only have been cut by gigantic floods. Pioneer Venus and the Soviet Veneras made it clear that our "sister" planet was a very odd relative indeed, and the Voyagers were off on their historic tours of the outer Solar System.

Before long it was clear that a planet-by-planet course organization could no longer work. It would be tediously repetitious to talk about craters and volcanoes on the Moon, then later to talk about craters and volcanoes on Mars, and then craters and volcanoes on Mercury, adding a new "craters and volcanoes" block for each new planet. While the number of planets kept multiplying, the number of different geologic processes did not. Pretty much the same processes, modified a bit for local conditions, act on every body we have investigated so far. So the modern course organization emphasizes processes, not individual planets. Furthermore, the body of information about each planet has multiplied to the point that it is no longer possible to comprehensively cover all that is known about even *one* planet within the confines of a one-semester class. If you doubt this, go to a library and look at the shelf of books about planets in just the University of Arizona's Space Science Series. The total collection occupies about two meters of shelf space, and it grows by a few tens of centimeters (or more!) every year.

This practical limitation accounts for the "process" orientation of this book. Beyond this, I had to make decisions about which processes to treat and in what order. Textbooks on terrestrial geomorphology abound and "process orientation" is a buzzword that most modern books respect, but fluvial processes dominate terrestrial geomorphology. Fluvial processes, however, are rare in the larger Universe and must take a back seat to more universal processes, such as impact cratering, in a planetary context. In teaching this class I have long used an approach that follows the planetary exploration mantra of "first flyby, then orbit, land, and finally return samples." I start with those aspects of a planet that you can see from the greatest distance, even telescopically (a level that we have just attained for extrasolar planets). Thus, we can ask: what determines a planet's shape and the topography of its surface? Deviations from a spheroidal shape must be supported by internal strength, which motivates a discussion of what strength is and how topographic variations can be supported.

If topography is limited by strength, then what happens when it is exceeded? The answer is tectonics: faults and fractures. As we approach ever closer to a planet, the next things we might notice are craters, just as the first features on Mars and Mercury imaged by spacecraft were cratered terrains. I had thus planned to make impact craters the subject of Chapter 5, followed by volcanism in Chapter 6. However, one of the anonymous reviewers of my original book proposal cogently argued that volcanism is most closely linked to tectonics so that the order of these two chapters should logically be reversed. I agree, and so the order is as you now have it – after all, the first things that Mariner 9 saw looming out of the global dust storm were the summits of Mars' four great shield volcanoes. The last five chapters are organized around the principle of most-to-least universal processes. All bodies have regoliths, although the regolith of airless bodies such as the Moon or asteroids differs profoundly from the agricultural soil of Earth. Regoliths do not need slopes to form, but mass movement is a process that acts only on slopes, so that is the subject of Chapter 8. Chapters 9, 10, and 11 are, in the broadest sense, about the processes that involve wind, water, and ice, even though the "wind" may be blowing carbon dioxide, the "water" liquid methane, and the "ice" solid carbon dioxide or methane. These chapters are really about transport by atmospheric gases (universal for large enough planets and moons), liquids (fewer bodies possess flowing liquids on their surfaces), and solids warm enough to flow at measurable rates (that is, very close to their melting points, which must be pretty unusual on a planet's surface).

In teaching this course I try to get through the entire set of processes in one semester (15 weeks of three hours of lecture per week). As my former students well know, I often do not succeed. New discoveries come up, someone asks a lot of deep questions about some topic, and I end up spending more time on one topic than the syllabus allows. The result is that I usually have to rush through the last sections. I have often said, "If only there were a text for this course, I could have the students read up on this topic and not miss out on an important idea." Well, here is the text. Maybe it will solve this problem.

Another note about how I teach this class: I typically assign challenging homework problems that are meant to encourage the students to think. There are sometimes no strictly right or wrong answers, just reasonable ones that admit of a lot of interpretation (there are also some easy problems that just involve substitutions, but I hope the answers are enlightening). I also ask the students to write a research paper on some topic that interests them, and I base much of the final grade on these research papers. These papers are about ten pages long and I encourage the students to think independently, not just regurgitate what they may have read in some published paper. New calculations or even small-scale experiments and field investigations are strongly encouraged. I do not penalize the students, gradewise, if some initially promising line of research does not work out. Many of these papers have turned into abstracts presented at the annual *Lunar and Planetary Research Conference*. Some have turned into papers published in the scientific journals and a few have become Ph.D. theses.

Because paper-writing becomes more intense as the semester proceeds, I ease off on the amount of homework assigned to allow the students time to explore their own ideas. This has often resulted in surprising bursts of creative activity that I do not wish to smother under too much "set work." For that reason you will find that the number and difficulty of the exercises associated with each chapter falls off toward the end of the book. I do this in the hope that the early part of the course will serve as a kind of "launch pad" for independent investigation of this fascinating field.

Anyone who teaches this subject must realize that planetary science is an active and ever-changing subject. New discoveries are constantly being made. I have tried to incorporate some of the latest discoveries in this text, but I fully realize that by the time this book appears in print some things I have written will be obsolete (indeed, in my own research I am doing my best to make that happen). So it is important to supplement this text with readings from the current literature and even news stories and NASA data releases.

Ah yes, one last piece of advice (and my former students would not forgive me if I failed to mention this!): The stories. Some of them are here in the book, cleaned up a bit and properly referenced. Not all of them (some of the good ones I was unable to verify in this way – they are in a file labeled "dubious stories" until I can find a reliable reference). Stories about people, about ideas, about what motivated whom to do what and how some great idea came from something that seemed wholly unrelated. Some of it is the usual scuttlebutt of science, told over coffee or around campfires. But most of my stories are different: Like Aesop's fables, they all have a moral. Like all teachers, I am often distressed by how little students seem to remember about some topic after the lapse of even one semester, let alone a few years. So I try to wrap the really important ideas into a really good story about someone or something. I think that makes the idea easier to remember and hope that the idea might remain mentally accessible long after the equation or intricate train of reasoning has passed beyond recall. I am not sure this works, but I do meet students who, after many years, still retain the story, if not the point that it was meant to illustrate. Not everyone who teaches this course will want to emulate this particular technique, but I do ask you not to drain the human interest from the science. Science is done by humans, and for humans to continue to do it they must realize how quirky and illogical the course of discovery can be.

   With that, I invite you to move on into this book and make your own discoveries. I hope you have as much fun learning this stuff as I have had.

October 2010
West Lafayette, Indiana

Acknowledgments

I am grateful, most of all, to the many students who participated in my classes and whose questions, answers to homework problems, responses to challenges, insightful research projects, and field trip presentations added immeasurably to my appreciation and knowledge of the surfaces of the various strange objects that inhabit our Solar System. Most of this work, particularly the field component of my classes, would have been impossible without the support of the Department of Planetary Sciences and the Lunar and Planetary Lab of the University of Arizona. There were some tense moments and challenging situations that developed in the field, but overall this support has been exemplary. I had long dreamed of distilling my class lectures into a book, while adding more material that I never had time to cover in a semester course, but it was Susan Francis, of Cambridge University Press, who finally persuaded me to take on this monumental task. The actual writing of this book has stretched over more years than I like to remember, during which time I received aid from a large number of people. Virgina Pasek drafted nearly all of the figures from my rough sketches. Her artistic ability and sense of graphic design shines through in every chapter, except 1 and 6. She has been a steady and patient collaborator and I am most grateful that she agreed to join me in this effort and persisted up to the last moment. As the book neared completion, Francis Nimmo read Chapters 3 and 4 and made numerous helpful suggestions that clarified the accuracy and precision of the material. Several former students contributed figures, particularly Jason Barnes, Eric Palmer, Ralph Lorenz, and Ingrid Daubar-Spitale. Steve Squyres contributed the beautiful cover image of Duck Bay, Mars, through several iterations of re-processing. I am also most grateful to my wife, Ellen Germann-Melosh, who has borne my preoccupation with writing on too many nights and weekends with patience and understanding. She is surely tired of hearing about "The Book" and I am delighted to finally send it forth to whatever fate awaits it in the larger world of science.

1

The grand tour

> The planets are no longer wandering lights in the evening sky. For centuries man lived in a universe which seemed safe and cozy – even tidy. The Earth was the cynosure of creation and Man the pinnacle of mortal life. But these quaint and comforting notions have not stood the test of time. … No longer does "the World" mean the Universe. We live on one world among an immensity of others.
>
> Carl Sagan (1970)

From the seventeenth to the middle of the nineteenth century it was customary for the scions of affluent British families to make a long tour of all the capitals of Europe to acquaint them with the architecture and culture of their larger world. In the twentieth century NASA planned a "grand tour" of our Solar System that would visit every planet outside the orbit of Mars. That tour never happened. Nevertheless, we have just about accomplished its goals, with the final New Horizons encounter with Pluto scheduled for 2015.

Scientific exploration of the Solar System can be said to have started around 1610, when Galileo Galilei (1564–1642) applied the newly invented telescope to investigate the world beyond the Earth. Telescopes have increased greatly in both size and sophistication since the days of Galileo, but even the best ground-based telescopes are unequal to the task of detailed exploration of the planets. The beginning of the Space Age, opening with the launch of Sputnik 1 in 1957, was the next leap forward in planetary exploration. Spacecraft, carrying instruments and humans, have greatly expanded our knowledge of the planets and moons around us.

The usual course of exploration of a planetary body, after remote astronomical observations, has been, first, flyby spacecraft, followed by orbiters, then landers and finally sample returns and human *in situ* visitation. The majority of bodies discussed in this book are still in the orbiter or flyby stage of exploration: Humans have returned multiple large samples only from our Moon, so far. In addition, very small amounts of material from comet Wild 2 and asteroid 25143 Itokawa have been returned by NASA's Stardust and Japan's Hayabusa missions, respectively. Nevertheless, a great deal has been learned about the other bodies orbiting our Sun. As this is written, the astronomical exploration of the planetary systems around other stars has been underway for about a decade.

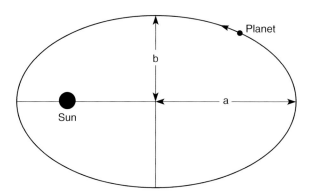

Figure 1.1 Planets orbit about the Sun in elliptical orbits that lie in a plane. The Sun lies at one focus of the ellipse. The size of the ellipse is defined by the semimajor axis a and the semiminor axis b.

1.1 Structure of the Solar System

Planets move around the Sun along paths that are, to a very good first approximation, elliptical (Figure 1.1). Johannes Kepler (1571–1630), using tables of planetary positions, deduced that the Sun lies at one focus of each planet's ellipse and that the square of the period P of each planet's orbit is proportional to the cube of its semimajor axis a, a relation now known as Kepler's third law (the first law is that the planets move in elliptical orbits with the Sun at one focus and the second is that, in moving along its orbit, planets sweep out equal areas in equal times). Isaac Newton (1642–1727) extended Kepler's empirical observations with his law of universal gravitation, writing that the period of a small body moving around a much larger body of mass M is given by:

$$P^2 = \frac{4\pi^2}{GM} a^3 \tag{1.1}$$

where G is Newton's gravitational constant, equal to $(6.67259 \pm 0.00085) \times 10^{-11}$ m^3/(kg s^2). The detailed position of a planet in its orbit is defined by six numbers, of which we shall be concerned with only three. The first is the semimajor axis a, which is conventionally expressed in Astronomical Units (AU), equal to the semimajor axis of the Earth's orbit around the Sun. One AU is approximately equal to 1.496×10^8 km. The second is the eccentricity e, defined by the ratio between the semimajor axis and semiminor axis b of the ellipse:

$$e = \sqrt{\frac{a^2 - b^2}{a^2}}. \tag{1.2}$$

The eccentricity is zero for a circular orbit and equal to one for a parabolic orbit. Eccentricities larger than one describe unbound, hyperbolic orbits. The last orbital parameter of interest to us is the inclination of the ellipse to a reference plane, conventionally chosen to be the plane of the Earth's orbit around the Sun, the ecliptic plane (another often-used reference plane is the invariable plane of the Solar System, which is mostly defined by Jupiter's orbit). The other three orbital parameters describe the orientation of the ellipse's axis in space (two numbers, the longitude of the ascending node and the argument of the perihelion) and the location of the planet in its orbit (the true anomaly).

1.1.1 Major facts of the Solar System

There are a number of important facts about the Solar System, not all of which are yet explained by models of Solar System formation, although most are at least consistent with the current solar nebula model of Solar System formation. Table 1.1 lists the orbital characteristics of the major planetary objects.

Planets and asteroids all formed at the same time. To the best of our ability to date them, the Earth, Moon, Mars, and the most ancient meteorites all have the same formation age, about 4.57 Gyr before the present. To less precision, the age of the Sun is also the same.

Planetary orbits are nearly coplanar. The planetary objects with the largest inclinations to the ecliptic plane are Mercury, inclined at 7.0°, Venus, inclined at 3.4°, and Saturn, inclined at 2.5°. Pluto, which is inclined at 17.1° is now believed to be a member of a family of similar *Trans-Neptunian Objects* or TNOs with similarly large inclinations (Eris, the largest TNO, is somewhat larger than Pluto and is inclined at 44° to the ecliptic).

Planetary orbits lie near the plane of the Sun's equator. The Sun's equator is inclined to the ecliptic at about 7.5°, so that its orbital momentum vector is similar in direction to those of the planets.

Planetary orbits are nearly circular. The three most eccentric planetary orbits are those of Mercury, $e = 0.2065$, Mars, $e = 0.0934$, and Saturn, $e = 0.0560$. Pluto, with $e = 0.2482$, seemed less like the other planets, until the discovery of many other similar TNO objects confirmed that it belongs to a different population of objects than the inner planets (Eris' eccentricity is 0.44, larger than Pluto's).

The planets all revolve around the Sun in the same direction. There are no exceptions to this rule, although many long-period comets do travel on retrograde orbits.

The rotation direction of most *planets is direct.* The obliquity (the inclination of the axis of rotation to the orbital plane) of most planets is small, with the following major exceptions: Venus, whose rotation is retrograde, is inclined at an angle of 177.3° to its orbital plane. Uranus is also retrograde, although with an obliquity of 97.86° it nearly lies on its side with respect to its orbit. Neptune is prograde, with an obliquity of 29.6°.

The spacing of planetary orbits follows a rough logarithmic series. Called the *Titus–Bode* law, this relation states that the semimajor axis of the *n*th planet, a_n, is given by:

$$a_n = c_1 + c_2 \, 2^n. \tag{1.3}$$

Table 1.1 *Planetary orbits*

Planet	Period (years)	Semimajor axis, a (AU)	Semimajor axis a by Titus–Bode	Eccentricity, e	Orbital inclination, i (degrees)
Mercury	0.241	0.387	0.55	0.2056	7.004
Venus	0.615	0.723	0.7	0.0068	3.394
Earth	1.000	1.000	1.0	0.0167	0.000
Mars	1.881	1.524	1.6	0.0934	1.850
Ceres	4.60	2.766	2.8	0.0739	10.585
Jupiter	11.862	5.203	5.2	0.0483	1.038
Saturn	29.458	9.539	10	0.0560	2.488
Uranus	84.01	19.191	19.6	0.0461	0.774
Neptune	164.79	30.061	*	0.0097	1.774
Pluto	248.54	39.529	38.8	0.2482	17.148
Eris	557	67.67	77.2	0.4418	44.187

* Advocates of the validity of this *law* frequently skip Neptune and list Pluto in its place, noting that the "agreement is better." Readers are encouraged to judge for themselves.

If the constants c_1 and c_2 are chosen to be 1 for Earth ($n = 3$) and 5.2 for Jupiter ($n = 6$), this rule states:

$$a_n = \frac{2}{5} + \left(\frac{3}{40}\right) 2^n. \tag{1.4}$$

Table 1.1 lists the predictions of this "law" and shows that it does agree roughly with observation. So far, there is no fundamental understanding of why this relation should be true.

1.1.2 Varieties of objects in the Solar System

Besides the major planets, many different types of object make up the Solar System. In order of decreasing size, these comprise the satellites of major planets: The larger ones in the Solar System include Ganymede and Callisto (satellites of Jupiter, 5262 and 4800 km diameter, respectively), Titan (satellite of Saturn, 5150 km diameter), Triton (satellite of Neptune, 2700 km diameter) and our Moon (3476 km diameter). Ganymede and Titan are larger than Mercury and, as bodies in themselves, can be classed as planetary objects. For the purposes of this book, we shall discuss these large satellites as varieties of planetary object and make no distinction between objects that orbit the Sun and those that orbit other planets.

In addition to satellites, we recognize asteroids, the largest of which is Ceres (diameter 950 km), but whose sizes range downward to a few kilometers, grading into objects that would be classed as meteoroids (there is no universally recognized size that divides asteroids from meteoroids: current authors set the dividing line anywhere between 1 km and a few meters). The total mass of all the asteroids is small, only about 4% of the Moon's mass, most of which resides in the largest asteroids, Ceres, Vesta, and Pallas. By definition an asteroid orbits the Sun and appears "star-like" in a telescopic image: That is, it does not display a "coma" or regularly emit gas and dust like a comet. Unfortunately for classifications, there are a few objects that do not possess comas but in every other respect are comet-like, while other objects long recognized as asteroids have suddenly acquired comas.

Comets are objects that, upon approaching the Sun, emit gas and dust to produce a *coma* (literally, "hair" in Latin: Comets are "hairy stars"). The diameters of cometary nuclei range from about a kilometer up to several tens of kilometers. They contain ices that, upon warming near the Sun, create their characteristic tails of gas and dust as the ices evaporate. There are several classes of cometary orbit, ranging from low inclination orbits typical of short-period comets to long-period comets that approach the Sun from the depths of the *Oort cloud* (which ranges out to a large fraction of the distance from the Sun to the nearest star).

In addition to these macroscopic objects, the Solar System also contains dust particles, whose diameter ranges down to submicron sizes, as well as individual atoms in the form of plasma and cosmic rays, most of which are emitted by the Sun, although a small component comes from interstellar space, as do some tiny dust particles.

1.2 Classification of the planets

The planetary-scale objects in our Solar System can be grouped into three general classes, with a number of important subclasses.

Terrestrial planets. Planets similar to the Earth are of most direct interest to us as inhabitants of such a planet. The main mass of these planets is composed of silicate minerals, although most have a metallic core rich in iron and nickel. Their densities range from 3000 to 6000 kg/m³ and they all have well-defined surfaces. Members of this class may or may not possess atmospheres. This class includes many satellites and asteroids as well as major planets. Examples are Mercury, Venus, Earth, our Moon, Mars, Io, and Ceres.

Icy satellites. Confined to the outer Solar System, this class consists of bodies mainly composed of water ice, with a possible component of other ices such as carbon dioxide, ammonia or, in the extreme outer Solar System, methane and nitrogen. These objects may have cores of silicate minerals that include hydrated silicates and, possibly, metallic inner cores. Densities of these objects range from 1000 to more than 2000 kg/m³.

Jovian (gas giant) planets. Jupiter and Saturn are composed of nearly the same materials as the Sun, mainly hydrogen plus helium and an admixture of heavier elements such as carbon and oxygen typical of "ices." These objects do not possess definite surfaces although they may have dense rocky or metallic cores. Their densities fall in the range of 700 to 2000 kg/m³. These planets are often divided into Jovian Planets (Jupiter and Saturn) and ice-rich Neptunian Planets (Uranus and Neptune), which contain a higher proportion of carbon and oxygen than the Sun.

1.2.1 Retention of planetary atmospheres

All of the large planets and the largest moons possess atmospheres. Planetary atmospheres are of great interest in themselves, but this book lacks space to discuss them: References to good sources of information on them are given in the *Further reading* section at the end of the chapter. However, the presence of an atmosphere strongly affects surface temperatures and, when present, permits a wide variety of surface processes to operate that would not be possible in its absence.

Whether or not a planetary object possesses a substantial atmosphere is dependent upon two main factors: The planet's escape velocity and the temperature of its exosphere. The escape velocity is the minimum speed required for a body initially on the surface to escape to infinity. If the mass of a planet is M and its radius is R, the escape velocity is given by:

$$v_{esc} = \sqrt{\frac{2GM}{R}} = \sqrt{2gR} \qquad (1.5)$$

where g is the surface acceleration of gravity. The second form of this relation can be used to compute the escape velocity of the objects listed in Table 1.2. The escape velocity of the Earth is 11.2 km/s, faster than almost any molecule near its surface (except hydrogen and helium) and so it retains a substantial atmosphere. The Moon, on the other hand, has an escape velocity of only 2.4 km/s and it has no permanent atmosphere.

The other factor in atmospheric escape is the temperature of the atmospheric gases. Kinetic theory tells us that the most probable velocity of a gas molecule of mass m at temperature T is given by:

$$v_T = \sqrt{\frac{2kT}{m}} \qquad (1.6)$$

where k is Boltzmann's constant. Note the dependence on the inverse molecular mass: Light molecules escape more easily than heavy ones, which accounts for the near absence of hydrogen and helium in the atmospheres of the terrestrial planets. There are two subtleties of this relation. The first is that, although Equation (1.6) is the most probable velocity, the velocity distribution has an exponential tail at high velocities and so a small fraction of gas molecules moves many times faster than v_T. Over geologic time a large fraction of the atmosphere may slowly leak away even though the most probable velocity is several times smaller than the escape velocity. The other subtlety is that the relevant temperature is not the surface temperature but the temperature high in the atmosphere, where the atmospheric gases are so thin that a molecule moving upward has a good chance of escaping into interplanetary space without colliding with another molecule. This portion of a planetary atmosphere is called the exosphere. Solar UV radiation tends to heat the upper reaches of planetary atmospheres to temperatures much higher than the surface, so that gases that might seem to be stable on the basis of the surface temperature can, in fact, leave the planet.

Table 1.2 *Physical data on terrestrial planets and major moons*

Planet or moon	Rotation period (days)	Equatorial radius (km)	Mean density (kg/m³)	Surface gravity (m/s²)	Inclination of equator to orbit (degrees)	Maximum surface temperature (K)
Mercury	58.6	2439	5430	2.78	2	700
Venus	243	6051	5250	8.60	177.3	735
Earth	1.00	6378	5520	9.78	23.45	311
Moon	28	1738	3340	1.62	6.68	396
Mars	1.03	3393	3950	3.72	25.19	293
Ceres	0.38	487	2080	0.27	~3	~239
Io	1.77	1821	3530	1.80	2.2	130
Europa	3.55	1569	3010	1.31	0.1	125
Ganymede	7.15	2634	1936	1.43	0–0.33	152
Callisto	16.69	2410	1834	1.24	0	165
Titan	15.95	2576	1880	1.35	0	94
Triton	5.88	1353	2061	0.78	0	38
Pluto	6.4	1153	2030	0.66	119.6	55
Eris	> 0.33 ?	1300	2250	~0.8	?	55

In addition to this thermal escape mechanism, other erosion processes such as impingement of the solar wind on atmospheres that are not defended by a planetary magnetic field, or ejection of atmospheric gases by impacts, may play important roles. One or both of these mechanisms may have depleted the Martian atmosphere over time. An early phase of strong UV emission by the newborn Sun may have ejected atmospheric gases from the early planets in a process known as hydrodynamic escape.

1.2.2 Geologic processes on the terrestrial planets and moons

The gamut of geologic processes that act on the surface of a planet or moon is the principal subject of this book. In subsequent chapters each process is considered in some detail. The chapters are arranged in approximate order of the universality of each process, ranging from processes that act on all planetary objects to processes that may affect only a few special bodies. Different processes act on different bodies, but if there is any overall organization to how different processes affect different planets (and there are many exceptions to any rule one might try to make), it is that planetary complexity increases with planetary size. Small bodies are home to only a limited variety of processes, while large planets are much more diverse.

Figure 1.2 is an attempt to capture this progression in a simple diagram that lists process as a function of planetary diameter, with a few examples added for definiteness.

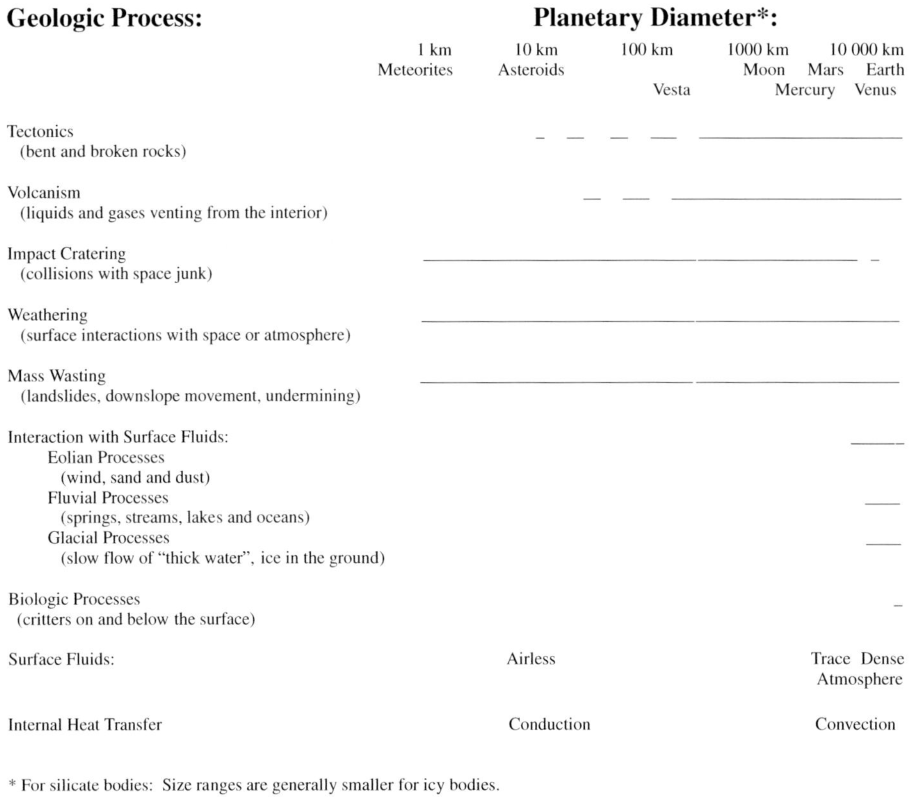

Geologic Process: **Planetary Diameter*:**

	1 km Meteorites	10 km Asteroids	100 km Vesta	1000 km Moon Mercury	10 000 km Mars Earth Venus
Tectonics (bent and broken rocks)		– — — —		────────	
Volcanism (liquids and gases venting from the interior)		– —		────────	
Impact Cratering (collisions with space junk)	───────────────────────────				–
Weathering (surface interactions with space or atmosphere)	────────────────────			────────	
Mass Wasting (landslides, downslope movement, undermining)	────────────────			────────	

Interaction with Surface Fluids:
  Eolian Processes
    (wind, sand and dust)
  Fluvial Processes
    (springs, streams, lakes and oceans)
  Glacial Processes
    (slow flow of "thick water", ice in the ground)
				──────
				
				────
				
				────	

Biologic Processes (critters on and below the surface)				–	
Surface Fluids:		Airless		Trace Dense Atmosphere	
Internal Heat Transfer		Conduction		Convection	

* For silicate bodies: Size ranges are generally smaller for icy bodies.

Figure 1.2 The activity of different geological processes is a function mainly of the size of the planetary body. The horizontal lines in this figure indicate the importance of each of the processes listed along the left side of the figure.

Impact cratering is probably the most universal process, although it is less important on large bodies than on small ones, mainly because other, more rapid, processes are more effective. Mass movement and surface modification of various kinds are similarly universal, although for small bodies both are strongly coupled with impact cratering. Tectonics, the process of rock deformation and fracture, is also important across the entire scale of sizes, but it is somewhat more effective on larger bodies where stresses can more easily approach the limit of rock strength. Volcanism might seem to be a mostly large-body process, yet evidence of melting has been found on even the smallest objects in the Solar System, remnants of an early era in which radioactive heat sources were more effective than now.

The cluster of processes that require active fluids on the surface of planets, including "wind" (the movement of any atmospheric gas), flowing "water" (which could be any liquid, such as methane on Titan), and "ice" (again, any highly viscous material near its melting point) are exclusively large-planet processes, because it requires a large body to hold an atmosphere. Finally, biologic processes are, so far as we know, confined to the largest and most diverse planet in the Solar System, Earth.

1.3 Planetary surfaces and history

Not all of the planets in the Solar System have well-defined surfaces. By surface, I mean a thin zone (thickness under a centimeter or less) across which physical and mechanical properties such as density, strength, sound speed, etc., show an abrupt change. The gas giant planets Jupiter, Saturn, Uranus, and Neptune probably do not show such abrupt changes and so can be considered to be without surfaces in this sense.

In this book we are mostly concerned with the surfaces of the terrestrial planets, silicate and icy moons. Much of what we will learn can also be applied to the surfaces of asteroids and comets. The surfaces of silicate-rich bodies are mostly composed of the refractory oxides of Si, Mg, Fe, Al, and Ca. These minerals have melting points far above ambient temperatures and so the physical state of the terrestrial planet's surfaces is solid: For most purposes we can treat the crust of these planets as a brittle elastic solid (something we loosely call *rock*). Because rocks do not deform until applied stresses exceed some yield point, it is difficult to remove all traces of past events that may have acted upon them. Rocks can record many aspects of the history of their planets. This fortunate aspect also makes planetary surfaces complex. A given planetary surface not only shows the effects of forces acting on it at the present time, but its structure also carries traces of forces that may have long ceased to exist. Similar considerations apply to the icy moons, so long as they remain well below the melting point of their ices. In the outer Solar System, beyond the orbit of Jupiter, temperatures are far below the melting point of water ice, and so water ice behaves in most respects like rock.

The gas giant planets are different in this respect from the solid planets: Although the atmospheres of the giant planets show complex structures, these structures are the consequence of presently acting forces: History is not a major player in their form. Even though the Great Red Spot on Jupiter has been visible for centuries, it would dissipate within hours if the vortex maintaining it were to die out.

The Earth's surface presents a minor exception to our emphasis on solid surfaces: Roughly ¾ of the Earth's surface is underlain by liquid water. The sea surface cannot maintain a record of the forces acting upon it for more than a few hours or days. The study of the sea surface is thus one of current or very recent forces acting upon the liquid: There is no paleontological aspect to sea surfaces. Of course, beneath the sea we find rocks that do maintain a record (although still of limited length due to plate tectonic recycling), so that we can learn much about the Earth's history by studying the sea floor.

Our emphasis in this book will be more on the response of rocky silicate or ice surfaces to applied forces than on the historical goal of using planetary surfaces to unravel the history of the planet. A clear understanding of these physical processes is a prerequisite to the interpretation of some particular surface. Moreover, we do not yet have sufficient data about most planets to make unambiguous statements about their history. Thus, at the present time, the most fruitful approach is the study of physical processes.

We begin our study of planetary surfaces with a review of the gross properties of the terrestrial planets and moons, and undertake a brief description of their present surfaces and what is known of their past. Table 1.2 lists some of the basic physical properties of a selection of Solar System objects with solid surfaces.

1.3.1 The Moon

Two easily recognizable terrain types dominate the surface of the Moon: The maria and the terrae (or highlands). The *maria* are dark (with a normal albedo of about 0.08), smooth plains that are found predominantly on the Moon's nearside. They are lightly cratered and are mantled with a layer of comminuted rock (the regolith) to a depth of a few meters. Lava flow fronts are observed in them, with scarps up to 100 m high (many times higher than typical for terrestrial lava flows). Samples returned by the Apollo astronauts show that they are composed of fine-grained basalt. The maria are volcanic plains formed by the extrusion of large volumes of highly fluid basalt. These flows may be equivalent to terrestrial flood basalts such as those that form the Columbia Plateau or the Deccan Traps. The mare basalts sampled so far were extruded over a ca. 700 Myr interval from about 3.2 to 3.9 Gyr ago. Some small central volcanic features are also observed, although there is little indication of silica-rich explosive volcanism.

The *terrae* represent an older surface than the mare. The typical terra is a rugged-looking hilly surface in which many circular, rimmed pits (impact craters) form at many different size scales. The terrae surfaces are typically bright (normal albedo about 0.16) and are best developed in the Moon's southern latitudes on the nearside. They are the predominant terrain type on the farside. The terrae are covered with a layer of broken rock tens of meters thick that may itself overlie a *megaregolith* of broken rock ten or more kilometers thick. All lunar samples that have been returned from terrae are breccias (that is, composed of broken angular rock fragments that may have been cemented by heat and pressure from impacts). The terrae may represent areas where the original crust of the Moon is exposed: They have, however, been subjected to such intense meteoritic bombardment that crater has obliterated crater to the point that no remnant of the original surface remains. The entire surface is in a "saturated" or "equilibrium" condition with respect to cratering. Although the terrae were also once composed of igneous rock, they are richer in Al than mare basalts. The composition of the terrae is often described as "anorthositic gabbro" (70% plagioclase, 20% orthopyroxene, 9% olivine, 1% ilmenite). Anorthositic gabbro is mafic (Fe and Mg rich and poor in SiO_2) compared to terrestrial granite.

Because the terrae all formed since the crust of the moon crystallized at about 4.6 Gyr, their high crater density, compared to that on the more lightly cratered mare, implies that an era with an especially high cratering rate must have occurred in the interval between 4.6 and 3.9 Gyr. A controversy is currently raging about whether this era of heavy bombardment was the tail end of a high cratering flux extending from the time of the Solar System's origin, about 4.6 Gyr ago, or whether it was part of a relatively short spike in the cratering rate, the "Late Heavy Bombardment." Current models of Solar System formation relate such a spike to a resonant interaction between Jupiter and Saturn that destabilized orbits in the asteroid belt and led to a shower of asteroids that affected all of the terrestrial planets, including the Moon.

Multiring basins dominate the regional geology of the Moon. The type example of such basins is Orientale, just over the Moon's western limb. Two or more inward-facing ring scarps surround its inner mare basin, each 2 to 8 km high and 600 to 900 km in diameter. Outside the rings a radially lineated and grooved terrain is composed of terra material (the Hevelius Formation). Associated with the lineated terrain are chains of secondary impact craters. These features, along with direct sampling of the lineated unit surrounding the Imbrium basin (the Fra Mauro Formation) imply that these basins were formed by very large impact events. Imbrium, with a ring diameter of 1340 km (nearly equal to the radius of the Moon itself) is the largest fresh basin. On the farside, the heavily degraded South Pole-Aitken basin is even larger, with a diameter of 2600 km and a depth of about 13 km. Grimaldi, the smallest multiring basin, is only about 200 km in diameter.

The surface of the Moon seems to be saturated with multiring basins (Figure 1.3), just as the terrae are saturated with craters. Every point on the Moon's surface is thus either inside an old multiringed basin or on the ejecta blanket of one. The surface of the Moon is thus layered to a depth perhaps exceeding 10 km, the stratification consisting of overlapping ejecta blankets from nearby basins. These basins are so large that the formation of each of them affected the entire Moon.

There is a widely held misconception (which has even prompted the writing of papers to "explain" it) that there are more large impact basins on the nearside than on the farside. Figure 1.3 should dispel this notion. It is not even true that the largest basin is on the nearside: South Pole-Aitken, on the southern farside, is nearly twice as large as Imbrium.

The source of this misconception is probably the very real asymmetry in maria between the nearside and farside. A multiring basin that has been flooded with basalt becomes a very striking circular feature, of which there are many on the nearside and essentially none on the farside. The coincidence between mare basalt flooding and a multiring basin also produces an anomalously large gravity field over the basin, a "mascon" that may locally increase the acceleration of gravity by 0.003 m/s^2 (or about 0.2% of the total field), comparable to the largest gravity anomalies observed on Earth.

Although there is no nearside–farside asymmetry of impact basins, there is a strong asymmetry in crustal thickness between the nearside and farside. The farside crust is almost twice as thick as that on the nearside. The major consequence of this asymmetry is the lack of dark maria on the farside, perhaps because of the difficulty lavas experienced in rising through the thicker crust. Another obvious consequence of this asymmetry is that the Moon's center of figure is displaced several kilometers away from its center of mass. This displacement, as might be expected, is away from the Earth, roughly along the line connecting the centers of the Earth and Moon.

In summary, the most common landforms on the moon are impact craters (Figure 1.4), circular rimmed pits that range in size from submicroscopic to multiring basins with diameters comparable to the Moon's own diameter. Each size range of crater has its own distinctive

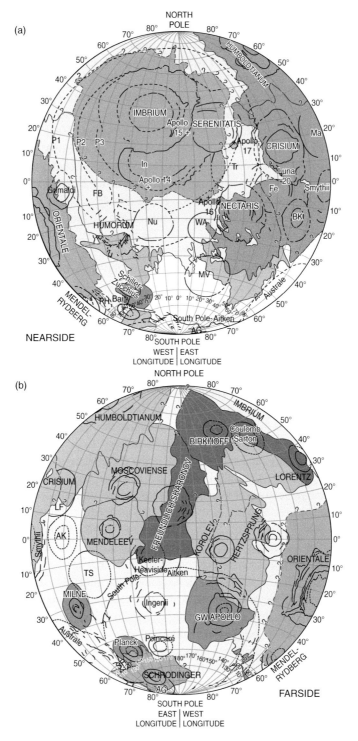

Figure 1.3 The major multiring basins of the Moon and the extent of their ejecta deposits are indicated. Curved lines indicate major rings. Panel (a) is the Moon's nearside and (b) is its farside. See color plate section for full detail. Blue indicates the deposits of the youngest (Imbrian) basins, yellow-orange Nectarian, dark brown Pre-Nectarian. After Plates 3A and 3B in Wilhelms (1987).

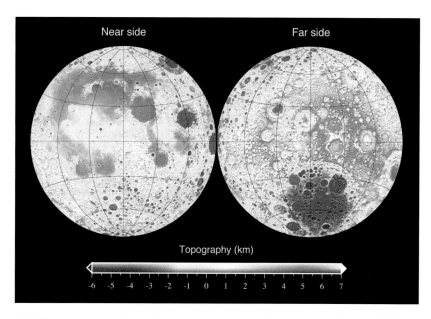

Figure 1.4 The topography of the Moon referenced to a sphere with a radius of 1737.4 km. Data were obtained from the Lunar Orbiter Laser Altimeter (LOLA) that was flown on the mission Lunar Reconnaissance Orbiter (LRO). The color-coded topography is displayed in two Lambert equal area images projected on the near and farside hemispheres, see color plate section. Courtesy Mark Wieczorek, August 7, 2010.

morphology and in each size range we see a gradation from "fresh," recently formed craters to *degraded*, barely recognizable battered remnants of the pristine form. While the large impact craters dominate the tectonics and stratigraphy of the lunar surface, the small to medium-size impacts make their contribution in forming the lunar soil or "regolith."

The regolith is a mantle of broken rock and glass which is 4 to 10 m thick on the mare and which covers the entire surface of the Moon. The regolith is slowly churned (*gardened*) by small meteorite impacts, which are also responsible for the bulk of lateral transport of material both on the plains and down slopes.

Except for the basaltic mare plains, there are few indications of internal tectonic activity on the Moon. The maria, especially the mascon basins, are cut by tectonic faults, with mainly extensional graben concentric to the basins and compressional mare ridges within the basins. These features are linked to subsidence of the Moon's lithosphere under the immense weight of the lava filling the basins. Lunar seismicity is weak by terrestrial standards, and slaved to alternating tidal stresses. Recent high-resolution observations by the Lunar Reconnaissance Orbiter do reveal a few small compressive thrust faults in the lunar highlands, suggesting a small amount of geologically recent global contraction. The major possible exception to this lack of tectonics is the lunar "grid," a preferential trend of lineations on the Moon's surface that may reflect ancient fracturing of the Moon's crust as the Moon receded from the Earth.

Figure 1.5 Oblique view of the surface of Mercury from the MESSENGER spacecraft showing volcanic flows of the intercrater plains. NASA/JPL/APL/Carnegie image PIA11773.

We can thus look upon the Moon as an essentially primitive body, chemically differentiated, but carrying a nearly unchanged record of the heavy meteorite bombardment that affected all of the planets inside the orbit of Jupiter. The Moon's surface has changed only slightly since about 3 Gyr ago, after the last lava flows had cooled. Only a few large young craters such as Copernicus (ca. 1 Gyr) and Tycho (ca. 100 Myr) were added to complete the present picture. Interestingly, the Moon's surface is only slightly younger than the oldest rocks preserved on the Earth's surface, so that the Moon preserves a record of events for which we have no comparable information in terrestrial history.

1.3.2 Mercury

Upon first glance, Mercury appears to be very Moon-like (Figure 1.5). Although its mean density of 5430 kg/m^3 indicates that it is internally different from the Moon (bulk density 3340 kg/m^3), the surface seems nearly indistinguishable from that of the Moon. Mercury's dominant landform is impact craters, ranging from the limit of resolution (a few 100 m at present) to the 1540 km diameter Caloris Basin.

Mercury possesses mare-like plains that are probably volcanic in origin, although Mercury's mare are not dark like the Moon's. The crater density on Mercury, however, is lower than that of the lunar highlands. There are extensive light intercrater plains separating individual large craters.

Unlike the Moon, Mercury shows some signs of internal tectonic activity early in its history. Long thrust fault scarps, up to a few kilometers high, wind hundreds of kilometers across the surface of the planet, cutting craters and plains alike. These scarps may have formed as Mercury's interior cooled and contracted after an elastic lithosphere developed. Recent images from NASA's MESSENGER mission flybys reveal at least one volcanic center located on the periphery of the large, volcanically flooded Caloris Basin.

Mercury's rotation is locked into a resonance with its orbital period: It orbits the Sun in 88 days as it rotates on its axis once every 58.6 days, in a 3:2 resonance that appears stably maintained by mass anomalies within the body of the planet. This resonance coordinates with its highly elliptical orbit so that one point on its equator (the *hot pole*) alternately faces toward and away from the Sun at perihelion. The *warm poles*, 90° away from these scorching locations (reaching about 700 K), are slightly more temperate but still very hot by terrestrial standards. On the other hand, radar reflections seem to indicate the presence of large masses of ice in permanently shadowed craters near Mercury's north pole.

In spite of its slow rotation Mercury possesses a small magnetic field, suggesting that its core is still molten and convecting. The magnetic field prevents the solar wind from impinging directly on its surface, which may have important consequences for regolith processes that are still not fully understood. It is unclear why Mercury should have such a large metallic core: It makes up a much larger fraction of Mercury's total mass than the core of any other terrestrial planet. Perhaps a large, nearly head-on collision early in Solar System history ejected a large fraction of its silicate mantle. In spite of its large iron-rich core, remote chemical measurements of Mercury's surface rocks show a surprising deficiency of iron oxide, which may account for its uniformly bright surface compared to the Moon.

The MESSENGER spacecraft is about to enter its orbital phase as this book goes to press, at which time a great increase in our understanding of Mercury's geology is anticipated. Readers should watch the news and subsequent papers for more information on this surprisingly interesting body.

1.3.3 Venus

Although Venus, of all the terrestrial planets, is the most similar in size to the Earth, it is most emphatically *not* Earth's twin. Venus' surface is searingly hot, baking at a nearly constant temperature of 730 K, hotter than the surface of Mercury. It is crushed under a CO_2 atmosphere 100 times more massive than the Earth's, which maintains the high surface temperature through a runaway greenhouse effect. Although wind velocities in this massive atmosphere are low, aeolian dune fields and wind streaks are abundant. Venus may once have possessed oceans, but it has lost its water through dissociation and preferential loss of hydrogen to space.

Venus' dense atmosphere supports a thick layer of high clouds composed mainly of a very stable sulfuric acid aerosol that reflects 67% of the incident sunlight and completely obscures the surface from visual imaging. Only radar from Earth-based or spacecraft platforms has finally revealed the details of its surface. Venus' rotation is retrograde, with a period of 243 days. This period puts is very close to a near-resonance with the Earth, such that Venus shows the same face to the Earth whenever conjunction occurs, which happens every five Venusian days.

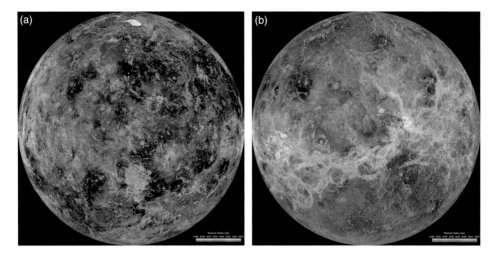

Figure 1.6 Topographic elevations from the Magellan radar altimeter. Panel (a) is centered on 0° Longitude, panel (b) is centered on 180°. The surface of Venus is occupied by seemingly randomly spaced rises and plains with a few highlands such as the Lakshmi Plateau in panel (a), near Venus' north pole. Maxwell Montes, the highest point on Venus, rises above the plateau. Note the extensive chain of circular coronae extending across the lower half of panel (b). This chain ends in the large incomplete circle of Artemis Chasma. Very few impact craters are visible. Panel (a) is NASA/JPL/ USGS PIA00157 and panel (b) is PIA00159. See also color plate section.

Venus does not lose internal heat through plate tectonics, but may suffer episodic global spasms of volcanic resurfacing. The most recent such event obliterated any pre-existing surface of the planet about 700 Myr ago, covering the entire surface with thick sheets of fluid basaltic lava. Large volcanic/tectonic complexes called coronae are the most characteristic feature (Figure 1.6), along with central volcanoes, folded mountain belts and extensional plains. Recent evidence from ESA's Venus Express mission suggests that volcanism may be currently active. The surface is warped and broken by myriads of faults of all varieties: Venus is a tectonic paradise. A small number of craters (about 940) punctuate the surface, the largest of which, Mead, is about 280 km in diameter: There is no sign of the ancient large basins that sculpture the crust of the Moon, Mercury or Mars.

The most surprising fact about Venus is that it possesses substantial topography: Its highest mountain, Maxwell Montes, rises 11 km above its mean elevation. It is a major puzzle, how rocks at the high temperatures of the Venusian surface can support such high topography for geologic periods. Perhaps a total absence of water in the Venusian crust adds an unearthly degree of strength to the near-surface material.

1.3.4 Mars

Mars has long been the center of attention of spaceflight enthusiasts and is still considered one of the most likely planets, other than the Earth, to harbor life. It appears Earth-like

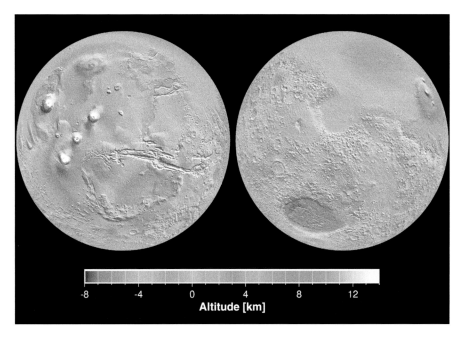

Figure 1.7 False color topography of Mars from the MOLA instrument aboard the Mars Global Surveyor spacecraft. The left hemisphere is dominated by the Tharsis Rise with its enormous volcanoes. Olympus Mons rises to the upper left. The gigantic trough of Valles Marineris extends to the right center. The northern lowlands and the Borealis plains dominate the upper half of the right hemisphere. The deep circular basin to the lower left is Hellas and the smaller basin near the center is Utopia. NASA/JPL image PIA02820. See also color plate section.

in its possession of an atmosphere (albeit very thin, with a surface pressure only 0.6% of Earth's and composed almost entirely of CO_2), polar caps, and clouds. Although Mars does not presently possess a magnetic field, detection of a magnetized ancient crust suggests that it once did.

Like the Moon, the Martian crust is divided into asymmetric halves: The crust beneath the northern plains is thinner than that beneath the southern highlands: A relatively narrow escarpment separates the two. Mars is heavily cratered. Its surface is dominated by several large impact basins, the largest of which, Hellas, is 2300 km in diameter and 7 km deep (Figure 1.7).

Mars stands as an intermediate between Moon-like bodies and the Earth. In addition to heavily cratered surfaces that presumably date from the time of heavy bombardment, Mars has been resurfaced by much younger lava flows and shows abundant tectonic activity. Olympus Mons, the largest volcano in the Solar System, sits astride the crustal dichotomy between the northern lowlands and southern highlands, as does the large volcanic Tharsis Rise that supports another three major volcanoes. The Tharsis volcanic system dominates the volcanic and tectonic framework of the entire planet. The grand canyon of Mars, Valles Marineris, is radial to the oval Tharsis uplift.

Figure 1.8 The Galilean satellites of Jupiter as imaged by the Galileo spacecraft. In order from left to right are Io, Europa, Ganymede, and Callisto. Volcanoes dominate Io, Europa is covered with an ice shell, Ganymede's surface is a patchwork of bright young and dark old terrain, and Callisto is an undifferentiated mixture of ice and rock. NASA/JPL/DLR image PIA01400. See also color plate section.

The most intriguing aspect of Mars is evidence of fluvial activity on its surface. Ancient valley networks appear to be fossil stream valleys, some of which may have been formed by rain running off the surface. Enormous channels cutting thousands of kilometers across the surface imply huge catastrophic floods in the distant past, while recent gullies may indicate contemporary fluvial activity. Features suggestive of shorelines now seem to line up at the same elevation, perhaps indicating an ancient global ocean, and glacial landforms suggest that extensive icecaps once covered parts of the southern highlands.

Evidence for water now appears to be everywhere: It is frozen in the polar caps and in permafrost beneath the high-latitude plains. Fluvially deposited sediments discovered by the MER rovers suggest shallow lakes on the surface, although concretions in the same sediments indicate highly acidic sulfur-rich waters. The puzzle in all of these observations is that liquid water is not presently stable on the surface of Mars: Not only is its surface too cold for liquid water, the present atmospheric pressure is below the triple point of water so that liquid water is never stable, except perhaps on hot summer days at the bottom of the Hellas Basin. If Mars did once possess oceans and icecaps, where did all that water go? And if Mars was once warm and wet, how were those high temperatures maintained?

Mars today is a cold, dry planet whose surface is dominated by aeolian activity, but the evidence for a very different climate in its past grows stronger with every new investigation. How this early climate was maintained and what happened to bring Mars to its current inhospitable state are major questions that remain to be answered.

1.3.5 Jupiter's Galilean satellites

Although Jupiter possesses many satellites, the four large ones first discovered by Galileo are the most interesting from a planetological perspective. Although satellites, they are comparable in size to Mercury and can be treated as individual worlds (Figure 1.8). They are also surprisingly dissimilar, although their peculiarities seem to be strongly correlated with their distance from Jupiter. All of these satellites are tidally locked to Jupiter and they, thus, rotate on their axes with the same period as they orbit about Jupiter. Their orbits are also regularly spaced and they interact strongly with each other.

Io, the innermost Galilean satellite, has a density suggestive of a silicate planet. It has a small orbital eccentricity, which is maintained by gravitational interactions with Europa. Because of this eccentricity its distance from Jupiter alternates with every orbit, flexing Io by the alternating tidal stresses. This continual flexing dissipates large amounts of heat in Io, whose surface heat flow is approximately 25 times larger than the Earth's. Io is the most volcanically active body in the Solar System. Its surface is mantled with volcanically extruded sulfur and sulfur dioxide, producing a wide range of lurid yellow and orange colors dotted with occasional black pools of molten sulfur. No impact craters are known anywhere on Io, whose surface is renewed at the rate of about 1 cm per year. Instead, volcanic calderas dot the surface, from which flows of sulfur-rich and silicate lavas proceed over the surface while plumes of sulfur dioxide spray more than 100 km upward into space before raining back onto the surface. The ultimate source of this volcanism appears to be exceedingly hot silicate lavas that mobilize the more volatile sulfur compounds.

In addition to the volcanic calderas, Io also possesses about one hundred mountains, the tallest of which towers 17 km above the surface. Sulfur is not strong enough to support such high topography, so these mountain masses indicate the presence of stronger silicate rocks in Io's thin lithosphere.

Europa is also a mostly silicate body, but its surface is covered by water ice, not rock or sulfur, although some albedo variations on the surface may be caused by sulfur implanted from Jupiter's magnetosphere. Europa has a few large impact craters, but its principal surface features are ridges and long curving fracture systems of several types. Like Io, Europa derives its internal heat mainly from tidal flexing, but it is almost twice as far from Jupiter as Io and its tidal heating is much less intense.

In addition to the ubiquitous double ridges, which reach heights of a few hundred meters and widths of a few kilometers, a substantial fraction of Europa's surface is pocked with irregular shallow depressions known as chaos regions, where the crust has apparently been disrupted and broken into polygonal blocks that were once mobile. The crust of Europa is believed to be mostly water ice that floats on a briny water ocean that may be 100 km deep. The thickness of the ice shell is currently controversial: Estimates of its thickness range from about 20 km to as little as a few kilometers. Topographic elevations or depressions seldom exceed 1 km anywhere on Europa.

The small number of craters on its surface indicates an average age of about 60 Myr for the surface, which must thus be continually renewed by internal activity. Europa, possessing a deep water ocean beneath a thin ice shell, has fueled speculation about its potential for supporting life, which may be nurtured by warm hydrothermal plumes ejected from the silicate floor of the subsurface ocean.

Ganymede is the next outer Galilean satellite. It circles Jupiter with a period twice that of Europa. Its density of 1940 kg/m^3 implies a body composed of about 60% rock and 40% ice, which is substantially more ice-rich than Europa. Ganymede possesses two principal terrain types: dark terrain that is heavily cratered and apparently very old, and light terrain that is much younger. Ganymede appears to be differentiated and may also have a small metallic core, as it surprisingly possesses a small magnetic field of its own. It also shows an induced magnetic field that suggests the presence of a thin liquid ocean buried some 170 km below the surface.

Craters on Ganymede show both fresh and viscously relaxed morphologies. A crater type endogenous to Ganymede is the palimpsest crater, a relaxed circular form that may either have been completely relaxed by viscous flow in relatively warm ice, or have formed in a fluid, slushy surface. Central pit craters are also abundant on Ganymede and Callisto. The younger bright regions are traversed by several generations of kilometer-wide furrows and grooves that seem to be extensional in origin, because all of the recognizable faults are extensional normal types. A substantial population of small craters overlies even the bright terrain, so these surfaces are all older than the surface of Europa. Although evidence for tectonic resurfacing is abundant on Ganymede, there is little sign of volcanic activity, although some smooth low areas and circular pits called paterae may be of cryovolcanic origin. Evidence for small-scale mass wasting and sublimation is abundant. Ganymede also possesses polar caps that extend down to latitudes near 40° in both hemispheres. These polar caps are evidence for the mobility of some water ice on the surface.

Callisto is the outermost Galilean satellite. Unlike the other three, Callisto does not appear to have completely differentiated into a crust, mantle and core. Tidal heating is not important and its surface seems to date from ancient times. Callisto is mainly notable for a number of very large multiring basins in which a central bright patch containing the crater and its ejecta blanket is surrounded by dozens of rings. Valhalla, 1800 km in diameter, is the type example of this class. Callisto's low density suggests that it is mainly composed of ice, with only a small rocky component, although its dark surface seems to be largely mantled by hydrated silicate dust.

Callisto apparently lacks tectonic features other than those associated with the multiring basins. There is also no sign of volcanism. At high resolution, however, its surface shows a bizarre alternation of bright, steep-sided peaks and mesas and dark, level plains that is attributed to the dominance of sublimation and mass wasting of the mixture of silicate dust and ice.

1.3.6 Titan

Titan, Saturn's largest satellite, has rapidly become one of the most interesting objects in the outer Solar System. It has long been known that, unique among moons, it possesses a dense atmosphere (Triton, the large satellite of Neptune, does have a thin N_2 atmosphere). Titan's atmosphere is mostly composed of N_2 with a small admixture of methane. The pressure at its surface, however, is about 1.6 times larger than the Earth's. Titan's upper atmosphere is filled with organic aerosols that obscure the surface almost completely at visible wavelengths, although infrared wavelengths do penetrate to the surface and reveal broad surface details (Figure 1.9).

Figure 1.9 Global view of Titan's surface from the VIMS spectrometer aboard the Cassini spacecraft. This false-color composite is constructed from three wavelengths in the infrared that penetrate Titan's hazy atmosphere (1.3 μm is shown in blue, 2 in green and 5 in red). The dark region in the center of the image is named Xanadu and may be the site of a large ancient impact. NASA/JPL/University of Arizona image PIA09034. See also color plate section.

Titan's density indicates a large admixture of ice in its composition and the surface is mainly composed of water ice, which at Titan's surface temperature of 94 K is as hard and strong as silicate rock on the surface of the terrestrial planets. Methane can exist as either a gas in the atmosphere or as a liquid on the surface, so Titan possesses a "hydrologic" cycle based on methane. It is apparent that methane rain occasionally beats down on its surface. Broad, shallow methane lakes accumulate in low areas near its poles. Radar images from the Cassini spacecraft show an eerily Earth-like landscape with lakes, river valleys, and mountain ranges. Images from the Huygens lander could easily be mistaken for a stream-bed on the Earth.

Titan does have a few large impact craters, but small craters are nearly absent: Like the Earth, erosional processes on Titan are active enough to remove most craters. Titan possesses an extensive equatorial belt of "sand" dunes that resemble the Libyan sand seas of the Sahara.

There are many puzzles concerning Titan, such as what kind of weathering processes act to disintegrate its frigid ice bedrock and how active are the methane rivers that have carved out deep valleys into its crust. Titan provides many fine examples of how the same processes can act on vastly different materials to produce similar landforms. The abundant organic chemistry acting in its atmosphere, lakes, and below its surface also offers an

Figure 1.10 Topographic map of the Earth from NOAA. ETOPO1 is a 1 arc-minute global relief model of Earth's surface that integrates land topography and ocean bathymetry. It was built from numerous global and regional data sets (Amante and Eakins, 2009). See also color plate section.

attractive laboratory for the study of whether some kind of carbon-based life could have gotten a start in an environment so different from our familiar Earth.

1.3.7 The Earth

The Earth presents the opposite extreme to the Moon and rounds out our brief tour of the planets (home is always the last stop on any long journey). It is the largest terrestrial planet. Probably as a result of its size, the Earth cannot rid itself of heat generated by the decay of long-lived radioactive elements (chiefly U and Th at the present time) by means of conduction alone. Only the convective transfer of heat by currents deep in the Earth's interior can cool the planet effectively. The surface of the Earth is profoundly affected by this internal convection and a slow (1–10 cm/yr) relative motion of its large surface structures is the result. Thus, the Earth's surface structure is largely generated by internal tectonics.

The process of plate tectonics determines the large-scale structures of the Earth (Figure 1.10). The surface of the Earth is divided into about a dozen large, semi-rigid plates that slowly glide over hotter, more fluid rock at depths of about 100 km. New basaltic crust is created at mid-ocean spreading centers. This crust is only 5 to 10 km thick, compared to the continental rafts that range from 30 to 70 km thick. The dense basaltic crust stands low, an average of about 4 km below the level of the continents that are composed of less dense silica-rich rocks. Earth's free surface water fills these low basins and, thus, exposes a large area of mostly dry land to the atmosphere above.

The oceanic crust is almost neutrally buoyant, so it is easily created and later reabsorbed in subduction zones where it sinks back into the mantle. In the process of being recycled, however, it partially melts and differentiates, creating long lines of volcanoes that rise over the subducting crust. The continental crust is too buoyant to subduct to great depths. It thus remains near the surface and has accumulated a history nearly as long as the Earth's own existence, although it is continually fragmented and the pieces rejoined to create a patchwork of amazing and often baffling complexity.

Linear mountain ranges rise where continental fragments collide, while erosion quickly reduces their elevations (at rates averaging about 10 cm/1000 yr) and distributes their material to lower elevations. Geologists divide continental rafts into young mountain ranges at the sites of recent slow collisions (tens to 100 Myr), platforms of middle age (less then about 1 Gyr), and ancient shields (more than 1 Gyr) that have not suffered fragmentation or collision during this long interval. Old continents that have escaped breakup may, in fact, possess cold and strong roots that resist fragmentation and thus preserve such undisturbed regions intact for long intervals.

This continental process of erosion, deposition and re-collision to raise new mountains led early geological theorists like James Hutton (1726–1797) to posit a universal rock cycle in which the rocks of the Earth are eroded, deposited as sediments, metamorphosed after burial, melted to form igneous rocks, uplifted, then eroded again to continue the cycle. This rock cycle has been one of the foundations of terrestrial geology, but it now appears to be a uniquely terrestrial process, one that requires both plate tectonics and active fluvial erosion. There is little hint of such a cycle on any other planet in the Solar System (although Io may undergo a weird variant of crustal recycling).

Free water is present on the Earth's surface under conditions where it can change phase freely from solid to liquid to gas. This creates a hydrologic cycle that can readily transport solid silicate debris from place to place. Even more damaging to the integrity of surface rock, life has produced an atmosphere of which 20% consists of a toxic and corrosive gas – oxygen (even living systems go to great lengths in their internal biochemistry to avoid being damaged by this reactive substance). Oxygen, in conjunction with water, can corrode and disintegrate nearly every silicate, except quartz, which is otherwise stable in the crust, leading to rapid degradation (weathering) of rocks that are brought to the surface. Thus, on Earth, probably more so than on any other planet, crustal structures are weathered and eroded by atmospheric processes, so further affecting the surface morphology and engendering relatively rapid changes of the landscape.

It is, thus, no surprise that the Earth's crust contains essentially no trace of the early meteoritic bombardment to which the Moon was subjected. Indeed, 20 km craters only a few tens of millions of years old are obliterated almost beyond recognition unless special circumstances preserve them. The giant impact basins that form the most ancient structures on most of the other planets have vanished from the Earth without leaving any presently recognized record.

The study of the Earth's surface therefore requires a wholly different orientation from the study of the Moon's surface. On Earth, geologists are primarily concerned with internally generated tectonics and its modification by weathering and erosion. The effect of external agencies is so slight that for many years geologists refused to accept the opposite situation on the Moon and other planets.

Because the surface of the Earth is treated at great length in other books on geology and geomorphology, in this book we emphasize those processes which are either universal, acting on all the planets including the Earth, or those that hardly affect the Earth at all, such as impact cratering. Compared with other texts this one says little about fluvial processes or even plate tectonics. On the other hand, as Carl Sagan said, the Earth is not the center of the Universe and its geology is not that of our neighboring planets.

Further reading

Readers wishing to learn more about orbital mechanics and its application to the Solar System should consult the book by Bate *et al.* (1971), which is my personal favorite for learning orbital mechanics in spite of its use of English units, or the comprehensive treatise by Murray and Dermott (1999). The best general overview of the planetary system is still the excellent collection of topical articles in Beatty *et al.* (1999). The subject of planetary atmospheres has lacked an up-to-date text for many years. However, a recent, well-reviewed book on this subject has just appeared (Sanchez-Levega, 2010).

The individual planets of the Solar System are well treated in the many volumes of the University of Arizona's Space Science Series. The series includes a book about Mercury, but it is badly out of date. Instead, I refer you to Robert Strom's summary article of what was learned from the Mariner 10 mission (Strom, 1984) or his popular but authoritative book (Strom and Sprague, 2003). The MESSENGER mission will shortly make anything now in print out of date. For Venus, see the collection of papers in Bougher *et al.* (1997). There are many geology books that describe the Earth and its tectonics. The latest edition of an old, reliable standard that has gone through many editions is Grotzinger *et al.* (2006a). A personal favorite of mine (I often recommend this as an introduction for non-geologists) is the visually oriented book *Geology Illustrated* (Shelton, 1966). Michael Carr has been writing the best summaries of Martian geology for years. His latest effort is Carr (2006). Asteroidal bodies are treated well in Bottke *et al.* (2002). The Jovian system after the Galileo mission is well summarized in Bagenal *et al.* (2004). Titan is the subject of a book that just appeared (Brown *et al.*, 2009), although Cassini's investigation of the Saturnian system is ongoing. In the outer Solar System, Neptune (Cruikshank, 1995) and Uranus (Bergstralh *et al.*, 1991) have not been visited since Voyager and so there is little change to the summary articles on each in the books cited. Pluto (Stern and Tholen, 1997) will still be something of an enigma until the New Horizons spacecraft visits that system in 2015. Trans-Neptunian objects have been recently reviewed (Cruikshank and Morbidelli, 2008), as have comets (Festou *et al.*, 2004).

2

The shapes of planets and moons

> The equal gravitation of the parts on all sides would give a spherical figure to the planets, if it was not for their diurnal revolution in a circle ….
>
> I. Newton, *Principia*, Theorem XVI

Modern space exploration has made everyone familiar with the idea that planets are mostly spherical. From a great distance a casual observer might not even notice that rotating planets and moons are not quite perfect spheres. However, careful examination reveals departures from perfection. Rotating planets are slightly oblate spheres, while tidally locked satellites are triaxial. Furthermore, once these bodies are approached closely, it becomes clear that nearly every planet and moon possesses topographic variations. Mountains, valleys, plains, and craters create landscapes that, up close, can challenge attempts to traverse them by mechanical rovers or human explorers.

The forces that create and maintain the topography of planetary bodies depend on the scale of the feature. The gravitational self-attraction that tends to make planets spherical operates differently on the scale of individual mountains. It is thus useful to distinguish several *orders of relief* that categorize different scales of topographic feature. This notion can be made mathematically precise through the use of spherical harmonics, a concept that will be discussed later in this chapter.

The tendency of large masses of material to take on a spherical shape was first recognized by Isaac Newton (1643–1727) in 1686. His brilliant insight into universal gravitation showed that, in the absence of other forces, the attraction of matter for other matter tends to mould all bodies into spheres. Gravity is weak compared to other forces so, on a human scale, bodies must be very large for gravity to dominate the electromagnetic forces that give atomic matter its strength to resist deformation. The conflict between strength and gravitation is the subject of Chapter 3. This chapter concentrates on the largest-scale features of planetary topography and its geometric properties.

Table 2.1 *Physical characteristics of large bodies in the Solar System*

Name	Equatorial radius (km)	Mass (kg)	Mean density (kg/m^3)	Equatorial acceleration of gravity (m/s^2)	Sidereal rotation period
Mercury	2 439	3.303 x 10^{23}	5430	2.78	58.75 days
Venus	6 051	4.870 x 10^{24}	5250	8.60	243.01 days
Earth	6 378	5.976 x 10^{24}	5520	9.78	1.00 days
Moon	1 738	7.349 x 10^{22}	3340	1.62	27.322 days
Mars	3 393	6.421 x 10^{23}	3950	3.72	1.029 days
Ceres	424	8.6 x 10^{20}	1980	0.32	9.08 hours
Vesta	234	3.0 x 10^{20}	3900	0.37	5.34 hours
Jupiter	71 492[a]	1.900 x 10^{27}	1330	22.88	9.925 hours[b]
Io	1 815	8.94 x 10^{22}	3570	1.81	1.769 days
Europa	1 569	4.80 x 10^{22}	2970	1.30	3.551 days
Ganymede	2 631	1.48 x 10^{23}	1940	1.43	7.155 days
Callisto	2 400	1.08 x 10^{23}	1860	1.25	16.689 days
Saturn	60 268[a]	5.688 x 10^{26}	690	9.05	10.675 hours[b]
Titan	2 575	1.35 x 10^{23}	1880	1.36	15.945 days
Uranus	25 559[a]	8.684 x 10^{25}	1290	7.77	17.240 hours[b]
Neptune	24 764[a]	1.02 x 10^{26}	1640	11.0	16.11 hours[b]
Triton	1 350	2.14 x 10^{22}	2070	0.78	5.877 days
Pluto	1 150	1.29 x 10^{22}	2030	0.658	6.3872 days
Charon	604	1.52 x 10^{21}	1650	0.278	6.3872 days

[a] Measured at the 1-bar pressure level.

[b] Internal rotation period derived from the magnetic field.

2.1 The overall shapes of planets

2.1.1 Non-rotating planets: spheres

The lowest order of relief, which in this book we call the "zeroth order" because it corresponds to the zeroth-order spherical harmonic, is the overall diameter of the body. The shape of a non-rotating, self-gravitating mass of material that has no intrinsic strength is a perfect sphere. We can describe such an object by its radius or diameter, and with this single datum our description of its shape is complete. Of course, the smaller bodies in the Solar System, such as asteroids or small satellites, may depart considerably from a perfect spherical shape, but it is still useful to describe them by the radius of a sphere of equal volume. Most tabulations of the physical properties of the planets and moons give their diameter, along with their mass (see Table 2.1). From the mass and diameter other physically interesting parameters, such as density or the surface acceleration of gravity, can be computed.

2.1 The overall shapes of planets

The volume V of a sphere of radius r is, as everyone learns in high school, $V = \frac{4}{3}\pi r^3$. Fewer readers may remember that the volume of a triaxial ellipsoid with semiaxes a, b and c is given by the similar formula $V = \frac{4}{3}\pi\, abc$. Thus, the mean radius (that is, the radius of a sphere of equal volume) of a triaxial body is given by $r_{mean} = \sqrt[3]{abc}$. In many cases a, b, and c differ only slightly from the mean. When this happens we can use the approximate formula $r_{mean} \approx (a + b + c)/3$. These simple formulas will prove useful in interpreting the relationships between various radii commonly encountered in tabulations.

2.1.2 Rotating planets: oblate spheroids

Newton himself recognized the first level of deviation from a perfectly spherical shape. Rotating bodies are not spherical: They are oblate spheres with larger equatorial radii than polar radii. In the language of spherical harmonics this is the second order of relief. Historically, the Earth's oblateness was not at all obvious to Newton's contemporaries and sparked a debate between Newton and contemporary astronomer Jacques Cassini (1677–1756) that resulted in one of the first major scientific expeditions, a project to precisely measure and compare the length of a degree of latitude in both France and Lapland. The result, announced in 1738, was the first direct evidence that the Earth is shaped like an oblate spheroid, whose equatorial radius is about 22 km longer than its radius along its axis of rotation. This apparently tiny deviation, of only 22 km out of 6371 km, is expressed by the *flattening* of the Earth, f, defined as:

$$f = \frac{a - c}{a} \tag{2.1}$$

where a is the equatorial radius of the Earth and c is its polar radius (Figure 2.1). The currently accepted value for the Earth's flattening, 1/298.257, is substantially different from Newton's own theoretical estimate of 1/230.

Newton derived his estimate of Earth's oblateness from a theory that assumed that the Earth's density is uniform. Under this assumption he was able to show that the flattening is proportional to a factor m, the ratio between the centrifugal acceleration at the equator (a consequence of rotation) and the gravitational acceleration at the equator. This ratio expresses the tendency of a planet to remain spherical: Smaller m implies that rotation is less important and the planet is more spherical.

$$m = \frac{\omega^2 a^3}{GM} \cong \frac{3}{4\pi} \frac{\omega^2}{G\bar{\rho}}. \tag{2.2}$$

In this equation ω is the rotation rate of the Earth (in radians per second), G is Newton's gravitational constant, 6.672×10^{-11} m³/kg-s², M is the mass of the Earth and $\bar{\rho}$ is its mean density. Even in Newton's time it was known that $m \approx 1/290$. Newton used a clever argument involving hypothetical water-filled wells drilled from both the pole and equator that join at the center of the Earth. Supposing that the wellheads are connected by a level canal

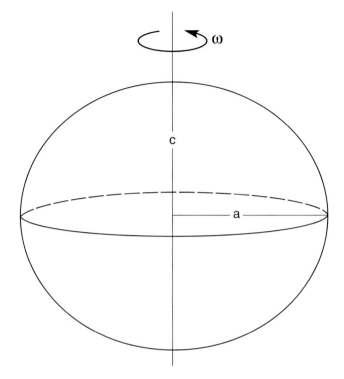

Figure 2.1 Oblate spherical shape of a rotating fluid planet with no strength. The equatorial radius is *a*, the polar radius is *c*, and the rotation rate is ω.

at the surface, he used the impossibility of perpetual motion to argue that the pressure at the bottom of the water columns had to be equal at the Earth's center, from which he derived the expression:

$$f = \frac{5}{4}m. \tag{2.3}$$

Although not quite correct for the Earth, this formula gives an excellent first approximation to the flattening.

Implicit in Newton's derivation is the idea that water tends to assume a *level surface*. That is, the surface that water naturally attains coincides with a surface on which the gravitational potential energy is constant. If the surface of a body of water, or any other strengthless fluid, did not follow a constant gravitational potential, it could gain energy by flowing downhill. Thus, in the absence of currents or other imposed pressure gradients, the surface of a fluid must coincide with an *equipotential*. This is sometimes called an *equilibrium surface*. This proposition holds equally well for planetary atmospheres, which also tend to have equal pressures on equipotential surfaces. It is important to note that on a rotating planet the equilibrium surface is not a sphere but an oblate spheroid (on very rapidly rotating bodies this surface becomes still more complex, but such surfaces do not seem to be important on any currently known planets). In real planets, lateral variations in density lead to further distortions of the equilibrium surfaces. An arbitrarily chosen equilibrium surface, called the *geoid*, forms the primary reference from which topographic elevations are measured on Earth. Deviations of a planet's surface from an equilibrium surface must be supported by strength.

Several hundred years of further effort by mathematical physicists were needed to extend Newton's simple flattening estimate to a more comprehensive form. In 1959 Sir Harold Jeffreys established a formula for the flattening of a rotating, strengthless body in hydro-static equilibrium (for which contours of constant density coincide with equipotentials at all depths within the body). This more complicated formula is:

$$f = \frac{\frac{5}{2}m}{1 + \frac{25}{4}\left[1 - \frac{3}{2}\left(\frac{C}{Ma^2}\right)\right]^2} \tag{2.4}$$

where *C* is the planet's moment of inertia about the polar axis. The moment of inertia is defined as the integral:

$$C = \int_{o}^{M} r^2 dm \tag{2.5}$$

where *r* is the radial distance of the infinitesimal mass element *dm* from the axis about which *C* is computed. In the case of the polar moment of inertia *C* this is the rotation axis.

The dimensionless moment of inertia ratio, C/Ma^2, expresses the concentration of mass toward the center of the planet. This ratio is equal to zero for a point mass, equals 2/5 (= 0.4) for a uniform density sphere and is measured to be 0.33078 for the Earth. Earth's moment of inertia ratio is less than 0.4 because mass is concentrated in its dense nickel–iron core. In the uniform density case, $C/Ma^2 = 2/5$, and Equation (2.4) reduces exactly to Newton's estimate.

Newton's contemporaries also noted the rather large flattening of the rapidly rotating planets Jupiter and Saturn. Modern measurements of the flattening of the planets in our Solar System are listed in Table 2.2.

Mars is a special case. Its observed shape flattening, $f_{Mars} = 1/154$ is considerably larger than that estimated from Jeffreys' formula above, which gives $f = 1/198$. This is because Mars is far from hydrostatic equilibrium, and so violates the assumptions of Jeffreys' der-ivation. The Tharsis Rise volcanic complex is so large and so massive that it dominates the gravitational field of the planet and warps its shape well out of hydrostatic equilibrium. Its equatorial radius varies by almost 5 km, depending on longitude. Geophysical models of Mars have yet to fully separate the effects of Tharsis from the radial concentration of mass towards its core. The shapes of small bodies in the Solar System – comets, asteroids and

Table 2.2 *Deviations of Solar System bodies from spheres*

Name	Moment of inertia factor, C/MR^2	Topographic flattening	Dynamical flattening	Center of mass – center of figure offset (km)
Mercury	0.33	~1/1800.	$1/1.03 \times 10^6$	?
Venus	0.33	0	$1/1.66 \times 10^7$	0.280
Earth	0.33078	1/298.257	1/301	2.100
Moon	0.394	1/801.6	$1/1.08 \times 10^5$	1.982
Mars	0.366	1/154.	1/198.	2.501
Jupiter	0.254	1/15.42[a]	1/15.2	–
Saturn	0.210	1/10.2[a]	1/10.2	–
Uranus	0.225	1/43.6[a]	1/50.7	–
Neptune	0.24	1/58.5[a]	1/54.4	–

Data from Yoder (1995).

[a] At 1-bar pressure level in the atmosphere.

small moons – do not obey Jeffreys' formula for similar reasons; their inherent strength produces large departures from hydrostatic equilibrium.

Mercury is another interesting planet from the flattening perspective. Although we do not know its flattening very accurately (not to better than 10%), it seems to be very small, about 1/1800. This is consistent with its current slow rotation period of about 57 days. However, Mercury is so close to the Sun that it is strongly affected by solar tides. It may originally have had a much faster rotation rate that has since declined, placing the planet in its current 3/2 spin-orbital resonance with the Sun (that is, Mercury rotates three times around its axis for every two 88-day trips around the Sun). If this is correct, then Mercury's flattening may have changed substantially over the history of the Solar System. Chapter 4 will discuss the tectonic consequences of this global shape change.

2.1.3 Tidally deformed bodies: triaxial ellipsoids

After rotation, the next degree of complication in planetary shapes includes the effects of tidal forces. Tides are the result of the variation of the gravitational potential of a primary attractor across the body of an orbiting satellite, along with the centripetal potential due to its orbit about the primary. The consequence of these varying potentials on a mostly spherical satellite is that it becomes elongated along the line connecting the centers of the primary and satellite, compressed along its polar axis, and compressed by an intermediate amount along the axis tangent to its orbit (Figure 2.2).

If the satellite (and this includes the planets themselves, which are satellites of the Sun) spins at a rate different than its orbital period, then the elongation of the equipotential surface in a frame of reference rotating with the satellite varies with time. A point on the

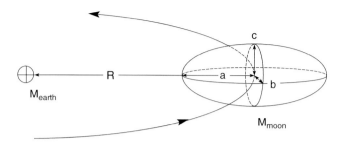

Figure 2.2 Tidal deformation of a synchronously rotating satellite such as our Moon, orbiting about its primary at a distance R. The tidal forces stretch the satellite along the line connecting its center to that of its primary such that the radius a along this line is larger than any of its other principal radii. The smallest radius c is perpendicular to the orbital plane and the intermediate radius b is parallel to the orbit's tangent.

equator is alternately lifted and dropped as it rotates from the line between its center and that of the primary, to the line tangent to the orbit. These periodic equipotential changes, which on Earth we recognize as the force responsible for oceanic tides, create motions that dissipate rotational energy and, for strong enough tides or long enough times, may eventually slow the satellite's rotation. The Earth is subject to both solar and lunar tides, which have gradually lengthened our day from 18 hours, 1.3 billion years ago, to its present 24 hours. The Moon, being less massive than the Earth, long ago lost any excess rotation and is now synchronously locked to the Earth, rotating once on its axis for each orbit around the Earth. Many other satellites in the Solar System are similarly synchronously locked to their primaries, including the four large Galilean satellites of Jupiter and Saturn's large satellite Titan. Mercury itself is an exception. Although astronomers long believed that Mercury is synchronously locked to the Sun, we now understand that its highly elliptical orbit led it to be trapped in the present 3/2 spin-orbit resonance.

Tidal forces produce a characteristic pattern of deformation on synchronously rotating satellites. Harold Jeffreys, in his famous book *The Earth* (Jeffreys, 1952), showed that the equipotential surface of a tidally locked body is a triaxial ellipsoid with three unequal axes $a>b>c$. The lengths of these axes are:

$$a = r_{\text{moon}}\left(1 + \frac{35}{12}\Omega\right)$$

$$b = r_{\text{moon}}\left(1 - \frac{10}{12}\Omega\right)$$

$$c = r_{\text{moon}}\left(1 - \frac{25}{12}\Omega\right) \tag{2.6}$$

where r_{moon} is the mean radius of the Moon. It is clear that the average of these three axial distances just equals the mean radius. These distortions depend on the dimensionless factor Ω, given by:

$$\Omega = \frac{M_{earth}}{M_{moon}} \frac{r_{moon}^3}{R^3} \tag{2.7}$$

where M_{earth} and M_{moon} are the masses of the Earth and Moon, respectively, and R is the distance between the Earth and the Moon.

Naturally, for satellites other than the Moon orbiting about some other primary, the analogous quantities must be inserted in Ω. The distortion is larger for a larger ratio between primary mass and satellite mass and smaller for increased distance between the two bodies. Jeffreys' equations above are strictly valid only for a Moon of uniform density. The book by Murray and Dermott (1999) describes how to treat the general case where the Moon's density is not uniform. Table 2.3 lists the ratios between the different axes of the satellites in the Solar System for which they are known.

According to the formulas (2.6) and (2.7), the maximum difference between the axes, $a - c$, on Earth's Moon is presently about 66 m. The Moon is 10 to 20 times more distorted than this, a fact that was known even in Harold Jeffreys' day. The current best estimates indicate that the Moon is at least 10 times more distorted than the hydrostatic prediction. This observation prompted Jeffreys to propose that the present figure of the Moon is the fossil remnant of a formerly much larger distortion. The Moon is presently receding from the Earth at the rate of about 3.8 cm/year. This recession rate was surely higher in the past when the Moon was closer to the Earth and tides were, therefore, higher (the full story of the evolution of the Moon's orbit is a complicated one, involving changes in tidal dissipation over the age of the Earth as continents and seas shifted). Nevertheless, it is clear that the Moon was once much closer to the Earth than it is now. Because the lengths of the axes depend on the Earth–Moon distance to the inverse cube power, Jeffreys postulated that the present figure could have been frozen-in at a time when the Moon was about 1/2.7 times its present distance from the Earth. At this distance the Moon would have circled the Earth in only 6.1 days and, by angular momentum conservation, a day on Earth would have lasted only 8.2 hours.

The major problem with Jeffreys' proposal (which he recognized himself) is that the hydrostatic formulas (2.6) make a definite prediction that that the ratio $(a - c)/(b - c) = 4$, independent of the Earth–Moon distance. The latest value of this ratio from Japan's Kaguya mission (Araki *et al.*, 2009) is 1.21, far from the hydrostatic value of 4. The general opinion at the present time is that, although the Moon clearly departs from a hydrostatic shape, the present shape is more a consequence of geologic forces that sculpted the lunar surface rather than the remnant of a former hydrostatic figure.

In examining Table 2.3 the ratio of axes differences in the last column is more often quite different from the theoretical value of 4 than close to it. The larger satellites approach most closely to this ideal ratio, while the smaller ones are clearly dominated by strength rather than gravitational forces.

Table 2.3 *Triaxial shapes of small satellites*

Name	Density (kg/m³)	Mean radius (km)	(a–c) (km)	(a–c)/ (b–c)
Earth's Satellite				
Moon[a]	3340	1737	0.713	1.21
Martian Satellites				
Phobos[b]	1900	11.1	4.0	2.2
Deimos[b]	1760	6.3	2.1	2.6
Jovian Satellites				
Metis[c]	3000	21.7	13.0	4.3
Adrastea[c]	3000	8.2	3.0	3.0
Almathea[c]	862	83.6	61.0	6.8
Thebe[c]	3000	49.2	16.0	2.3
Io[d]	3528	1818.1	14.3	4.1
Europa[d]	3014	1560.7	3.0	3.8
Ganymede[d]	1942	2634.1	1.8	4.5
Callisto[d]	1834	2408.3	0.2	2.0
Saturnian Satellites				
Mimas[e]	1150	198.1	16.8	2.7
Enceladus[e]	1608	252.1	8.3	2.7
Tethys[e]	973	533.0	12.9	3.6
Dione[e]	1476	561.7	3.5	5.0
Rhea[e]	1233	764.3	4.1	−6.8
Titan[f]	1881	2574.5	0.410	3.8
Iapetus[e]	1083	723.9	35.0	n/a
Uranian Satellites				
Miranda[g]	1200	235.7	7.1	5.5
Ariel[g]	1670	578.9	3.4	17.0

Data from:
[a] Araki *et al.* (2009)
[b] Thomas (1989)
[c] Thomas *et al.* (1998)
[d] Davies *et al.* (1998)
[e] Thomas *et al.* (2007)
[f] Iess *et al.* (2010)
[g] Thomas (1988)

Most of the other moons in the Solar System depart substantially from a hydrostatic shape. However, it appears that the shape of Jupiter's large, tidally heated moons Io and Europa may be close to equilibrium ellipsoids. If that is the case, then the maximum distortion, *a–c*, should be 14.3 km for Io and 3.9 km for Europa. The Galileo spacecraft confirmed these expectations with an accuracy of 0.12 and 0.65 km, respectively (Davies *et al.*, 1998).

2.1.4 A scaling law for planetary figures?

William Kaula (1926–2000), who contributed extensively to understanding the figure of the Earth from satellite measurements, proposed a scaling law that seems to approximately give the deviations of planetary figures from a hydrostatic shape. The law is based on the idea that all planets have about the same intrinsic strength, and that stress differences are proportional to the surface acceleration of gravity, g (Kaula, 1968, p. 418). In the absence of any other information about a planet, this law may give a useful first estimate of how far a "normal" planet's figure departs from the hydrostatic shape.

This departure is given by Kaula's first law:

$$\left.\frac{C-A}{Ma^2}\right|_{\text{non-hydrostatic}} \propto \frac{1}{g^2} \tag{2.8}$$

where C and A are the moments of inertia about the shortest and longest axes of the figure, respectively. The difference in moments of inertia is proportional to the normalized difference of the lengths of the axes, $(a\text{-}c)/a$, so that this rule also implies that the deviations of this ratio from its hydrostatic value should depend on the inverse square of the gravitational acceleration. For constant density, g is proportional to the planetary radius, so this ratio should equally depend on the inverse square of the planetary radius. Looking ahead to Chapter 3, we will see in Section 3.3.3 and Figure 3.5 that this relation does seem to hold approximately in our Solar System for the larger bodies, but it fails badly for small objects for which strength is controlled by frictional forces that depend on pressure.

The possibility that planets may have substantial non-hydrostatic contributions to their figures plays an important role in studies of rotational dynamics and the tidal evolution of bodies in the Solar System. For example, the present orientation of the Moon with respect to the Earth may be partly due to the distribution of dense lavas in the low-lying basins on the nearside. The orientation of Mercury may be controlled by mass anomalies associated with the Caloris Basin. And if Europa has too large a non-hydrostatic figure, then its putative slow non-synchronous rotation cannot occur.

2.1.5 Center of mass to center of figure offsets

One of the widely publicized results of the Apollo missions to the Moon was the discovery that the Moon's center of mass is about 2 km closer to the Earth than its center of figure. For many years this offset was known only in the Moon's equatorial plane, as all of the Apollo flights circled the Moon's equator. Now, as a result of the unmanned Clementine mission, we know more precisely that the offset is 1.982 km.

Although, in hindsight, it is not really surprising that such an offset should exist, the possibility that the center of figure of a planet might not correspond with its center of mass was never considered in classical geodesy. All harmonic expansions of the gravity field of the Earth are made about its center of mass and so the reference geoid and all other equipotential surfaces are also centered about this point. In fact, the Earth itself has a substantial center of mass – center of figure offset if the water filling the ocean basins is neglected. The floor of the Pacific Ocean is about 5 km below sea level, whereas the opposite hemisphere is dominated by the continental landmasses of Asia, Africa, and the Americas. The waterless Earth's center of figure is thus offset from its center of mass by about 2.5 km at the present time. Of course, as the continents drift around over geologic time this offset gradually changes in both direction and magnitude. The fundamental reason for the offset is the difference in density and thickness between oceanic crust and continental crust. Table 2.2 lists these offsets for the bodies where they are known.

For reasons that are still not understood, most of the terrestrial planets show striking asymmetries on a hemispheric scale. The nearside of the Moon looks quite different from the farside, and lies at a lower average elevation with respect to its center of mass. It is generally believed that this is due to a thicker crust on the farside, although what caused the thickness variation is unknown. Mars also possesses a strong hemispheric asymmetry. The northern plains of Mars lie an average of 5 km lower than the southern highlands. Here again the immediate cause may be a difference in crustal thickness or composition, but the ultimate reason for the difference is presently unknown.

2.1.6 Tumbling moons and planets

Most rotating bodies in the Solar System spin about an axis that coincides with their maximum moment of inertia, the C axis in our terminology. The moment of inertia is actually a second-rank tensor, written I_{ij}, which can be defined for any solid body as a generalization of the definition (2.5) for C:

$$I_{ij} = \int r_i r_j \, \mathrm{d}m \qquad (2.9)$$

where the subscripts i and j run from 1 to 3 and denote, respectively, the x, y, and z axes of a Cartesian coordinate system. The symbol r_i for $i = 1$, 2, and 3, thus, denotes the x, y, and z coordinates of a mass element dm with respect to the origin around which the moment of inertia is computed, generally taken to be the center of mass of a body.

A fundamental theorem of tensor mathematics states that a suitable rotation of the coordinate axes can always be found in which the tensor (2.9) is diagonal (that is $I_{ij} = 0$ unless $i = j$) about three perpendicular axes. The moments of inertia about these special axes are called the principal moments of inertia and are labeled C, B, and A for the maximum, intermediate, and minimum principal moments, $C \geq B \geq A$. The lengths of the corresponding principal axes are conventionally written in lower case and, perhaps confusingly, the c axis is usually the *shortest* of the three: $c \leq b \leq a$. The reason for this inverse order is that the moment of inertia is largest about that axis for which the mass elements are most distant and, thus, the r_i perpendicular to the axis are largest.

The reason that the C axis is special is that a body that rotates about this axis has the lowest kinetic energy possible for a fixed angular momentum. Angular momentum is conserved for an isolated body, but kinetic energy can be converted into heat. A body rotating about the A axis, say, spins relatively quickly but this is the highest-energy rotational state, so if it exchanges kinetic energy for heat it must spin about another axis, or a combination of axes in a complex tumbling motion. When the body finally spins about its C axis it attains a minimum energy configuration and cannot change its rotation further (unless some external torque acts on it).

This lesson was learned the hard way in the early days of space exploration when the first US satellite, Explorer 1, which was shaped like a long narrow cylinder, was stabilized by spinning it around its long axis. A broken antenna connection flexed back and forth as the satellite rotated, dissipating energy. Within hours the satellite was spinning about its C axis – the short axis, perpendicular to the cylinder. Because the spacecraft was not designed to operate in this configuration radio contact was soon lost.

Most Solar System bodies, therefore, spin about their C axes. A few small asteroids have been discovered that are in "excited" rotation states in which the object is not in its minimum-energy rotational state and thus tumbles, but these exceptions are rare because nearly every object has some means of dissipating energy internally and thus eventually seeks out the lowest energy configuration.

Mars is a prime example of the importance of this process. Mars' major positive gravity anomaly, the Tharsis Rise, is located on its equator. This location puts the excess mass as far as possible from its rotational axis, maximizing the moment of inertia. The opposite extreme is illustrated by the asteroid Vesta, which suffered an impact that gouged out a crater (a mass deficit) nearly as large in diameter as Vesta itself. The central peak of this crater is now located at its rotational pole (which happens to be the south pole), the most stable configuration.

2.2 Higher-order topography: continents and mountains

2.2.1 How high is high?

As silly as this question may seem, it highlights a common assumption that underlies our thinking when we consider the elevation of some topographic feature. An elevation is a number or contour that we read off a map (or, more commonly in modern times, a color code on an image). But what is that elevation *relative* to? Does it directly give the distance from the center of the planet? Or the height above an arbitrary spheroid? Or, more commonly on Earth, the elevation above mean sea level? The answer to all these questions is, "it depends."

The fundamental reference surface for elevations, or *geodetic datum*, is established empirically. Historically, it has varied with the technology for measuring topography, completeness or accuracy of information and, sometimes, pure convenience. The era of space exploration has brought great changes in this type of measurement, as well as bringing new planets under close scrutiny. Each new planet has presented unique challenges in the apparently simple task of measuring the elevations of its surface features.

In the case of the Earth, detailed mapping began in the eighteenth century with trigonometric surveys utilizing telescopes, carefully divided angle scales, rods, chains, levels, and plumb bobs. The primary referent for elevations was mean sea level, already a problematic concept in view of changes of water level in response to tides, currents, and meteorological pressure changes. Individual countries established topographic surveys and used astronomical measurements to locate prime coordinate points. Nations with access to an ocean established elaborate gauging stations from which a time-average or *mean* sea level could be defined. Survey crews could carry this elevation reference inland by the use of levels and plumb bobs. Elevations of high and relatively inaccessible mountains were estimated barometrically, by comparing the pressure of the air at the top of the mountain to that at sea level. All of these methods, on close examination, amount to referring elevations to an equipotential surface. The equipotential that corresponds to mean sea level is called the geoid, and all elevations are, ideally, referred to this surface.

The geoid is quite hard to measure accurately. Although it roughly corresponds to a flattened sphere, as described previously, slight variations in density from one location to another gently warp it into a complex surface. Determination of the geoid thus requires precision measurements of the acceleration of gravity, as well as accurate leveling. Much of both classical and satellite geodesy is devoted to determining the geoid and, thus, permitting elevations to be defined with respect to a level surface (level in the sense of an equipotential, down which water will not run). Historically, each nation with a topographic survey created its own version of the geoid, although thanks to satellite measurements these are now knit into a consistent global network.

Elevations referred to the geoid are very convenient for a variety of purposes. Besides engineering applications, such as determining the true gradient of a canal or railway line, they are also essential to geologists who hope to estimate water discharges from the slope of a river system. The "upstream" ends of many river systems (the Mississippi is one) are actually closer to the center of the Earth than their mouths. And yet the water still flows from head to mouth because water flows from a higher to a lower gravitational potential.

The modern era of the Global Positioning System (GPS) is ushering in new changes. The orbiting GPS satellites really define positions with respect to the Earth's center of mass, so converting GPS elevations to elevations with respect to the geoid requires an elaborate model of the geoid itself. It has become common to refer elevations to a global average datum that locally may not correspond to the actual geoid.

The determination of elevations on other planets is becoming nearly as complex as that on Earth, thanks to a flood of new data from orbiting spacecraft. The first body to be orbited by a spacecraft capable of determining elevations precisely was the Moon. The Apollo orbiters carried laser altimeters that measured the elevations of features beneath their orbital tracks. Although it was many years before the Clementine spacecraft expanded this method beyond the equatorial swaths cut by the Apollo orbiters, a prime reference surface for lunar elevations had to be chosen. Rather than trying to use the very slightly distorted equipotential surface, the reference surface was chosen to be a perfect sphere of radius 1738 km centered on the Moon's center of mass. All lunar topographic elevations are relative to this sphere. Since this sphere is not a geoid (although it is close enough for many purposes) care must be taken when, for example, estimating the slopes of long lava flows over the nearly level plains of the maria.

The topographic reference levels chosen for different planets varies depending on the rotation rate (and direction) of the planet, its degree of deviation from a sphere, and the technology available. The reference surfaces of Mercury and Venus are spheres. The mean diameter of Mercury has yet to be established: At the moment, the only global data set comes from radar measurements of equatorial tracks, but the MESSENGER laser altimeter should soon make a better system possible. The reference surface for Venusian elevations is a sphere of diameter 6051 km, close to the average determined by the Magellan orbiter.

Mars offers mapmakers serious problems when it comes to elevations. The Martian geoid is far from a rotationally symmetric spheroid, thanks to the large non-hydrostatic deviations caused by the Tharsis gravity anomaly. The geoid is intended to coincide with the level at which the mean atmospheric pressure equals 6.1 mb, the triple point of water. However, the atmospheric pressure varies seasonally by a substantial fraction of the entire pressure, so locating this point is not straightforward.

Some Martian elevation maps are referenced to a spheroid with flattening 1/170, a system recommended by the International Astronomical Union (IAU) (Seidelmann *et al.*, 2002). However, much of the high-precision data currently available is referenced to a more complex and realistic geoid, so that the user of such information must be alert to the system in use. Elevations with respect to a geoid are most useful in determining what directions are really downhill (that is, toward lower gravitational potential), which determines the expected flow direction of water or lava. Because geoids improve with time, no map of elevations is complete without a specification of the reference surface in use.

The surfaces of small asteroids such as Eros or Ida, comet nuclei such as Tempel 1, and many other small bodies that will be mapped in the future, present new problems. They are too irregular in shape to approximate spheres. At the moment their surface elevations are defined in terms of the distance from their centers of mass.

The gas giant planets in the outer Solar System lack solid surfaces and so elevations are especially difficult to define. The convention is now to refer elevations on these bodies to a spheroid at the 1-bar pressure level, which is a good approximation to an equipotential surface on such planets.

2.2.2 Elevation statistics: hypsometric curves

Elevation data can be processed and interpreted in many ways. A map of elevations, a topographic map, is certainly the most familiar and contains a wealth of data. However, one can extract more general features from such data that tell their own stories. On a small scale, roughness is important for safely landing on and roving over planetary surfaces. Some of the methods for analyzing roughness are described in Box 2.1. On the large scale, an elevation plot known as the hypsometric curve has proven useful in highlighting general properties of planetary crusts.

A hypsometric curve is a plot of the percentage of the area of a planet's surface that falls within a range of elevations. Curves of this type have been constructed for the Earth ever since global topographic data sets became available and they show one of the Earth's major features. Figure 2.3c illustrates the Earth's hypsometric pattern, binned into elevation intervals of 500 m. The striking feature of this curve is the two major peaks in areas that lie between the maximum elevation of 7.83 km and the minimum of −10.376 km below mean sea level (note that these are not the highest and lowest points on the Earth – they are the highest and lowest elevations averaged over 5′ x 5′, 1/12-degree squares). A sharp peak that encompasses about 1/3 of the area of the Earth lies close to or above sea level. The second peak lies about 4 km below mean sea level and accounts for most of the rest of the Earth's area. These two peaks reflect the two kinds of crust that cover the surface of our planet. The low level is oceanic crust, which is thin (5–10 km), dense (about 3000 kg/m^3), basaltic in composition and young, being created by mid-oceanic spreading centers. The second principal topographic level is continental crust, which is much thicker (25–75 km) than oceanic crust, less dense (about 2700 kg/m^3), granitic in composition and much older than the ocean floors. Plate tectonics creates and maintains these two different crustal types.

Although Figure 2.3a shows a hypsometric curve for Mercury, the data set from which this is derived is sparse at the moment, consisting of a number of mostly equatorial radar tracks. At least with this data, however, there is no indication of an Earth-like dichotomy of crustal thickness.

Figure 2.3b illustrates the hypsometric curve of Venus, for which an excellent data set exists from the Magellan radar altimeter. This curve is an asymmetric Gaussian, skewed toward higher elevations. There is no indication of a double-peaked structure, from which we must infer that, whatever processes are acting to create the crust of Venus, they must differ profoundly from those that affect the Earth.

The Moon's hypsometric curve is illustrated in Figure 2.3d. Like Venus, the Moon lacks a dichotomy of crustal types, although there are important differences in elevation between the nearside and farside, illustrated by the thin lines that show separate hypsometric curves for the two hemispheres. This is generally attributed to systematic differences in crustal thickness between the nearside and farside, rather than compositional differences. The role of the large basins, especially the gigantic South Pole-Aitken basin, is still not fully understood.

Finally, we come to Mars in Figure 2.3e. Mars, surprisingly, shows a double-peaked distribution similar to that of the Earth. As shown by the light lines, the lower peak is accounted for by the Northern Lowlands, while the high peak represents the contribution of the Southern Highlands. The two terrains are divided by the Martian Crustal Dichotomy, an elliptical region tilted with respect to the north pole that may represent the scar of an

Box 2.1 **Topographic roughness**

The roughness of a surface is a concept that everyone is familiar with. Surprisingly, although there are many ways to measure roughness, there is no standard convention. Intuitively, rough surfaces are full of steep slopes, while smooth surfaces lack them. One measure of roughness would, thus, cite some statistical measure of the frequency of occurrence of a particular slope, extended over a range of slopes. For example, Root Mean Square (RMS) measures have often been used because they can be easily extracted from radar backscatter data. This statistic implicitly assumes that the distribution of slopes approximately follows a Gaussian curve, an assumption that needs to be tested. One might also cite mean or median slopes.

These simple statistical measures, while perhaps capturing our intuitive idea of *roughness*, do miss an important aspect of the concept, and that is the scale of the roughness. A surface that is smooth on a scale of 100 m might be very rough on the scale of 10 cm, a difference that is of overwhelming importance when one is trying to set a 1 m lander down safely onto a planetary surface. We thus need to define a baseline, L, in addition to the slope of a surface and to describe the statistics of slopes with respect to a range of baselines.

We could, for example, define the slope $s(L)$ of a surface for which we measure the elevations $z(x, y)$, where x and y are Cartesian coordinates that define a location on the surface. The surface slope is then given by:

$$s(L) = \frac{z(x,y) - z(x',y')}{L}. \tag{B2.1.1}$$

The points where the slope is evaluated, (x, y) and (x',y'), are separated by the distance $L = \sqrt{(x-x')^2 + (y-y')^2}$. On a two-dimensional surface we need some understanding of how to locate the different points at which elevations are evaluated, and many methods have been devised for this. At the present time, elevation data is often acquired by laser altimeters on orbiting spacecraft. The tracks along which elevations are measured are linear or gently curved, simplifying the decision process. New techniques may make this problem more acute, however: The laser altimeter aboard the Lunar Reconnaissance Orbiter now collects elevation data simultaneously from an array of five non-collinear spots, so that a full two-dimensional array of slopes can be defined.

A promising statistic is derived from fractal theory, the Hurst exponent, and has been applied to the analysis of Martian slopes in the MOLA dataset (Aharonson and Schorghofer, 2006). This statistic, at present, is limited to linear sets of elevation data, $z(x)$, in which the y coordinate is ignored. The variance $v(L)$ of the elevation differences along the track is computed for a large number N of equally-spaced locations:

$$v(L) = \sqrt{\frac{1}{N} \sum_{i=1}^{N} \left[z(x_i) - z(x_i + L) \right]^2}. \tag{B2.1.2}$$

The roughness is thus given by the slope $s(L)$:

$$s(L) = \frac{v(L)}{L} = s_0 \left(\frac{L}{L_0} \right)^{H-1}. \tag{B2.1.3}$$

Box 2.1 (cont.)

It is often found that the variance $v(L)$ is a power function of the baseline L, so that if the variance is compared to the slope s_0 at some particular baseline L_0, a relation expressed by the second term in (B2.1.3) is found with an exponent H known at the *Hurst exponent*. Whether this exponent is constant over a broad range of baselines or depends in some simple way upon the scale is not yet known, nor is it understood how surface processes are related to this exponent, although some proposals have been made (Dodds and Rothman, 2000). More progress can be expected as more closely spaced elevation data is collected on a number of different planets.

A disadvantage of the definition (B2.1.2) is that a long, smooth slope also contributes to the variance because the elevations $z(x)$ on a straight slope differ by a constant amount along the track. One would really like to filter out all elevation differences except those close to the scale L. One solution to this problem is the median differential slope, which is based upon four points along the track. The two extreme points are used to define a regional slope, which is then subtracted from the elevation difference of the two inner points to achieve a measure of slope that is not affected by long straight slopes, but responds to short wavelength variations on the scale of the distance between the inner points (Kreslavsky and Head, 2000). The array of four points is located at $-L$, $-L/2$, $L/2$, and L around an arbitrary zero point that slides along the spacecraft track. This differential slope s_d is given by:

$$s_d(L) = \frac{z(L/2) - z(-L/2)}{L} - \frac{z(L) - z(-L)}{2L}. \tag{B2.1.4}$$

The first term is the slope between the inner points and the second term is the larger scale slope. The differential slope is zero for a long straight slope, as desired.

In the past, the study of roughness tended to focus on landing-site safety, but current efforts are also making progress on extracting information on the surface processes creating the roughness. One may expect to hear more about this in the future.

ancient giant impact. Do these two peaks represent two types of crust, as on the Earth, or are these just areas with very different crustal thickness? We do not believe that Mars possesses plate tectonics, although it has been suggested that some plate processes may have acted in the distant past.

The Martian hypsometric curve offers an interesting lesson in the importance of referencing elevations to the geoid. Earlier plots of Mars elevations showed a Gaussian-like distribution of elevations similar to that of Venus or the Moon. Only after a good gravity field was measured and elevations referenced to a true geoid did the double-peaked character of Martian elevations become apparent.

2.2.3 Where are we? Latitude and longitude on the planets

Latitudes and longitudes are the conventional means for locating features on the surface of a planet. However, before such a system can be defined, the pole of rotation must be established. All systems of latitude and longitude are oriented around the north pole, which must

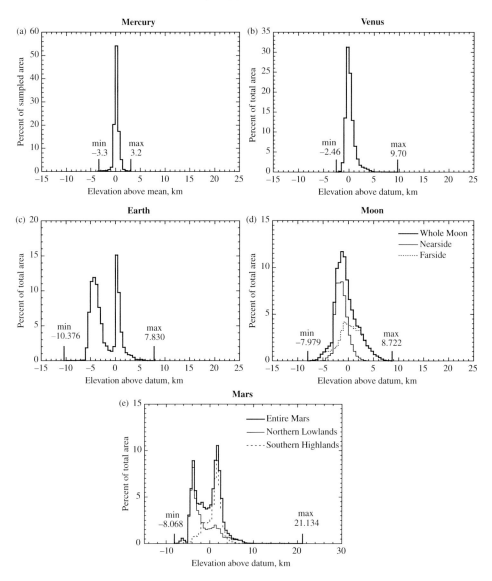

Figure 2.3 Hypsometric curves of the terrestrial planets and the Moon. Maximum and minimum elevations are also shown. These elevations refer to the maximum and minimum elevations averaged over square sample areas of different sizes, not the highest and lowest points on the planet's surface. (a) Mercury data from 9830 radar elevations in PDS file allMerc.txt. (b) Venus hypsometric curve derived from a 1° x 1° Magellan map of Venus from PDS dataset MGN-V-RDRS-5-TOPO-L2-V1.0, file TOPOGRD-DAT. (c) Earth, data is binned in 1/12 degree squares, from National Geophysical Data Center file TBASE.BIN. (d) Moon, data is ¼ degree data from Clementine LIDAR from PDS dataset CLEM1-L-LIDAR-5-TOPO-V1.0, file TOPOGRD2.DAT. Light lines show separate curves for the nearside and farside. (e) Mars, ¼ degree MOLA gridded data from PDS dataset MGS-M-MOLA-5-MEGDR-L3-V1.0, file MEGT90N000CB.IMG. Light lines show data separately for the Northern and Southern crustal provinces. For this purpose, the planet was divided into two hemispheres by a great circle whose pole is located at 53° N and 210° E longitude.

be determined in some absolute system of coordinates. The locations of the Earth's poles are determined astronomically, by the position of the extrapolated rotation axis among the stars in the sky. Although the Earth's axis precesses slowly with a period of about 26 000 yr, this gradual change is predictable and can be taken into account when referring to the pole position by citing the time, or epoch, at which the position is cited. The pole positions of the other planets are defined in a similar way, in terms of the celestial coordinates, the declination and right ascension, of their projected northern axis of rotation.

The choice of prime meridian for different planets is entirely arbitrary, but must be a definite location. On the Earth, we use the longitude of a point in Greenwich, UK, as the zero of longitude. On Mercury a crater known as Hun Kal defines the location of the 20° longitude. The choice of the 0° longitude on Venus fell to the central peak of a crater known as Ariadne, while on Mars the 0° longitude passes through a small crater known as Airy-0. Pluto's 0° of longitude (at present) passes through the mean sub-Charon point. As new bodies are mapped and their rotation axes determined, new choices for the prime meridian have to be made.

The prime meridians of the fluid gas giant planets in the outer Solar System are much harder to define and are based on the rotation rates of their magnetic fields rather than the shifting patterns of clouds in their atmospheres. Because the clouds rotate at different rates depending on latitude, they do not yield definite rotation rates for the entire planet. For these planets an accurate rotation rate must be determined and that rotation rate, plus the epoch at which it was established, defines the prime meridian.

Venus presents an interesting cartographic problem because its spin is retrograde. Its north pole, nevertheless, lies on the north side of the ecliptic by convention and, also by cartographic convention, longitudes increase eastward from 0° to 360°.

The International Astronomical Union adopted a convention in 2000 that defines the latitude and longitude coordinate system used in locating features on the surface of planets. The first principle defines the north pole of a Solar System body:

(1) The rotational pole of a planet or satellite that lies on the north side of the invariable plane will be called north, and northern latitudes will be designated as positive.

The second principle is more controversial and there is some disagreement between geophysicists and cartographers about the most sensible way to present longitudes:

(2) The planetographic longitude of the central meridian, as observed from a direction fixed with respect to an inertial system, will increase with time. The range of longitudes shall extend from 0° to 360°.

Thus, west longitudes (i.e., longitudes measured positively to the west) will be used when the rotation is prograde and east longitudes (i.e., longitudes measured positively to the east) when the rotation is retrograde. The origin is the center of mass. Also, because of tradition, the Earth, Sun, and Moon do not conform with this definition. Their rotations are prograde and longitudes run both east and west 180°, or east 360° (Seidelmann *et al.*, 2002).

This convention means that Mars presents a left-handed coordinate system, a consequence not favored by geophysicists. This debate is not resolved at the present time, so the user of cartographic data must carefully check what conventions are in use – it is very easy to download geophysical data for Mars and find that one is working on a mirror image of the actual planet Mars.

2.3 Spectral representation of topography

Maps showing contours of elevation are not the only way of codifying topographic information. Just as any function of spatial coordinates can be broken down into a Fourier series in an inverse space of wavenumbers, topography can be represented as a sum of oscillating functions on a sphere. This mode of representation is known as harmonic analysis or spectral analysis and for many data sets is preferred over a purely spatial representation.

Spectral analysis averages elevations over all positions on a sphere and presents the information as a function of wavelength, not position. No information is lost in this process: With the appropriate mathematical tools one can freely transform from space to wavelength and back again.

The full details of harmonic analysis are too specialized for full presentation in this book. The interested reader is referred to a fine review of the entire subject by Wieczorek (2007). In this book it is enough to note that any function of latitude φ and longitude λ, such as elevation $H(\varphi,\lambda)$, can be expressed as:

$$H(\varphi,\lambda) = \sum_{l=0}^{\infty} \sum_{m=-l}^{l} H_{lm} Y_{lm}(\varphi,\lambda) \qquad (2.10)$$

where l and m are integers and the $Y_{lm}(\varphi,\lambda)$ are spherical harmonic functions of order l and degree m. They are given in terms of more standard functions as:

$$Y_{lm}(\varphi,\lambda) = \begin{cases} \bar{P}_{lm}(\sin\varphi)\cos m\lambda & \text{if } m \geq 0 \\ \bar{P}_{l|m|}(\sin\varphi)\sin|m|\lambda & \text{if } m < 0 \end{cases} \qquad (2.11)$$

where the $\bar{P}_{lm}$ are normalized associated Legendre functions given by:

$$\bar{P}_{lm}(\sin\varphi) = \sqrt{(2-\delta_{0m})(2l+1)\frac{(l-m)!}{(l+m)!}} P_{lm}(\sin\varphi) \qquad (2.12)$$

where δ_{ij} is the Kronecker delta function and P_{lm} are the standard, unnormalized Legendre functions. These functions are tabulated in standard sources and, more importantly, can be computed with readily available software. There are many issues about the conventional normalizations of these functions, which are not standardized across all scientific disciplines.

The functions Y_{lm} are the spherical analogs of sines and cosines for Fourier analysis. They are simple for small-order l and become more complex, with more zero crossings, as l increases. They possess $2|m|$ zero-crossings in the longitudinal direction and $l - |m|$ in the latitudinal direction. Thus Y_{00} is a constant, Y_{10}, Y_{1-1}, and Y_{11} correspond to the displacement of a sphere from the center of mass (for example, a center-of-figure, center-of-mass offset) and the five $l = 2$ terms describe an oblate or triaxial tidally distorted sphere. In general, as l increases the wavelength of the feature that can be represented by these harmonics decreases. This is made more precise by an approximate relation between the wavelength w of features that can be represented by spherical harmonics of order l: $w \approx 2\pi a / \sqrt{l(l+1)}$, where a is the planetary radius.

The expansion of topography in spherical harmonics makes the idea of orders of relief precise: The zeroth-order harmonic is just the radius, the first is the center-of-mass center-of-figure offset, the second is the rotational or tidal distortion, etc. Harmonic coefficients for the topography of the planets to degree and order 180 are becoming common, and still higher degrees exist for the Earth and are planned for the other planets as sufficiently precise data becomes available. In addition to topography, the geoid and gravity fields are also represented by spherical harmonics, a format that makes many computations of, for example, global isostatic compensation, much simpler than it is for spatial data.

The spherical harmonic functions are orthogonal after integration over the complete sphere, so that the harmonic coefficients H_{lm} can be obtained from the topography $H(\varphi, \lambda)$ by:

$$H_{lm} = \frac{1}{4\pi} \int_0^{2\pi} \int_{-\pi/2}^{\pi/2} H(\varphi, \lambda) \, Y_{lm}(\varphi, \lambda) \cos\varphi \, d\varphi \, d\lambda. \qquad (2.13)$$

One can freely pass from the spatial representation of topography to the harmonic representation and back again. There is no loss of information, nor any saving in the amount of data to be stored for one or the other representation.

Just as the hypsometric function attempts to distil useful global information from the map of elevations, a similar extraction of data is often made from the harmonic coefficients. This data contraction is called the spectral power and is a measure of how much of the topography is due to a particular wavelength. The RMS spectral power density collapses a full set of l^2+2l+1 numbers for harmonic coefficients to degree and order l down to a set of only l numbers by summing over all orders for each degree:

$$S_l = \sqrt{\sum_{m=-l}^{l} H_{lm}^2}. \qquad (2.14)$$

The RMS spectral power densities given by Equation (2.14) are plotted in Figure 2.4 for each of the terrestrial planets (excluding Mercury) and the Moon. Except for the Moon, most of the RMS power densities on this plot rise with increasing wavelength, so each body

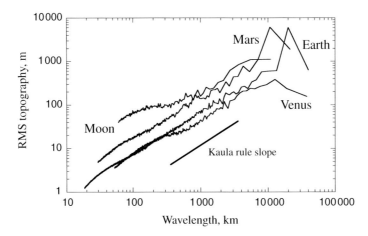

Figure 2.4 Topographic power spectra of the terrestrial planets and the Moon, excluding Mercury for which the necessary data does not yet exist. Lunar spectral data are from the Kaguya data set (Araki *et al.* 2009). Spectral data on Earth, Venus, and Mars are from Mark Wieczorek's website, http://www.ipgp.fr/~wieczor/SH/SH.html, files SRTMP2160, VenusTopo719.shape and MarsTopo719. shape, respectively.

has more power in longer-wavelength topography. Another way of saying this is that the slopes of the surfaces are approximately independent of their scale. It is popular to call this a fractal relationship, but it is unclear, at present, exactly what this means. The Earth and Venus are comparably smooth at short wavelengths, while Mars is rougher than both and the Moon is rougher still. The Moon's roughness does not rise with increasing wavelength as fast as that of the larger planets.

William Kaula worked extensively on harmonic representations of topography and his studies of the Earth's topography led him to formulate what we will call here "Kaula's second law," which is that the RMS topography depends on the inverse order, $1/l$. Because wavelength depends on the inverse order as well, his law states that the power is directly proportional to the wavelength. The prediction of this law is shown on Figure 2.4 and it does seem to hold fairly well for the major planets, but not for the Moon. The deviation shown here for the Moon is relatively new: It was not known before the data from the Kaguya laser altimeter were analyzed.

Spectral representations are, at present, difficult to interpret (see, e.g. the discussion by Pike and Rozema, 1975). Spectral data at a very small scale is widely used for computations of "trafficability" of vehicles across terrains and for designing vehicle suspension systems, but its use in geologic interpretation has been limited. The reason for this may be that the spectral method averages over a wide variety of different terrain types that are shaped by different processes and so loses the signatures characteristic of individual processes. Whatever the reason, it is currently an analysis technique in search of an interpretation, although some suggestive models have provided more insight into the interpretation of such data (Dodds and Rothman, 2000).

Further reading

Newton's Principia is tough going, but full of surprising results. It is astonishing how far Newton went with the theory of the Earth's figure (Newton, 1966). You do not need to know Latin to read the translation of the Principia, but you do need to be patient and resourceful. Harold Jeffreys was deeply interested in the figure of the Earth, the Radau approximation, and the theory of the Moon's triaxial figure. There are six editions of his famous book *The Earth*, but the third (Jeffreys, 1952) and fourth present the apex of his insight into this problem. The problem of determining the shape of the Earth has been of major interest to astronomers and mathematicians since Newton. The early history of investigation of the figure of the Earth is exhaustively told in the full language of mathematics by Todhunter (1962). More modern extensions are well covered in Chandrasekhar (1969) and Jardetzky (1958). Kaula's book (Kaula, 1968) is now very dated in its facts, but he covered many of the methods of planetary geophysics, particularly geodesy, in great detail. The nature of the geoid on Earth and its determination are well discussed in Lambeck (1988). The details of modern planetary cartography are described in book form by Greeley and Batson (2000). Spectral analysis of both topography and gravity are the subjects of a very recent and very clear review by Mark Wieczorek (2007) that offers the simplest introduction to spherical harmonics that I am aware of. He also goes to some trouble to explain the different normalization conventions in the geophysical literature.

Exercises

2.1 A whirling moon

Saturn's moon Iapetus is currently synchronously locked to Saturn, with a rotation (and orbital) period of 79.3 days. In spite of its slow rotation, Iapetus has a considerable equatorial bulge, $a-c \approx 35$ km (Table 2.3). Iapetus' density is not very different from that of water ice, so it can be treated as an approximately homogeneous body. If Iapetus' equatorial bulge is a fossil remnant from a time when it was spinning faster than at present, estimate the minimum initial period of Iapetus' rotation (explain why this is a minimum estimate). What do you think may have happened to Iapetus?

2.2 Hot Jupiteus shaped like water melons

Planet WASP-12b circles a Sun-like star about 600 light years from Earth in the constellation Auriga. It is a *hot Jupiter* planet, with a mass equal to 1.41 times that of Jupiter, radius 1.83 times larger than Jupiter, but circles only 0.0229 AU (Astronomical Units) from its star with a period of 1.0914 days. Use Equations (2.6) and (2.7), suitably generalized for a planet orbiting a star, to compute the tidal distortion of this planet, assuming that it is synchronously locked to its star (which is almost certainly true). Tabulate the lengths of the three principal axes a, b, and c. What do you think this implies for the planet? For more on this system, see Li *et al.* (2010).

2.3 The axis of least effort

The kinetic energy of rotation of a body with a principal moment of inertia I about some axis is given by $E = \frac{1}{2}I\omega^2$, where ω is the angular rotation rate (radians/s). The angular momentum L of a rotating body is given by $L = I\omega$. For fixed angular momentum, show that the kinetic energy of a rotating body is a minimum if it rotates about the axis with the maximum moment of inertia C of the three principal moments $C \geq B \geq A$.

Extra Credit: If a body is rotating stably about its C axis and some internal process in the body redistributes its internal mass and switches the C and B principal axes, what happens to this body? Note that a process of this kind has been proposed for, among others, Enceladus (Nimmo and Pappalardo, 2006).

3

Strength versus gravity

The existence of any differences of height on the Earth's surface is decisive evidence that the internal stress is not hydrostatic. If the Earth was liquid any elevation would spread out horizontally until it disappeared. The only departure of the surface from a spherical form would be the ellipticity; the outer surface would become a level surface, the ocean would cover it to a uniform depth, and that would be the end of us. The fact that we are here implies that the stress departs appreciably from being hydrostatic; …

H. Jeffreys, *Earthquakes and Mountains* (1935)

3.1 Topography and stress

Sir Harold Jeffreys (1891–1989), one of the leading geophysicists of the early twentieth century, was fascinated (one might almost say obsessed) with the strength necessary to support the observed topographic relief on the Earth and Moon. Through several books and numerous papers he made quantitative estimates of the strength of the Earth's interior and compared the results of those estimates to the strength of common rocks.

Jeffreys was not the only earth scientist who grasped the fundamental importance of rock strength. Almost fifty years before Jeffreys, American geologist G. K. Gilbert (1843–1918) wrote in a similar vein:

If the Earth possessed no rigidity, its materials would arrange themselves in accordance with the laws of hydrostatic equilibrium. The matter specifically heaviest would assume the lowest position, and there would be a graduation upward to the matter specifically lightest, which would constitute the entire surface. The surface would be regularly ellipsoidal, and would be completely covered by the ocean. Elevations and depressions, mountains and valleys, continents and ocean basins, are rendered possible by the property of rigidity.

G. K. Gilbert, *Lake Bonneville* (1890)

By *rigidity* Gilbert meant the resistance of an elastic body to a change of shape. He was well aware that this *rigidity* has its limits, and that when some threshold is exceeded Earth materials fail to support any further loads. We call this threshold *strength* and recognize that this material property resists the tendency of gravitational forces to erase all topographic variation on the surface of the Earth and the other solid planets and moons.

The importance of strength is highlighted by a simple computation that Jeffreys included in his masterwork, *The Earth* (1952). This computation is summarized in Box 3.1, where it is shown that, without strength, a topographic feature of breadth w would disappear from the surface of a planet in a time $t_{collapse}$ given by:

$$t_{\text{collapse}} = \sqrt{\frac{\pi}{8} \frac{w}{g}} \tag{3.1}$$

where g is surface gravitational acceleration. Without strength, a mountain 10 km wide on the Earth would collapse in about 20 seconds, and a 100 km wide crater on the moon would disappear in about 3 minutes. Clearly, such features can and do persist for much longer periods of time.

Planetary topography, and the material strength that makes it possible, lend interest and variety to planetary surfaces. However, when seen from a distance, it is clear that the shapes of planets are, nevertheless, very close to spheroids. Only very small asteroids and moons (Phobos and Deimos are examples) depart greatly from a spheroidal shape in equilibrium with their rotation or tidal distortion. Thus, although the strength of planetary materials (rock or ice) is adequate to support a certain amount of topography, it is evidently limited. Such things as 100 km high mountains do not exist on the Earth because strength has limits. The ultimate extremes of altitude on a planet's surface are regulated by the antagonism between the strength of its surface materials and its gravitational field.

Although everyone has an intuitive idea of *strength*, the full quantification of this property is both complex and subtle. Many introductory physics or engineering textbooks present strength as if it were a simple number that can be looked up in the appropriate handbook. This impression is reinforced by handbooks that offer tables of numbers purporting to represent the *strength* of given materials. But further investigation soon reveals that there are different kinds of strength: crushing strength, tensile strength, shear strength, and many others. Strength sometimes seems to depend on the way that forces or loads are applied to the material, and upon other conditions such as pressure, temperature, and even its history of deformation. The various strengths of ductile metals, like iron or aluminum, typically do not depend much on how the load is applied, or how fast it is applied, but common planetary materials behave quite differently.

Quantitative understanding of the relation between topography, strength, and gravity requires, first, some elementary notions of stress and strain and, second, a more detailed understanding of how apparently solid materials resist changes in shape. This chapter introduces the basic concepts of stress, strain, and strength before failure, and applies them to the limits on possible topography. It also introduces the role of time and temperature in limiting the strength of materials and the duration of topographic features. The next chapter examines deformation beyond the strength limit and the tectonic landforms that develop when this limit is exceeded.

Box 3.1 **Collapse of topography on a strengthless planet**

Consider a long mountain ridge of height h, width w and effectively infinite length L standing on a wide, level plain. For simplicity suppose that the profile of the mountain is rectangular, with vertical cliffs of height h bounding both sides (Figure B3.1.1). The surface gravitational acceleration of the planet on which this mountain lies is g, and ρ is the density of the material from which both the mountain and planetary surface are composed.

The weight of the mountain is $\rho ghwL$. If there is no strength, this weight (force) can only be balanced by the inertial resistance of material accelerating beneath the surface, according to Newton's law $F = ma$. The driving force F equals the weight of the mountain, $F = \rho ghwL$. The acceleration a is equal to the second time derivative of the mountain height, $a = \dfrac{d^2 h}{dt^2}$. The mass being accelerated is less easy to compute exactly, but it is approximately the mass enclosed in a half cylinder of radius $w/2$ beneath the mountain (this neglects the mass of the mountain itself, which is not strictly correct, but if h is small compared to w, the mountain mass is only a small correction). The mass is then $m \approx \dfrac{\pi}{8} w^2 L \rho$. This yields a simple, second-order differential equation for the mountain height h as a function of time, t:

$$\frac{d^2 h(t)}{dt^2} = \frac{8}{\pi} \frac{g}{w} h(t). \tag{B3.1.1}$$

This equation has the solution

$$h(t) = h_o e^{-t/t_{\text{collapse}}} \tag{B3.1.2}$$

where h_0 is the initial height of the mountain and the timescale for collapse is given by:

$$t_{\text{collapse}} = \sqrt{\frac{\pi}{8} \frac{w}{g}}. \tag{B3.1.3}$$

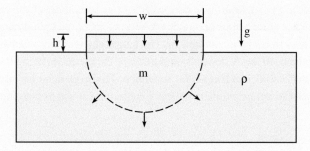

Figure B3.1.1 The dimensions and velocity of a linear collapsing mountain of height h and width w on a strengthless half space of density ρ that is compressed by the surface gravity g on a fluid planet. As the mountain collapses vertically it drives a plug of material of mass m underneath it that flows out through the dashed cylindrical surface.

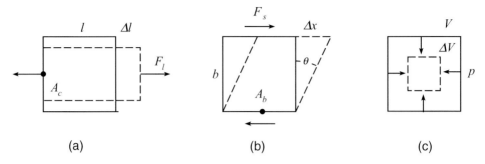

Figure 3.1 Three varieties of strain. (a) Longitudinal strain, in which a block of material of original length l and basal area A_c is extended an amount Δl by a force F_l. (b) Shear strain, in which the top of a block of height b is sheared a distance Δx relative to its base (to an angle θ) by a differential force F_s. (c) Volume strain, in which a block of original volume V is compressed an amount ΔV by a pressure p.

3.2 Stress and strain: a primer

A full exposition of the continuum theory of stress and strain is beyond the scope of this book. For the intimate details, the reader is referred to sources such as Turcotte and Schubert's excellent book *Geodynamics* (2002). A few simple concepts will suffice for a general understanding of planetary surface processes, although the actual computation of stresses under the different loading conditions illustrated later in this chapter requires an application of the full theory of elasticity.

3.2.1 Strain

Strain is a dimensionless measure of deformation. It is a purely geometric concept that is meaningful only in the limit where solids are approximated as continuous materials: All relevant dimensions must be much larger than the atoms of which matter is composed. Historically, the concept of strain was derived from measurements of the change in length of a rod that is either stretched or compressed. When a force is applied parallel to a rod of length l, its length changes by an amount Δl. The length change Δl is observed to be proportional to the length l itself, so Δl depends on the size of the specimen being tested. A measure of deformation that is independent of the specimen size is obtained by taking the ratio of these two quantities to define a dimensionless *longitudinal strain* as (see Figure 3.1a):

$$\varepsilon_l = \frac{\Delta l}{l}. \tag{3.2}$$

A full description of extensional strain in a three-dimensional body requires three perpendicular longitudinal strains, one for each direction in space.

In addition to stretching or compression, a solid can also be deformed by shear, in which one side of a specimen shifts in a direction parallel to the opposite side. In the special case of *simple shear* the top of a layer of thickness b is displaced by a horizontal distance Δx from the bottom, while its thickness b remains constant. In this case the *shear strain* is defined as (Figure 3.1b)

$$\varepsilon_s = \frac{\Delta x}{b} \approx \theta \tag{3.3}$$

where θ is the slope angle of the sheared material. This angle becomes exactly equal to $\Delta x/b$ as Δx approaches zero. Again, because space is three-dimensional there are three independent shear strains.

Mathematically sophisticated readers may note that the six strains are not vector quantities, but form components of a 3×3 symmetric tensor. The three perpendicular longitudinal strains are the diagonal components and the shear strains are the off-diagonal components. An important theorem states that the coordinate axes can always be rotated to a system in which the strain tensor is diagonal. In this coordinate system all strains are longitudinal, although some may be compressional while others are extensional. A general 3×3 matrix has 9 components, not 6. The extra three (which form an antisymmetric tensor) correspond to pure rotations, which, because they do not cause distortions of the material, are wisely excluded from the definition of the strain tensor.

Finally, if all the dimensions are shrunk or expanded equally, the shape is preserved, but the volume V changes, and the resulting deformation is described by the *volume strain* (Figure 3.1c):

$$\varepsilon_V = \frac{\Delta V}{V}. \tag{3.4}$$

There is only one volume strain and it depends entirely on the longitudinal strains, because it can be expressed as the sum of the three perpendicular longitudinal strains.

3.2.2 Stress

Stress is a measure of the forces that cause deformation. In the limit of small deformations it is linearly proportional to strain for an elastic material. Just as the strain is expressed as a ratio of the change in length divided by the length, to make it independent of the size of the test specimen, stress is expressed as the ratio between the force acting on the specimen and its cross-sectional area. Defined in this way, stress is independent of the size of the test specimen and has dimensions of force per unit area, the same as pressure. Thus, if the cross-sectional area of a rod is A_c, and a force F_l is acting to stretch or compress it, the *normal stress* in the rod is defined as:

$$\sigma_l = \frac{F_l}{A_c}. \tag{3.5}$$

Similarly to longitudinal strain, there are three normal stresses, one for each perpendicular direction of space.

Stress is defined as positive when a rod is extended. This makes stress proportional to strain times a positive number. This is a sensible procedure and is used without further comment in engineering texts, in which positive stress is tensional. However, in geologic applications stresses are nearly always compressional. Even when stretching does occur, it is often under conditions of an overall compressional background stress, so that the stress in the extended direction is simply less compressive than the other directions (in this case, the stress is often said to be extensional as opposed to tensional). For such applications it would obviously be simpler if compressional stress is taken as positive. However, such a convention complicates other simple relations in the full theory of stress and strain. Various geological authors have tried special definitions to deal with this problem, although few have gone so far as to make the constants relating stress and strain negative. Turcotte and Schubert, in their otherwise excellent book, actually switch conventions halfway through, and other authors recommend changing the sign of the strain definition. The least drastic convention, and the one followed in this book, is to define pressure as the negative of the average of the three perpendicular stresses, so that compressive (negative) stress always give rise to positive pressure. This means that a compressional stress acting on a rock mass is negative.

In close analogy to shear strains, the three *shear stresses* are defined as the ratio between a deforming force F_s and, in this case, the basal area of the sheared layer A_b:

$$\sigma_s = \frac{F_s}{A_b}. \tag{3.6}$$

Just as for strains, stresses are components of a 3×3 tensor whose diagonal components are the normal stresses and the off-diagonal components are the shear stresses. (The three antisymmetric components of the full 3×3 tensor are torque densities, which almost never arise in practice. We do not consider them further.) Stresses are not vectors: The forces are vectors, but because the forces are divided by an area that also has a direction in space, the stresses are components of a tensor. Stresses, thus, do not point in some direction in space. However, it is always possible to rotate the coordinate axes such that the off-diagonal shear stresses are zero in the new coordinate system, and stresses are sometimes graphically represented as triplets of arrows of different lengths pointing in perpendicular directions. But beware! Such arrows cannot be added or subtracted in the same fashion as vectors!

Finally, in the special case where the stresses are equal in three perpendicular spatial directions, the negative of the force per unit area (all directions are equivalent in this case) is defined as the pressure:

$$P = -\sigma_{\text{vol}} = -\frac{F}{A}. \tag{3.7}$$

Because stresses, and stress differences in particular, play a major role in determining the ability of a solid to resist deformation, it is often convenient to single out the three perpendicular normal stresses in the special coordinate system in which the shear stresses vanish. These special stresses are called *principal stresses* and are frequently denoted σ_1, σ_2, and σ_3 for the maximum (most tensional), intermediate, and minimum (most compressive) normal stress directions – but be careful of stress conventions here: in geologic applications the maximum stress is often taken as the most compressive. So long as this is understood, it causes little difficulty. In the case of hydrostatic stress (pressure) these principal stresses are all equal. When there are three unequal deviatoric stresses the definition of pressure in Equation (3.7) is generalized so that p is equal to the negative average of the three principal stresses. This quantity plays a special role in the tensor description of stress because it is a rotational invariant, the (negative) trace of the stress tensor, divided by 3.

Because of the qualitatively different dependence of strength on pressure and shear, the stress is often separated into a component that depends only on differential stresses, called the *deviatoric stress* (often written as σ' – thereby forming a test of the readers' attentiveness) plus the (negative) pressure. The principal stresses are then written as $\sigma_1'\text{-}p$, $\sigma_2'\text{-}p$ and $\sigma_3'\text{-}p$, whereas the shear stresses are the same as before.

The ultimate strength of many materials is often found to depend on the magnitude of the difference between the maximum and minimum principal stresses, $|\sigma_1 - \sigma_3|$, without any dependence on the intermediate principal stress. A somewhat more complicated measure of the total distortional stress that does take the intermediate principal stress into account is called the *second stress invariant* Σ_2 (pressure is the *first invariant*):

$$\Sigma_2 = \sqrt{\frac{1}{6}\left[\left(\sigma_1 - \sigma_3\right)^2 + \left(\sigma_1 - \sigma_2\right)^2 + \left(\sigma_2 - \sigma_3\right)^2\right]}. \tag{3.8}$$

The factor of 1/6 under the square root is a conventional part of the definition. There is also a *third invariant*, whose role in failure mechanics is more complex, and is not considered further in this text. These quantities are called invariants because their magnitude does not depend on the orientation of the coordinate system. Once their values are established in one coordinate system, they are the same in all.

It may seem surprising that there is no shear stress term in either of these formulas: after all, it is common experience that solids break more readily in shear than under compression. However, shear actually *is* incorporated, although this may not be apparent. The reason is that shear is one of those off-diagonal components that are intentionally eliminated by the coordinate rotation that brings the stress tensor to its diagonal form. It can be shown that a state of pure shear stress σ_s is equivalent to one in which the coordinate axes are rotated 45° and the principal stresses are $\sigma_1 = -\sigma_3 = \sigma_s$.

3.2.3 Stress and strain combined: Hooke's law

English scientist (and Newton's arch-rival) Robert Hooke (1635–1703) recorded some of the first observations of the relation between stress and strain in 1665. Working mainly with springs (Hooke was really interested in clocks) that produce visible deformations under relatively small loads, Hooke hypothesized a linear relation between longitudinal stress and strain, now known as *Hooke's law*:

$$\sigma_l = E\,\varepsilon_l \tag{3.9}$$

where the proportionality constant E has dimensions of pressure and is generally known as Young's modulus, after a much later researcher who studied the extension of elastic rods. Although it was once believed that a single elastic constant is sufficient to describe the stress–strain relation for a given material, it was finally demonstrated in the early 1800s that at least two constants are necessary to characterize an isotropic solid (in fact, for a single crystal, up to 21 elastic constants may be necessary, but here we consider only the minimum required). The second constant is often taken to be the shear modulus μ that relates shear stress to shear strain:

$$\sigma_s = 2\,\mu\,\varepsilon_s. \tag{3.10}$$

The factor of 2 is a conventional part of the definition that derives from the way shear strain is defined. Because there are two elastic constants they can be, and often are, combined in various ways. For example, pressure and volume strain are related by a constant K usually known as the bulk modulus:

$$p = -K\,\varepsilon_V \tag{3.11}$$

(note the minus sign because of the way pressure is defined). Because there are only two independent stress–strain constants, one of these three must obviously be a function of the others: It can be shown that $E = 9K\mu/(3K + \mu)$.

Another useful combination is called Poisson's ratio ν. In Figure 3.1a the extended rod is illustrated as having contracted in the direction perpendicular to its extension. This is a real, observed effect (indeed, the case of pure extension, without lateral contraction, is very difficult to realize in practice as it requires tensional loads perpendicular to the extension axis to maintain a constant cross section). The dimensionless Poisson's ratio is defined as the ratio between the amount of lateral contraction and the longitudinal extension of a laterally unconstrained rod. The deformation illustrated in Figure 3.1a actually involves both a volume change and shear (change of shape), so that the Young's modulus contains contributions from both the bulk modulus and shear modulus. In terms of Poisson's ratio, ν, the Young's modulus is $E = 2(1 + \nu)\mu$.

Relations between stress and strain are generally known as *constitutive relations*. Hooke's law was simply the first of what is now understood to be a large class of possible relationships between deformation (strain) and applied force (stress). Such relations may also involve time: We will shortly meet the concept of viscosity (invented by Newton) that relates the strain *rate* (the derivative of strain with respect to time) to applied stress. In modern times the study of the relation between deformation and stress has reached a high degree of sophistication. This field is now known under the name of *rheology*. Because the materials that make up planets are complex, the rheologic properties of materials as diverse as rock, air, ice, and lava are crucial for an understanding of how the surfaces of planets and moons formed and continue to evolve.

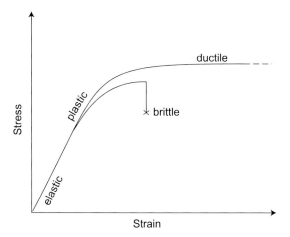

Figure 3.2 In a real solid, stress is linearly proportional to strain only for small stresses and strains (typically only up to a strain of about 0.001). Beyond this limit the relationship becomes non-linear. In this regime the flow deformation may be reversible (non-linear elasticity) or non-reversible (plastic). At even larger strains the material may fracture, losing its strength suddenly in a brittle fracture, or continue to deform to large strains in ductile flow.

The mathematically convenient linear relation between stress and strain does not hold in all, or even in most, real situations: Although stress and strain are always proportional for sufficiently small deformations, when the deformation becomes large enough (and *large* may be a strain of only 0.001 – not even visible to the human eye!) the relation becomes non-linear and catastrophic failure of various kinds may occur (Figure 3.2). Nevertheless, the combination of simple constitutive laws, such as that of Robert Hooke, and the requirement that both internal and external forces are in balance (often known under the name *stress equilibrium*) has been immensely fruitful in explaining the ability of planets to support topographic loads.

3.2.4 Stress, strain, and time: viscosity

Just as ideal elasticity is a useful limit describing the deformation of materials at small strains, so too is the concept of ideal viscosity. Isaac Newton first recognized viscosity on the basis of his extensive experimental studies, and proposed an ideal generalization of his experiments (in fact, Newton proposed this property mainly to undermine his rival Descartes' vortex theory of planetary motion). Ideal elasticity relates shear stress σ_s and shear strain ε_s by a linear equation. Similarly, ideal (or *Newtonian*) viscosity relates the shear stress and shear strain *rate $\dot{\varepsilon}_s$* through a single constant η, the *viscosity*:

$$\sigma_s = 2\eta\dot{\varepsilon}_s. \tag{3.12}$$

Viscosity has dimensions of stress × time, or Pa-s in SI units. The rules for viscous flow are somewhat more complicated than those of elasticity because the volume strain ε_V

cannot be a function of time: If it were, the volume of a viscous substance under pressure would gradually decrease to zero! Discussions of viscous flow must, therefore, pay careful attention to the difference between volume strain and shear strain. In most ideal models the volume strain is set equal to zero; this is called the incompressible limit. A more realistic, but mathematically more complex, approximation is to treat the volume strain as elastic and the shear strain as viscous.

3.3 Linking stress and strain: Jeffreys' theorem

3.3.1 Elastic deformation and topographic support

The earliest and simplest models of topographic support are derived from applications of the classic theory of elasticity. This theory combines the full tensor definitions of stress and strain with a linear Hooke-type relation between stress and strain (with just two elastic constants, the minimum number) and the stress equilibrium equations to derive a closed mathematical system. Within the context of this theory, one can show that, starting from an unstressed initial solid, the stress and strain throughout the solid are uniquely determined by the forces and displacements acting on its surface. Thus, if we approximate a planet, or some well-defined portion of it, as an elastic solid, and treat the weight of topography as a load acting on its surface, the stress differences induced by the topography can be accurately computed throughout its interior.

Of course, this is an unrealistically rosy picture of what is actually possible: The troubles come from the detailed conditions under which elastic theory is valid. Harold Jeffreys, to whom we owe many of the results that follow, was painfully aware of the limitations of the elastic model, and he devoted much effort to understanding both its successes and its failures. The first difficulty is the obvious limitation of elastic behavior to small deformations. Once failure or flow occurs, elastic theory becomes invalid. In principle this can be addressed by numerical methods and is thus inconvenient but not insurmountable. The second, more insidious difficulty stems from the condition of an *unstressed initial solid*. All planetary surfaces with which we are familiar exhibit a long history of change, of repeated events that certainly exceeded the limits of linear elasticity. So to what extent can the near-surface material be considered *initially unstressed*?

All planetary materials have mass and all are subject to gravity, so at a minimum, the rocks beneath the surface must develop sufficient stresses to support their own weight. However, even a liquid, without resistance to deformation (but still resisting volume change!) can support its own weight. It does this by compressing slightly and thus balancing the gravitational force of the overlying material against the much stronger quantum mechanical forces that resist the close approach of atoms (gravity eventually wins this struggle in the stellar collapse to a black hole, but this is far outside the range of planetary processes). The stresses are hydrostatic in this case, and the pressure p a distance h below the surface of a body with uniform density ρ and surface gravitational acceleration g is given by:

$$p = \rho g h. \tag{3.13}$$

Although such *lithostatic* pressures may be very large compared to the stress differences needed to cause rock failure, the large value of the bulk modulus K for most substances ensures that the associated volume strain is small. In this case, we can simply add the lithostatic stress and strain of the subsurface rock to that caused by other loads. This is a consequence of the linearity of the theory of elasticity: Two solutions can always be added to give a third solution, so long as the boundary conditions of the third solution are the sums of those of its components.

If the rock beneath a planet's surface crystallizes from a deep liquid mass, or is heated to such a high temperature that all differential stresses relax after some time, then the lithostatic stress state described above can be accurately considered to be the initial state and the response to any subsequent loads can be computed as elastic additions to this basic state. Unfortunately, most planets are not so cooperative: In most cases one cannot assume that all differential stresses were erased just before the latest episode of topographic loading.

Another elastic solution useful for describing an initial state is derived from the stresses that develop in an initially unstressed and very wide elastic sheet that is suddenly subjected to the force of gravity. The elastic sheet cannot expand laterally; it can only compress vertically. In this case the principal stresses are not all equal (lithostatic), but the vertical stress σ_V and horizontal stresses σ_H differ in magnitude:

$$\sigma_V = -\rho g h$$
$$\sigma_H = -\frac{\nu}{1-\nu} \rho g h \tag{3.14}$$

where ν is Poisson's ratio, which can be no larger than 0.5. Poisson's ratio for most solid rocks is close to 0.25, although it can approach 0.0 for loosely consolidated sediments. In this solution the magnitude of the horizontal stress is smaller than the magnitude of the vertical stress. The difference between the horizontal stresses and the vertical stress increases linearly with depth and so, at some large enough depth failure must occur, but this is often so deep that the solution has great practical value.

Alert readers may wonder that this solution has any practical value at all: the idea that a mass of rock might be assembled in the absence of gravity, which is afterwards magically turned on, seems so artificial that it could not apply to any real situation. However, as demonstrated by Haxby and Turcotte (1976), this is precisely the stress state that develops in a rock mass assembled from the gradual accumulation of a stack of thin, broad and initially stress-free layers. Thus, the stresses that develop in a thick pile of lava flows, or in an accumulating sedimentary basin, are well described by this model. Compilations of vertical and horizontal stress measurements in the Earth (McGarr and Gay, 1978) show that, in many places, such as southern Africa or in sedimentary basins in North America, stresses are bounded between the lithostatic and infinite-layer results (this is not true everywhere: In Canada and much of Europe horizontal stresses are much larger than suggested by these solutions).

Although the two basic states just described are frequently useful, they are certainly not unique: Through all six editions of *The Earth*, Jeffreys invariably emphasized that, due to the generally unknown history of previous deformation, there are an infinite number of stress and strain configurations that are compatible with the presently observed topography. So why did he devote so much time and effort to obtaining elastic solutions when he did not believe that such solutions could be accurate? Jeffreys frequently cited a theorem he called *Castigliano's principle*, which asserts: "Of all states consistent with given external forces, the elastic one implies the least strain energy" (Jeffreys, Ed. 6, Appendix C). Thus, to the extent that the forces acting below a planetary surface tend toward a minimum of energy, the elastic solution delineates the favored minimum. A second reason is that, although a given elastic solution may not represent the complete stress state, it does often indicate how the stresses *change* in response to a small change in the applied loads. For example, the formation of a distant impact crater or a change in planetary spin rate or tidal stresses may cause stress changes that are accurately described by an elastic deformation. In either case, the elastic solutions are of greater significance than the limitations of the strictly conceived elastic model would suggest.

3.3.2 Elastic stress solutions and a limit theorem

Using the full theory of elasticity, stresses can be computed beneath various surface loads, assuming an initially hydrostatic initial state. Contour plots of the second invariant Σ_2 for four of these configurations are shown in Figure 3.3a–d. Figures 3.3a–c apply to long loads intended to represent idealized mountain profiles, originally computed by Jeffreys. Figure 3.3d shows the stress differences underneath an axially symmetric idealized impact crater with a depth/diameter ratio of 0.3.

Although the patterns illustrated by these various solutions are diverse in detail, there are a number of similarities. Most obvious is that the maximum stress differences are not at the surface, but occur some distance below. Thus, most of the weight of a sinusoidal series of mountain ridges is not supported by the strength of the material in the mountains them-selves, but by material some distance below. This is an important lesson (one ignored by the builders of the Tower of Pisa): Foundations are important! The second important lesson is that the maximum stress difference is about 1/3 of the total load itself for all four cases illustrated. These results are summarized in Table 3.1, where the depth to the maximum stress and the maximum stress differences for Figures 3.3a–d are listed.

The first lesson from these solutions, the isolation of the maximum stress region below the surface, is not strictly valid outside the domain of elastic solutions. More sophisti-cated analyses, using the theory of plasticity described below, show that, although first failure upon loading does, indeed, occur where the elastic solution predicts the max-imum stress differences, once this failure has occurred the failure zone may work its way toward the surface, especially if the load has sharp edges, as for a cliff or steep surface slope. The final, visible failure may, thus, involve a surface landslide localized at one of these sharp edges. However, the region over which the strength of the material is

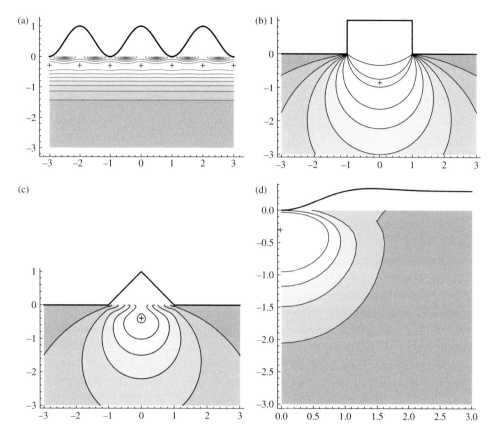

Figure 3.3 Stresses below various loads placed on an originally unstressed elastic half space. Contours are of the second invariant Σ_2 and are drawn at intervals of 0.05, 0.1, 0.15, 0.2, 0.25, 0.3, 0.35, and 0.4 of the maximum load. These plots were constructed by summing the Fourier components of the Airy stress function that satisfies the load boundary conditions. (a) Shows the differential stress magnitudes beneath a series of very long mountains with sinusoidal hills and valleys. (b) Stresses beneath a vertical-sided strip mountain. (c) Stresses beneath a long mountain with a triangular profile and (d) Stresses beneath a circular impact crater with depth/diameter ratio 0.3. Plots are not vertically exaggerated; horizontal dimensions are in units of the load width. The + sign marks the position of the stress maximum in each plot.

exceeded is far broader than such a surface manifestation and is well delineated by the elastic solution.

The second lesson from the elastic analysis is more enduring. Generations of structural engineers have devoted their ingenuity to ways of extending their ability to analyze the maximum stresses that develop in any given structure. The results of this effort (and the subject of a huge literature of its own) are the so-called *limit theorems*. Although theorems of this type do not give the user the detailed distribution of stresses in some complex structure (this must be done on a case-by-case basis using a full knowledge of the structure and its history of loading), they do give some overall constraints on how

Table 3.1 *Elastic stress differences, Poisson's ratio* $v = 0.25$

Load shape	Maximum stress difference $\Sigma_z/\rho gh$	Depth of maximum below surface
Sinusoidal strip, wavelength λ	0.384	$0.289\ \lambda$
Rectangular strip, width w	0.352	$0.865\ w$
Triangular strip, basal width w	0.305	$0.388\ w$
Axisymmetric crater, depth/diameter=0.3, diameter D	0.359	$0.305\ D$

strong materials must be to support some given load, independent of structure and history of construction.

As summarized by Jeffreys, structural limit theorems assure us that to support a surface load of order ρgh, *somewhere* in the body stresses between ½ and ⅓ of this load must be sustained. Furthermore, this stress is generally supported at a depth comparable to the load width (exceptions to this depth rule, such as loads supported by strong, thin plates, usually imply stresses greatly in excess of the minimum).

This fundamental theorem is so important (and so often overlooked in the planetary literature!) that I set it out by itself for emphasis:

Jeffreys' Theorem: The *minimum* stress difference required to support a surface load of ρgh is ($^1/_2$ to $^1/_3$) ρgh. This stress is usually sustained over a region comparable in dimensions to the load.

Of course, this theorem does not prevent much larger stresses from developing in specific situations, but a given topographic load cannot be supported by any smaller stress difference. The value of this theorem is that it can be linked to specific strength models to obtain quick estimates of the maximum topographic variation to be expected on any given Solar System body, even when the specifics of interior structure and history are unknown. An example of this procedure is given in the next section.

3.3.3 A model of planetary topography

Consider a generic planetary body (Figure 3.4) of mass M, average radius $\bar{R}$ and average density $\bar{\rho}$. The surface acceleration of gravity g is:

$$g = -\frac{G\,M}{\bar{R}^2} = -\frac{4}{3}\pi\,G\,\bar{\rho}\,\bar{R} \tag{3.15}$$

where G is Newton's gravitational constant.

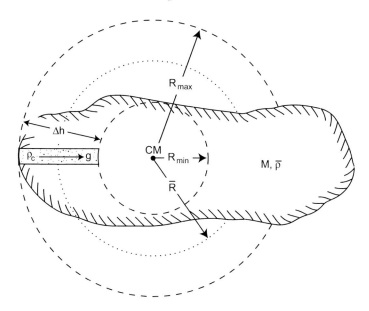

Figure 3.4 A simple model of the gravitational forces in an irregular self-gravitating body such as an asteroid. The average radius is $\bar{R}$ and the maximum and minimum radii for points on the surface are R_{max} and R_{min} from the center of mass CM. The mean density of the object is $\bar{\rho}$.

This relation is exact for a spherical body, and approximate for any other shape. If the surface has topography of order Δh, and its material is of density ρ_c, the surface load imposed by this topographic variation is about $\Delta \sigma = \rho_c g \Delta h$. Applying Jeffreys' theorem, a minimum stress of magnitude Y must be present somewhere in the body's interior:

$$Y \approx \frac{1}{2}|\Delta \sigma| = \frac{2}{3}\pi G \bar{\rho} \rho_c \bar{R} \Delta h. \tag{3.16}$$

Rearranging, we obtain an equation that relates the maximum topographic variation, Δh, to some measure of *strength*, Y.

$$\Delta h \approx \frac{3}{2\pi} \frac{Y}{G \rho_c \bar{\rho}} \frac{1}{\bar{R}}. \tag{3.17}$$

Applying this equation to the Earth, take $\bar{\rho} = 5200$ kg/m³, $\rho_c = 2700$ kg/m³, $\bar{R} = 6340$ km. We find:

$$\Delta h_{\text{Earth}} (m) \approx 80.4\, Y\, (\text{MPa}). \tag{3.18}$$

Taking $Y \approx 100$ MPa, which is about the crushing strength of granite, we see that the Earth can support abut 8 km of topography – not far off the 8850 m height of Mount Everest or the 11 000 m depth of the Marianas trench, when the buoyancy of submerged rock is taken into account. However, the dependence of Δh on $1/\bar{R}$ means that, if Y is the same for all the terrestrial planets, we should expect 8 km high mountains on Venus, 24 km high mountains on Mars and 50 km high mountains on the Moon. As shown in Figures 2.3b and 2.3e, this is not far off for Venus and Mars, but is more than twice the observed topographic range on the Moon in Figure 2.3d. Evidently strength is not the major factor limiting the Moon's topography: History must play a role, too.

Applying this model for topography to the smaller bodies of the Solar System, such as Phobos, this rock strength limitation leads to ridiculous conclusions about the topographic ranges on these bodies (see Problem 3.1 at the end of the chapter). One might be tempted simply to give up and look for factors other than strength that limit topography. However, as we shall see in the next section, a better appreciation of the concept of strength lets us go considerably farther down the strength limitation path. In particular, we need to appreciate the laws that govern the strength of broken rock.

3.4 The nature of strength

3.4.1 Rheology: elastic, viscous, plastic, and more

Rheology is the study of the response of materials to applied stress. Although stemming from roots in prehistory, E. C. Bingham (of whom we will learn much more in Chapter 5) first established it as a scientific discipline in the 1930s. It is not a simple science: Real materials are complex and so is their detailed description. However, much of this complex behavior can be understood in terms of the properties of a number of simple *ideal materials*, which are then compounded to approximate real substances. We have already described ideal elastic and viscous substances. A third ideal behavior is implicit in the idea of *strength*: An ideal *plastic* substance is one which does not undergo any strain at all until the strength reaches some limiting value, after which the strain increases to any extent consistent with other constraints on the material. Of course, no real material behaves in this way, but many materials do not undergo any very large strains until some limiting stress is reached, after which strain increases rapidly. A slightly more realistic model is to compound elastic behavior with plastic yielding to arrive at an elastic-plastic substance that responds to applied stress as an ideal elastic material until the stress exceeds some limit, after which its strain is limited only by system constraints. Then we could add materials whose elastic strain depends on a non-linear function of stress. We can add time dependence by coupling elastic and viscous behavior. And so on.

This section explores some examples of such compound behavior relevant to understanding planetary topography and its long-term evolution. The first topic we examine is the ultimate limits to topographic heights, after which we will look at more realistic limits.

3.4.2 Long-term strength

The ultimate strength of atomic matter. A full understanding of the strength of matter was achieved only in the mid-twentieth century. Despite the triumphs of quantum mechanics in explaining the bulk properties of matter in the early twentieth century, an explanation of strength came much later. The earliest modern attempt to compute the strength of materials from basic principles was a mitigated disaster: Yakov Frenkel (1894–1952), in 1926 (Frenkel, 1926), constructed a simple model of shear resistance (see Box 3.2 for his derivation) that relates the ultimate strength, Y_{ultimate}, of a material to its shear modulus μ:

$$Y_{\text{ultimate}} = \mu/2\pi. \tag{3.19}$$

Box 3.2 The ultimate strength of solids

The first estimate of the theoretical upper limit to the strength of a solid was formulated by Yakov (a.k.a. Jacov or James) Frenkel (1926). Frenkel started from the fact that atoms in a crystal lattice are uniformly spaced at the interatomic distance a. When a solid is subjected to shear strain, each plane of atoms parallel to the direction of the strain shifts a small distance u with respect to the plane immediately above or below. The net shear strain is thus given by $\varepsilon_s = u/a$, and is numerically the same at both the atomic and macroscopic scales (see Figure B3.2.1). The force resisting this deformation increases as one plane of atoms shifts over the adjacent plane, because the length of the bonds between each atom and its neighbor increases. However, when the deformation becomes so large that the atoms of adjacent planes are midway between lattice sites (that is, at a strain ε_s equal to ½), the attraction to the next atom in the adjacent plane equals the attraction from the shifting atom's previous neighbor and the resistance to deformation drops to zero. Further deformation brings each atom into closer proximity to its new neighbor. New bonds form: The atomic plane snaps into a new position, jumping forward by one atomic step.

The force between adjacent atomic planes of a strained crystal is thus periodic, with a repeat distance equal to the interatomic spacing. Frenkel assumed that this periodic function would be the simplest that he could think of: A sine function. He set the force resisting deformation equal to a constant times sin $(2\pi u/a)$. Because the maximum value of the sine function is 1 (when $u = a/4$), the constant equals the ultimate strength of the crystal, Y_{Frenkel}. Thus, he supposed that the shear stress is given by:

$$\sigma_s = Y_{\text{Frenkel}} \sin\left(\frac{2\pi u}{a}\right) = Y_{\text{Frenkel}} \sin(2\pi\varepsilon_s). \tag{B3.2.1}$$

To determine the constant, he noted that very small deformations are elastic, and in this limit $\sigma_s = \mu\varepsilon_s$. Expanding the sine function for very small arguments yields Frenkel's relation for the ultimate strength of a solid in terms of the shear modulus μ,

$$Y_{\text{Frenkel}} = \frac{\mu}{2\pi}. \tag{B3.2.2}$$

Although defect-free solids such as fine whiskers and carbon microtubules can approach this limit, Table 3.2 shows that Frenkel's limit greatly overestimates the strength of real materials, even for rocks at high confining pressures.

Accurate computation of the actual strength of materials is not yet possible, so that measurement and empirical estimates are still necessary to determine the strength of a real substance under conditions of interest to planetary science.

Box 3.2 (**cont.**)

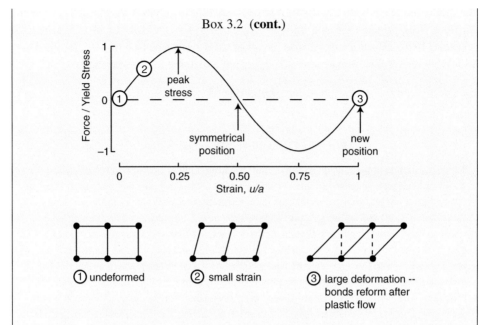

Figure B3.2.1 The theoretical limit to the strength of a solid, based on the model of Yakov Frenkel. The graph on the top shows the sinusoidal dependence of shear force on shear strain, indicating that it is a periodic function of lattice displacement. The lower part of the figure shows the deformation of a lattice at three different strains, correlated with points on the force–strain plot above by the circled numbers: (1) is the undeformed solid, (2) has been subjected to a small strain, while (3) indicates a strain so large that the atoms in the solid are again in register with their neighbors, so that the shear force vanishes.

The shear modulus has been measured for a large variety of materials. It is a bulk property that can now be computed from first principles for many single crystals. Although Frenkel's formula is elegantly simple, it is also grossly inadequate: As shown in Table 3.2, the actual measured strength of most materials is a factor of 100 or more smaller than the Frenkel limit. Nevertheless, the Frenkel limit is not wholly wrong or useless: The strength of a few materials, such as carefully prepared single crystals or fine carbon fibers, does approach this limit. However, the Frenkel limit clearly does not capture the factors controlling the strength of the materials we are likely to meet in planetary interiors.

The principal shortcoming of Frenkel's strength estimate is its neglect of *defects*. Rocks are composed of crystals of individual minerals. While the crystals themselves might be strong, they are bonded through weaker surface interactions. Most igneous rocks, such as granite or basalt, have cooled through a large range of temperatures and, because of the different thermal expansion coefficients of their constituent minerals, tiny grain-boundary cracks develop in abundance. Sedimentary and metamorphic rocks also contain vast numbers of microscopic cracks and weak bonds between individual grains. All rocks contain

Table 3.2 *Theoretical vs. observed material strength*

Solid material	$Y_{ultimate}$ $= \mu/2\pi$ (GPa)[a]	$Y_{observed}$ At $p = 1$ and 5 (GPa)[b]
Iron, Fe	13.0	0.11–1.0
Aluminum, Al	4.14	0.10–0.30
Corundum, Al_2O_3	25.9	0.26–0.92
Periclase, MgO	20.9	0.14–1.07
Quartz (Opal), SiO_2	7.08	0.35–1.8
Forsterite, Mg_2SiO_4	12.9	1.13 ($p = 0.5$ GPa)[c]
Calcite, $CaCO_3$	5.09	0.27–0.84
Halite, NaCl	2.34	0.09–0.29
Ice, H_2O	0.54	0.20–1.0[d]

[a] Elastic moduli from Bass (1995).
[b] At 23°C from Handin (1966) Table 11–9, except as noted.
[c] At 24°C Handin (1966), Table 11–3, Dun Mtn., NZ, peridotite.
[d] At 77–115 K; extrapolated from Beeman *et al.* (1988).

macroscopic cracks in the form of joints. In addition to cracks between mineral grains, the minerals themselves inevitably contain arrays of a peculiar sort of strength-related line defect called *dislocations*. First described in the 1950s by engineers studying the creep elongation of turbine blades in high-temperature jet engines, dislocations flow under stresses far below the Frenkel limit. It is only by studying the properties and interactions of entities such as cracks and dislocations that progress has been made in understanding the practical limitations on the strength of materials.

Although the strength of materials is a large field of endeavor in itself, one too vast to cover in this book (references for this literature are provided at the end of this chapter), the basic take-away lesson is that defects rule the macroscopic strength properties of materials. One cannot expect planetary materials to be stronger than a small fraction of the Frenkel limit. And, in spite of a half-century of progress in understanding the fundamental basis of strength, there are so many complex contributing factors that the strength of a particular material under given conditions of pressure, temperature, and chemical environment is still best determined by experiment.

Traditional material science focuses on the strength properties of metals. Only recently have the much more complex problems presented by the strength of ceramics and geologic materials, such as rocks, become amenable to rational explanation. Naturally, experimenters did not wait for theoreticians to make up models of the strength of rock, so that much of our present understanding is based upon empirical observations.

Built upon sand: The strength of broken rock. Most experts on asteroids now believe that all but the very smallest asteroids (bigger than a few tens of meters in diameter) are better described as fragmented *rubble piles* than as solid chunks of rock. Unlike solid rock, rubble piles have no tensile strength. Their entire ability to resist changes in shape depends on the frictional forces acting across the rock–rock contacts between their components.

Coulomb in 1785 first formulated the laws governing the mechanical behavior of a mass of broken rock (or a pile of sand). Because the frictional resistance at a rock–rock contact is proportional to the force pushing the rocks together, the strength of a mass of broken rock is proportional to the pressure. This fact was first clearly stated by Leonardo da Vinci (1452–1519) in the fifteenth century, but not published by him. Guillaume Amontons (1663–1705) in 1699 resurrected this relation from da Vinci's codices. This behavior is in stark contrast to the strength of ductile metals, such as aluminum or steel, which is nearly independent of pressure. Many experimental studies of the strength of sand or soil show that the mass begins to yield when the applied shear stress σ_s reaches a constant fraction of the overburden pressure p:

$$|\sigma_s| = f_f p = \tan \phi_f \, p \qquad (3.20)$$

where f_f is the *coefficient of friction* and ϕ_f is the related *angle of internal friction*. This angle is also closely related to ϕ_r, the angle of repose, which is the maximum steepness of a slope composed of this material (See Section 8.2.1 and Table 8.1 for more on internal friction). This coefficient is typically about 0.6 for most geologic materials (including water ice well below its freezing point), making ϕ_f about 30°.

Applying this formula to a model of small-body topographic support, the most obvious evidence of topography on small bodies is the difference between their longest and shortest dimensions, $R_{max} - R_{min}$ (refer back to Figure 3.4) This *out of roundness* corresponds to a load of breadth comparable to the mean radius of the body itself, $\bar{R}$. The stress supporting this load is, thus, localized deep within the body. The average pressure in the center of a homogeneous body ($\rho_c = \bar{\rho}$) is $p_{ctr} = \frac{1}{2} \bar{\rho} g \bar{R}$, so that the *strength*, Y, or resistance to yield, is $Y \approx f_f p_{ctr}$. Inserting this into the equation for Δh, we find that a small-body model of strength implies:

$$\Delta h_{smallbody} \approx f_f \bar{R}. \qquad (3.21)$$

Another way of deriving the same result is to note that a constant coefficient of friction implies a constant angle of repose, which is nearly equal to the angle of internal friction. Imagine a hypothetical, maximally out-of-round asteroid constructed in such a way that every slope on its surface is at the angle of repose in its local gravitational field (such a shape has now been constructed by Minton, 2008). Although the precise shape is complex, it is clear that, in traversing the surface of the asteroid from equator to pole, a distance of $(\pi/2)\bar{R}$, up (or down) a constant slope of angle ϕ_r, an elevation change of the order of $(\pi/2)\bar{R} \tan \phi_r$ must take place. This yields essentially the same $\Delta h_{smallbody}$ as above.

This small-body topography model predicts that the maximum fractional deviation from sphericity, $(R_{max} - R_{min})/\bar{R}$, is actually *independent* of size. This is in strong contrast to the constant-strength model derived for the Earth, which suggests that, as a body becomes larger, its shape becomes relatively closer to a spheroid because $(R_{max} - R_{min})/\bar{R} \propto 1/\bar{R}^2$, so that the ratio decreases as $\bar{R}$ increases.

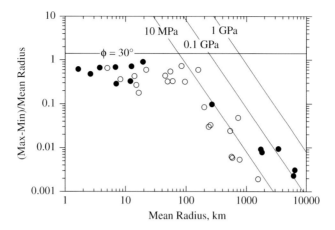

Figure 3.5 The ratio of the maximum elevation difference to the radius for various Solar System bodies as a function of diameter. Up to a diameter of about 200 km, this ratio is nearly constant, as expected for rubble piles supported only by frictional strength. Above this diameter the ratio falls off, consistent with an ultimate planetary crustal strength of about 0.1 GPa. The solid dots are silicate bodies and the open circles are icy. The data suggests that icy bodies are weaker than silicate objects although they have similar friction coefficients.

How do these model predictions fare against reality? Figure 3.5 plots the maximum fractional deviation from sphericity against mean radius for a variety of Solar System objects. It is clear that the topography of the smaller bodies does, indeed, follow a law that suggests the dominance of frictional strength. There is no obvious tendency for the fractional topographic deviation to decrease with increasing size. However, at a radius of about 200 km the frictional relationship breaks off and the maximum topographic deviations of the larger planets and moons decrease sharply with increasing diameter, following an approximate $1/\bar{R}^2$ dependence on the log–log plot. For these large objects greater size does imply greater smoothness. The trend of the curve for larger planetary objects suggests that the ultimate strength of planetary crusts is about 0.1 GPa.

The constancy of the maximum fractional deviation for small objects is a direct consequence of the ability of pressure to increase the strength of broken rock materials. Obviously, however, this frictional increase in strength has its limits. This fact is also clear from laboratory measurements of rock strength: As shown in Figure 3.6, the frictional regime holds up to some maximum stress, generally a few GPa, when the intrinsic strength of the rock is reached and yielding occurs in spite of increasing overburden pressures. As in the large–planet topography model, it seems that the ultimate limit to topography lies in the ultimate ability of matter to resist deformation. It is thus worth inquiring just what determines this resistance.

David Griggs and the strength of rocks. The most obvious feature of the rocks outcropping on the surface of the Earth is that they are pervaded by fractures at all scales. How these fractures actually form, however, is much less obvious. It took many years before experimenters could reproduce the pressures and temperatures existing in the Earth's

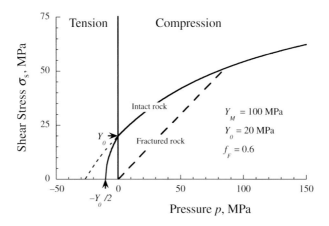

Figure 3.6 Yield stress of a *typical* intact rock specimen (heavy line) described by the Lundborg strength envelope, Equation (3.23). Note the substantial tensional strength (equal to $Y_0/2$ by the Brace construction, which is, nevertheless, weaker than the extrapolation of the Lundborg strength envelope, shown by the dotted line, would suggest) indicated on the negative pressure axis. Shown also as a heavy dashed line is the yield curve for a fractured rock specimen for which the shear resistance is entirely due to friction.

interior and come to an understanding of how rocks break. Indeed, this is still an active area of research in the earth sciences. David Griggs (1911–1974) was one of the first people to systematically investigate rock fracturing under high pressures and temperatures. Griggs' interest in geologic processes began as a boy, when he accompanied his father, geologist Robert Griggs, on a National Geographic expedition to study the deposits of the famous 1912 eruption of Mount Katmai in Alaska (Griggs, 1922). From his experience in the field, David decided to study how rocks break deep within the Earth. He sought out Percy Bridgeman at Harvard University and signed on as his graduate student in 1933. Bridgeman's laboratory was one of the few places in the world where pressures approaching those deep in the Earth's crust could be attained.

Griggs eventually perfected an apparatus widely known as a *Griggs' rig* that could both compress and heat a small rock sample, typically a cylinder a few centimeters in length and diameter, while subjecting it to controlled differential stresses. Continuing his work after World War II at UCLA, and accompanied by a growing number of similarly motivated experimenters, he showed that, unlike metals, the fracture strength of rock is a strong function of both pressure and temperature.

It has long been known that metals and alloys, such as iron or steel, fail at similar stresses under both compression and tension. Ideal plasticity is a useful approximation to metal failure, in which half the stress *difference* at failure (equivalent to the shear stress through a coordinate rotation) is assumed to be a constant Y, the *yield stress*:

$$|\sigma_s| = \frac{|\sigma_1 - \sigma_3|}{2} = Y. \tag{3.22}$$

Table 3.3 *Lundborg strength parameters for representative rocks*

Rock	Friction coefficient, f_f	Cohesion, Y_0(MPa)	Von Mises plastic limit, Y_M, (MPa)
Granite I	2.0	60	970
Granite II	2.5	50	1170
Quartzite	2.0	60	610
Gray slate	1.8	30	570
Black slate	1.0	60	480
Limestone I	1.2	30	870
Limestone II	1.0	20	1020
Sandstone	0.7	20	900

Data from Lundborg (1968).

The yield stress of metals is, to a good approximation, independent of pressure and strain, although it declines with increasing temperature. Because of its utility in engineering, the theory of failure of ideally plastic materials is highly developed, in spite of serious mathematical difficulties that stem from this very lack of dependence on strain (Hill, 1950).

Experimental studies of rock fracture show, however, that the strength of rock depends very strongly on pressure, at least up to pressures approaching 5 GPa (50 kilobars). Many analytic representations of the failure strength of rock have been proposed; among them, one that seems to fit many materials was suggested by Lundborg (1968) for unfractured rock:

$$|\sigma_s| = Y_0 + \frac{f_f\, p}{1 + \left(\dfrac{f_f\, p}{Y_M - Y_0}\right)} \tag{3.23}$$

where Y_0 is the strength at zero pressure, often called *cohesion*, and Y_M is known as the von Mises plastic limit of the material. Y_M limits the maximum stress that can be achieved at arbitrarily high pressure. The Lundborg form of the failure law is illustrated in Figure 3.6 and some representative values of the parameters are listed in Table 3.3.

Although the Lundborg law, and others like it, gives a good description of the failure of rock over the full range of pressures from very low to very high, much more data has been collected in the low pressure regime where a linear version is generally adequate. Thus, when $p \ll Y_M$,

$$|\sigma_s| \approx Y_0 + f_f\, p \quad \text{for} \quad p \ll Y_M. \tag{3.24}$$

Table 3.4 lists representative values of Y_0 and f_f for a small number of materials, ranging from a hard igneous rock (at crustal temperatures) to weak sedimentary rock.

Table 3.4 *Low-pressure failure envelope for representative rocks*

Rock	Friction coefficient, f_f	Cohesion, Y_0(MPa)
Westerly granite @ 500°C	0.6	50
Pennant sandstone @ 25°C	0.97	35
Limestone @ 25°C	0.75–1.6	3.5–35
Siltstone @ 25°C	0.55	21
Chalk @ 25°C	0.38	0.9

Data from Handin (1966).

The sloping, low-pressure portion of the failure law illustrated in Figure 3.6 is superficially similar to that of sand. However, in this case the pressure coefficient f_f is less obviously related to friction, although it is often referred to as a coefficient of *internal friction*, presumably because it is dimensionless and relates strength linearly to overburden pressure, as does the true friction coefficient. Numerically, it is also similar to the coefficient of rock-on-rock friction, although the reader should not confuse the two: f_f is the (approximate) linear slope of the strength envelope that defines the stress conditions under which intact rock fails, whereas f_B is the (static or starting) coefficient of friction of a pre-existing planar rock fracture sliding over another. The difference between these two curves is responsible for the *brittle–ductile* transition that gives rise to discrete faults in rock, as will be discussed in more detail in Section 4.6.1.

Extensive tables of the strength envelopes of rocks under various conditions can be found in Handin (1966) and Lockner (1995). The ultimate strength limit of about 0.1 up to 1 GPa for real rocks is in fair agreement with the observed trend of topographic deviations on the larger planets illustrated in Figure 3.5. It, thus, appears that we presently have a good first-order understanding of the strength properties of planetary bodies, although many details remain to be worked out.

The presence of pre-existing fractures in most large rock masses greatly complicates analyses of the strength of rock. The actual strength of a large volume of rock generally lies somewhere between that of intact rock and that defined by the coefficient of friction (the dashed line in Figure 3.6). A constant value of the friction on a pre-existing fracture, $f_B \simeq 0.85$ (up to a mean pressure p of about 100 MPa; the slope is somewhat less at larger pressure) is often known as *Byerlee's law* after the researcher who showed that this value describes the friction of a wide variety of rock surfaces (Byerlee, 1978). In its exact form Byerlee's law states:

$$\sigma_s = \begin{cases} 0.85\,\sigma_n & \sigma_n < 200 \text{ MPa} \\ 50 + 0.6\,\sigma_n & \sigma_n \geq 200 \text{ MPa} \end{cases} \qquad (3.25)$$

where σ_n is the normal stress across a fracture, σ_s is the shear stress and all stresses are in megapascals.

Note that the *mean pressure*, p, in Equation (3.23) is somewhat confusingly equal to the negative of either one-half of the sum of the maximum and minimum principal stresses, or (more correctly, if less frequently seen) to one-third of the sum of all three principal stresses. A similar equation is often written in which, in the location occupied by the term p in Equation (3.23), a term for the normal stress acting across the failure plane appears instead. Byerlee's law is strictly valid only for this normal stress. The disadvantage of this formulation is that the failure plane must be known before the equation can be applied. Thus, for the present goal of defining a strength envelope, a formulation in terms of stress invariants (pressure and shear stress) is preferable. The wary user of data tables is careful to make sure which definitions are in use before accepting a given *coefficient of internal friction* at face value!

The *mean pressure*, p, in the Equation (3.23) must be modified by subtracting the pore fluid pressure, $p \rightarrow p - p_f$, when the rock is pervaded by a fluid that itself is at some hydrostatic pressure p_f. This modification is very important when a fluid such as water or oil on Earth, or methane on Titan, is present. It was first introduced by Terzaghi (1943) for soils, and by Hubbert and Rubey (1959) for rocks. Its detailed implications are the subject of a large literature. It will be discussed further in Section 8.2.1, but suffice it to say now that high fluid-pore pressures cause substantial weakening of rock through this pressure subtraction effect.

The coefficient Y_0 in Equation (3.23) is the zero-pressure strength or cohesion. Mathematically, it is the intercept of the strength envelope with the zero-pressure axis (see Figure 3.6). Physically, it represents the adhesion of crystals in the rock to one another and can range from only a few megapascals for weak sedimentary rocks to several tenths of a gigapascal for intact granite. It is strongly affected by pre-existing cracks in the rock and drops to zero in a fully fractured rock mass. An extrapolation of this line to negative values of p intercepts the pressure axis (zero shear stress) at $p_T = -Y_0/f_f$. This intercept corresponds to the tensile strength of the rock. The linear extrapolation yields an overestimate of the actual yield stress by a factor of two to three: More sophisticated models based on crack theory (Brace, 1960) give a different, and more accurate, analytic form for tensile stresses that is indicated by the heavy yield curve on Figure 3.6.

The slope of the failure curve decreases at large values of the average pressure, and the maximum shear stress that the rock can sustain approaches a constant Y_M, independent of pressure. This rollover occurs when the frictional stress of sliding on inter- and intracrystalline cracks approaches the intrinsic strength of the individual crystals. A full understanding of this process is still under development, but the general outlines are now in fairly good agreement with observations (Ashby and Sammis, 1990). This change in the dependence of the strength on pressure is known as the *brittle–ductile* transition, for reasons that will be discussed in more detail in the next chapter, Section 4.6.1. It occurs at, or near, the point where the failure curve for fractured rock crosses that for intact rock in Figure 3.6.

The ultimate yield stress Y_M in Equation (3.23) is, as shown in Figure 3.6, still far below the Frenkel limit because of intra-crystalline defects such as dislocations. Although independent of pressure, by definition, it does depend strongly on temperature. There is no universal law for this temperature dependence, which must be determined empirically, but it is clear that the strength must vanish at the melting temperature, T_m. Using this hint, a widely used approximation to the temperature dependence is to multiply both Y_0 and Y_M by the same factor:

$$F_T = \left(\frac{T - T_m}{T_m} \right)^2 \qquad (3.26)$$

which assures that the strength falls to zero as the temperature approaches the melting point. The exponent in this relation is purely empirical, chosen to fit a large body of data on both metals and rocks.

3.4.3 Creep: strength cannot endure

David Griggs and the flow of rocks. When David Griggs began his now-classic work in 1933 he was already the veteran of many geologic field excursions and knew from personal experience that the rocks of the Earth's crust often show signs of large amounts of deformation *without* fracture. This fluid-like deformation had long been attributed to the high pressure and temperature within the inaccessible depths of the Earth, but no one understood the rates or conditions under which this flow occurred. Griggs began his lifework with a relatively simple apparatus that measured the slow deformation of rocks under an applied load as a function of time, initially working at room temperature and pressure (Figure 3.7). Although he found that most rocks deform elastically only for periods of time less than a year, he discovered a few that exhibited slow *pseudoviscous flow* or creep according to a simple law relating the strain ε and time t:

$$\varepsilon = A + B \log t + C t \qquad (3.27)$$

where the constant A represents *instantaneous* elastic deformation, B a kind of decelerating creep now often called *primary* creep, and C is the rate of steady, long-term flow. Although the primary creep term is important for short-term flow processes, such as the response to fluctuating tidal stresses or the small strains that accompany planetary reorientation and spin changes, most geologic interest centers on the third, steady-state term, because it represents deformation that increases steadily with increasing time, apparently without limit. In this respect the flow of rocks resembles that of more familiar viscous liquids, such as honey, motor oil or tar.

Sixty years of subsequent research by Griggs and a large cadre of laboratory geologists who recognized the importance of this research has shown that the rate of steady-state creep is a function of stress, temperature, and pressure, as well as rock composition, grain size, presence or absence of water, trace elements, and a host of other factors. Most creep experiments can be fit by a formula of the form:

$$C = \dot{\varepsilon}_{\text{steady}} = A_c \, \sigma^n \, e^{-\frac{Q^*}{RT}} \qquad (3.28)$$

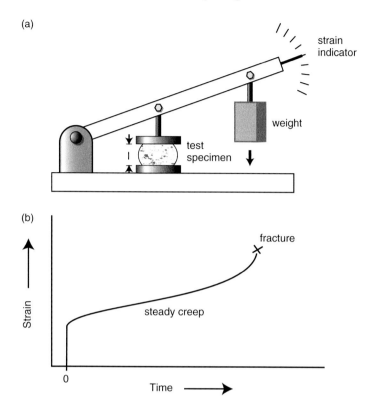

Figure 3.7 Schematic representation of a creep experiment on rock, similar to Griggs' 1933 room-temperature measurements. (a) The test specimen, of original length *l*, is mechanically loaded (by a weight and a lever) while its deflection is measured on a sensitive scale. (b) Schematic creep curve, showing strain as a function of time after loading. The curve shows three distinct portions after the initial elastic deflection: A period of decelerating creep, a long period of steady creep and, for lab specimens, a final acceleration just before rupture.

where A_c is a constant with dimensions (stress)$^{-n}$ time^{-1}, σ is deviatoric stress, n a dimensionless constant, Q^* is *activation enthalpy* (this term incorporates most of the pressure dependence because $Q^* = E^* + pV^*$, where p is pressure and E^* and V^* are constants), R the gas constant, and T is absolute temperature. The dot over the strain ε, following Newton's *fluxion* notation, indicates differentiation with respect to time.

It is often convenient to express the rate of steady-state creep, Equation (3.28), in terms of an *effective* viscosity, even though it depends on the stress level. Adapting the definition of viscosity, Equation (3.12), the effective viscosity η_{eff} is defined as:

$$\eta_{\mathrm{eff}} = \frac{\sigma_s}{2\,\dot{\varepsilon}_{\mathrm{steady}}} = \frac{e^{\frac{Q^*}{RT}}}{2\,A_c\sigma^{n-1}}. \tag{3.29}$$

This definition of viscosity generalizes Newton's original definition, which applies to the case $n = 1$. It has now become common to refer to the case $n = 1$ as "Newtonian viscosity" and to use the term "viscosity" in the broader sense for any value of n, as long as it refers to a flow law in which the strain rate is a function of stress.

Unlike viscous liquids, the power n relating stress and strain rate is usually larger than 1 for creeping rocks and minerals, justifying the use of the term "pseudoviscous" for this kind of flow. Doubling the stress on materials such as ice or olivine may cause the creep rate to increase by a factor of 10, in strong contrast to ideally viscous materials in which the creep rate only doubles. It is also important to realize that creep rate depends exponentially on the temperature. Although rocks deform very slowly at low temperatures, as the temperature climbs toward the melting point the creep rate increases rapidly (by as much as a factor of 10 for each 100°C increase in temperature for many rocks). A useful approximation is that for most materials, creep rates become important over geologic time periods (millions of years, which implies $\dot{\varepsilon}_{\text{steady}} \approx 10^{-13}$ s^{-1} or less) when the temperature reaches one-half the melting temperature, $T \sim 1/2T_m$. A useful simplification of the temperature dependence of the creep rate is to absorb the activation energy and melting temperature into a constant g and express the temperature as the dimensionless ratio T/T_m, the *homologous temperature*:

$$C = \dot{\varepsilon}_{\text{steady}} = A_c \, \sigma^n \, e^{-g\frac{T_m}{T}}. \tag{3.30}$$

Table 3.5 gives typical values for A_c, n, Q^*, T_m, and g for a few materials of geologic and planetary interest.

Extensive tables, such as that of Kirby and Kronenberg (1987a, b) and Evans and Kohlstedt (1995), have been compiled to categorize the creep of rocks, and theoretical models have been developed to explain this flow behavior in terms of diffusion and dislocation motion (e.g. Evans and Kohlstedt, 1995; Poirier, 1985). However, for the purposes of this book the principal concept to remember is that at high temperatures rocks can flow like liquids over geologic timescales.

J. C. Maxwell and the viscosity of "elastic solids." Observation and experiment have taught us that cool materials (that is, materials at temperatures well below their melting point) deform elastically under applied loads, while hot materials gradually flow. Elastic behavior is mostly recoverable: that is, when the load is removed the deformation reverses itself; while viscous flow is not recoverable: when the load is removed the deformation remains. The alert reader might wonder how these very different types of behavior can be reconciled at intermediate temperatures: At what point does the elastic response stop and viscous flow take over?

This important question received a definitive answer from an unlikely source. Most people who recognize the name of nineteenth-century physicist J. C. Maxwell (1831–1879) think immediately of Maxwell's equations that describe electric and magnetic fields, or perhaps of his contributions to thermodynamics and statistical mechanics. In fact, it was during his 1867 study of the viscosity of gases that Maxwell faced the puzzling dichotomy

Table 3.5 *Creep properties of selected materials*

Material	A_c (MPa^{-n} s)	n	$Q*$ (kJ/mol)	T_m (K)	$g = Q/RT_m$
Olivine				2200	27
Dry	1.2×10^2	3.0	502		
Wet	2.0×10^3	3.0	420		
Diabase				1100	53
Dry[a]	5.4–347	4.7	485		
Wet[b]	6×10^{-2}	3.05	276		
Quartz				1996	8.1
Dry	1.3×10^{-6}	2.7	134		
Wet	2.0×10^{-2}	1.8	167		
Granite (Westerly)				1320	12.7
Dry[c]	2.5×10^{-9}	3.4	139		
Wet[c]	2.0×10^{-4}	1.9	137		
Anorthosite	3.2×10^{-4}	3.2	238	1400	20.5
Halite, NaCl	6.3	5.3	102	1074	11.4
Water ice, Ih,[d] T > 258 K $\sigma > 1$ MPa	6.3×10^{28}	4	181	273	80
Solid CO_2, 150 < T < 190[e]	4.4×10^3	4.5	31	217	17
Limestone, Dry, Solenhofen ls	2.5×10^3	4.7	298	1520	23.6

Data is from Evans and Kohlstedt (1995), except as noted:
[a] Mackwell *et al.* (1998)
[b] Caristan (1982)
[c] Kirby and Kronenberg (1987b)
[d] Durham and Stern (2001)
[e] Durham *et al.* (1999)

between the elastic and viscous behavior of solids (Maxwell, 1867). His insight came from what might seem like an annoying detail: The steel wire supporting the torsion pendulum he was using to measure gas viscosity exhibited viscous behavior of its own. He invented a theory of what are now known as viscoelastic materials to separate the viscosity of the pendulum wire from that of the gas.

Maxwell proceeded by postulating that the total deformation of his wire is the simple sum of the elastic plus the viscous strain, $\varepsilon_{total} = \varepsilon_{elastic} + \varepsilon_{viscous}$. He supposed that each strain would develop under the influence of the same stress, obeying the equations previously stated for ideal elastic and viscous behavior. His equation, however, suffers a serious mathematical problem, because the viscous strain is not determined directly from the stress: The stress determines only the strain *rate*. It is possible to write the viscous strain as the time integral of the strain rate, but it is more straightforward to differentiate both sides of Equation (3.10) with respect to time and sum the result to obtain the fundamental equation for a *Maxwell viscoelastic* substance, $\dot{\varepsilon}_{total} = \dot{\varepsilon}_{elastic} + \dot{\varepsilon}_{viscous}$. Inserting the definitions of each term:

$$\dot{\varepsilon}_{total} = \dot{\sigma}/2\mu + \sigma/2\eta. \qquad (3.31)$$

This equation embodies both an elastic response for loads applied quickly and viscous flow for long sustained loads. Its full solution is complex because volume strain and shear strain must be treated differently in each term of the full tensor equation. However, it is not necessary to actually solve this equation to attain an insight of major importance. Simple dimensional analysis shows that the ratio of the viscosity η to the elastic shear modulus μ has the dimensions of time. This ratio is known as the *Maxwell time* τ_M and it plays a fundamental role in the transition from elastic to viscous behavior. Its definition is:

$$\tau_M \equiv \frac{\eta}{\mu}. \qquad (3.32)$$

If a load is applied instantaneously to a Maxwell viscoelastic material, then held constant, the Maxwell time is equal to the length of time that passes before the accumulated viscous strain equals the instantaneous elastic strain. Thus, for times shorter than the Maxwell time, the material response is dominated by the elastic deformation. For times longer than the Maxwell time, the response is essentially viscous. Maxwell supposed that even water must act as an elastic material on a short enough timescale, but he computed this time as about 10^{-13} s – unobservably small in the late 1800s. However, he did later succeed in observing both elastic and viscous behavior in Canada balsam (pine tree sap).

Although Equation (3.32) was derived from the equations for ideal elastic and viscous substances, a generalization of the idea of Maxwell time can be applied even to pseudoviscous materials that do not obey the equation of ideal viscosity: The generalized Maxwell time is the length of time over which creep must act for the total creep strain to equal the elastic strain. In the form of an equation:

$$\tau_M = \frac{\text{(elastic strain)}}{\text{(creep strain rate)}} = \frac{\varepsilon_{elastic}}{\dot{\varepsilon}_{creep}}. \qquad (3.33)$$

The Maxwell time is often surprisingly short. This is because the elastic strain in most geologic materials is invisibly small – typically only about 0.0001, even for stresses near fracture. For this reason Australian geologist S. Warren Carey invented a term, which he called *rheidity* (Carey, 1953), and for which he proposed a timescale of exactly 1000 τ_M. Although this rheidity concept adds nothing fundamental to the idea of Maxwell time, it does give an estimate of the time necessary for viscous or pseudoviscous flow to become *visible* to the human eye.

Most children are familiar with the high-polymer material known as *Silly Putty*™, which behaves as a brittle elastic material on a short timescale – it can be fractured by a hammer blow – but flows like a liquid when left undisturbed for a long period. It is less widely appreciated that *all* materials behave this way, if only the timescale is chosen appropriately. Water ice is another example: ice cubes in common experience are brittle elastic materials,

Table 3.6 *Maxwell time and rheidity time for various materials*

Material	Shear modulus, μ (GPa)	Viscosity, η (Pa–s)	Maxwell time, τ_M	Rheidity time, τ_R
Soda-lime glass @ 250°C	25	4.3×10^{11}	17 s	4.8 hr
Glacier ice @ 0°C	4	$\sim 10^{13}$	42 min	29 days
Halite @ 200°C	20	3×10^{16}	17 days	48 yr
Earth mantle from glacial rebound	50	10^{20}	66 yr	66 000 yr

but it is obvious from glaciers that ice flows like a liquid over long timescales. Table 3.6 lists the Maxwell and rheidity times for a number of geologic materials.

Maxwell viscoelasticity neatly resolves other apparent paradoxes of earth science. William Thomson, later Lord Kelvin, used the difference between solid Earth tides and ocean tides to show that the Earth's elastic modulus is similar to that of steel. Kelvin himself, and Harold Jeffreys after him, never accepted the idea that over long intervals of time the Earth's mantle could flow like a liquid (England *et al.*, 2007). However, our modern understanding of mantle convection and plate tectonics requires just that. The resolution of this conundrum is through Maxwell viscoelasticity: Table 3.6 shows that the Earth's mantle (which is mainly composed of the mineral olivine) has a Maxwell time of about 100 years. Thus, the mantle behaves as an elastic solid with respect to the month-long tidal deformation (or even the 22-month Chandler wobble of its axis), and yet flows like a liquid during the 100 Myr timescale of mantle convection.

Ironically, Lord Kelvin himself provided one of the most graphic illustrations of the role of viscoelastic flow in the Earth and other planets. Kelvin loved mechanical models, often stating that he could "never satisfy myself until I can make a mechanical model of a thing" (Kargon and Achinstein, 1987). In his famous *Baltimore Lectures* of 1884, Kelvin described a classroom model in which he floated a layer of "Scottish shoemaker's wax" on a beaker of water. He submerged a number of corks underneath the wax and set a few lead bullets on top (Figure 3.8). Over the course of a semester, the bullets sank into the viscoelastic wax while the corks burrowed upward into it. By the semester's end, the bullets had dropped to the bottom of the beaker and the corks had emerged on top. While he could not have found a better analogy for the geologic behavior of the Earth, Kelvin himself used this model to illustrate his concept of the hypothetical aether, to show how the Earth could move through the all-pervading aether apparently without friction, while light waves traveled like elastic waves in this universal substance.

Maxwell's model of viscoelastic flow turns out to be only one of many possible variations. Depending upon how the elastic and viscous strains combine (coupled, more generally, with the possibility of plastic flow), a variety of viscoelastic (or elasto-visco-plastic) responses to stress are possible. Kelvin himself proposed a model in which elastic and viscous stresses are summed and the strains are then set equal. Now known as the

Figure 3.8 Lord Kelvin's class demonstration. Over the course of a semester, Kelvin showed that dense bullets would sink and light corks would rise through a layer of viscoelastic wax on top of a beaker of water. Although Kelvin himself did not intend it as such, it provides an apt illustration of the long-term flow properties of a planetary mantle.

Kelvin–Voigt model, it provides a better description of short-term flow, such as primary creep (the B term in Equation (3.27)), than does the Maxwell model (the C term in Equation (3.27)). The Kelvin–Voigt model may provide a good description of flow under short-term oscillatory stresses, such as tidal flexing, while the Maxwell model is more appropriate for steady, long-term deformation in which the total strain can increase without limit.

3.4.4 Planetary strength profiles

It should now be clear to the reader that the "strength" available to support topographical features on a planet is a complex issue. Topographic loads can be supported by the resistance to deformation exerted by cold solids, as well as by slow viscous or pseudoviscous deformation of warm solid materials. "Strength," thus, depends on pressure, temperature, and the duration of the load, among many other modifying factors such as the pressure of included fluids, presence or absence of chemical weakening agents such as water, and even the history of previous deformation.

   Given this complex response of "solid" materials to differential stresses, can one make *any* simple generalizations at all about the ability of planets to support topographical features? One simple observation is that strength generally decreases as temperature rises (and vanishes at the melting temperature). Most planets are warmer inside than on their outsides, although exceptions to even this apparently obvious situation occur during planetary accumulation and very large impact events. Thus, a simple generalization is that most of a planet's strength resides near its surface. This observation gives rise to the idea of a *lithosphere*, a relatively thin shell near the surface of planets large enough to have hot interiors, which embodies most of its long-term strength. The outermost part of the lithosphere is usually cool enough to exhibit brittle strength, while deeper portions resist loads by slow deformation (this definition of the lithosphere is oversimplified: It will be made more precise in the next chapter when the concept of Maxwell time is applied).

   These ideas are used to construct *strength profiles*; envelopes that show the maximum differential stresses that can be supported as a function of depth in any given planet. Besides

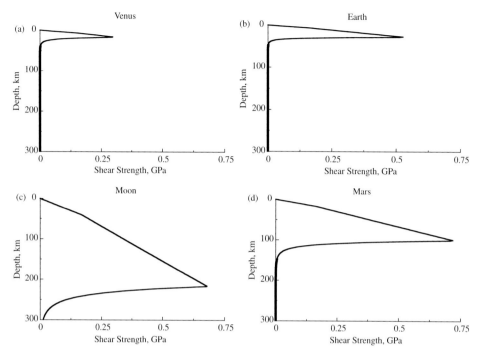

Figure 3.9 Strength profiles for the lithospheres of (a) Venus, (b) the Earth, (c) the Moon, and (d) Mars. The upper parts of the curve are controlled by friction on pre-existing fractures and, thus, follow Byerlee's law, Equation (3.25). The lower portions are cut off by creep in olivine, with parameters listed in Table 3.5. Temperatures are computed from mantle heat flow on the Earth and by assuming an average chondritic composition for the other planets. Thermal conductivity is taken to be 3.0 W/m-K. The strain rate is 10^{-13} s^{-1}, although the curves are only slightly different at 10^{-15} s^{-1}.

the failure laws themselves, these profiles require knowledge (or estimates) of the temperature and pressure as a function of depth. Because many factors influence strength, these curves are oversimplifications of the actual facts, and are useful only for general guidance to the strength levels available. There are also many variants of this kind of curve, each designed to show some especially pertinent relationship. To show the effect of strain rate, the strength profiles constructed here are assumed to reflect deformation at some particular strain rate. The cohesive strength of cold rocks is neglected because it is assumed that beyond some small strain the rock will fracture, so that only frictional strength continues to act.

Figure 3.9 shows computed strength profiles for Venus, the Earth, the Moon, and Mars. In all of these curves the upper cold portion is assumed to follow Byerlee's law, while the lower portion is controlled by the rheological properties of the common mantle mineral, olivine. Each assumes that the lithosphere is stretched at a strain rate of 10^{-13} s^{-1}, a typical plate-tectonic strain rate on the Earth. Very similar curves would result for lithospheric compression, with slightly higher frictional strengths. Lower strain rates decrease the stresses in the lower part of the lithosphere and push the cusp marking the transition between friction and pseudoviscous flow to shallower depths. The sharpness of the cusp is artificial: In reality the transition is probably gradual, but the flow laws are not known well enough to represent this accurately.

The main lesson from these curves is that the maximum strength in a planet's interior resides neither at its surface, due to the pressure dependence of rock friction, nor at great depths, due to the weakening effect of high temperatures. The maximum strength is at an intermediate depth, and it is at this depth that most of the forces that support long-term topography are exerted.

3.5 Mechanisms of topographic support

3.5.1 *Plastic strength: Jeffreys' limit again*

Short-wavelength loads on a planetary lithosphere are supported by plastic strength, as described in Section 3.3.2. Stress differences reach approximately 1/3 of the vertical load and are supported at a depth comparable to the width of the load. The meaning of "short wavelength" is defined by reference to the thickness of the lithosphere. If the breadth of the load is comparable to or larger than the lithosphere's own thickness, then new factors come into play and more sophisticated models, such as the flexural models discussed later, in Section 3.5.5, must be brought into play. These new factors generally decrease the ability of the lithosphere to support the load: Jeffreys' theorem must always hold, but it does not guarantee that the stresses are not much larger than the minimum given by his limit. Direct support of a load by a strong material right underneath is always the most effective way to carry the weight of a topographic feature.

3.5.2 *Viscous relaxation of topography*

Just as a mound created on the surface of a dish of honey gradually relaxes to a flat surface, so topographic features formed on the surface of a planet whose interior materials obey a viscous or pseudo-viscous flow law will eventually relax to a flat plain. Because of the complexity of the full non-linear pseudoviscous creep law determined for real rocks, most analyses of the viscous relaxation of topography approximate the actual flow law as Newtonian (at present, numerical methods are rapidly superseding such crude approximations, but there is still much to be learned from "back of the envelope" computations using Newtonian viscosity). The viscosity determined from such an analysis is then termed an "effective viscosity," η_{eff}, and its value must be accompanied by an estimate of the stress at which it is determined. Although such a procedure is not exact, and in some special cases may be seriously misleading, it often yields useful insights into the mechanical behavior of a planetary body, so long as the user understands what the effective viscosity really is, and does not mistake it for what it is not.

Viscous relaxation acts to gradually erase any deviation from a "level" planetary surface (that is, from a surface coinciding with a gravitational equipotential surface). Thus, both elevations and depressions will gradually fade away with time. How much time this requires depends on the viscosity. If the viscosity is large enough, even a few billion years is not enough to erase the topography and we can speak of the surface elevations as "permanent," even though, in principle, there is no such thing as a solid and all materials eventually creep to relax their deviatoric stresses.

The first estimates of the Earth's viscosity derived from the early nineteenth-century observation by Swedish naturalist Celsius that some shorelines around the Baltic Sea are rising as rapidly as one meter per century. Hotly contested at the time, it is now accepted that central Scandinavia, formerly depressed by the weight of continental ice sheets, is gradually rebounding to its pre-ice age position. A still larger area in North America is currently rebounding from the former weight of the Laurentide ice sheets, which melted away about 11 000 yr ago. Although detailed analyses of the implications of this uplift have been ongoing for the past 60 yr, it is easy to perform a first-order estimate of the viscosity of the Earth's interior that gives a value for its effective viscosity close to the most sophisticated modern determinations.

Following in the spirit of Jeffreys' computation in Box 3.1, it is possible to balance the stress created by a depression (or elevation: the analysis is identical except for the sign) of time-dependent depth $h(t)$ against the rate of deformation implied by the ideal viscous stress relation. Jeffreys' theorem tells us that this stress difference is of order $0.3\rho gh$. The strain rate $\dot{\varepsilon}_s$ is of order $\dot{h}/w$, where w is the breadth of the depression. Inserting these factors into the definition of viscosity, Equation (3.12), yields a first-order differential equation for $h(t)$, $\dot{h}(t) = -[0.3\,\rho ghw/\eta_{eff}]h(t)$, whose solution is:

$$h(t) = h_0 \exp[-t/\tau_R]$$

$$\text{where } \tau_R = \frac{\eta_{\text{eff}}}{0.3\rho gw}. \tag{3.34}$$

In this equation h_0 is the initial depression due to the weight of the ice and τ_R is the timescale for relaxation. It is this relaxation timescale that yields an estimate of the effective viscosity, which is thus given by:

$$\eta_{\text{eff}} = (0.3\,\rho gw)\,\tau_R. \tag{3.35}$$

As one might intuitively expect, the relaxation time grows longer as the viscosity increases, and it decreases as the crustal density or gravity increases. Perhaps the least intuitive result is that the relaxation time depends inversely on the width of the load w. A physically intuitive way of appreciating this result is to realize that as w increases, the depth over which the flow occurs also increases. For a given pressure gradient, the flow is faster in a wider channel, leading to a faster relaxation rate. Thus, for a given viscosity, broad loads relax faster than narrow ones. The important implications of this result will shortly be highlighted in more detail.

In addition to the viscous half-space assumed in the relaxation computation just outlined, a second important limit is that of a thin viscous channel underlying the load. Following through a derivation similar to that above yields an equation similar to (3.34), except that the inverse load width $1/w$ is replaced by w^2/d^3, where d is the depth of the thin channel (this derivation requires the equation for the parabolic velocity profile driven by a pressure gradient in a thin layer, a topic of so-called lubrication theory). In this case the relaxation time *is* proportional to the load width, squared. A more general analysis of the relaxation of an axisymmetric crater of arbitrary profile on a substrate whose viscosity is a more complex function of depth can be found in Section 8.4 of Melosh (1989).

Performing an actual estimate of the viscosity beneath the Canadian shield, take 2700 kg/m^3 as the average crustal density, 9.8 m/s^2 as the acceleration of gravity, suppose the load is 3000 km across and that it relaxes over a timescale of 6000 yr. This yields an order-of-magnitude viscosity estimate of 5 x 10^{21} Pa-s, nearly identical to the current best estimate for the Earth's lower mantle. To interpret this estimate, remember that most of the stress generated by a broad load is supported at a depth of about 1/3 the load width; that is, about 1000 km deep in this case, or near the top of the Earth's lower mantle. Furthermore, this is an effective viscosity that applies to a stress level of about 0.3 $\rho g h_0$ or around 20 MPa, assuming that h_0 was about 2 km.

Although the idea of using the duration of topographic support to estimate planetary viscosity was first applied to the Earth, planetary geologists were quick to apply this idea to the planets. Ralph Baldwin, in his epochal 1963 book, *The Measure of the Moon*, made the first estimates of the Moon's viscosity based on the persistence of its non-hydrostatic tidal bulge and on the depths of lunar basins. In 1967 Ron F. Scott, a soil mechanics engineer at Caltech, was inspired to show how lunar surface viscosities could be estimated from the shape of relaxed lunar craters. He created a number of model crater shapes in a pan of viscous tar and allowed them to relax, recording how their shapes changed with time. Three of his time steps are shown in Figure 3.10. The most prominent characteristic of these changes is the dependence of relaxation rate on the size scale of the feature. Thus, large craters relax faster than small ones, so long as the viscous substrate is deeper than the diameter of the crater. Furthermore, the small-scale crater rims persist long after the larger-scale crater bowls have relaxed, just as Equation (3.34) suggests (other factors, such as the presence of a shallow lithosphere, may account for the persistence of crater rims on real planets, as opposed to craters in pans of uniform-viscosity tar).

Although Scott and others thus showed how viscous relaxation affects crater morphology, it has not yet been conclusively demonstrated that viscous relaxation has actually occurred in craters on any of the terrestrial planets or moons. Processes such as impact erosion or lava infilling often obscure any depth changes caused by viscous flow. The absence of relaxation does give useful lower limits to the viscosity, but this does not constitute a numerical measurement. However, the icy moons of the outer Solar System tell a different story. Figure 3.11 shows a 500 km wide crater, Odysseus, on the Saturnian satellite Tethys, contrasted with an unrelaxed 130 km wide crater, Herschel, on Mimas. Odysseus' floor has clearly relaxed to conform to the equipotential surface of the satellite, while its still-sharp

Figure 3.10 Viscous relaxation of model craters produced in asphalt of viscosity about 10^5 Pa-s. The largest crater is about 10 cm in diameter and the smaller craters about 2 and 0.2 cm. (a) 0.1 minute after the craters were molded into the surface. (b) After 30 minutes the larger crater floor has rebounded and the middle-sized crater floor is beginning to rise. All of the crater rims are still sharp. (c) After 18 hours the large and middle crater and their rims have relaxed, while the smallest crater is still evident. Image selection from Scott (1967).

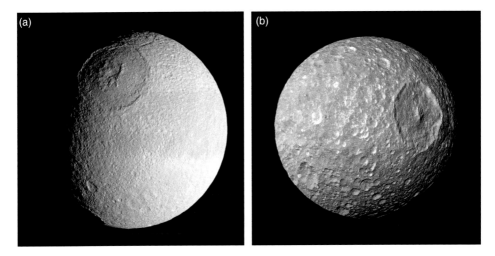

Figure 3.11 Large craters on moons of Saturn (a) The floor of the 500 km diameter crater Odysseus on Tethys has mostly relaxed to conform with the spherical shape of the satellite. NASA Cassini image PIA 08400. (b) The 130 km diameter crater Herschel on Mimas shows little sign of viscous relaxation. NASA Cassini image PIA 12570.

rim and central peak attest to the size-dependence of viscosity. If the age of the crater were known, these two observations would produce a tight bound on the viscosity of the moon, and by implication (from laboratory measurements of the flow law of ice) give an estimate of the internal temperature of Tethys. Much effort by planetary scientists is currently being expended on viscosity estimates of this kind, with the ultimate goal of estimating internal temperatures and even temperature gradients.

An important, but often overlooked, point refers back to Jeffreys' theorem: Topographic loads are typically supported at a depth comparable to the width of the load. Thus, the viscosity deduced from the relaxation of a feature of breadth w applies to a depth comparable to w itself (or, slightly better, about $w/3$, as indicated in Table 3.1). The fact that a narrow crater rim relaxes more slowly than the crater bowl itself is, thus, due to both the scale dependence of the relaxation time in Equation (3.34) and also to a possibly different (generally larger) viscosity at shallower depths below the narrow rim. In effect, topography of breadth w "probes" the viscosity at a depth of about $w/3$. When sufficiently detailed data on crater relaxation profiles are available this effect can be used to invert for the depth dependence of viscosity and, by inference from creep measurements on the (presumed known) underlying material, for the subsurface temperature gradient.

One of the major surprises of the past few decades is the existence of substantial topography on the planet Venus. Shortly after its 730 K surface temperature was discovered, but before its surface had been imaged by spacecraft-borne radar systems, material science expert J. Weertman (1979) predicted that any mountains on Venus would have long since relaxed away and that its surface must be a vast, gently undulating plain. On Earth, the temperature contour that defines the bottom of the elastic oceanic lithosphere is similar to

Venus' surface temperature, so this prediction seemed very reasonable. The discovery of large topographic variations on Venus, first by the Soviet Venera 15 and 16 radar missions and then by the US Magellan mission, was thus greeted with consternation. To this day we do not fully understand why the crust and upper mantle of Venus are so strong. The most common assumption is that high temperatures have cooked all of the water out of its near-surface rocks, thus eliminating the major weakening agent affecting terrestrial and Martian rocks. However, even the total elimination of water and the assumption of a low thermal gradient can barely explain the existence of the 13 km high Maxwell Montes, the highest elevation on Venus.

3.5.3 The topographic advantages of density differences: isostatic support

Most of the long-wavelength topography on the silicate planets and moons is a direct consequence of the difference in density between a crust and underlying mantle. Where no density differences exist, such as on the icy moons of Jupiter or Saturn, elevation differences tend to be of short wavelength. The resulting concept of *isostasy* has long been a staple of geological explanation on Earth. It has found broad application to the Earth-like planets. The basic idea of isostatic equilibrium is that high topography is high because it is underlain by rocks that are less dense than average. Elevation correlates with either density itself (Pratt isostasy) or with the thickness of a layer of lesser density (Airy isostasy). The crust is supposed to be in floating equilibrium, so that at some depth below the surface (the depth of isostatic compensation) the pressure of the overlying rock layers is the same along an equipotential surface.

A key component of the idea of isostatic equilibrium is that at the depth of isostatic compensation, deviatoric stresses vanish and pressure is the only force available. This concept accords well with the observational facts indicating that as the temperature rises, rock strength declines and creep rates increase. Initial stresses, even those applied nearly instantaneously by, say, the formation of an impact crater, relax rapidly on a geologic time-scale and bring topography into a state of isostatic equilibrium. This idea puts a premium on determining the depth of this level of compensation. If its depth can be determined, for example, using the methods discussed in the next section, this information can be converted to an estimate of the planet's interior temperature.

Geodesist Colonel George Everest accidentally initiated the discovery of isostasy in 1847, while he was triangulating the "Great Arc" in India. As he approached the massive Himalayan mountains he found that he could not get good agreement between his triangulated positions and astronomical measurements of latitude. J. H. Pratt, archdeacon of Calcutta, who was familiar with Newton's law of universal attraction, suggested that the mountains deflected the vertical, although the observed deflection was much less than what he first calculated. Pratt then supposed that the rocks underlying the Himalayas might be less dense than those underlying the Indian peninsula. Pratt announced his conclusions in 1855, the same year that G. B. Airy, the Astronomer Royal of Great Britain, suggested that variations in the thickness of a low-density crust floating on a denser substratum could account for Everest's observations. Four years later Pratt published his own theory of isostasy in which he attributed variations in the elevation of surface features to lateral variations in the density of the crust above a level of "compensation," at which the density is uniform.

Geodetic observations in the late 1800s could not discriminate between the Pratt and Airy models. In the early twentieth century the US Coast and Geodetic survey officially adopted the Pratt model because of its computational simplicity, but when the developing field of seismology revealed deep roots beneath the Alps and Himalayas, the weight of opinion swung in favor of Airy isostasy for most of the twentieth century. Most recently, however, it has been shown that Pratt isostasy dominates California's southern Sierra Nevada. The elevation of the western US's Colorado Plateau now appears to be due to low densities in the mantle, not the crust. Furthermore, precise gravity measurements from the Magellan spacecraft have shown that the Pratt mechanism, with the low densities supplied by some combination of high temperature and a low-density mantle residuum, may support the volcanic uplands of Venus (Smrekar *et al.*, 1997). Evidently, the Pratt and Airy mechanisms are end members of a continuum and the determination of crustal and mantle density and thickness must be pursued independently, insofar as that is possible.

The application of the idea of isostasy to planetary topography is simple, which is part of its appeal. Figures 3.12a and 3.12b illustrate the idea of isostatic balance between two crustal columns in both the Pratt and Airy limits. For simplicity, these examples assume constant densities for both the crust and mantle, but it is easy to generalize these examples by integrating a depth-dependent density from the surface down to the depth of compensation.

For the Pratt hypothesis, Figure 3.12a, the pressure at the depth of compensation, d_c, beneath the plains and highlands crustal blocks is given by:

$$[\rho_{cp}t + \rho_m(d_c - t)]g = [\rho_{ch}(t + h_P) + \rho_m(d_c - t)]g \qquad (3.36)$$

The contribution from the depth of compensation, as well as the mantle density and the acceleration of gravity all cancel out, so that the topographic elevation on the Pratt hypothesis, h_P, is given in terms of the crustal thickness below the plains, t, by:

$$h_P = \left(\frac{\rho_{cp} - \rho_{ch}}{\rho_{ch}}\right)t. \qquad (3.37)$$

For the Airy hypothesis, Figure 3.12b, the density of the crust ρ_c is the same in both crustal columns, but the thicker crust sinks with respect to the thinner crust and produces a root beneath the highlands of thickness t_R. Performing the same type of pressure balance as for the Pratt case,

$$[\rho_c t + \rho_m(d_c - t)]g = [\rho_c(t + h_A + t_R) + \rho_m(d_c - t - t_R)]g. \qquad (3.38)$$

Again benefiting from many cancellations, including the crustal thickness itself, the final expression for the elevation in terms of the depth of the root is:

$$h_A = \left(\frac{\rho_m - \rho_c}{\rho_c}\right)t_R. \qquad (3.39)$$

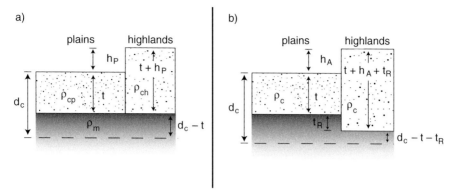

Figure 3.12 Isostatic compensation of topography is possible where a low-density crust overlies a higher-density mantle. (a) Pratt isostatic compensation, in which highlands are underlain by less dense crustal material than lowland plains. (b) Airy isostasy, in which the crust is the same density everywhere, but is thicker under highlands than plains. The dimensions defined in this figure are used in equations described in the text. The horizontal dashed line near the bottom of both figures is the depth of compensation, below which no stress differences are postulated to exist.

Because of these cancellations, the depth of isostatic compensation does not contribute directly to the topography for either Pratt or Airy isostasy. This is both a blessing and an annoyance: Insofar as the depth of compensation cannot be directly measured, we lose an important piece of information about the flow in a planet's interior. We must thus resort to indirect methods to learn about internal temperatures.

Isostatic equilibrium, although often assumed to be the final state of topographic relaxation in much of the geological literature, is not, in fact, the most stable end state: The minimum energy of any self-gravitating body is attained only when the density decreases monotonically outwards from the center of mass. Thus, even when the topography is fully compensated there is a tendency for low-density material to spread over adjacent denser rocks, and for mountain roots to spread laterally. Stress differences, thus, still exist in a state of full isostatic compensation. Calculation of these stress differences and their implications for flow in the crust of the Earth is now a part of terrestrial geodynamics (Sonder and Jones, 1999).

Although isostasy does contribute greatly to topographic support, it cannot evade the principal part of Jeffreys' theorem: Stresses of order $\rho g h$ must still develop somewhere beneath the load. The advantage of the isostatic support mechanism is that it shifts the location of the maximum stress differences from a depth comparable to the width of the topographic load to the region above the depth of compensation, where the rocks are cooler, stronger and, thus, more capable of bearing the load. Vertical topographic loads are converted into horizontal loads acting near density or thickness gradients. It is even possible to convert topography into detailed horizontal stress maps, assuming that isostasy is strictly valid (Artyushkov, 1973; Fleitout and Froidevaux, 1982; Molnar and Lyon-Caen, 1988). Thus, planet-wide elevation differences, such as the hemispheric dichotomy of Mars or the center of mass-center of figure offset of the Moon can be supported by modest stresses in the cooler, stronger outer layers of the planet, rather than by long-term strength at great depths.

3.5.4 Dynamic topography

Slow viscous or pseudoviscous deformation can do more than just eliminate pre-existing topographic features. Stresses generated by forced flow can actually create topography. When this occurs, the resulting elevations and depressions are referred to as *dynamic topography*. Slow flows in the interior of planets may be generated by a number of processes, but the most common are driven by density differences, where the density deviations from the mean are due either to temperature differences (the process is then referred to as thermal convection) or to compositional differences (compositional convection). The slow convective flows that drive plate tectonics in the Earth are a combination of both types of difference.

The most striking dynamic topographic features on the Earth are the deep submarine trenches that mark the sites of subduction zones. At subduction zones the cold, dense, and relatively stiff tectonic plates sink into the warmer, less dense mantle at rates up to about 10 cm/yr. As the plates sink, they undergo a sharp bend, changing their attitude from nearly level to plunging at angles that may exceed 45°. The cool, highly viscous material of the plates thus undergoes a large amount of stretching on the upper part of the bend, counterbalanced by compression at depth. This stretching creates stresses that literally suck the overlying surface downward, resulting in the observed topographic troughs. The depth of the trough is readily estimated from the definition of the effective viscosity, Equation (3.29), along with Jeffreys' theorem:

$$h_{\text{dynamic}} = \left[\frac{\eta_{\text{eff}}}{0.3\,\rho g}\right]\dot{\varepsilon} = \left[\frac{\eta_{\text{eff}}}{0.3\,\rho g}\right]\frac{v}{w} \tag{3.40}$$

where v is the velocity of motion (subduction, in this case) and w is the distance scale over which bending occurs.

Unfortunately, we cannot accurately determine the effective viscosity from first principles, but we can invert the formula and determine how large it must be to give the observed ca. 5 km of trench depth as the plate bends through a radius $w \sim 200$ km. The result, about 2×10^{21} Pa-s at a stress of about 50 MPa, is at least reasonable – it is almost two orders of magnitude greater than that of the underlying asthenosphere and in moderately good agreement with extrapolated laboratory measurements of the creep rate of olivine at this stress level.

One of the major complications in making this kind of estimate precise for subduction zones is that the much stiffer brittle-elastic plate that tops the tectonic plates interferes with the viscous flow deeper within the plate. Indeed, many models for subduction zones focus exclusively on the elastic plate and neglect viscous flow entirely. Such models, which were among the earliest explanations of subduction zone topography, suffer from the prediction of enormous extensional stresses in the strongly bent elastic plate (up to 5 GPa, far in excess of any measured rock strength) and neglect seismic data that indicate that the elastic plate is extensively fractured and, thus, unlikely to support any extensional loads at all. Nevertheless, this kind of elastic-viscous coupling is common in planetary tectonics and will be discussed in more detail in the next chapter.

The other likely source of dynamic topography on the Earth (with its unique plate tectonics) and the other planets is associated with rising (or descending) convective plumes. Arising from deep within a planetary interior, buoyant plumes approach the surface and exert viscous stresses on the overlying cool rock layers. These stresses account for a substantial portion of the uplift associated with the plume's arrival (the rest is associated with the plume's low density). Equation (3.40) can also be used to estimate the dynamic portion of the topography associated with a plume of horizontal dimension w rising at a velocity v, provided an estimate of the effective viscosity can be made.

Because dynamic topography can develop even when temperatures are too high to permit much static rock strength, it has been suggested that a vigorous plume rising from deep within the Venusian mantle might cause the astonishingly high elevations of Maxwell Montes and Beta Regio on Venus. The estimated plume velocities must be quite high, on the order of meters per year, and if the flow fluctuates with time, one might expect to see the elevation of Maxwell Montes fluctuate in concert with the flow. Pursuing this idea, the Magellan radar altimeter repeatedly measured the height of Maxwell throughout the duration of the mission, seeking for measurable fluctuations. Unfortunately, none were found and the reason for Maxwell's high elevation remains unresolved.

3.5.5 *Floating elastic shells: flexural support of topographic loads*

A small, cool planet or moon may possess considerable long-term strength right down to its center. However, as interior temperatures rise, strength declines and the ability of a large planet to support long-duration, non-hydrostatic loads comes to reside exclusively near its surface. This gives rise to the concept of a lithosphere, a cool outer rind whose strength is controlled by increasing pressure near its top and by slow viscous creep near its base. The mechanical behavior of such a lithosphere can be very complex: Its upper portion responds to loads both by elastic deformation and plastic failure, while its underside flows on long timescales. However, a drastic but surprisingly effective approximation neglects the viscous deformation altogether and treats the lithosphere as an elastic plate floating on a perfectly fluid substratum. Loads on the surface flex the lithosphere downward and are supported by a combination of elastic stress from the lithosphere itself and the buoyancy of the displaced fluid below. The lithosphere thus supports loads in the same way that a skater on a frozen pond is supported by the flexure of the layer of ice. Indeed, Heinrich Hertz, otherwise renowned for his discovery of radio waves, first published the equations describing the effect of a point load on floating ice in 1884 and so initiated the mathematical study of lithospheric support.

Flexural models of topographic support were first proposed around 1900, when most scientists supposed that the interior of the Earth is literally molten and that the continents simply float on a liquid interior like ice on a frozen pond. Although it is now clear from the propagation of seismic shear waves and the slow rate of post-glacial rebound that the Earth's interior is actually a hot, viscoelastic solid, the relaxation of differential stresses at high temperatures still makes the elastic flexure approximation a good one for loads of long duration.

The bottom of the elastic lithosphere is now understood to be the depth at which the Maxwell time equals the duration of the load to be supported. Surface rocks behave elastically above this depth and flow gradually below it. Because the duration of the load enters into the lithosphere thickness, this concept is a bit fuzzy: For a load lasting only a few minutes, as might be applied by a meteorite impact, the entire mantle of the Earth is the lithosphere. For glacial rebound over 10 000 yr, the effective lithosphere is about 100 km thick, whereas for a mountain chain built over 100 Myr the lithosphere thickness might be only a few tens of kilometers. However, because the creep rate of most rocks is a strong function of temperature, the effective lithosphere thickness varies only by a small amount for loads lasting from a few million to a few billion years. Under these circumstances the lithosphere can be approximated as having a constant thickness determined by its composition and the near-surface thermal gradient.

The equations describing the response of such a floating elastic shell are very complex. In their simplest form, for a thin flat plate of uniform thickness, they obey a fourth-order partial differential equation called the biharmonic equation. However, these equations need not be solved to attain a qualitative idea of how topographic loads are supported by an elastic lithospheric shell. The most important concept deriving from these equations is embodied in a factor with dimensions of length called the *flexural parameter*, α, which is defined as:

$$\alpha = \left[\frac{1}{3\,(1-\nu^2)} \frac{Et^3}{\rho_m g} \right]^{1/4} \tag{3.41}$$

where t is the thickness of the lithosphere and ρ_m is the density of the mantle underlying the lithospheric plate. E is Young's elastic modulus, ν is Poisson's ratio, and g is the acceleration of gravity. A representative value of α for the Earth's oceanic lithosphere is about 53 km, derived from the deflection caused by the Great Meteor Seamount in the Atlantic Ocean (Watts *et al.*, 1975). It is generally a few times larger than the lithosphere thickness itself, which in this case is about 16 km.

The flexural parameter describes the tradeoff between elastic flexure and buoyancy in supporting a concentrated load. Loads of breadth smaller than the flexural parameter are mainly supported by elastic stresses that develop in the warped lithosphere, whereas broader loads must be supported by buoyancy; that is, by isostatic forces. Flexure thus fills the gap between topographic loads much narrower than the thickness of the lithosphere, which are supported essentially on an elastic half-space, and very broad topographic loads that are supported by isostasy.

There is, however, a price to be paid for the advantages of flexural support. Plate flexure creates bending stresses, and for broad loads these stresses are usually much larger than the minimum required by Jeffreys' theorem. In the absence of isostatic support, for a sinusoidal load of wavelength λ elastic plate flexure theory gives a maximum stress of:

$$\sigma_{\max} = \frac{3}{2\pi^2} \frac{\lambda^2}{t^2} \rho_c g h. \tag{3.42}$$

Thus, because of the factor λ/t, squared, stresses build rapidly when the width of the load becomes substantially broader than the plate thickness. This equation suggests that as the load breadth increases, the flexural stress increases without limit. However, this does not occur when a low-density crust overlies a denser mantle: At long wavelengths isostasy takes over and the stresses actually decline as $1/\lambda^2$ after the stress peaks at a wavelength of $\sqrt{2}\,\pi\alpha$.

The surprisingly high stresses that develop in an elastically flexed plate are reduced somewhat when plastic yielding occurs and spreads the stresses over a larger volume, but the lesson is that flexure cannot support very broad loads. Indeed, in locations where the topography suggests that flexural support is important, it is common to observe tectonic evidence of rock failure.

The flexural parameter α is often directly observable. The size of the region depressed by a concentrated load is governed by the flexural parameter. Thus, the island of Hawaii is surrounded by a broad shallow moat where the elastic lithosphere of the Pacific Ocean floor is flexed downward by the weight of the volcanic pile. Similarly, the ice shell of Europa is flexed downward by the weight of the ridges crisscrossing its surface, creating shallow troughs flanking the ridges (Figure 3.13). The gigantic Artemis Corona on Venus is partially surrounded by a moat similar to that around Hawaii, and also may be due to a flexing (or, perhaps, subducting and flexing) lithosphere.

One of the goals of planetary surface studies is to find evidence for such flexural depressions flanking topographic loads and, from their breadth, use Equation (3.41) to determine the thickness of the lithosphere. This thickness, in turn, can be used to estimate the near-surface temperature gradient, and, hence, planetary heat production.

3.6 Clues to topographic support

With all of the different mechanisms that can contribute to topographic support, the question naturally arises, how can we tell which mechanism, or what combination of mechanisms, is actually supporting the topography of a given planet? Some first-order guesses are easy: Long-wavelength loads are generally supported by isostasy, short-wavelength loads by flexure. Limits to the strength of materials provide some clues. However, how can we know, in a particular case, what mechanism is actually in play?

Box 3.3 **Flexure of a floating elastic layer**

Determining the deformation of the surface of a floating elastic plate under a given load is both an old problem and a difficult one. Heinrich Hertz (of radio-wave fame) first offered a solution in 1884 (Hertz, 1884). His interest was not planetary lithospheres, but rather the form of the surface of a frozen pond under the weight of an ice skater. The weight of the skater (what we would now call a concentrated load) is supported both by the bending of the ice layer itself and by the buoyancy of the underlying water displaced by the deflection of the lower surface of the ice layer. Hertz's solution relied on centuries of mathematical study of the deflection of beams and the creation of an effective theory of elasticity. His work on the deflection of a floating plate found immediate application to railway engineering, where the bed of the tracks forms a support similar in many ways to a dense liquid layer.

In more recent times, the well-known geophysicist Don Anderson relates (personal communication, 1971) that his first introduction to plate tectonics took place courtesy of the United States Air Force, which required him to use Hertz's theory to determine how close together airplanes might be parked on an ice floe before the ice ruptured. Currently, the theory of plate flexure is widely applied in geodynamics to investigate the structure and evolution of planetary lithospheres. It is the principal subject of at least one modern monograph (Watts, 2001) and is discussed as part of thousands of papers on both terrestrial and planetary geophysics.

A serious student of terrestrial or planetary geophysics should, thus, be familiar with both the derivation and many applications of the flexural equations. However, the passage of time has not made this subject much easier than it was for Hertz, and a full derivation would be out of place in a broad overview (indeed, the correct application of the lower boundary condition between elastic and fluid materials presents a subtlety so obscure that, of all the books I know, only one (Cathles, 1975) treats it correctly!). The equations are, happily, linear for small vertical deflections w of the centerline of the plate. They are, however, *fourth*-order differential equations that, thus, have four parameters that must be determined from the boundary conditions. The most frequently used version of the full equations assumes that both the plate and the load are uniform in the y direction, hence, the solution depends only on the horizontal distance x:

$$D\frac{\partial^4 w}{\partial x^4} + N\frac{\partial^2 w}{\partial x^2} + \rho_a g\, w = q(x) \qquad (B3.3.1)$$

where N is a horizontal force applied in the x direction (taken to be positive in compression), ρ_a is the density of the underlying fluid layer, g is the acceleration of gravity and q is an applied surface load (force per unit area). D is the flexural rigidity, defined in terms of the Young's modulus E, Poisson's ratio v and plate thickness t as:

$$D = \frac{E\,t^3}{12\,(1 - v^2)}. \qquad (B3.3.2)$$

A typical solution to this equation is given by Figure B3.3.1, which shows the deflection of the lithosphere under a load of uniform thickness a that extends arbitrarily far to the left of the center, $x = 0$ and infinitely far perpendicular to the page. This might represent the edge of a very broad plateau with a straight edge. The density of both the load and the flexed plate is ρ_l, while that of the underlying fluid layer is ρ_a. The formula for the vertical deflection of the *center of the lithosphere* in this case is:

Box 3.3 **(cont.)**

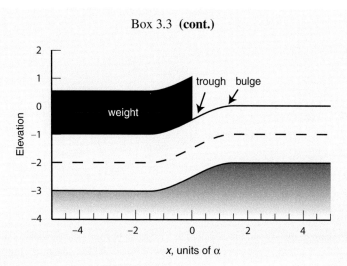

Figure B3.3.1 Deflection of a floating elastic plate by a sharp-edged load with the same density as the plate. The horizontal dimension is in units of the flexural parameter α, while the elevation is in units of half of the lithosphere thickness. The dashed line is the neutral sheet of the plate. The load uniformly depresses the plate on the left, but near the edge of the load the plate is flexed down with a curvature comparable to the flexural parameter. A very low bulge develops beyond the main flexure. Equation (B3.4.3) in Box 3.3 describes the deflection of the neutral sheet.

$$w(x) = \begin{cases} \dfrac{\rho_l}{\rho_a}\dfrac{a}{2}\left\{2 - e^{x/\alpha}\cos(x/\alpha)\right\} & x \le 0 \\[2ex] \dfrac{\rho_l}{\rho_a}\dfrac{a}{2}e^{-x/\alpha}\cos(x/\alpha) & x > 0. \end{cases}$$ (B3.3.3)

In these equations the horizontal scale of the deflection is determined by the flexural parameter, α, defined in the text. To find the actual topography one must be careful to add in the thickness of the overlying half of the plate, plus the load. The lithosphere is deflected downward by a distance of $\dfrac{\rho_l}{\rho_a}\dfrac{a}{2}$ right under the edge of the load, while far to the left it achieves the deflection required by isostatic equilibrium, $\dfrac{\rho_l}{\rho_a}a$.

Note, in Figure B3.3.1, the very slight reversal of the vertical deflection that crests at a distance of $3\pi\alpha/4$ (labeled "bulge" in the figure). Noted by Hertz in his solution, this small-amplitude reverse deflection is characteristic of flexural solutions. On Earth, this slight bulge is readily apparent on topographic maps of the great oceanic trenches where the oceanic lithosphere is subducted into the mantle. A few hundred kilometers seaward of every trench there is a small rise, termed the "outer rise," that seems to represent the flexure of the oceanic lithosphere. The flexural trough surrounding each of the Hawaiian islands is likewise accompanied by a slight outer rise farther from the island loads.

The above results are valid only for a flat elastic plate. When the lithosphere has a substantial curvature it is technically called a shell and the solutions for topographic support must include membrane stresses from the stretching or compression of the plate in addition to flexural stresses. This case is examined in some detail by Turcotte *et al.* (1981).

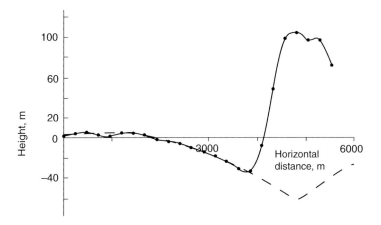

Figure 3.13 Flexure of the lithosphere adjacent to a ridge on Europa. The plot shows the topography determined by the method of photoclinometry applied to a Galileo image of Europa. Note the prominent trough flanking the ridge and the low bulge beyond the trough. This trough and bulge topography is the expected shape for a loaded floating elastic plate of thickness 350 ± 50 m. After Figure 4 in Hurford *et al.* (2005).

   This question would be easy to answer if we could directly determine the stresses acting in a planet's interior. Unfortunately, stresses are difficult to measure even in the laboratory. *In situ* stress measurement techniques do exist, however, and are sometimes used at shallow depths in the Earth (Engelder, 1993), but such data do not yet exist for any other planet. Earthquakes also give clues about stress magnitudes and directions, but even in the Earth their interpretation is presently somewhat controversial, and for other planets the necessary seismic data sets do not exist (apart from some intriguing, but limited-term, data on lunar seismicity). Associations of tectonic features, to be discussed in more detail in the next chapter, may give clues about regional stress distributions and this is an important source of information. However, the most direct information on topographic support comes from measurements of the acceleration of gravity.

3.6.1 Flexural profiles

The most direct method of determining lithospheric thickness is to observe a topographic flexural profile. As discussed above, such profiles are commonly observed around volcanic islands in the Earth's oceans (Fig 3.14), and have been noted on Venus and Europa. Figure 3.13 is a good example of such a profile, first discovered in 2005 (Hurford *et al.*, 2005). Upon observing new images of the surface of a terrestrial planet or satellite, one of the first efforts of any geophysically oriented planetary scientist is to seek evidence of such surface deformation. Once found, the wavelength of the observed topographic flexure is readily expressed in terms of the flexural parameter α, Equation (3.41), and with a few additional assumptions it yields the thickness of the elastic lithosphere. Still more assumptions give an estimate of the planet's heat flow, a number that is difficult to determine remotely in almost any other way. The amplitude of the flexural deflection yields the magnitude of the load via the full flexural equations described in Box 3.3.

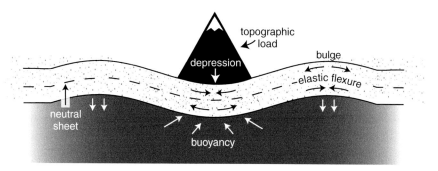

Figure 3.14 Flexure of a floating elastic plate subjected to a topographic load. The weight of the load is supported by a combination of flexural stresses developed by bending of the plate and buoyancy generated by the depression of the lithosphere into the fluid mantle below. This schematic drawing indicates the neutral sheet in the plate by a dashed line.

3.6.2 Anomalies in the acceleration of gravity

To first order, the gravitational acceleration of a large planet or satellite is directed toward its center of mass and varies as the inverse square of the distance from that center, outside of the body itself. However, small but significant deviations from an ideal spheroidal field (for a rotating body) are readily detected either on the surface or from orbiting satellites. This is not the place for a detailed discussion of this large area of research (see, e.g. Lambeck, 1988), but a few important results require brief discussion.

Deviations of the gravitational acceleration of a body from an ideal assumed field are known as *anomalies*. Free-air anomalies are probably the easiest to understand: they are simply the difference between the expected gravitational acceleration and the observed acceleration at some conventional elevation above the surface. Gravity anomalies are generally reported in units of *gal* (short for galileo: 1 gal = 1 cm/s^2). The excess gravitational acceleration produced by an infinitely wide sheet of density ρ and thickness h is given by:

$$\Delta g = 2\pi \, G\rho h. \tag{3.43}$$

Thus, the gravitational acceleration exerted by a broad sheet of basalt 1 km thick is about 125 mGal. Although this is a very small acceleration compared with the surface acceleration of the Earth (9.8 m/s^2, or 980 000 mGal), such anomalies are readily detected by the orbital perturbations of low-flying satellites. Indeed, accuracies of 0.1 mGal are presently achievable.

The importance of free-air gravity anomalies is that, to first order, isostatically compensated topography has no anomaly. Because the extra mass of surface topography is compensated by a low density at depth, the net excess mass below the surface of isostatically compensated topography is zero, and so is the gravity anomaly. This simple conclusion is not quite true for loads of limited lateral extent, in which case detailed modeling is required, but it works as a general rule of thumb. Topography supported by strength in the form of flexure or supported dynamically is not compensated by underlying low densities and so exhibits free-air gravity anomalies, which may be quite substantial. For example, free-air gravity anomalies over the dynamic topography of subduction zone trenches may drop below –300 mGal. Flanked by smaller positive anomalies along the volcanic arc, they constitute the largest free-air gravity anomalies on Earth.

Anomalies in the acceleration of gravity (both excesses and deficits) can be interpreted as equivalent topographic loads (positive or negative) and, thus, sources of stress in the lithosphere. A rough-and-ready method of estimating such stresses is to replace the gravity anomaly, Δg, by a broad layer of sufficient mass to create the same anomaly, then to equate the weight of this extra load to the applied stress. This yields an equation for the equivalent stress σ_{grav} of the gravity anomaly:

$$\sigma_{\text{grav}} = \frac{g\,\Delta g}{2\,\pi G} \tag{3.44}$$

where G is Newton's gravitational constant. In the case of the 300 mGal anomalies observed in terrestrial oceanic trenches, this corresponds to a stress of about 70 MPa – close to the crushing strength of rock. On the moon, the same anomaly would only imply a stress of about 12 MPa: Although the mass anomaly is the same on the moon, its lower acceleration of gravity implies a smaller stress.

An important episode in the history of lunar exploration was the unexpected (and, at the time, unwelcome!) discovery of strong mass concentrations on the lunar nearside. Termed "mascons" after they were discovered in 1968, they are circular anomalies of up to about 500 mGal that are associated with basalt-filled impact basins. The Apollo mission planners went to great trouble to compensate for the effects of these anomalies in the manned Apollo orbits. The mascons' effect on satellite orbits is so strong that they crashed the Apollo 16 subsatellite PFS-2 onto the lunar surface only a month after the astronauts released it.

It is now understood that isostatically uncompensated lava within the circular nearside impact basins creates most of the mascon anomalies, with an additional contribution from an uplifted mantle plug underneath the basins (Neumann *et al.*, 1996). The effect of the lava's enormous weight is clearly visible in the Humorum basin, where adjacent craters tilt inward toward the sagging basin center and the stresses generated by the load have fractured the crust in great circumferential faults.

A second commonly employed type of gravity anomaly is the Bouguer anomaly. It is computed from the free-air anomaly by subtracting an estimate of the gravitational attraction of the topography above (or below) the reference geoid. The resulting anomaly reflects the mass deficit (or excess) below (or above) the topographic elevation (or depression). For more details on this and other types of gravity anomaly, see the book by Garland (1965). Isostatically compensated mountainous terrain, such as the Alps in Europe, exhibits small free-air anomalies and strong negative Bouguer anomalies, reflecting the low-density root compensating the weight of the mountains. If it is further assumed that the density anomalies are entirely due to variations in the thickness of a constant-density crust, then gravity anomalies can be inverted to create maps of crustal thickness. Such maps have been produced for the Earth, Moon, and Mars, but it should be understood that a great many assumptions enter into such maps unless they are constrained by seismic data.

Because isostatic compensation by a floating lithosphere depends on the breadth of the load, a useful approach is to compute the Fourier transforms of both gravity anomalies and topography. Comparison of the amplitude of the gravity anomaly to topographic height as a function of wavelength on a so-called admittance diagram may then be used to estimate the thickness of the elastic lithosphere. Unfortunately, subsurface loads easily confuse this method; more recent work focuses on correlations between gravity anomalies and topography as a function of wavelength. This kind of study yields maps of lithosphere thickness, which have been compiled to date for the Earth, Venus, and Mars.

3.6.3 Geoid anomalies

A somewhat different type of gravity anomaly is the ratio between geoid height and topographic elevation (or depression). The geoid, by definition, coincides with a gravitational equipotential surface, whereas the acceleration of gravity is proportional to the potential gradient perpendicular to this surface. The geoid and gravitational acceleration anomalies, thus, contain different types of information. It can be shown that geoid height variations depend on the near-surface density *gradient*, rather than the density itself. Thus, the Geoid to Topography Ratio, or GTR, allows estimates of the depth of isostatic compensation as well as of the type of compensation, whether of the Airy, Pratt, or mixed type.

Over the Earth's oceans the geoid can be measured directly by precision observations of the shape of the sea surface. Over the land areas of the Earth and over planets with solid surfaces it can be constructed from careful tracking of gravitational perturbations of satellite orbits.

The geoid of Mars is utterly dominated by the huge Tharsis dome, which affects a region about 5000 km in diameter and appears to overlie an enormous lens of low-density material in the Martian mantle. Because of the uncompensated load, the Tharsis dome causes such a large geoid distortion that it obscures efforts to determine that planet's moment of inertia and thus the size and mass of its core (Neumann *et al.*, 2004). On Venus the GTR of highland features is much larger (tens of m/km; Smrekar *et al.*, 1997) than those of terrestrial features (typically less than 5 m/km in the oceans). These large ratios indicate much larger depths of isostatic compensation than observed on the Earth (a paradox, considering the much thinner lithosphere on Venus because of its high surface temperature!). Furthermore, the GTR varies widely from one highland feature to another, suggesting a highly variable depth of compensation. The major geoid anomalies on Earth are associated with subduction zones. However, unlike the acceleration anomalies, which are negative in the trenches and positive over the flanking volcanic arcs, the geoid anomalies are broad positive welts that follow the trend of the subduction zones. These anomalies are generally attributed to the cold, dense subducted slabs slowly sinking into the mantle.

The great importance of gravity anomalies for understanding topographic support is the principal reason that planetary geophysicists are eager to establish polar orbiters about as many planets and satellites as possible. The Earth, Venus, and Mars have been well covered

---

Box 3.4 **The ambiguous "lithosphere"**

The term "lithosphere," as introduced in this chapter, refers to that part of a planet's interior that responds elastically to applied loads. Because of the time-dependent response of rock materials to applied stress, the lithosphere's size and location is somewhat ill defined, as it depends on the duration of the load under consideration and the rheology of the material of which it is composed. It is, nevertheless, a useful concept because the extreme variations in the effective viscosity of most planetary materials make the uncertainties in the lithosphere's boundaries small in relation to the size of the elastic region itself for timescales of geologic duration.

However, the ambiguities of the term "lithosphere" only begin with this definition. Numerous geophysicists over the past 70 years have complained that the same term is promiscuously applied to three disparate concepts (Anderson, 1995), but to little avail: The word "lithosphere" is employed by large segments of the geophysical community to mean either the elastic portion of a planet's interior (the "elastic lithosphere"), the portion of the Earth above the seismic low-velocity zone (the "seismic lithosphere"), or the cold boundary layer of a thermally convecting cell (the "thermal lithosphere"). It is wise to be cautions when encountering the term "lithosphere" and to ask oneself which usage is intended!

Another frequent confusion is between "lithosphere" and "crust." The outer regions of planets are frequently differentiated into an outer, less dense crust that overlies a deeper interior zone often called the "mantle," in analogy to the divisions of the Earth's interior. The distinction between the crust and mantle is purely chemical: They are composed of materials with different average densities. In contrast, the elastic lithosphere is a mechanical division. In the Earth's ocean basins, the elastic lithosphere comprises *both* the oceanic crust and upper mantle, while on the continents the elastic lithosphere may include only the upper portion of the crust (and this is often underlain by a *second* lithosphere at the top of the mantle).

---

by such orbiters. Our Moon is not so well understood because we cannot directly measure the range to a satellite over its farside. High-precision lunar gravity measurements of the farside will require missions that incorporate at least two satellites tracking one another. Recently, the successful Japanese Kaguya mission has produced the best and most complete lunar gravity field, including the farside. At the time of writing, the MESSENGER mission is on its way to orbit Mercury in 2012 and we are eagerly awaiting the data on Mercurian gravity that will result from successful completion of that mission. In the future we can hope for missions that orbit, and track, spacecraft around the major moons in the outer Solar System.

Further reading

G. K. Gilbert achieved a great deal of insight into the relationship between temporary loads on the Earth's surface and the viscous response of the underlying mantle in his famous monograph on Lake Bonneville (Gilbert, 1890), which can still be read with profit. The history of the investigation of the strength of the Earth and gravity anomalies, among other things, is well told by Greene (1982). Long a classic in the field, the book *Fundamentals of Rock Mechanics* by J. C. Jaeger has gone through many editions. It was out of print for many years, but a new edition has recently appeared that updates its notation and includes many new measurements (Jaeger *et al.*, 2007). The ideas behind our modern understanding of material strength are engagingly told in a semi-popular book (Gordon, 2006), while the details of brittle fracture theory are explored by Lawn and Wilshaw (1993) and for rocks by Paterson and Wong (2005). The nature and theory of dislocations is well described in Hull and Bacon (2001) and Weertman and Weertman (1992). Gilman (1969) applies dislocation mechanics to the plastic deformation of solids. Harold Jeffreys devoted much of his life to understanding the relation between strength and topography of the Earth. He wrote a fine, although now somewhat dated, popular book (Jeffreys, 1950), but his enduring masterpieces are the third and fourth editions of *The Earth* (Jeffreys, 1952, 1962). Later editions of this book exist, but by the fifth edition, the aging Jeffreys was on a campaign to stamp out the upstart theory of plate tectonics and these later editions are rather polemic. The best treatment of the relation between gravity, the geoid, and the shape of the Earth is Lambeck (1988). A clear, detailed discussion of rheology applied to the Earth is Ranalli (1995), although a more recent book focused on the detailed mechanisms of deformation and flow is Karato (2008). The now-classic book on the application of theories of elasticity and viscosity to geodynamic problems is Turcotte and Schubert (2002). This book has become a standard text for advanced courses in geodynamics. The flexure of the lithosphere and its relation to isostatic support is now well covered at book length by Watts (2001).

Exercises

3.1 Strength vs. gravity

a) Phobos, the innermost satellite of Mars, is an irregular, potato-shaped body with extremes of radius, $r_{min} \approx 10$ km and $r_{max} \approx 14$ km, and mean density 1900 kg/m³. If these extremes are the maximum that Phobos' strength could support, how large is the strength of its rock? If the strength of Phobos' rock is similar to that of the Moon's, about 10 MPa, how large could the extremes of Phobos' radii be? What do you think this means?

b) If asteroids are incoherent "rubble piles", the maximum slope that can exist on their surface is the angle of repose for rock debris, about 30° for most types of rock. Estimate the *maximum* difference in elevations possible on a non-rotating rubble-pile asteroid and compare this to the actual difference in dimensions of known asteroids.

Extra Credit: Suppose the asteroid is rotating at the limit for breakup. Now estimate (crudely: to do this exactly is a *very* hard problem) the maximum possible difference in the asteroid's dimensions. For a more sophisticated approach to this problem see Minton (2008).

3.2 Viscous flow

Use the formula, similar to the one derived in Section 3.5.2, for the relaxation time τ of a disk-shaped load on a viscous half-space with viscosity η:

$$\tau = 5\eta/\rho g R$$

where ρ is the rock density, g is the surface acceleration of gravity, and R is the radius of the disk.

a) The Imbrium mascon ($R = 500$ km) is not isostatically compensated (relaxed). The last lavas on its surface are ca. 3×10^9 yr old. Derive a lower limit for the Moon's present-day viscosity.

b) Lake Bonneville (Ancestral Great Salt Lake in Utah) relaxed almost completely in the 1500-year interval between the Bonneville and Provo stages, when much of its water drained out to the northern Columbia River drainage. Its radius $R \approx 100$ km. What is the viscosity of the mantle beneath Utah? Compare this to the viscosity (10^{21} Pa-s) of the average mantle. What does this mean?

c) Over the last 3×10^9 yr (probably), 100 km diameter crater basins on Ganymede have relaxed completely, but their 10 km wide rims are still clearly visible. What does this imply about the viscosity of Ganymede? Is there more than one possible interpretation?

3.3 Warmed-over Uranian moons

Tiny Miranda, radius 236 km and surface temperature about 70 K, has a shape that is indistinguishable from that of an equilibrium tidal ellipsoid, with a maximum tidal bulge of about 7.1 km. Using the fact that Miranda must have relaxed into this shape over the past 4.5×10^9 yr, derive an upper limit for the viscosity of its interior (you may need to know $G = 6.67 \times 10^{-11}$ Nm^2/kg^2 and the mean density of Miranda is 1200 kg/m^3).

3.4 The ultimate limit to core formation

Use Frenkel's estimate of the ultimate strength of a solid, $Y_F = \mu/2\pi$, to estimate the maximum radius, r_{max}, of an iron sphere that can be supported in the Earth's mantle. Some relevant data are ρ(iron) = 8000 kg/m^3 (at mantle pressures), ρ(mantle) = 5000 kg/m^3, and $\mu = 2.5 \times 10^{11}$ Pa.

Reference (consult this *after* you have solved the problem!): G. F. Davies (1982).

3.5 Global isostasy

a) Suppose that the Moon's center of mass (CM) is 1.6 km closer to the Earth than its center of figure (CF), as was determined by the Clementine mission. Model the Moon's interior as a mantle of density 3300 kg/m^3 and a crust of density 2800 kg/m^3. If the crustal thickness on the nearside is 60 km (determined by the Apollo seismic experiment), how thick must the farside crust be to explain the CM–CF offset?

b) Another way to estimate the Moon's crustal thickness is to note that the floor of the gigantic, 2600 km diameter, South Pole-Aitken basin lies about 8 km below the best-fitting sphere representing the lunar mean elevation. If we make the reasonable assumption that this impact cleared away *all* of the overlying crust, leaving bare mantle on the floor of the basin, use isostasy to estimate the Moon's mean crustal thickness. If this answer differs from (a) above, what may be the reason?

3.6 Supporting Maxwell

Maxwell Montes is the highest elevation on Venus, rising 11 km above the planet's mean radius and extending over a 500 km diameter region. It is possible that Maxwell is supported dynamically by viscous stresses induced by a rising mantle plume impinging on the overlying lithosphere. If this is correct, use order-of-magnitude estimates to deduce the *velocity* of the plume necessary to support Maxwell. You may assume the mean viscosity of the mantle is 10^{19} Pa-s, similar to the Earth's asthenosphere. If this velocity fluctuates by 10% over the year that Magellan observed the altitude of Maxwell's summit, how large would the variations in elevation be? Do you think these elevation changes would be detectable?

3.7 Flexed Venusian lithosphere

The northern edge of Ishtar Terra on Venus is an enormous scarp, 4 km high, that stands near 30°, the angle of repose. Just north of the plateau edge is a deep trough that is bounded still farther away by a low rise that crests about 50 km away from the deepest part of the trough. Use the theory of a sharp-edged load on a floating elastic plate to estimate the Venusian elastic lithosphere thickness. How does this agree with other estimates of lithosphere thickness?

You may need to recall that the acceleration of gravity on Venus is 8.6 m/s^2, and may assume that the elastic constants of the Venusian crust are approximated by $E = 1.6 \times 10^{11}$ Pa and $v = 0.25$.

4

Tectonics

Though the primary direction of the force which thus elevated them [the strata] must have been from below upwards, yet it has been so combined with the gravity and resistance of the mass to which it was applied, as to create a lateral and oblique thrust, and to produce those contortions of the strata, which, when on the great scale, are among the most striking and instructive phenomena of geology.

John Playfair, Illustrations of the Huttonian Theory of the Earth
(1802, p. 45)

4.1 What is tectonic deformation?

During NASA's Mariner 10 spacecrafts first encounter with the planet Mercury on March 29, 1974, planetary scientists at both Caltech and JPL (the Jet Propulsion Laboratory in Pasadena, California) were initially disappointed by the Moon-like appearance of Mercury's surface. They had hoped for something more exotic than a monotonous, gray, heavily cratered planetary landscape. But closer examination revealed a major difference: dozens of kilometer-high scarps wound hundreds of kilometers across the surface, cutting plains and craters alike. These "lobate scarps" indicated a planetary history quite distinct from that of our moon and implied that powerful compressive stresses fractured Mercury's lithosphere sometime after the period of heavy meteorite bombardment.

Planetary surface features created by internal stresses that fracture or deform the lithosphere are collectively termed tectonic. Not all tectonic features are as dramatic as Mercury's lobate scarps: subtle long-wavelength topographic warping is also classified as tectonic. On Earth, such warping was first recognized by changes in the position of the land and the sea. The word "tectonic" is derived from the Greek word for "builder" and has long been identified with the forces that produce mountains. Terrestrial geologists use this term to describe earth movements and constructional landforms not created by volcanism or impact.

Analysis of the scarps, ridges, and troughs created by tectonic processes reveals the nature of the forces that have acted within a planet's lithosphere. Stress directions and magnitudes can often be inferred from topography coupled with knowledge of the rheology of the surface materials. Supplemented by information on the age of such features, valuable insights into the internal workings and history of a tectonically deformed planet may be gained.

4.1.1 Rheologic structure of planets

Small planetary bodies lacking a substantial internal heat source are rheologically diverse: Depending on the condition of the materials composing them, they may be described as rubble piles, monolithic blocks or fractured remnants of larger bodies. However, if the internal temperatures rise to about half the melting temperature of the planet's principal constituents, a simple generalization becomes possible. Inside such an internally heated body creep relaxes differential stresses over geologic time, while its exterior remains cold. The actual numerical values of the temperatures of the near surface and interior depend upon the heat sources available, size of the body, mode of heat transport, surface temperature, and many other factors. However, for present purposes, it is sufficient to characterize a large planet as a roughly spherical mass of material with a soft, stress-free interior enclosed within a brittle outer shell. Gradations of the strength and response of the shell to stress certainly exist, and are functions of depth, composition, and temperature, as will be discussed later in this chapter. Nevertheless, to first order, the vision of a large planet or satellite as a soft, strengthless mass beneath a thin brittle (or sometimes ductile) shell captures much of the mechanical behavior of the rocky and icy objects in our Solar System.

The concept of an elastic lithosphere, discussed in the last chapter, is central to understanding the tectonic processes that are manifested in the basins, plateaus, scarps, and troughs that diversify planetary landscapes. The thickness of the lithosphere determines the size scale for many of these phenomena: The horizontal scale of lithospheric warping is determined by the flexural parameter α, which itself depends more sensitively on the lithospheric thickness t than on any other parameter, satisfying the functional relation $\alpha \sim t^{3/4}$, Equation (3.41). Similarly, the width of fault-bounded troughs is controlled by the thickness of the broken elastic layer and, thus, often reflects the thickness of the underlying lithosphere.

In addition to providing long-term topographic support to a planet, the lithosphere transmits horizontal stresses to long distances away from the actual source of applied loads. On the Earth, stresses arising from the negative buoyancy of sinking slabs in subduction zones are broadcast throughout planetary-scale tectonic plates. On smaller planets, such as Mercury, changes in the rotation rate that affect the shape of the equator alter stresses acting at the poles. On Mars, the development of the Tharsis Rise affected tectonic patterns on the opposite side of the planet. The lithosphere is, thus, not only the seat of long-term strength in planetary interiors, but also an effective stress guide that may integrate the tectonic fabric of an entire planet. Understanding the nature and origin of the lithosphere is thus of prime importance to understanding the tectonic features observed on the surface of a planet.

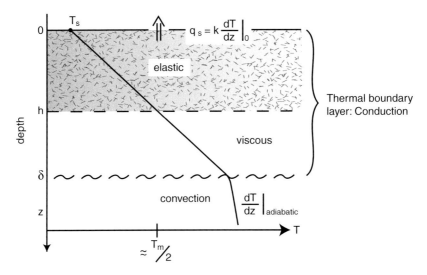

Figure 4.1 Schematic illustration of temperature in the lithosphere and below. The temperature near the surface of a planet grows approximately linearly with depth, increasing from its surface temperature T_S to approximately half the melting temperature of the planet's dominant material, $T_m/2$. Material above this depth responds elastically over geologic time to applied stress. This region defines the lithosphere. At greater depths viscous creep relaxes stress differences, but the temperature gradient is still usually linear until the viscosity becomes so low that thermal convection alters the temperature gradient from a conductive profile to a less-steep adiabatic gradient where convection dominates.

To first order, the thickness of the elastic lithosphere is determined by its composition and the planet's surface temperature and heat flow q_S. If the surface temperature of the planet is T_S, and the temperature at the base of the lithosphere is about half its melting point, $T_m/2$, then the thickness of the lithosphere t is given by (Figure 4.1):

$$t \approx \frac{k\left(\dfrac{T_m}{2} - T_S\right)}{q_S} \qquad (4.1)$$

where k is the thermal conductivity of the material that makes up the lithosphere. For rocks, k is about 3 W/m-K, while for ice it is roughly 2.2 W/m-K (see Table 4.2 for other values). Note that temperatures must be expressed in degrees K! Equation (4.1) assumes that the temperature gradient is linear in the lithosphere. This equation must be amended if there are layers of varying thermal conductivity or if heat is generated internally. Nevertheless, this simple formula gives a good first approximation to the thickness of the elastic lithosphere on most planets. For example, the thickness of the Earth's oceanic lithosphere varies widely, from near zero at spreading centers to about 35 km in the oldest plates, but its bottom coincides nearly everywhere with the 750 K isotherm, which is just about half of

Table 4.1 *Elastic lithosphere thickness estimates for the terrestrial planets*

Planet or moon	Surface heat flow (mW/m^2)	Surface temperature (K)	Elastic lithosphere thickness (km)
Mercury	13	440	172
Venus	56	730	22
Earth	81	273	34
Moon	27	273	103
Mars	23	218	126

Assumes $k = 3.3$ W/m-K, $T_m = 2200$ K. Surface heat flow for Venus and Mars assumes a bulk heat production equal to carbonaceous chondrite material. Bulk heat production for Mercury is reduced by a factor of 0.55 to account for the fact that its core is about 65% of its total mass, whereas the Earth's core is only 31.5% of its total mass.

the melting point of depleted peridotite, its primary constituent (Watts, 2001). Table 4.1 lists estimates of the lithosphere thickness for the terrestrial planets and the Moon, derived from Equation (4.1).

4.1.2 One- and multiple-plate planets

The simple rheologic structure discussed so far applies well to planetary bodies such as our Moon, Mercury, Mars, and many of the larger icy satellites of Jupiter and Saturn. A single lithospheric shell encloses all these objects. This shell transmits stresses horizontally from one part of its surface to every other part. Such objects are aptly termed "one-plate planets." Their lithospheric shells are relatively strong and they respond globally to applied stresses. Geographic features on such planets are fixed with respect to one another and often reflect events early in Solar System history.

On the other hand, the lithospheres of larger objects such as the Earth and perhaps Venus and Europa are segmented into multiple lithospheric plates. On Earth, these tectonic plates are composed of cooler, stronger domains separated by narrow zones of hotter and, thus, weaker material. Stresses cannot easily cross these boundaries and each plate acts, to some extent, independently of the others. This structure, probably a consequence of Earth's high surface heat flow (and perhaps the weakening effect of water), is known as plate tectonics. Much of Earth's tectonic history can be understood in terms of the motion and interactions of these lithospheric shells. Tectonic plates are created at spreading centers, glide past each other, and sink or are "subducted" back into the mantle in regions of negative buoyancy. This arrangement enhances heat transfer through the recycled oceanic lithosphere, while the more buoyant continental rafts are pushed to and fro, alternately colliding and being rifted apart in the slow dance that has given Earth a highly variable geography and geology.

Venus, Europa, and Io also have high heat surface flows, but these objects do not seem to respond to this heat flow in the same manner as the Earth. Venus appears to have been catastrophically resurfaced about 700 Myr ago, when all of its then-existing impact craters were erased. Tectonic and volcanic processes have heavily altered its surface, but it is not organized into recognizable plates similar to those of the Earth. This may be a consequence of its high surface temperature, producing a very thin elastic lithosphere that deforms so readily that it cannot transmit stresses any appreciable distance. Europa and Io may also possess such thin lithospheres that recognizable plates do not form, although Europa possesses what appears to be spreading centers in some of its broad ridge bands. Even on Earth there are some regions, such as the Mediterranean and perhaps eastern Asia, that must be described in terms of a multitude of *microplates*, for which the plate tectonic paradigm is not particularly apt.

Perhaps we should recognize a third category of "deformable-plate" planets for objects similar to Venus, Europa or Io, but is not clear that such an awkward designation would be of much use. A better organization might focus on the horizontal distance to which stress perturbations can be transmitted: globally on one-plate planets, regionally on Earth and only locally on Venus and similar objects. Such a "coherence length" is closely linked to a combination of lithospheric thickness, strength, and viscosity, which themselves are functions of the temperature gradient, mean temperature, and composition.

4.2 Sources of tectonic stress

Tectonic deformation cannot occur without forces to drive it. There are many sources of tectonic stress: Nearly every possible source that has been suggested theoretically has by now been found (or is believed to be) acting on some planetary body. A widely used classification divides these forces into either external or internal. As in most attempts at exclusive classification, however, this is not an absolutely clean separation, but it is at least useful for a starting point.

4.2.1 External sources of tectonic stress

Forces that act on a planetary object from outside and lead to a change in shape may produce tectonic stress. Such forces include tides induced by adjacent massive bodies, changes of rotation rate, and spin-axis reorientation due to true polar wander. Internal temperature variations, although they might ultimately be caused by external changes in heat received from a star or nearby planet, are classed as internal (such is the limitation of exclusive classifications!).

The relation between planetary shape and tidal forces or rotation rate was discussed in Chapter 2. Translation of these shape changes into stresses requires a rheologic model, which itself depends on the particular circumstances of each planet. Nevertheless, a Dutch geophysicist, Felix Vening-Meinesz (1887–1966) created a very useful approximation (Vening-Meinesz, 1947). Although Vening-Meinesz made major contributions to understanding subduction zones, he lived in the era before plate tectonics and believed that the Earth could be described as a one-plate planet, with a very thin elastic lithosphere floating on a strengthless substratum. He postulated that global shear patterns would arise as the Earth's day lengthened due to tidal friction. To support this idea, he derived an equation for the stresses that develop in a thin, initially unstressed, elastic shell as the Earth's flattening changes by an amount Δf. Because the shell is thin, there are no vertical stresses σ_{rr}(other than lithostatic stresses, which must be added to the following equations to get the total stress). The horizontal stresses in the east-west direction $\sigma_{\theta\theta}$ and north–south direction $\sigma_{\phi\phi}$ are given by:

$$\sigma_{\theta\theta} = \frac{\Delta f}{3} \mu \left(\frac{1+\nu}{5+\nu}\right)\left[5+3\cos(2\theta)\right] \tag{4.2}$$

$$\sigma_{\phi\phi} = -\frac{\Delta f}{3} \mu \left(\frac{1+\nu}{5+\nu}\right)\left[1-9\cos(2\theta)\right] \tag{4.3}$$

where μ is the shear modulus of the lithosphere, ν is its Poisson's ratio and θ is the colatitude. Note that these stresses are equal, as they must be, at the pole ($\theta = 0°$) and are mostly extensional for a decrease in flattening (spindown). Their largest difference is at the equator, where the east–west stress is more compressive than the north–south stress for spindown. These stresses are uniform through the depth of the lithosphere in the thin lithosphere approximation.

Although it may seem that a change in rotation rate is a very special situation, Equations (4.2) and (4.3) can be applied to both tidal deformation and to planetary reorientation by adding together multiple sets of stresses obeying similar equations after rotating the solutions about an appropriate axis (Melosh, 1980a, 1980b). Such solutions have been applied to compute the stresses in the Moon, Mercury (Melosh, 1977), Europa (Greenberg *et al.*, 1998; Helfenstein and Parmentier, 1983), Mars, and, most recently, in Enceladus (Matsuyama and Nimmo, 2008) from spindown, tidal deformation or spin-axis reorientation.

4.2.2 Internal sources of tectonic stress

Any load, positive or negative, added to the surface of a planet creates stress in its lithosphere. These stresses are required to support the additional topography and were discussed in the last chapter. It should be obvious that the emplacement of volcanic plains applies loads to the underlying lithosphere, as for example, did the lavas that created the lunar mascons. Magma that does not erupt on the surface is frequently intruded into the lithosphere and creates tectonic deformation as it displaces the surrounding rock layers. Impact craters, likewise, remove material and serve as an immediate source of negative loading. It may be less obvious that on planets with an atmosphere or hydrosphere, such as the Earth, Mars or Titan, erosion and sedimentation by surface fluids also redistribute mass and produce varying surface loads. On the Earth, loading by sediments deposited in enclosed basins, on continental shelves, and by large river deltas has been extensively analyzed in the geophysical literature (Watts, 2001). Similarly, erosion of rising mountain masses, such as the Alps or Himalayas, unloads the crust and usually leads to substantially larger uplift in the core of the mountain ranges than would otherwise occur. The tectonic changes induced by the growth and decay of Pleistocene ice sheets on the Earth provide unique insights into the rheologic structure of our planet.

The stresses produced by loads on or beneath a planetary surface are as various as the loads and the underlying lithospheric structures themselves. There are no simple rules for computing the exact sizes and directions of the stresses: Each case must be addressed on its own terms. Fortunately there are now many examples that have been examined and there are methods, including sophisticated numerical techniques such as finite element analysis, for deducing tectonic stresses and rheologic response from given initial conditions (Turcotte and Schubert, 2002). Although these methods yield results that often cannot be anticipated by simple rules, the theory of floating elastic plates described in Section 3.5.5 does provide some guidance. Immediately beneath a positive load the surface stresses are generally compressive as the lithosphere bows downward under the load. The stresses at the bottom of the lithosphere are correspondingly extensional due to stretching. Outside the boundary of the load itself, at a distance comparable to the flexural parameter α, these stresses are much smaller, and reverse in sign. Thus, the centers of lava-filled mare basins on the Moon are generally deformed in compression, while extensional structures surround them. Naturally, the same description applies to uplifted regions (negative loads), but with all signs reversed.

A very important, but often overlooked, role is played by the curvature of the planetary surface. Most published analyses of stresses in floating plates adopt a flat-plate approximation, in which the curvature of the surface is ignored. This is often an excellent approximation and only small errors are incurred by adopting it. However, when curvature of the plate is important (in this case the "plate" is called a "shell" in the engineering literature) the directions of the principal stresses may shift 90°, changing, for example, the radial extension surrounding an uplift to concentric extension (Janes and Melosh, 1990). The effect of planetary curvature is appreciable for loads of surprisingly small dimensions. A simple criterion for estimating the breadth of the load for which curvature is important was derived by Landau and Lifshitz in their famous book on elasticity (Landau and Lifshitz, 1970, p. 62). If the breadth of the load is L and the radius of the planet is R, then shell curvature is important when:

$$L > \sqrt{R\,t}. \tag{4.4}$$

Thus, on the Moon, $R = 1738$ km, if the lithosphere is 100 km thick, then loads broader than about 400 km will be strongly affected by planetary curvature. In particular, this means that stresses from mascon loads cannot be correctly computed using a flat-plate model (Freed *et al.*, 2001).

Aside from surface loads, any internal process that changes the shape or volume of a planet can produce tectonic stresses. The largest scale change is an alteration in planetary volume. This can be caused by heating or cooling (which is probably the source of Mercury's crustal compression), phase changes such as the crystallization of ice from water in the icy satellites or of liquid iron to solid iron in the cores of the terrestrial planets, or even rearrangements such as the separation of iron from silicates during core formation. All changes in planetary volume cause a change in radius (moreover, this also alters the rotation rate slightly and thus produces stresses that we have just classified as "external"). For a thin elastic shell, the stresses due to a change in radius ΔR are computed on the same basis as Vening-Meinesz's estimate of stresses caused by flattening changes. Both horizontal stresses are equal, and the vertical stress change is, again, zero to first order. Thus:

$$\sigma_{\theta\theta} = \sigma_{\phi\phi} = 2\,\mu\left(\frac{1+\nu}{1-\nu}\right)\frac{\Delta R}{R}. \tag{4.5}$$

Expansion produces isotropic extensional stress and contraction produces isotropic compression, both of which are uniform through the lithosphere and thus cannot cause preferred orientations of tectonic features.

In addition to global changes in radius, local processes produce stresses through volume changes. Those most important on the Earth today are probably temperature variations, although changes in chemical composition or mineral phase are also often cited as reasons for uplift or downwarping. Such changes are often linked to internal convection and so cannot be easily separated from convective stresses. The buoyancy stresses due to a local change in volume are easy to estimate: If the local density alteration is $\Delta\rho$ and the vertical dimension of the region over which the alteration occurs is L, then the buoyancy stress is $\Delta\sigma \sim gL\Delta\rho$. The density alteration due to a temperature change ΔT is simply $\Delta\rho = -\rho\alpha_V\,\Delta T$, where α_V is the volume coefficient of thermal expansion when pressure is maintained constant:

$$\left.\frac{\Delta V}{V}\right|_P = -\left.\frac{\Delta\rho}{\rho}\right|_P = \alpha_V\,\Delta T. \tag{4.6}$$

For substances that expand uniformly in all directions the volume coefficient of expansion is exactly three times larger than the linear coefficient of expansion. Because of the importance of this parameter for tectonic deformation, Table 4.2 lists the size of the volume expansion coefficients for a number of substances important in planetary interiors.

Simple density estimates can be made for chemical and phase changes, such as may occur during hydration of silicate minerals, freezing of liquids, or solid-state phase changes due to alterations of pressure or temperature in a planetary interior. Many liquids, upon freezing, decrease their volume by about 10% (water is an important exception!). Such volume changes do not, however, translate directly into tectonic stress in the lithosphere, as we must know how the volume or density change interacts with the lithosphere. If the volume of the elastic lithosphere itself changes, then the change is resisted by the elastic modulus (Turcotte and Schubert, 2002, Section 4–22). Such changes usually give rise to enormous stresses because of the large size of the elastic modulus compared with the

Table 4.2 *Selected thermal conductivity and volume thermal expansion coefficients at constant pressure and standard temperature and pressure*

Material	k (W/m-K)	$\alpha_V \times 10^5$
Rocks:		
Sandstone	1.5–4.2	3
Dolomite	2–3.4	2.4
Granite	2.4–3.8	2.4
Gabbro (basalt)	1.9–4.0	1.6
Peridotite (Earth mantle)	3–4.5	2.4
Minerals:		
Water ice, H_2O	2.2	5
Enstatite, $MgSiO_3$	4.47[b]	2.3[a]
Forsterite, Mg_2SiO_4	4.65[b]	2.5[a]
Metallic iron, Fe	0.84[c]	1.5[a]

After Turcotte and Schubert (2002, Appendix II.E) unless otherwise noted:
[a] Poirier (1991) p. 23
[b] Clauser and Huenges (1995)
[c] Weast (1972)

stresses caused by topographic loads. However, in most cases the volume change occurs in the more mobile material underlying the lithosphere and then the nature of coupling to the lithosphere becomes important, particularly whether it is by viscous shear stresses or by vertical uplift of the surface.

When heat is transferred by convection in the interior of a planet, the subsurface convective flow exerts forces on the overlying lithosphere and may manifest itself by tectonic deformation of the surface. This is the principal cause of tectonic deformation on the Earth and probably Venus. Convection, whether continuous or episodic, may also have affected the surfaces of Europa, Ganymede, and Enceladus, to name just a few. The size of the stresses can be crudely estimated from the equations above: For thermally driven convection, substitution of the equation relating density and temperature into the buoyancy equation tells us that $\Delta\sigma \sim \rho g \alpha_V L \Delta T$, where d is a relevant distance scale, often the thickness of the convecting layer. But what determines the temperature difference ΔT? And what is the spatial pattern of the convecting region and the pattern of stresses that convection induces in the overlying lithosphere? A great deal of effort in both planetary science and geophysics is currently being expended to answer these questions, but these answers seem to require elaborate numerical simulations of convection and stress coupling. Nevertheless, we can make a few simple statements about the onset of convection in the first place and about its rate and ability to transfer heat out of the interiors of planets, as well as the approximate size of ΔT. This will be discussed in the next section in the context of internal heat sources and thermal convection.

4.3 Planetary engines: heat sources and heat transfer

4.3.1 Accretional heat

The tectonic engine that wracks the surfaces of planets and satellites is powered by heat. Internal motions and differential stresses in the lithosphere are created by the flow of internal heat toward the cooler surface. A fundamental understanding of planetary tectonics thus requires understanding the fundamentals of heat deposition and transport in the interiors of planetary-scale bodies.

A potent source of internal heat, at least for larger planets, is their gravitational energy of formation. When an object of mass m strikes the surface of a planet at velocity v, its kinetic energy $\frac{1}{2}mv^2$ is released, a portion of which may remain buried in the planet (most of the balance is radiated rapidly to space). Some of this impact velocity is due to the gravitational attraction of the planet, for which the energy increment is GMm/a, where a is the radius of the planet, M is its mass and G is Newton's gravitational constant. If the planet grows from the accumulation of much smaller bodies, the total gravitational energy released upon accretion is a constant of order one times GM^2/a. For a uniform density sphere the "gravitational binding energy" is:

$$E_{\text{bind}} = \frac{3}{5}\frac{GM^2}{a} = \frac{16}{15}\pi^2\bar{\rho}^2Ga^5 \tag{4.7}$$

where $\bar{\rho}$ is the mean density. This binding energy clearly grows rapidly as planetary size increases. The Earth's accretional heat, if all of it were retained, would raise the entire planet's temperature by about 38 000 K. For the Moon, this figure is only 1700 K. These numbers make it clear that the initial heat of formation may play a major role in the thermal evolution of planetary-sized bodies.

The original heat of accretion is usually negligible for asteroidal bodies, but it is of major significance for the larger planets and moons. The principal uncertainty in its effectiveness, however, is how large a fraction of this heat is lost by radiation and how much is buried deep within the body. At the present time a full accounting of the heat deposited by impacts of various sizes has not yet been established, largely due to large contributions from factors of opposite sign. It has become common practice to assume a "burial efficiency" for impact heat, to which various writers have assigned values between 0.3 and 0.5. Most recently, however, the recognition of the role of truly giant impacts, in which bodies of comparable size collide, has greatly complicated this type of computation and much work still remains to be done on the heat retained during accretion (Melosh, 1990a).

One of the peculiar consequences of accretional heating is that, in the earliest stages of accretion when the mass of the growing planet is small, collisions are gentle and little heating takes place. But as the planet grows and its mass increases, collisions release correspondingly more heat. This produces a nascent planet with a cold center whose internal temperature increases roughly as the square of distance from its center. Such an "upside-down" thermal profile is stable against thermal convection until other heat sources, such as radiogenic heating, compositional convection, or planetary-scale impacts rearrange the interior into the more common pattern of a hot core overlain by a mantle whose temperature decreases towards the surface. The inverted temperature profile caused by accretional heating has occasioned much discussion over when and how planetary cores form (Stevenson, 1981).

4.3.2 Tidal dissipation in planetary interiors

One of the major discoveries in planetary science is that the tectonic, and even volcanic, evolution of many satellites in the Solar System is driven by heat created from tidal dissipation. Just as vigorous flexing of a squash ball warms it and renders it more flexible, so the inexorable tidal flexing of satellites, such as Jupiter's Io, dissipates energy in their interiors and may strongly affect their surfaces. Tidal dissipation also causes orbital changes, such as the circularization of elliptical orbits, the expansion of our Moon's orbit, or the gradual de-orbiting of Phobos (currently scheduled to impact Mars' surface in about 100 Myr). Tides have slowed the rotation of many planets or satellites, leading to the synchronization of rotation and orbital periods or capture into stable resonances, such as the 3:2 spin state of Mercury. Iapetus' oblate figure, for example, suggests a primordial 16-hour period, instead of its current 79-day synchronous rotation about Saturn.

A full discussion of tidal energy dissipation and its orbital consequences is beyond the scope of this text (for a full account, see the text by Murray and Dermott, 1999), but a few simple ideas will aid in understanding how tidal energy is dissipated in planets. Just as in our humble squash-ball analogy, tidal flexing of a planet dissipates energy by internal friction. The type of friction varies a great deal from one body to another. In very small objects that are solid throughout, a variety of solid-state deformation mechanisms dissipate energy by shear between atoms or atomic structures such as dislocation arrays. This type of dissipation is generally inefficient and energy losses from such mechanisms are relatively small. In our Earth, tidal currents in shallow oceans dissipate most of the energy from lunar and solar tides. The energy dissipated by these tides is not small: It amounts to about 10% of the total surface heat flow of the Earth. If this energy were dissipated deep within our planet (as was once believed) it would contribute more than enough energy to drive the geodynamo. This energy comes ultimately from the Earth's rotation and its loss causes the day to gradually lengthen (900 Myr ago the day was only 18 hours long: Sonnett *et al.*, 1996) and, through the conservation of angular momentum, causes the Moon to recede from the Earth at about 3.7 cm/yr.

Planets or moons with internal liquid–solid, or even slipping solid–solid interfaces can also lose large amounts of tidal energy at these junctures, a mechanism that is currently the subject of much calculational effort. Entirely gaseous planets such as Jupiter, Saturn or even stars like the Sun, have fluid-dynamical loss mechanisms that cause tidal dissipation, although such mechanisms are usually even less efficient than solid dissipation.

The rate of energy loss in a periodically disturbed system is conventionally parameterized by a quantity called simply "Q," short for "quality factor." The concept of *Q* originated in the study of oscillating electrical circuits, but it is an apt way of describing the loss of energy in any periodic system. Q is defined as the inverse ratio between the energy lost per cycle ΔE, to the periodically oscillating energy present in the system, E_{max}, times the conventional factor of 2π (this seemingly arbitrary factor converts the energy loss from per cycle to per radian and usually eliminates many 2π factors elsewhere):

$$Q \equiv \frac{2\pi E_{max}}{\Delta E}. \tag{4.8}$$

As the energy dissipated per cycle decreases, the value of Q rises – the less loss, the higher the "quality" of an oscillator. Another way of thinking of Q is that it is the number of cycles it takes for the energy to decrease by 1/e, so that the higher the Q, the larger the number of cycles that pass before the oscillation is damped.

The energy stored in a tidally distorted body is not easy to compute: For a solid body it requires a coupled application of full elasticity theory and gravitational potential theory for a distorted body. However, the result of such a computation for a uniform, elastic, incompressible satellite distorted by an adjacent massive primary around which it travels in a circular orbit is easy to state (Goldreich and Soter, 1966):

$$E_{max} = \frac{9\pi}{2} \frac{GM^2 a^5}{\left(1 + \frac{19\mu}{2g\rho a}\right)} \frac{1}{r^6} \tag{4.9}$$

where G is Newton's gravitational constant, M the mass of the primary, and r is the distance between the primary and the orbiting object. The tidally distorted satellite is characterized by its radius a, mean density ρ, surface acceleration of gravity g, and average shear modulus μ. Note that the distortional energy is a very strong function of distance from the primary: It falls off as the sixth power of r, so that tidal dissipation is generally only important for very close satellite–primary pairs. The factor in parentheses describes the balance between gravitational and elastic distortional energy. If gravity alone dominates, as in very large planets, this factor is essentially 1, whereas in small objects dominated by elastic distortion the second term dominates.

In general, Q must be computed from detailed models of dissipation. As might be expected, it depends on the frequency of flexing. For extremely strong tides it also depends on the amplitude of the flexing, but this is usually neglected in current models. Q is readily computed for idealized rheologic models, such as the Maxwell model or the Kelvin–Voigt model described in Chapter 3. The dependence of Q on angular frequency ω is very different for these two models (Knopoff, 1964):

$$Q = \frac{\omega\eta}{\mu} \qquad \text{Maxwell solid}$$

$$Q = \frac{\mu}{\omega\eta} \qquad \text{Kelvin–Voight solid} \tag{4.10}$$

where η is the viscosity and μ is the shear modulus of the material. Dissipation in liquids follows the Kelvin–Voigt equation quite accurately, whereas it is expected that the slow deformation of hot solids should approximate the Maxwell rule. The dissipation of seismic waves in the Earth enigmatically shows very little frequency dependence (Knopoff, 1964). This is generally attributed to the simultaneous operation of many different mechanisms of seismic energy loss, but the subject is still controversial (Karato, 1998).

The time rate of energy dissipation is related to Q and E_{max} through the frequency:

$$\frac{dE}{dt} = \frac{\omega\, E_{\mathrm{max}}}{Q}. \tag{4.11}$$

In a Maxwell solid the rate of energy dissipation is independent of frequency, whereas in a Kelvin–Voigt solid it depends on the square of the frequency. The situation is more complicated than this discussion suggests when applied to planets, because the maximum stored tidal energy E_{max} in a Maxwell solid is itself frequency-dependent: The overall dissipation depends on $1/\omega$ at high frequencies and ω at low frequencies.

The observed rates of energy dissipation derived from rates of orbital evolution can be used to estimate Q for various Solar System bodies. Q is known only for a few bodies at present, but its value serves as a useful probe of the internal state of planetary objects. For example, the surprisingly low Q of 27 for the Moon suggests frictional losses at the solid–liquid interface of a small liquid core, whereas the much higher $Q = 86$ of Mars suggests that its core is solid. It is common practice, when a value for Q is not known for a solid or icy body, to assign it a nominal value of 100. Table 4.3 lists known values of Q for bodies in the Solar System.

4.3.3 Heat transfer by thermal conduction and radiogenic heat production

It is currently believed that planets and satellites are born hot, heated by the gravitational energy of infalling material as they accreted from smaller bodies in the planetary nebula around their nascent star, or by heat generated from the decay of radioactive elements. When growing planets and satellites are not so hot as to actually melt, their accretional heat is initially lost by solid-state conduction to their cooler surfaces. As they cool, the associated volume changes can drive tectonic deformation.

Joseph Fourier (1768–1830) established the laws governing heat conduction around 1810. Fourier had a deep interest in geophysics and spent much time and effort descending into deep mines or visiting geothermal spas to measure the Earth's internal heat. Fifty years before Fourier, his countryman Leclerc de Buffon measured the cooling rate of different-sized balls of white-hot iron in an effort to determine the age of the Earth (Sigurdsson, 1999). For a long time it was believed that the Earth's age could, in fact, be determined by the rate at which it cooled by conduction from a presumed incandescent initial condition (Burchfield, 1990). Nineteenth- and early twentieth-century geologists thought that

Table 4.3 *Q in the Solar System*

Planet or satellite	Approximate tidal period	Dissipation location	Q
Moon[a]	1 month	Core-mantle boundary?	27
Earth[a]	1 day	Shallow oceans	12
		Solid Earth	300
Mars[a]	1 day	Solid interior?	86
Io[b]	1.8 days	Hot mantle?	~36
Jupiter (Io)[c]	1.8 days	Atmosphere	~3 x 10⁴
Saturn (Mimas)[c]	0.9 day	Atmosphere	≥ 1.8 x 10⁴
Uranus (Ariel)[c]	2.5 days	Atmosphere	≥ 2 x 10⁴

[a] Yoder (1995)
[b] Murray and Dermott (1999)
[c] Segatz *et al.* (1988) Model A

terrestrial tectonics was caused by compression of the crust as the Earth gradually cooled (Greene, 1982). Folded mountain chains were likened to the shriveled skin of a shrinking, drying apple (Bucher, 1933).

Fourier (1955) derived a simple, linear differential equation for heat conduction and devised a clever series method for its solution. In modern vector notation, he started with the equation governing the flow of heat down a temperature gradient:

$$q = -k\nabla T \tag{4.12}$$

where q is the vector heat flux and k is the thermal conductivity. Adding the conservation of energy to this equation (actually, Fourier wrote his equation for the hypothetical fluid "caloric," but we now recognize that energy is the conserved quantity), he obtained the general equation governing the temperature in a solid body:

$$\frac{\partial T}{\partial t} = \nabla \bullet \left(\kappa \nabla T\right) + \frac{H}{c_P} \tag{4.13}$$

where H is internal heat generation per unit mass, and κ is the thermal diffusivity. The diffusivity is related to the thermal conductivity k by $\kappa = k/\rho c_P$, where c_P is the heat capacity of the solid at constant pressure (c_P is typically about 1000 J/kg-K for both rock and ice). Equation (4.13) includes internal heat generation and the form written above, with the diffusivity κ inside the divergence operator, also applies to the case where the diffusivity is a function of position.

Fourier and many authors after him devoted extensive efforts to finding solutions to his equation. A widely used reference containing analytical solutions to many problems of heat conduction is the monograph by Carslaw and Jaeger (1959).

One of the simplest, but widely useful, solutions to Fourier's equation is for the steady-state temperature in an infinite plane slab with uniform conductivity k and no internal heat generation. If the surface temperature is T_S and the surface heat flux is q_S, then the temperature is a linear function of depth in the slab z:

$$T(z) = T_S + \frac{q_S}{k} z. \tag{4.14}$$

This solution was used above in Equation (4.1) for the thickness of the elastic lithosphere. Adding a slightly higher level of complication, if the slab contains a uniform source of heat H, the solution acquires a third, quadratic, term:

$$T(z) = T_S + \frac{q_S}{k} z - \frac{H}{2c_P} z^2. \tag{4.15}$$

According to this equation, internal heat generation actually *lowers* the subsurface temperature compared to a slab without a source of heat. This is because heat from sources within the slab does not pass through the entire thickness of the slab, thus finding a shorter path to the surface and lowering the temperature at depth. This is one of the principal mechanisms by which the interior temperatures of planets are reduced: By differentiation of heat-producing radioactive elements upward toward the surface, internal temperatures are lowered.

Because the current major sources of radioactive heat in the Solar System, ^{238}U, ^{235}U, ^{232}Th, and ^{40}K, are large lithophile ions, they tend to concentrate near the surfaces of silicate planets and thus cause less internal heating than if they were uniformly distributed. Table 4.4 illustrates this effect by listing the current radiogenic heat generation for primitive Solar System material (carbonaceous chondrites), differentiated rocks, and typical mantle rocks. This cooling strategy does not help the icy satellites, however. Because differentiation of ice-rich bodies puts their heat-generating silicates beneath their icy mantles, their interiors are strongly heated. This may partially explain why the small icy satellites are so tectonically active compared to their rocky compatriots. Similarly, in the early Solar System the short-lived radioactive isotope ^{60}Fe was active for a few million years. Differentiated into the cores of planetesimals, this heat source would have been particularly effective in convulsing the interiors and surfaces of these small bodies, and ^{26}Al was briefly but fiercely effective in heating their mantles. Table 4.5 lists the rate at which heat is produced by the decay of various radioactive species important in our Solar System.

A crucially important aspect of thermal conduction is its control over the rate at which planets can heat or cool. At the present time, nearly any problem that can be posed involving heat conduction can be readily solved on a computer to any desired degree of accuracy; many computer programs are designed to do just this. Such numerical methods can also incorporate temperature-dependent thermal diffusivity as well as many other more realistic aspects of planetary materials. However, despite the present-day ease and sophistication for solving problems in heat conduction, one simple consequence of Equation (4.13) stands out and can be widely applied without the use of any computers at all. This result follows

Table 4.4 *Radiogenic heat production in silicate rocks, current Solar System*

Material	Heat production H (10^{-12} W/kg)
Bulk Earth (does not include primordial heat)	4.7
Bulk Moon (based on only two heat flow measurements)	~14
Average granite	1050
Alkali basalt	180
Venus surface rocks (Vega 1 and 2, Veneras 8, 9, 10)[a]	117–531
Tholeliite (Earth oceanic basalt)	27
Eclogite	9.2
Peridotite (Earth mantle)	1.5
Ordinary chondrite	5.85
Carbonaceous chondrite	5.23

Data from Stacey (1992), except:
[a] Schubert *et al.* (2001) Table 14.3

Table 4.5 *Radiogenic heat production in the Solar System*

Isotope	Half-life (years)	Heat generation (exclusive of neutrino losses) (W/kg)	Concentration in CI chondrites 4567 Myr before present (10^{-9} kg/kg)
^{238}U	4.47×10^9	9.46×10^{-5}	1.93
^{235}U	7.04×10^8	5.69×10^{-4}	0.610
^{232}Th	1.40×10^{10}	2.64×10^{-5}	4.47
^{40}K	1.25×10^9	2.92×10^{-5}	583.
^{26}Al	7.4×10^5	3.55×10^{-1}	430.
^{60}Fe	1.5×10^6	6.3×10^{-2}	840.

Data on heat generation and half-lives of U, Th, K from Van Schums (1995), element abundances from Newsom (1995), ^{26}Al energetic data from Schramm *et al.* (1970), initial abundance from McPherson *et al.* (2010). ^{60}Fe data from Quitté *et al.* (2005).

from dimensional analysis of this equation in the case of no internal heat generation and uniform thermal diffusivity. In this case the order of magnitude of the cooling (or heating) time of an initially hot (or cold) body of dimension L is simply given by:

$$\tau_{\text{cool}} = \frac{L^2}{\kappa}. \tag{4.16}$$

This simple formula has powerful implications for planetary timescales and is even crucial for understanding thermal convection. The fact that the size of the cooling body is squared in Equation (4.16) means that increasing the size of a body leads to a seemingly

Table 4.6 *Conductive thermal time constants*[a]

Object dimensions	Typical example	Conductive cooling/ heating time
1 mm	Fire fountain droplet	1 second
1 cm	Large chondrules	2 minutes
10 cm	Lava flow chilled crust	3 hours
1 m	Thin pahoehoe flow	12 days
10 m	Typical lava flow	3.2 yr
100 m	Rhyolite lava flow	320 yr
1 km	Impact crater ejecta blanket	32 000 yr
10 km	Basin ejecta blanket, planetesimal	3.2 Myr
100 km	Typical lithosphere, asteroid	320 Myr
1000 km	Planetary mantle, large moon	32 Gyr
10 000 km	Entire terrestrial planet	3200 Gyr

[a] Assumes a typical thermal diffusivity for rock, $\kappa = 10^{-6}$ m²/s.

disproportionate increase in the cooling time. To emphasize this important point, Table 4.6 lists the cooling times for rocky bodies of a variety of dimensions of planetary interest, from 1 mm (small chondrules or droplets from volcanic fire-fountains) to 10 000 km (an entire planet), assuming a thermal diffusivity appropriate for rock.

Thermal conduction timescales quickly become longer than the age of the Solar System (indeed, even longer than the age of the universe!) for planetary-size bodies. However, for objects the size of the larger asteroids or the smaller moons of Saturn, conductive cooling probably determined the pace of their thermal evolution. Volume changes due to conductive cooling are typically small: The volume expansion coefficient α_V is around 3 x 10^{-5} K^{-1} for most materials. Nevertheless, if a 100 km diameter moon cools by, say, 1000 K (over a timescale of a few 100 Myr, according to Table 4.6), its volume decreases by about 3% and, hence, it circumference decreases by about 6 km, a change that, if translated to offsets on compressional faults, would be quite visible in planetary images. Because of the large size of elastic moduli, this strain implies compressional stresses far in excess of the strength of any planetary material, so material failure is inevitable.

Although strains due to temperature changes may seem small, the most consistent topographic variation on Earth is almost entirely due to this effect. Thus, oceanic plates are born hot at mid-ocean ridges and cool, following Equation (4.16), as they move away from the ridge at nearly constant velocity. As the plates cool, they contract and the sea floor subsides from a depth of about 2.5 km at mid-oceanic ridges to about 6 km for the oldest plates (about 1/3 of this 3.5 km subsidence is due to isostatic adjustment to the loading by the weight of ocean water). The resulting age–depth relation is one of the most regular topographic relations in the Solar System. The depth of the sea floor plotted versus the square root of its age accurately follows a straight line over about 100 Myr of cooling, before it levels out when convective heat transfer overcomes conduction.

The cooling and contraction of Earth's oceanic plates nicely illustrates the point that planetary heating and cooling events need not affect the entire body: Local thermal events that affect only a limited region often occur. In all cases, however, Equation (4.16) provides guidance to the tempo of these changes, given only that L is chosen appropriately.

4.3.4 Thermal convection and planetary heat transfer

On sufficiently large bodies the density variations caused by temperature differences cause bulk motions deep below the surface. Heat is carried efficiently along with the material in a process known as advection. When advection is driven by temperature differences it is called thermal convection. In addition to the heat exchanged by this process, stresses develop whose tectonic consequences often provide the only visible manifestation of convection in a planetary interior. The velocity at which material is advected sets the tempo for these tectonic processes.

The mathematical analysis of the process of thermal convection has a huge literature and has been the subject of a large amount of sophisticated analysis. For the interested (and mathematically inclined) reader, the standard treatment is found in a book by Chandrasekhar (1961). However, the basic principles governing convection are not complex and a great deal can be captured using ideas that have already been presented, augmented by the numerical results of more exact analyses.

Rayleigh number. When an initially stagnant fluid is heated from below (or cooled from above), thermal expansion generally lowers the density of the deeper material. This produces a mechanically unstable configuration, where more dense material overlies less dense material. The natural tendency of the less dense material is to rise and displace the more dense material. The timescale for this exchange is governed by the viscosity η of the fluid. Referring to the last chapter, Equation (3.34), the timescale for relaxation of topography on a viscous substrate also gives the rate at which a density anomaly can rise through a viscous fluid. We need only replace the topographic density ρ by the density anomaly $\Delta\rho$ and write (ignoring the numerical constant):

$$\tau_{\text{overturn}} \approx \frac{\eta}{\Delta\rho\, g\, L} \tag{4.17}$$

where τ_{overturn} is the timescale for the density anomaly to turn over a mass of fluid of dimension L in the gravity field g. Because the density anomaly is caused by a temperature difference ΔT, a substitution using Equation (4.6) for thermal expansion gives the overturn time in terms of the temperature difference between the hot region and the overlying colder region:

$$\tau_{\text{overturn}} \approx \frac{\eta}{\alpha_v\, \Delta T \rho\, g\, L}. \tag{4.18}$$

Unlike the density anomaly due to a permanent, compositional, difference of materials, a thermal density anomaly lasts only as long as the temperature difference persists.

Thus, if the overturn time is so long that thermal conduction can eradicate the temperature difference, no motion will take place and the heat will be lost by conduction, in spite of the density anomaly. We can see that there must be a threshold below which heat is transported by conduction. Only if the threshold is exceeded will overturn occur and can heat be transported by convection. This threshold is defined by the ratio between the viscous overturn timescale and the cooling timescale, Equation (4.16). The criterion for the onset of convection was first derived by Lord Rayleigh (1916) and is, therefore, known as the Rayleigh number Ra:

$$\mathrm{Ra} = \frac{\tau_{\mathrm{cool}}}{\tau_{\mathrm{overturn}}} = \frac{\alpha_V \rho g L^3}{\kappa \eta} \Delta T. \qquad (4.19)$$

Because the temperature difference ΔT between the top and bottom of a planetary mantle is usually not known, a closely related formulation based on the surface heat flow q_S and Fourier's relation (4.12) is usually more practical:

$$\mathrm{Ra_q} = \frac{\alpha_V \rho g L^4}{\kappa k \eta} q_S. \qquad (4.20)$$

The definitions (4.19) and (4.20) are equal only at the onset of convection. Above the convection threshold they differ by a factor known as the Nusselt number, Nu: $\mathrm{Ra_q} = \mathrm{Nu}\,\mathrm{Ra}$. Nu, which is a measure of the net heat transport, will be defined more exactly below. For internally heated convecting systems, q_S in Equation (4.20) can be replaced by the equivalent heat generation per unit surface area, $\rho L\,H$, assuming that heat generation and loss through the surface are in balance and that the heat generation H is uniformly distributed.

If the conductive cooling time is much shorter than the overturn time, then Ra << 1 and convection will not occur. But in the opposite extreme, when Ra >> 1, convective overturn will occur and most heat will be transferred through material exchange between the underlying hot region and the surface cool zone. The "critical" Rayleigh number, at which convection just becomes possible, would thus seem to have a value close to one. However, sometimes quantities that physicists dismiss as "of the order of unity" involve high powers of π or other "small" numbers. The critical Rayleigh number $\mathrm{Ra_C}$ for convection is one of these: It ranges between 654 and 1709, depending on the boundary conditions for fluid flow at the top and bottom of the convecting region. Such precision is only possible for the restrictive case in which the viscosity and likewise the thermal expansion coefficient are constants, independent of temperature and stress, and the geometry is a simple plane layer.

Lord Rayleigh also determined that the convecting material organizes itself into cells, where sinking cool material separates upwelling warm regions. The horizontal breadth λ of these cells is roughly three times larger than the thickness of the convecting region at the onset of convection (Figure 4.2). At higher Rayleigh numbers the primary distance scale is set by the thickness of the boundary layers, as discussed below.

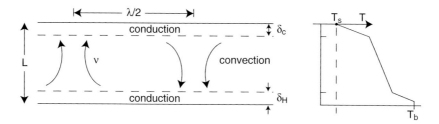

Figure 4.2 Schematic illustration of the flow pattern and temperature in a convecting layer of thickness L bounded between a cool surface layer at temperature T_S and a hot basal layer at temperature T_b. Fluid in the center of the layer rises and sinks in convective cells of horizontal dimension λ at velocity v. The upper and lower surfaces of the convecting layer are conductive boundary layers of thickness δ_c at the cool upper boundary and δ_H at the hot lower boundary. The temperature in these thin layers changes rapidly, as shown in the right panel, while that in the central convecting region changes slowly, approximately following an adiabatic gradient.

The perceptive reader may be dismayed by the simplifications introduced by the Rayleigh analysis and, in view of the fact that the viscosity of planetary materials is strongly dependent on both temperature and stress and that planetary mantles are hardly infinite planes, wonder whether this type of analysis has any practical use at all. Professional geoscientists have shared the same doubts and a great deal of effort has been expended over the past half-century to lift these restrictions. The answer is not simple: The interested reader should consult the massive (940 pages) monograph by Schubert *et al.* (2001) for a comprehensive survey. But the general conclusion is that, if the viscosity and other parameters are defined appropriately, then the simple Rayleigh criterion for the onset of convection does work. For example, in a fluid with a strongly temperature-dependent viscosity, a meaningful Rayleigh number is computed from the value of the viscosity at the fluid's mean temperature.

Furthermore, a number of simple generalizations give a good first-order description of convection well beyond the threshold. The thermal structure of a vigorously convecting region is divided into three regions (Figure 4.2): A relatively narrow, cool top zone where heat is transported only by conduction (the "cold thermal boundary layer"), a similar region at the bottom (the "hot thermal boundary layer") and an intervening region where the temperature variations are small. The hot thermal boundary layer may be absent if the layer is strongly heated internally, as by the decay of radioactive elements, and the cool boundary layer may be highly mobile, as in plate tectonic convection. The intervening zone is traversed by hot, buoyant masses rising to transport heat upward, which are counterbalanced by cool sinking masses that similarly transport heat upward by cooling the lower regions of the convecting zone. The average temperature in this central region approximates an "adiabat," where the temperature and pressure are related by the condition that the entropy of the material is constant. Although the derivation of the equation describing this gradient is deferred to Chapter 5 on volcanism, the equation for the adiabatic temperature gradient is:

$$\frac{dT}{dz}\bigg|_{\text{adiabat}} = \frac{\alpha_V g T}{c_P}.$$

(4.21)

This gradient is typically about 0.36 K/km in the Earth, which is very small compared to the average surface temperature gradient of 30 K/km. Because the adiabatic gradient scales with the gravitational acceleration g, it is smaller in smaller planets or moons. This relation must not be applied through phase changes, where the constant entropy constraint may require a temperature jump. Nevertheless, the temperature change across an adiabatic, vigorously convecting region is generally small compared to that across the top thermal boundary layer.

Nusselt number. The thickness δ of the boundary layers is related to the total amount of heat transported, given by Fourier's Equation (4.12). Because the boundary layer is thinner than the convecting region itself, the total heat transported by conduction through this layer is larger than if it were conducted through the thickness of the entire convecting region, by the ratio L/δ. A dimensionless number known as the Nusselt number expresses this fact:

$$\text{Nu} = \frac{\text{total heat transport}}{\text{heat transport by conduction}} \equiv \frac{q_S}{k\dfrac{\Delta T}{L}} = \frac{L}{\delta}.$$

(4.22)

The Nusselt number is equal to 1 if no heat is transported by convection, and it rises above 1 as convection becomes more vigorous. The mean convective velocity $\bar{v}_{\text{conv}}$ can now be estimated from the convective heat flux. The heat content of a unit volume of rising hot material is of order $\rho c_p \Delta T$, and the volume arriving per unit time and per unit area at the base of the cold boundary layer is just $\bar{v}_{\text{conv}}$. Add this to the rate at which heat is transported by conduction per unit area through the entire layer, $k\Delta T/L$, and use the definition of the Nusselt number,

$$\text{Nu} = \frac{k\dfrac{\Delta T}{L} + \bar{v}_{\text{conv}}\rho c_P \Delta T}{k\dfrac{\Delta T}{L}}$$

(4.23)

to derive the mean velocity of convection:

$$\bar{v}_{\text{conv}} \approx \frac{\kappa}{L}\left(\text{Nu} - 1\right)$$

(4.24)

Applying this equation to the Earth, the only planet for which seismology gives us independent information on the boundary layer thickness, $\delta_{Earth} \sim 100$ km (this is often stated to be the thickness of the tectonic plates, which is correct if the thermal boundary layer thickness is meant – it is much larger than the elastic plate thickness), yields a mean convective velocity of about 0.2 mm/yr. This is much smaller than the plate velocities themselves, but it does agree rather well with the relative velocities between the mantle plumes and, thus, reflects the speed of gross motions deep in the mantle. Equation (4.24) is only an average: Rising or sinking plumes may be much smaller than the convecting region as a whole and may, thus, move faster by the inverse ratio of their size to that of the convecting layer. Taking this smaller scale to be that of the boundary layer itself, we have

$$v_{\text{plume}} \approx \frac{L}{\delta} \bar{v}_{\text{conv}} = \frac{\kappa}{L} \operatorname{Nu}(\operatorname{Nu} - 1). \tag{4.25}$$

Parameterized convection. What is still missing is a means of estimating the Nusselt number, or equivalently, the boundary layer thickness. A simple means of doing this, which seems to hold up very well even in rather complex rheologies and with complicated planetary structures, is through parameterized convection models (Schubert, 1975). Such models relate the Nusselt number to the Rayleigh number through a simple power law:

$$\operatorname{Nu} = \left(\frac{\operatorname{Ra}}{\operatorname{Ra}_C} \right)^{\beta} \tag{4.26}$$

where Ra_C is the Rayleigh number at the onset of convection and β is a power that is close to 1/3 (the best measurements for high Rayleigh number suggest that it is actually about 0.29).

The convective velocity, Equation (4.24), allows us to estimate the rates of convective processes. In addition, we need an estimate of the stresses generated by convective motions. The buoyancy stress of a rising (or sinking) mass of material is given by $\rho \alpha_V \Delta T g d$, where d is some "appropriate" distance scale. For most tectonic processes this is the thickness of the boundary layer δ, not the depth of the convecting region itself. Furthermore, the temperature difference ΔT can be computed from the surface heat flow q_S (assuming that most of the temperature drop occurs across the boundary layer) to give an estimate of the stress generated by plume-scale convective processes:

$$\sigma_{\text{plume}} \approx \left(\rho \alpha_V \Delta T g \right) \delta = \frac{\rho \alpha_V q_S g}{k} \delta^2 = \left(\frac{\operatorname{Ra}_q}{\operatorname{Nu}^2} \right) \frac{\eta \kappa}{L^2}. \tag{4.27}$$

The crucial parameter missing before the above equations can be used to obtain numerical estimates of convective velocity, strain rate, and stress is the viscosity, η. Because the viscosity of hot, creeping solids is a strong function of composition, temperature, and stress, it might seem impossible to make a meaningful estimate of the viscosity of any given planetary body. On the Earth, some viscosity estimates are available from the rate of post-glacial rebound, from which it appears that the upper mantle, at least, has a viscosity in the neighborhood of 10^{21} Pa-s. However, aside from some rather crude bounds on viscosity from the rate of impact crater relaxation on a few extraterrestrial bodies, the mantle viscosity of other planets and satellites is not constrained by observations.

A recipe for planetary viscosity. Although this situation may seem dismal, there is an argument that may make meaningful viscosity estimates possible. This argument dates

Table 4.7 *Convection in the terrestrial planets*

Quantity:	Mercury	Venus	Earth	Moon	Mars
Mantle depth, km	640	3000	3000	1500	2000
Surface heat flow, mW/m^2	13[a]	56[a]	81	27	23[a]
Rayleigh number	7.1×10^4	5.3×10^7	7.7×10^7	9.3×10^5	3.9×10^6
Nusselt number	3.4	23.4	26.1	7.3	11.0
Thermal boundary layer thickness, km	186	130	110	210	180
Mean convection velocity, mm/yr	0.09	0.18	0.20	0.10	0.12
Plume velocity, mm/yr	0.31	4.2	5.3	0.72	1.3
Plume-scale convective stresses, MPa	38	190	240	43	66

Assuming: $k = 3.3$ W/m-K, $\kappa = 7.5 \times 10^7$ m^2/s, $\eta = 10^{21}$ Pa-s, $\rho = 3200$ kg/m^3.
[a] Estimated by assuming a bulk heat generation equal to that of carbonaceous chondrite material. This is multiplied by 0.55 for Mercury to account for its excess iron abundance.

back to the dawn of plate tectonics in the late 1960s when an eccentric geophysicist, David Tozer, was impressed that thermal convection in strongly temperature-dependent materials tends to be self-regulating (Tozer, 1965, 1970). Because of the exponential temperature dependence of viscosity, a small change in the mean temperature of a planetary mantle makes a big change in the rate of heat transport. Tozer reasoned that the internal temperature of a convecting planet adjusts itself to carry all internally generated heat to the surface, but that the necessary temperature variations are relatively small. He then concluded that the viscosity of *all* convecting planetary mantles is virtually identical: 10^{21} Pa-s. Although the logic behind this assertion is somewhat elusive, detailed modeling of convection in planetary mantles does exhibit strong self-regulation and the models show only one or two orders of magnitude variation in the mean viscosity (Davies, 1980; Schubert, 1975), either from one planet to another or over the course of planetary evolution.

Convecting planets. Provisionally accepting Tozer's prescription for viscosity, Table 4.7 summarizes the major features of mantle convection and the tectonic framework for the terrestrial planets and Earth's moon. Not unexpectedly Earth, the largest planet, convects the most vigorously, has the highest plume velocities, and its lithosphere is subject to the largest stresses. Perhaps this is the reason that Earth, apparently alone of the terrestrial planets, possesses plate tectonics, although the presence of a weakening agent, water, in its lithosphere may play a larger role (water is not taken into account in the estimates presented here). Plume velocities in Table 4.7 predicted by Equation (4.25) are about an order of magnitude lower than average plate velocities. This is because the crude convection model described in this section does not consider plate tectonics explicitly. The velocities of the plates are largely determined by a balance between the negative buoyancy of the subducted lithosphere and drag along the transform plate boundaries and are, thus, not correctly computed from simple convection models. Venus is very similar to Earth, but its elastic lithosphere is even thinner because of its high surface temperature: If Venus possesses plates, they are individually much smaller than the Earth's, and there is little evidence for subduction recycling. On Venus, volcanic resurfacing seems to be the rule. The interiors of Mercury, Mars, and the Moon are imprisoned within thick lithospheres, convective stresses are small and internal strain rates are low. These are one-plate planets whose tectonic framework is dominated by global stress systems that are apparently insensitive to internal convective processes.

An important discovery that derived from the parameterized convection models in the 1970s is that the Earth's internal heat generation and present heat loss through the surface are not in balance. Previous to this time it was assumed that because of the rapid turnover of its internal convective cells, the Earth's primordial heat had long since been dissipated. What was not realized is that the interiors of these cells exchange heat with the rising and falling plumes by conduction, for which the time constant is very long. Careful thermal modeling shows that about half of the Earth's present heat flow comes from heat buried at the time the planet formed (Schubert *et al.*, 2001). In the case of less vigorously convecting planets this time constant is even longer, so that it is doubtful that any of the terrestrial planets' heat flow today reflects the rate of internal heat generation from radioactive decay. This fact makes simple computations such as those reported in Table 4.7 somewhat dubious: Each planetary body must be treated on a case-by-case basis, with careful attention to the particulars of each planet and its history.

4.4 Rates of tectonic deformation

The tempo of planetary tectonic movements varies over the entire spectrum, from the near-instantaneous response to large asteroid impacts to very slow movements that are perceptible only when integrated over the age of the Solar System. Rates are important for understanding the mechanisms that form tectonic features: Stresses due to creep and viscosity are functions of the strain rate. The Maxwell time plays a dominant role in rate considerations. For example, on the timescale of an impact, earthquake, or even the monthly lunar tide, the Earth's mantle acts like a massive elastic body with a shear modulus larger than steel. However, on the much longer timescales associated with internal convection, the Earth's mantle is utterly strengthless and flows like warm honey under the small stresses caused by thermal expansion and contraction. The mantle's Maxwell time of about a century defines the watershed between these two contrasting types of behavior.

The slowest rates of tectonic deformation are those related to thermal conduction in bodies of planetary dimensions. The cooling and contraction timescale for an object the size of our Moon is measured in tens of billions of years. Even the larger asteroids, like Ceres or Vesta, are still gradually losing their primordial heat. Tectonic deformation of such objects is very slow and strains are small: Unless other processes, such as impacts, intervene, the surfaces of such objects should reflect their most ancient histories.

Early in Solar System history, however, radiogenic heating of planetary precursors was intense. Short-lived radioactive species, especially ^{26}Al and ^{60}Fe, were active. Planetesimals as small as 30 km in diameter, if formed early enough, could have melted and differentiated. We know little of such objects at present, aside from a small number of differentiated meteorites, but continued exploration of objects in the asteroid and Kuiper belts may reveal a few survivors whose tectonic and volcanic evolution was both short and intense.

The rates of tectonic processes caused by tidal heating vary enormously. Because tidal dissipation depends upon the inverse sixth power of the distance from the primary, tidal heating may vary from dominance, as for Jupiter's moon Io, to unimportance, as for Jupiter's Callisto, even within a single satellite system. Strain rates vary similarly. Because of volcanic resurfacing, Io's crust is continually sinking (and being regenerated) at the rate of about 1.5 cm/yr, leading to strain rates in the thin elastic crust $\Delta a/a \sim 3 \times 10^{-16}$ s^{-1}. On the other hand, Mercury lost its primordial spin by solar tidal friction over a period of perhaps 300 Myr, during which time its original equatorial bulge of, say, 20 km, relaxed away, implying a strain rate of about 10^{-18} s^{-1}.

Strain rates in planets large enough to support internal convection are controlled by the speed of convective overturn and, when convection is vigorous, by the impingement of rising and sinking plumes on the surface. In the Earth and Venus this occurs at velocities of a fraction of a centimeter per year and the associated strain rates are around 10^{-15} s^{-1}. Earth's plate tectonic regime implies much higher strain rates, up to a maximum near 10^{-13} s^{-1} near plate boundaries. Similar strain rates may have affected the crust of Venus during its latest episode of global resurfacing. Internal strain rates in Mars and Mercury are both much smaller and only dubiously manifested on their surfaces: On these bodies the dominant tectonic features seem to be related to surface loading by either impact or volcanic processes.

4.5 Flexures and folds

4.5.1 Compression: folding of rocks

The distortion of layered rocks into spectacular drape-like folds was one of the earliest observations in the history of geology. Even the earliest observers, such as James Hutton (1726–1797), recognized that folding is caused by compression. Although Hutton had no clear idea of the origin or even magnitude of the forces necessary, folding clearly indicates shortening of the rock layers and was quickly attributed to a cooling, and therefore shrinking, Earth by a series of "dynamical geologists" that extended from Lord Kelvin through Harold Jeffreys. Mapping of the Alps and Appalachian Mountains revealed that their rocks are thrown into enormous trains of folds whose wavelengths often reach tens of kilometers (Figure 4.3). The analogy between these mountain-scale folds and the more humble folds that develop in crumpled piles of paper or fabric is irresistible: It led to many elaborate experiments to reproduce mountain structure in the laboratory. USGS geologist Bailey Willis, for example, devoted years to re-creating the Appalachian Mountains' structure in a large wooden box filled with layers of carpeting and tar (Willis, 1893).

Visual analogies, however, are often poor guides to mechanics, and folding provides a particularly poignant illustration of this fact. The mathematical theory of folding by elastic

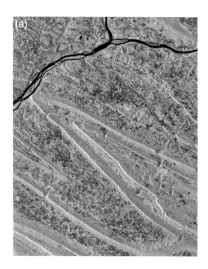

Figure 4.3 Folded mountains on Earth and Venus. (a) Shuttle synthetic aperture radar image of the area near Sunbury, PA, USA, in the Appalachian folded Valley and Ridge Province. The frame is 30.5 km wide and 38 km high. North is to the upper right. The large river is the Susquehanna and the tributary is the West Branch River. NASA image PIA01306. (b) Northern portion of Ovda Regio near the equator of Venus. This image shows compressional east–west ridges cut by orthogonal extensional fractures. Note that in these radar backscatter images created by Magellan synthetic aperture radar, dark is smooth (at 12.6 cm wavelength) and bright is rough. The frame is 300 km wide and 225 km high, approximately 10 times larger than the image in (a). NASA image PIA00202.

instability is briefly outlined in Box 4.1, and illustrated in Figure 4.4. Equation (B4.1.7) gives the minimum stress to initiate buckling in a floating elastic plate of thickness t. The minimum stress grows as the square root of the plate thickness. Inserting values of the elastic constants and density suitable for laboratory materials, one finds good agreement with the results of small-scale experiments, such as those of Bailey Willis, or the archetypical wrinkling skin of a drying apple. However, as soon as these results are scaled up to the dimensions of the Earth, trouble arises. Harold Jeffreys, through many editions of his famous book, *The Earth*, scoffed at the idea that elastic instability had anything to do with mountain ranges. He turned Equation (B4.1.7) around and asked the question: "What is the maximum thickness of an elastic layer that can buckle before the stress reaches the fracture limit?" His answer is surprising: 12 m is the maximum thickness of a granite layer that can buckle into folds by elastic instability before the stress exceeds its crushing strength! If instead one inserts values of the oceanic lithosphere's flexural rigidity deduced directly from seamount loading, $D = 6 \times 10^{22}$ N-m, corresponding to a 16 km thick elastic lithosphere, one finds that it would buckle into folds with a wavelength of about 240 km, but that the minimum buckling stress is 5.3 GPa – about 50 times larger than the crushing strength of basalt!

The theory of elastic folding, thus, offers a classic case of appealing small-scale model experiments whose extrapolation to the planetary scale runs into insuperable conflict with a well-tested mechanical analysis. Many efforts have been made to modify the

Box 4.1 **Elastic and viscous buckling theory**

Leonhard Euler (1707–1783) initiated the theory of elastic buckling in a treatise on elastic flexure in 1743. However, the theory was not applied to geological problems until the dawn of the twentieth century. The theory begins with the equation for the deformation of a floating elastic plate, discussed in Box 3.3, when no external load is applied, $q(x) = 0$, but in the presence of a compressive load N (per unit width of the plate) in the plane of the plate:

$$D\frac{\partial^4 w}{\partial x^4} + N\frac{\partial^2 w}{\partial x^2} + \rho_a g\, w = 0. \qquad (B4.1.1)$$

The buckling instability essentially involves the amplification of a small initial deformation to the point where it dominates the appearance of the plate (see Figure 4.4). It is assumed that any real planetary surface possesses some initial topography at all wavelengths (in the sense that Fourier analysis of even random irregularities contains non-zero components at all possible wavelengths). The vertical deflection of the plate, w, is thus divided into an initial deflection, w_0, and a component w_b representing the additional deflection due to the applied horizontal force N, $w = w_0 + w_b$. Assuming that the plate is in stress-free equilibrium when w_b is absent (that is, w_0 is a solution to (B4.1.1) when $N = 0$), substitution of w into (B4.1.1) yields an equation for w_b:

$$D\frac{\partial^4 w_b}{\partial x^4} + N\frac{\partial^2 (w_0 + w_b)}{\partial x^2} + \rho_a g\, w_b = 0. \qquad (B4.1.2)$$

Assuming that the Fourier component of the initial deflection w_0 is equal to $a_0 \sin k\,x$, where k is the wavenumber (equal to $2\pi/\lambda$, where λ is the wavelength), the solution to (B4.1.2) is found by supposing that w_b is of form $C \sin k\,x$, substituting, then solving the algebraic equation for the unknown C. The overall result of this process is that:

$$w = (a_0 + C)\sin kx = \left\{ \frac{Dk^4 + \rho_a g}{Dk^4 - Nk^2 + \rho_a g} \right\} a_0 \sin kx. \qquad (B4.1.3)$$

The main thing to notice about this solution is that the "amplification factor" in brackets becomes infinite if the denominator vanishes. This occurs when the compressive force N attains the special value:

$$N_b = Dk^2 + \frac{\rho_a g}{k^2}. \qquad (B4.1.4)$$

The buckling force N_b depends on the wavenumber k. Because the first term in Equation (B4.1.4) grows as k^2 for large values of k and the second term grows as $1/k^2$ for small values of k, there is some intermediate value of k at which the amplification factor grows toward infinity for a minimal applied compressive load. Setting the derivative of N_b with respect to k equal to zero, it is easy to show that this unique wavenumber is:

$$k_b = \left(\frac{\rho_a g}{D} \right)^{1/4} = \frac{\sqrt{2}}{\alpha}. \qquad (B4.1.5)$$

Box 4.1 (cont.)

where α is the flexural parameter defined in Chapter 3. The buckling wavelength that corresponds to this wavenumber is given by $\lambda_b = \sqrt{2}\,\pi\alpha$, so that the characteristic wavelength for buckling is directly proportional to the flexural parameter (by a factor of "order one" equal to 4.44).

Of course, the deformation at this unique "buckling wavelength" is not actually infinite, because we have used a flexure theory that is only valid for deformations that are small compared to the plate thickness. Nevertheless, numerical studies of buckling that incorporate realistic deformation models agree with this simple model that, when the buckling force is exceeded, large-amplitude sinusoidal deformation pops up rapidly at just the wavenumber computed from Equation (B4.1.5).

The minimum force needed to initiate buckling, N_{min}, is computed by inserting the wavenumber (B4.1.5) in Equation (B4.1.4):

$$N_{min} = \frac{4D}{\alpha^2} = 2\sqrt{\rho_a\,g\,D}. \tag{B4.1.6}$$

The corresponding stress, σ_{min}, is obtained by dividing the minimum buckling force by the lithosphere thickness t. Expanding the definition of the flexural rigidity D of a plate:

$$\sigma_{min} = \frac{N_{min}}{t} = \sqrt{\frac{\rho_a g}{3}\,\frac{E\,t}{\left(1-\nu^2\right)}} \tag{B4.1.7}$$

where ν is Poisson's ratio and E is Young's elastic modulus.

Buckling of compressed layers is not confined to floating elastic plates. The sinusoidal deformation of layers that we call folding arises from an antagonism between the elastic forces tending to relieve in-plane compression by lateral displacement and restoring forces tending to limit lateral motion. In Equation (B4.1.1) the restoring force is the weight of the rising folds, expressed by the third term on the left, $\rho_a gw$. However, this is not the only possible restoring force.

Another buckling scenario is a stiff elastic layer embedded in a surrounding medium of lower elastic modulus. In this case the restoring force in (B4.1.1) is replaced by an elastic reaction force, $E_0 kw/(1-\nu_0^2)$, where E_0 is Young's modulus of the soft layer and ν_0 its Poisson's ratio. Note the explicit appearance of the wavenumber k in this expression. After an analysis similar to that described above, the buckling wavelength is computed to be:

$$\lambda_b = 2\pi t \left[\frac{E(1-\nu_0^2)}{6E_0(1-\nu^2)}\right]^{1/3}. \tag{B4.1.8}$$

The most important result of this calculation is that the buckling wavelength is directly proportional to the layer thickness. This prediction agrees well with many observations of folds in rock layers, where the wavelength of the fold is clearly proportional to the layer thickness: Thin layers are buckled into a large number of short-wavelength folds, while thick layers are distorted in broad, open folds. An assembly of data on folds and layer thickness shows a good correlation over an astonishing range of wavelengths, ranging from centimeters to 30 km

Box 4.1 (cont.)

(Currie *et al.*, 1962). Over this range the wavelength seems to be approximately 27 times the layer thickness. The reason for this particular numerical value is presently obscure.

The stress required to initiate buckling in the layer in this case is given by a rather complicated formula:

$$\sigma_{min} = \frac{3D^{1/3}}{t}\left[\frac{E_0}{2(1-\nu_0^2)}\right]^{2/3}. \qquad \text{(B4.1.9)}$$

A very similar analysis can be performed for compression of layers of differing viscosity. Although the mechanics of elastic and viscous flow differs profoundly, elastic strain can often be simply replaced by the viscous strain rate in the mathematical equations and the same analysis carried through. This works well for compression of layers of differing viscosity. The result of such an analysis is an equation for the viscous buckling wavelength that closely resembles (B4.1.8):

$$\lambda_b = 2\pi t\left[\frac{\eta}{6\eta_0}\right]^{1/3}. \qquad \text{(B4.1.10)}$$

As for elastic buckling, the wavelength is directly proportional to the layer thickness. The major difference between elastic and viscous buckling is that for viscous layers there is no minimum buckling stress, thus avoiding the problem (discussed in the text) that elastic buckling often requires stresses far exceeding the fracture stress. Another difference is that, although the "buckling" wavelength corresponds to the fastest growing instability, the growth rate is not infinite but instead is limited by the viscosity of the layer and its surrounding medium. For more details see the books by Johnson *et al.*, cited below.

Elastic and viscous buckling are the subjects of several monographs. The interested reader will find discussions of elastic buckling applied to folds in terrestrial rocks in Johnson (1970). Viscous buckling is discussed at length by Johnson and Fletcher (1994) and Pollard and Fletcher (2005). Ramsay (1967) offers a comprehensive description of folded rocks on the Earth along with a clear general discussion of the theory of buckling.

conclusions of the mechanical analysis. One of the most plausible is to recognize that layered rocks are not simple monolithic masses but, like reams of paper or piles of carpeting, are composed of much thinner layers that can slide over one another. Because the flexural rigidity D of a layer is a function of the thickness of the layer cubed, if the layer is divided into n sublayers each of thickness t/n, and if the layers flex independently of one another, the flexural rigidity of the entire pile of n layers is proportional to $n (t/n)^3$, so that splitting a layer into n sublayers decreases its flexural rigidity by $1/n^2$. The buckling stress, Equation (B4.1.7), depends on the square root of D, so that dividing the layer into n sublayers decreases the buckling stress by $1/n$, and decreases the buckling wavelength by $1/\sqrt{n}$. Of course, this explanation depends on the layers' ability to slip freely over one another, and friction between layers greatly diminishes their independent motion.

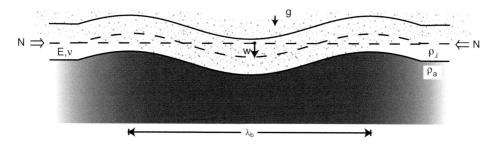

Figure 4.4 Buckling of a floating elastic plate under a horizontal load N. The deflection of the midplate neutral sheet (dashed line) is w, the wavelength of the buckled plate is λ_b, the Young's modulus of the plate is E, its Poisson's ratio is ν, the density of the plate is ρ_l, and that of the underlying asthenosphere is ρ_a. The surface acceleration of gravity is g.

Other efforts to preserve the major results of buckling theory appeal to alternation of soft and stiff layers or to buckling of viscous or even viscoelastic layers. It seems that buckling theory does apply in many cases to small-scale folds in rocks, and many monographs expound buckling theory for that reason, finding good qualitative and even quantitative agreement between small-scale observations of crumpled rocks and this theory.

Buckling theory has been applied to explain undulations observed on the surface of Europa (Prockter and Pappalardo, 2000), in this case appealing to the special rheology of ice to argue for a very thin elastic layer of small elastic modulus and thus a low buckling stress. Seismic reflection images of the Indian plate on the Earth reveal east–west ridges that strongly resemble compressional buckles consistent with the stress state in the plate (Weissel *et al.*, 1980). Although the geometry of these ridges suggests elastic buckling, the required stresses are again excessive, around 2.4 GPa, greatly exceeding the estimated crushing strength of rock. There is no evidence for thin layering in this oceanic plate. However, it has been suggested that the pervasive fracturing of the plate might lower its flexural rigidity by the required amount (Wallace and Melosh, 1994).

4.5.2 Folding vs. faulting: fault-bend folds

Folded mountain belts are apparent on both the Earth and Venus, as shown in Figure 4.3. Even the Moon and Mars possess smaller-scale folds in the form of mare ridges. If these are not the result of elastic buckling, then how do they form? Geologist John Suppe, who made a detailed study of the Pine Mountain structure in the southern Appalachians, discovered an answer to this question (Suppe, 1983). Suppe found that the large-scale folds in this area, and elsewhere in the Appalachians, developed as the layered rocks were pushed up and over sloping steps created by compressional thrust faults that cut through the rock layers. This model requires that the rock layers glide freely along nearly horizontal detachment faults between the rock layers until they meet a sloping step, up which they are forced to move by regional compressional stresses. The rocks then bend

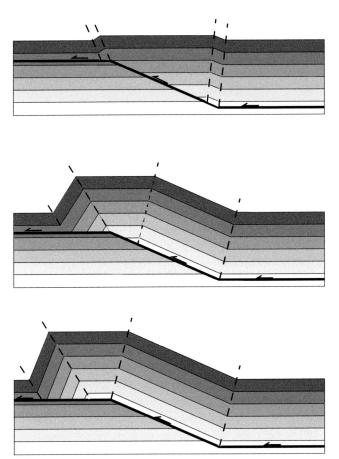

Figure 4.5 Formation of a fault-bend fold. The top frame shows a layered stack of rocks that are cut by a thrust fault dipping at about 30° from the horizontal. The fault becomes horizontal, following weaker beds, both above and below the dipping fault. The middle frame shows the evolution of the fault as the layered rocks are thrust up and over the fault ramp. Bending takes place in narrowly defined hinge zones, here represented as perfect planes (dashed lines). The lowest frame shows the further evolution of the compressed stack of rocks, indicating the formation of a ridge on the surface that overlies horizontal rocks below it. Concept after Suppe (1983).

to accommodate the distortion, as shown in Figure 4.5. Although this model does a good job explaining the geometry of the folds, it only shifts the stress problem from one of elastic buckling to one of explaining how rocks can glide apparently effortlessly along detachment faults. Nevertheless, the fault-bend fold model gives an excellent description of folds in a variety of circumstances (Woodward *et al.*, 1989), and is widely used in oil exploration. The stress problem may be partly resolved, at least on Earth, by the presence of pressurized fluids (water or oil) in the compressed rock mass, but this explanation cannot apply to Venus or the Moon.

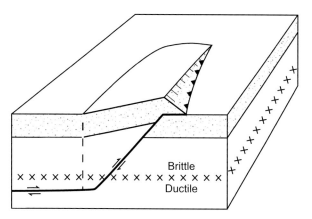

Figure 4.6 Schematic of mare-ridge formation on Mars. The steep scarp of these mare ridges occurs along a jump in surface elevation that reveals the presence of a deep-seated thrust fault. The width of the gently sloping surface above the dipping fault is about two times the depth of the 30° dipping fault. After Mueller and Golombek (2004).

The fundamental idea of fault-bend folding is consistent with the appearance of mare-ridge folds on Mars (Mueller and Golombek, 2004). As shown in Figure 4.6, the Martian ridges separate areas of different surface elevation, strongly suggesting that the surface folding is a manifestation of deep-seated detachment faulting in the Martian crust.

4.5.3 Extension: boudinage or necking instability

Whereas crumpling due to compression is a matter of daily experience, the formation of wavelike thickness variations in a stretched layer is much less familiar. Nevertheless, geologists have long recognized sinusoidal variations in the thickness of stretched rock layers. Because these pinch-and-swell variations resemble strings of sausages on the face of rock outcrops, they are commonly called boudins, after the French word for sausage. In three dimensions they do not resemble sausages at all, as they form in parallel waves on the surface of the stretched layers. They are easily distinguished from folds because the thickness of a folded layer remains nearly constant through the fold, whereas boudins vary greatly in thickness (Figure 4.7), to the extent that individual boudins may become detached, at which point the stretched layer comes to resemble a collection of parallel, elliptical cylinders.

Boudinage is unfamiliar because it requires special rheological conditions to form. A Newtonian viscous or elastic layer does not spontaneously break into sinusoidal waves as it is stretched: It merely becomes thinner. Mathematical analysis of stretching reveals the fundamental reason: Boudins form only in materials in which the stress–strain relation is non-linear. Sometimes called necking or pinch-and-swell instability, it occurs only in materials that flow more readily when subjected to larger stresses or strains. As a uniform layer is stretched, local regions that may be just a bit thinner than average concentrate stress and strain. If that concentration lowers the resistance to further strain, the layer becomes

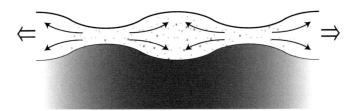

Figure 4.7 Schematic illustration of boudinage or pinch-and-swell deformation of a floating viscous or plastic lithosphere that is undergoing extension. Boudinage requires a non-linear relation between stress and strain and so is somewhat unfamiliar in daily experience.

still thinner than average and the process runs away, eventually producing regularly spaced variations in layer thickness.

Obscure as this process might seem, it is now believed to account for one of the most widespread terrain types on Ganymede (Figure 4.8), as well as the periodically spaced mountain ranges of the Earth's basin and range provinces (Fletcher and Hallet, 1983). Ganymede's grooved terrain exhibits a variety of dominant wavelengths, ranging from about 2 to 15 km. The prominent grooves, with amplitudes up to 500 m and lengths ranging from hundreds to thousands of kilometers, occur in regional sets that crosscut one another and evidently record a long history of extension (Bland and Showman, 2007; Collins *et al.*, 1998). Because there is little evidence for compression or subduction on Ganymede, they strongly suggest an era in which the satellite expanded, perhaps due to internal rearrangements or tidal heating.

4.5.4 Gravitational instability: diapirs and intrusions

In addition to the horizontal forces of compression or extension, vertical forces arising from density instabilities may warp planetary lithospheres. In actively convecting planets these could be generated by warm plumes of material rising from deep within the mantle. Even within a planetary crust, basal melting caused by intrusion of hot melts can mobilize rocks, which then rise toward the surface because of their lower density. On the Earth, this process was first recognized in mushroom-shaped structures produced by the rise of low-density salt buried under accumulating thicknesses of overlying sediment (Figure 4.9). Swedish geologist Hans Ramberg made this process the prime focus of his research career (Ramberg, 1967). Through mathematical analysis and extensive experimentation with centrifuged models, he elucidated the forms taken by gravitationally unstable configurations and correlated them with structures observed in the Earth's crust.

The slow upward movement of viscous masses of low-density rock material (whether it is hot rock, salt, warm ice or even mud) often disrupts and pierces through overlying rock layers. Such structures are called diapirs and the process by which they are emplaced is called diapiric intrusion. Because petroleum is often associated with salt diapirs, particularly in

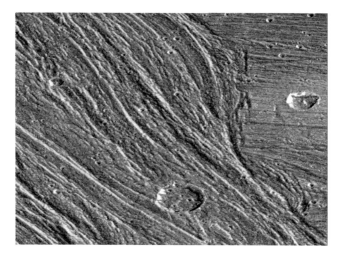

Figure 4.8 Grooved terrain in Nippur Sulcus on Ganymede. Groove spacing is 5–10 km. This Galileo image is approximately 79 km wide by 57 km high, and centered at 51° latitude and 204° longitude. North is toward the top. The largest crater in the image is 12 km in diameter. NASA image PIA01086.

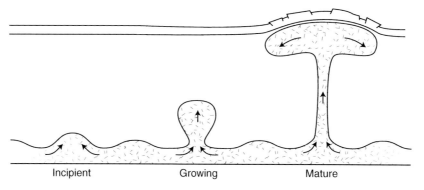

Figure 4.9 The initiation and growth of a viscous layer of density lower than that of the overlying material. An incipient instability forms, triggered by small accidental variations in layer thickness. This layer begins to thicken, bulging upward while the adjacent low-density material flows toward and into the rising column. Eventually this rising plume impinges on the surface. It then spreads laterally, warping the overlying surface upward and often fracturing it. The base of the low-density column is usually surrounded by a depressed moat, called a rim syncline.

the Middle East and the Gulf Coast of the United States, they are well studied and their structures at depth are well understood. As diapirs rise, adjacent rocks are broken and tilted upward and away from the core of the rising mass. Where they reach the surface, overlying rocks are uplifted and stretched. Radial extensional faults often mark their location, as does a local minimum in the acceleration of gravity. Where multiple diapirs reach the surface,

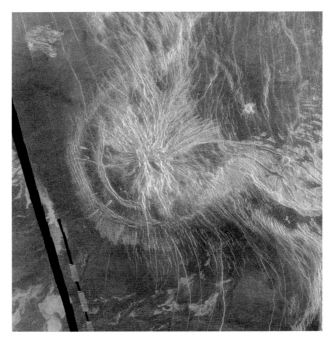

Figure 4.10 Selu Corona on Venus measures 350 km in diameter. It is an archetypical corona marked by both radial and concentric fractures, probably created as hot plumes from Venus' deeper mantle rose and uplifted the surface. The plumes released melt as they rose, feeding lava flows on the surface and perhaps injecting lava beneath the surface. Portion of NASA Magellan image C145S110.

their horizontal spacing offers clues to both the thickness of the buried source layer and the depth of the overlying rocks.

Planetary examples of diapiric surface modifications include coronae on Venus, which are crudely circular volcano-tectonic structures that range from about 100 to more than 1000 km in diameter (Figure 4.10). Venusian coronae are interpreted as the product of rising masses of hot mantle material that first approached the surface, bowed it upwards and thus generated an initial radial fracture system (Squyres *et al.*, 1992). The hot underlying material then spread out underneath the surface, generating melts and feeding lava flows in addition to subsurface intrusions. The region eventually cooled and subsided, producing a flexural trough and the associated concentric faults surrounding the surface load.

On a much smaller scale, the similarly named coronae on Uranus' tiny moon Miranda (480 km in diameter) are also believed to be the surface manifestations of large diapirs (Figure 4.11) that pushed toward the surface early in Miranda's history. The major difference is that upwarping of the surface on this small body created concentric, not radial, extensional faults. This is almost certainly due to the large curvature of Miranda's lithosphere and the flip in tectonic style that results from the curvature of the elastic shell (Janes and Melosh, 1990).

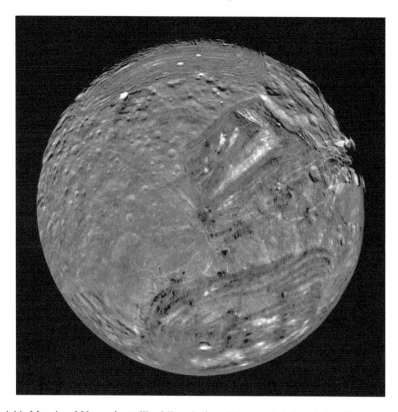

Figure 4.11 Mosaic of Uranus' satellite Miranda from Voyager 2. Miranda is 480 km in diameter. This image shows the striped and ridged terrain composing three large coronae, Elsinore at the top, Inverness to the right of center, and Arden at the bottom. The ridges are believed to represent normal faults, and the terrain is created as plumes rose from Miranda's interior to just under the surface. NASA image PIA01490.

4.6 Fractures and faults

4.6.1 Why faults? Localization

A casual glance at nearly any steep hillside or roadside outcrop reveals that nearly every rock mass is broken by fractures. Cracks in rocks form at all scales, from tiny microcracks along grain boundaries to enormous fractures that extend over much of a planet's circumference. When cracks cut through a rock mass without sensibly displacing the rocks on either side, they are called joints, whereas when the rocks are displaced relative to one another, the displacement is called a fault. It is common experience that an overstressed solid breaks along almost invisibly thin surfaces, leaving the adjacent material largely unchanged: Anyone who has dropped a glass or plate onto a hard surface can verify this. Rocks underlying the surfaces of planets are no different. Because this occurrence is so common, few people stop to ask themselves, why do fractures occur in such narrow zones?

Field geologists have a wider experience of rock fracture. In many places on the Earth uplift, tilting, and erosion have exposed expanses of rock that once extended deep below the surface. In these fortunate places one can sometimes find an ancient fault and trace it from the surface down into the former depths. One finds that the fault, while knife-sharp near the surface, gradually widened and blurred as it went deeper. At great depths the fault zone may have widened to the point of invisibility.

The transition between narrow fractures and broad zones of deformation is called the "brittle–ductile" transition. Brittle behavior implies cracks and narrow zones of deformation. The sudden loss of strength upon fracture may also cause "stick-slip" motion that may be related to earthquakes. Crockery and glasses, as well as rocks at the surface of our planet, all fracture in the brittle domain. Nevertheless, increasing the pressure or temperature in materials like rock can push them into the ductile flow domain. Ductile behavior is more characteristic of some metals and plastics. Ductile flow is not the same thing as pseudoviscous creep: Ductile materials retain considerable strength and their strain is not a function of time. Only at still higher temperatures do rock materials begin to flow under small stresses.

It is evident that materials such as rock can be either brittle or ductile, depending on external conditions of temperature or pressure. Figure 4.12 illustrates this transition for a particularly well-studied rock, Westerly granite (Stesky *et al.*, 1974), and shows how the observed pressure and temperature dependence of the brittle–ductile transition translates into mechanical behavior as a function of depth.

But we still have not explained what really underlies the formation of narrow fractures, the "brittle" behavior so familiar from daily experience. The basic reason for brittle fracture is simple, although a full understanding has taxed the mathematical ingenuity of a generation of material scientists. The key is to apply the concept of positive feedback. Positive feedback occurs when the response to some disturbance is enhanced by the disturbance itself, resulting in an overall response that quickly "runs away." In the case of brittle fracture, a small region of a stressed, heterogeneous solid may suffer more strain than adjacent areas. If this extra increment of strain weakens the region, it leads to a further increase of strain, and a strain runaway begins. Deformation quickly concentrates in the already-deformed region and a narrow zone of macroscopic displacement – a joint or fault – develops. This self-reinforcing response to stress is called strain softening and the process is called localization. Localization is typical of all brittle materials. Simply stated, a macroscopic crack forms when an initial increment of failure locally weakens the solid and causes more failure to concentrate around it, weakening it still further.

The above description of localization is oversimplified: Decades of experimental and mathematical analysis have considerably refined our understanding of when and how localization occurs. Readers interested in this fascinating topic must probe the original literature (for example, Hobbs *et al.*, 1990; Rice, 1976). Nevertheless, this idea provides the rationale for the observed complexities of rock fracture, and offers a qualitative explanation of why rock failure transitions from brittle to ductile as the pressure and temperature rise.

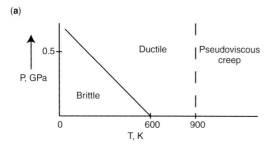

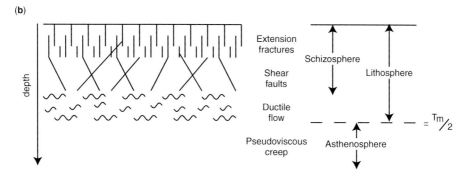

Figure 4.12 The brittle–ductile transition in rock. Panel (a) shows the three domains of mechanical behavior as a function of pressure and temperature. Rock is brittle, subject to sudden failure in narrow fracture zones at low pressures and temperatures. At higher pressures and temperatures it undergoes a transition to ductile fracture, in which it still retains considerable long-term strength, but failure is gradual and distributed over wide zones. Only at temperatures exceeding about half of the melting temperature does rock begin to flow like a highly viscous fluid with no long-term strength. This plot is constructed on the basis of data on Westerly granite (Stesky *et al.*, 1974). Panel (b) applies these domains of rock fracture to the crust of the Earth or planets whose surface conditions begin in the brittle domain. The outermost shell fails in an extensional regime, where fractures are parallel to the most compressive stress. At higher pressures this transitions to the shear failure regime, in which fractures form at an acute angle to the most compressive stress (refer to Figure 4.14 for more on this). Both regions are in the brittle fracture domain termed the schizosphere by C. Scholz (1990). Still deeper is the ductile flow domain. All of these domains that possess long-term strength are termed the elastic lithosphere. Still deeper, rising temperatures permit rock to flow under arbitrarily low stresses by pseudoviscous creep. This region is the asthenosphere.

4.6.2 Joints, joint networks, and lineaments

Joints are ubiquitous cracks in rocks across which no perceptible offset occurs. They frequently form in long parallel sets, periodically spaced at intervals that range from centimeters to hundreds of meters. Joints are found in all consolidated rock types, ranging from soft shales to hard igneous rocks. Special types that define hexagonal columns, called columnar joints, result from shrinkage of cooling basalt or volcanic tuff layers. Joints are clearly tensile fractures: Detailed surface features such as hackle marks and plumose patterns record the propagation of fast-running tensile cracks. Nevertheless, the origin of joints is not clearly understood, although the mechanics of many special cases has been worked out. Many circumstances conspire to produce tensile stresses: Cooling near the surface, uplift of deeply buried rock masses (with attendant cooling), or surface exposure by erosion can all produce tensile cracks in various orientations (a good description of joints can be found in Chapter 6 of Suppe, 1985).

Although it may seem that the formation of tensile cracks must be strictly limited to shallow depths, in the Earth, Mars, and Titan the presence of pressurized fluids in the crust (water, natural gas or, in Titan's case, liquid methane), may extend the tensile regime to many kilometers in depth. Moreover, experiments on the crushing of unconfined rock specimens indicates that tensile fractures, in which a crack forms in the plane of the compressional force, may occur in the regime where the average of the three principal stresses is slightly compressive.

Griffith crack theory. The mechanics of tensile failure is now well understood (Gordon, 2006; Lawn and Wilshaw, 1975). A young aeronautical engineer, A. A. Griffith (1893–1963), was the first to comprehend the relation between cracks and strength. Griffith supposed that all materials contain initial flaws, tiny cracks of various lengths l. When a tensile stress acts perpendicular to the plane of the crack, the crack may grow longer if the energy cost of lengthening the crack is less than the strain energy released by its extension. Equating these two terms, he found that if the surface energy of the crack is γ J/m^2, then the crack grows when the tensile stress exceeds:

$$\sigma_t = \sqrt{\frac{E\,\gamma}{l}} \qquad (4.28)$$

where E is Young's modulus. Tensile cracks extend in their own planes and, because of the $1/\sqrt{l}$ dependence in Equation (4.28), once started they continue to extend until either the stress is relieved or the crack reaches some impassible barrier.

Because tensile failure depends on the presence of initial flaws, the "strength" of a rock mass in tension depends on the history of the rock and its population of initial flaws. Moreover, since large rock masses are more likely to contain an especially vulnerable flaw than small rock samples, the tensile strength of a rock unit depends on its size. There is no necessary relation describing the dependence of strength on size, but many empirical studies suggest that tensile strength depends roughly on the inverse square root of the linear dimension of a rock.

On a planetary scale rocks are nearly strengthless in tension: When tensile forces develop, the rocks generally yield and joints, thus, form readily. The great lengths attained by many joints are due to the ability of tensile cracks to grow in their own plane: Once begun, a tensile crack just gets longer until the applied stress is relieved.

Because joints are weaker than the surrounding rock and offer ingress to corrosive fluids on planets with a hydrosphere, they are frequently expressed in the landscape. Joints often strongly control the course of rivers and the directions of valleys on the Earth and perhaps on Mars and Titan. Maps of the density, direction and regional extent of joints on the Earth are an important part of every geotechnical project, whether for building roads or evaluating sites for nuclear power plants. Joint directions are used to infer stress orientations in the lithosphere of the Earth and planets (Engelder, 1993), although care must be given to the time at which the joints formed. The orientations of impact crater walls and the head scarps of landslides are often visibly controlled by joint directions in the pre-existing rocks.

Lineament analysis. Joint analysis has long been an industry on the Earth and it is now frequently extended to the other planets. Because joints themselves are small-scale features, their presence and direction is usually inferred from their effects on the landscape. This is especially true in planetary science, where high-resolution surface investigations are currently rather limited. The art of inferring joints (including small-offset faults) from landscape patterns is called "lineament analysis," because the fractures that form major joint sets frequently occur in linear arrays. It is easy to pick out apparently straight features and alignments on planetary images, a technique at which the human visual processing system seems particularly adept. However, because of this very ability, great caution needs to be applied when pursuing this type of analysis.

Non-existent linear features have been reported on the surfaces of planets since Percival Lowell believed he saw artificial canals on Mars in 1895. More recently, the Apollo 15 astronauts reported layering in the face of Mount Hadley, 4.5 km from their landing site and took photographs of the "layers" (Figure 4.13a). Subsequent analysis showed that the apparent "layers" were all aligned with the direction of the Sun. As shown in Figure 4.13b, experiments on illuminating model hills with random surface textures demonstrate the illusion of layering (Wolfe and Bailey, 1972). These investigations indicate that lineament analysts must be conscious of such illumination effects (Howard and Larsen, 1972), and must avoid working near the limits of resolution, where the eye has a tendency to "connect the dots" (which was probably what fooled Lowell).

4.6.3 Faults: Anderson's theory of faulting

Faults on planetary surfaces betray their presence by sudden changes of elevation (scarps and troughs), offset surface features, or linear alignments. Miners and geologists first recognized them as abrupt breaks or shifts in rock layers. By definition, a fault is a narrow surface along which the rocks have moved relative to one another. Although faults have been recognized and described for several centuries, it was not until the late 1940s that experimenters, among them the ubiquitous David Griggs (Griggs and Handin, 1960), began to clarify the mechanics of how they form. Even now, the process of earthquake faulting, especially on large active faults, is shrouded in deep mystery, but progress has been made.

Griggs and similar experimenters crushed centimeter-sized rock specimens to failure under conditions of pressure and temperature approximating those in the Earth. They found that when the stresses in all directions are compressive, failure occurs by sliding along planar surfaces inclined at an acute angle, typically about 30°, to the direction of the maximum compressive load (Figure 4.14). No tensile stresses occur under these conditions, and the

Figure 4.13 Panel (a) is an Apollo 15 photograph taken from the lunar surface. It shows Mount Hadley 4.5 km away from the landing site, on which the astronauts reported prominent sloping layers. Top portion of NASA photograph AS15–90–12208. Panel (b) is a photograph of a pile of structureless cement powder, 15 cm high, taken under oblique illuminations similar to that at the Apollo 15 landing site, indicating how solar illumination can often produce the appearance of layered structure when none is present. From Figure 7 of Wolfe and Bailey (1972).

simple Griffith crack theory does not apply. The maximum shear stress in the specimen occurs on planes tilted at 45° to the direction of the applied load, so there is more to this failure mode than simple shearing.

For many years these results have been rationalized by a graphical stress representation invented by Otto Mohr (1835–1918). Mohr's circles enable one to correlate the failure angle with the shape of the yield envelope by resolving the stresses into normal and tangential forces acting across the eventual failure plane. Although this type of analysis, which is expounded in nearly every text and monograph on the subject of rock failure, is applicable to frictional sliding on a pre-existing fracture, it begs the question of how the inclined fracture surface develops in the first place.

Very recent work applies the theory of sheared microcracks (Paterson and Wong, 2005). The emerging picture is far from simple: Unlike cracks in tension, sheared cracks do not propagate in their own plane and thus cannot simply lengthen to create a fracture surface. Instead, sheared microcracks sprout secondary "wing" cracks that interact with neighboring

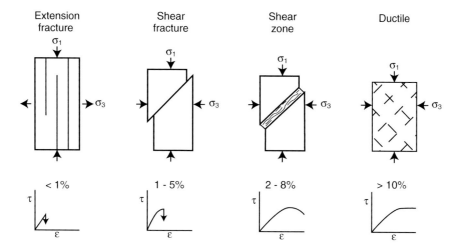

Figure 4.14 The effect of increasing pressure on the style of rock fracture. At very low pressures extensional cracks form parallel to the direction of compression. As pressure increases the failure surface becomes inclined to the direction of maximum compression at an acute angle, often close to 30°. Increasing pressure results in broader shear zones, until at very high pressures deformation is distributed throughout the failing rock specimen and the fracture is described as ductile. The upper row of figures schematically illustrates the failure mode and the lower row plots the relation between shear stress τ and strain ε. The arrows indicate sudden failure. After Griggs and Handin (1960).

cracks to eventually link up and form the observed inclined failure plane. This explains the generally irregular fracture plane that forms at low strains. Once a fault is formed and sliding progresses along the fault, the surface is ground smooth and characteristic "slicken-side" grooves develop in the direction of motion.

Although such detailed microscopic theories are greatly advancing our understanding of how faults work, these theories are fortunately not necessary for a basic correlation between fault type and stress orientation. E. M. Anderson (1877–1960) published a short but influential book on faults and stress directions (Anderson, 1951) that applied the premise that most faults are produced by shear. He recognized three basic types of fault and related each type to the stress field that produces it.

Anderson began by resolving the general stress tensor into its principal stresses and directions. He observed that, because the surface of a planet (supposed to be essentially horizontal) is free of shear stresses, one of these principal stresses must be perpendicular to the surface. There are then three possible kinds of fault, depending on whether that stress is the maximum, intermediate or minimum principal stress. Accepting the empirical observation that shear faults form at an acute angle to the maximum compressive direction, he supposes that the fault plane also contains the intermediate principal stress (Figure 4.15, left panels). There is an ambiguity, however, because there are two planes that satisfy this criterion. These are called conjugate planes and either one is a potential fault. Which one is more important is not determined by the theory: It is supposed that boundary conditions

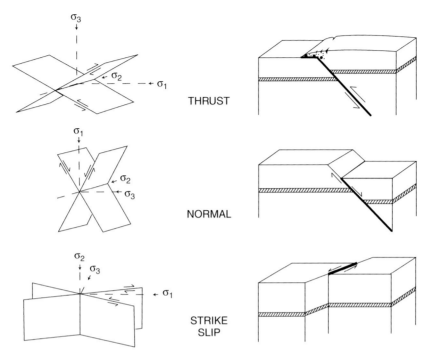

Figure 4.15 The relation between shear fractures and large-scale faults according to the theory of Anderson (1951). One of the three principal stresses must be perpendicular to the free surface, which results in the separation of possible fracture directions into thrust faults (top), normal faults (middle), or strike-slip faults (bottom). Right panel after Figure 11–22 in Hartmann (1972).

or geometric constraints finally determine the plane actually selected. Experimentally, in the initial stages of crushing both planes are often activated, producing X-shaped fracture patterns.

Normal faults. Anderson's theory predicts that so-called "normal" faults arise when the maximum principal compressive stress is vertical, strike-slip faults appear when it is intermediate, and thrust faults occur when it is minimum. Normal faults typically dip steeply, at an angle of about 60° with respect to the vertical (See Box 4.2). The name "normal" fault comes from the English Coal Measures, where the most common faults are of this type. The names *thrust* and *strike-slip* should be self-explanatory (Figure 4.15, right panels).

Thrust faults. Thrust faults arise when the maximum compressive stress is horizontal and the minimum stress is vertical. The dip angle of thrust faults is typically shallow, close to 30° (See Box 4.2). In a fault of this type a wedge of rock is thrust up and over a sloping ramp of underlying rock, uplifting the surface on the upthrust side and slightly depressing it on the opposite side of the fault. This type of faulting commonly produces asymmetric scarps across which the surface elevation suddenly changes. In plan, thrust faults are typically lobate. The most famous planetary thrust faults are the lobate scarps of Mercury (Figure 4.16), which may be several kilometers high and wind hundreds of kilometers

Box 4.2 **Dip angle of Anderson faults**

Normal faults and thrust faults are two of the most common types observed on planetary surfaces. Their appearances in images are strongly affected by the angle at which the faults dip below the surface. Thrust faults dip at shallow angles, typically about 30° with respect to the surface, and frequently present an irregular, wavy trace where they cut the surface. Normal faults dip more steeply, at angles near 60°, and their trace is typically straighter because the steep dip masks small variations in the depth of the fault surface.

The difference in the dip of thrust and normal faults is a simple geometric consequence of how shear and normal stresses are localized on the fault surface. Although the main text of this book has avoided dealing with the details of stress components, this box will show how this profound difference in fault dip comes about for either compressional horizontal stresses (thrust faults) or extensional stresses (normal faults).

The derivation begins with Coulomb's frictional law for sliding on a rock surface, Equation (3.20):

$$\sigma_s = \pm f_f \sigma_n \qquad (B4.2.1)$$

where σ_s is the shear stress on the potential plane of sliding, f_f is the coefficient of friction and σ_n is the "normal stress," or pressure perpendicular to the potential sliding plane. The alternate + and – sign indicates that sliding may occur either up the plane or down the plane when this shear stress is exceeded.

These stresses on the potential sliding surface, which we suppose dips at angle θ with respect to the horizontal (Figure B4.2.1), must now be related to the vertical stress, σ_V, and horizontal stress, σ_H, acting below the solid surface. The full derivation of this relation is

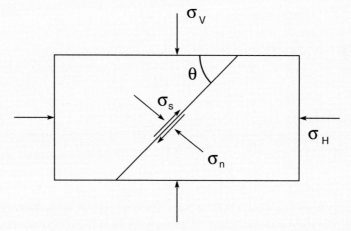

Figure B4.2.1 Schematic illustration of the stresses on a fault plane dipping at an angle θ with respect to a horizontal plane. Stresses σ_n and σ_s are stresses normal (perpendicular) to the fault plane and acting across it in shear, respectively. The vertical principal stress σ_V is nearly equal to the weight of the overlying rock, while the horizontal principal stress σ_H varies with tectonic processes until failure occurs in either compression or extension.

Box 4.2 (**cont.**)

presented in the book by Turcotte and Schubert (2002, Section 8.4). We simply quote the result here:

$$\sigma_n = \frac{1}{2}(\sigma_H + \sigma_V) - \frac{1}{2}(\sigma_H - \sigma_V)\cos 2\theta. \qquad (B4.2.2)$$

$$\sigma_s = \frac{1}{2}(\sigma_H - \sigma_V)\sin 2\theta. \qquad (B4.2.3)$$

The vertical stress σ_V is nearly equal to the weight of the rock overlying the fault plane, $\sigma_V \simeq \rho g z$, where ρ is the density of the rock, g is the surface acceleration of gravity, and z is the depth below the surface. Tectonic forces in addition to the confining pressure of the surrounding rocks generate the horizontal stress σ_H, which we treat as a variable in this computation. We, thus, insert the expressions for the normal (B4.2.2) and shear stresses (B4.2.3) into the failure Equation (B4.2.1) and solve for the horizontal stress, with the result that, at failure,

$$\sigma_H = \left[\frac{\sin 2\theta \pm f_f(1+\cos 2\theta)}{\sin 2\theta \mp f_f(1-\cos 2\theta)}\right]\sigma_V. \qquad (B4.2.4)$$

To find the dip θ of the plane most susceptible to failure, minimize the expression in square brackets with respect to this angle by differentiating it with respect to θ and then setting the derivative equal to zero. The result is:

$$\cot 2\theta_\pm = \pm f_f. \qquad (B4.2.5)$$

This can be simplified by defining the angle of internal friction $\phi_f = \arctan f_f$ and using trigonometric identities to obtain:

$$\theta_\pm = \frac{\pi}{4} \mp \frac{\phi_f}{2}. \qquad (B4.2.6)$$

The angle of internal friction is typically about 30° for most rocks. We can then deduce that for compressional stresses (upper sign) $\theta_+ \simeq 30°$, whereas for extensional stresses (lower sign) $\theta_- \simeq 60°$, in good agreement with observations.

across the surface, cutting craters in their path. The compressive deformation is also indicated by the distortion and shortening of initially circular craters cut by the fault. The prevalence of thrust fault scarps on Mercury indicates a dominantly compressional stress regime in Mercury's crust at the time that they formed, perhaps as a result of cooling and shrinking of the planet.

Strike-slip faults. Strike-slip faults can be easily recognized in planetary images because they produce horizontal offsets of surface features. On the Earth, strike-slip faults grow to enormous lengths because they form one of the three major types of plate boundary.

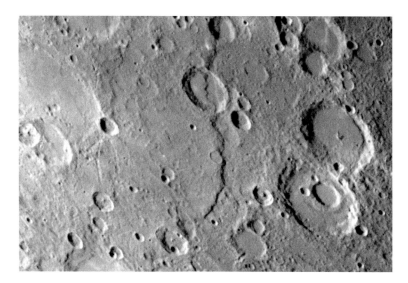

Figure 4.16 Discovery scarp on Mercury is a thrust fault roughly 350 km long that reaches a maximum height of 3 km. It transects two craters of 35 (top) and 55 km (middle) diameter. Note that these craters are shortened in the direction perpendicular to the scarp, as expected for a compressional thrust fault. NASA Mariner 10 image, PIA02446.

The San Andreas Fault in California and the Alpine Fault in New Zealand are examples. Strike-slip motion is not so evident on planets that lack plate tectonics, but careful mapping usually reveals some displacement of this type. Venus and Europa both exhibit strike-slip faults with a few kilometers of offset, ranging up to more than 40 km on Europa (Hoppa *et al.*, 1999). Global planetary grids, discussed below, may be networks of conjugate strike-slip faults with undetectably small offsets.

This simple scheme captures the appearance of many faults on the Earth and other planets and so is in wide use. Lest the reader think that this scheme is inevitable, a strange and little-documented episode in the history of American structural geology should serve as a warning. In the 1920s the first American textbook on structural geology by C. K. Leith (1923) expounded a geometrical theory of rock failure. Based on distortions of the strain ellipsoid, this theory predicted that failure occurs on planes making an *obtuse* angle to the direction of maximum compression. Structural geologists who accepted this idea were evidently unable to make any sense of field relations and an entire generation of American geologists grew to distrust any relationship between faults and stress directions. Their European counterparts, who did not accept this theory, were more fortunate and the literature of the time shows them more confident in their stress determinations. This error was not corrected until the 1940s when a new structural geology textbook by C. M. Nevin (1942) finally got things right. Of course, this confusion did not prevent the young David Griggs (1935) from weighing in on the right side of the argument.

Grabens. There is more to faulting, however, than Anderson's simple theory can describe. Normal faults have a strong tendency to form together in conjugate pairs, dropping a long,

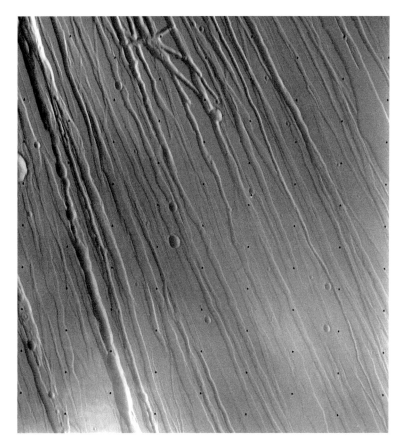

Figure 4.17 Dozens of parallel grabens form a corduroy-like terrain on parts of Mars. These grabens north-east of Tharsis at 25° N and 101° W in Ceraunius Fossae are spaced approximately 6–10 km apart and can be traced up to 300 km along their length. Image is 135 x150 km. NASA Viking frame 39B59.

flat-floored wedge down relative to its surroundings. Such fault-bounded troughs are known as grabens, from the German word for *ditch*, and often form in parallel and numerous sets (Figure 4.17). The reason for this common association is probably due to the fact that it minimizes distortional energy in the surrounding rocks (Melosh and Williams, 1989). Telescopic observers were the first to describe grabens on the Moon. Given names like "straight rilles" and "arcuate rilles," they form flat-bottomed troughs that stretch hundreds of kilometers across the surface of the Moon. In a now-classic study George McGill (1971), and later Matt Golombek (1979), recognized that the increases in the apparent width of these rilles as they cut through ridges reflect the dips of the bounding faults (Figure 4.18) and they showed that the normal faults bounding these lunar grabens dip at angles identical with those of fault-bounded troughs on the Earth. Referring to Figure 4.18, if the apparent

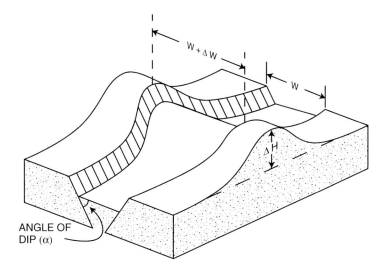

Figure 4.18 Schematic illustration of the interaction of a graben with a topographic ridge. Because the sides of the graben slope at a constant dip angle, when the graben cuts the ridge the apparent width of the graben expands from W to $W+\Delta W$. If the height of the ridge ΔH is known, the dip angle can be computed from Equation (4.29). After Figure 1 of McGill (1971).

width of the graben expands from W to $W+\Delta W$ and the height of the ridge is ΔH, then the dip angle α is given by:

$$\alpha = \tan^{-1}\left(\frac{\Delta H}{\Delta W/2}\right). \tag{4.29}$$

Grabens are found on virtually every planet and satellite in the Solar System and are one of the most recognizable indications of extensional strain. The floor depth and spacing of grabens makes it easy to compute the amount of extension quantitatively and, thus, to estimate the extensional strain, presuming that the terrain between grabens behaves as essentially rigid blocks. Grabens or vertical open cracks also form locally over the tips of vertical dikes that do not quite reach the free surface, and so are common features in volcanic terrains. Many of this type have been identified on Venus and perhaps Mars (Ernst *et al.*, 2001). Distinctive lines of nearly rimless conical drainage pits sometimes mark incipient grabens where extension may create gaping vertical "dilational" faults. Pit chains of this kind occur on Mars (Ferrill *et al.*, 2004) and maybe explain the distinctive crater chains on Phobos (Horstman and Melosh, 1989). The spacing between these pits, which require a loose regolith overlying a more competent substrate, is nearly equal to the thickness of the regolith layer, and, thus, provides an estimate of the thickness of the loose debris.

A much larger type of graben creates large rift zones on the Earth, such as the East African rift valleys. These grabens apparently originated with a single normal fault that cut the entire lithosphere. Flexure then produces extensional strain a distance from the first fault roughly equal to the flexural parameter and, when the stress is large enough, a second normal fault forms. Heiskanen and Vening-Meinesz (1958, Part 10D) first described flexural rift zones of this type.

Other combinations of stress produce distinctive structures: Simultaneous compression and shear create "transpression" structures (known as "flower structure" in the petroleum geology literature), while extension and shear create "transtension." Shearing of soft material between more rigid blocks produces "Riedel shears," while shearing of stiff material between soft layers generates "bookshelf" kinematics, so called because of its similarity to the motions between a laterally sheared row of upright books. Bends in faults produce characteristic associations of secondary faults. It would take an entire book to describe all of the known variations and varieties of faults. Fortunately, there are several, of which Mandl (1988) is one of the more comprehensive.

Mare ridges. Mare ridges, first observed telescopically on the Moon, where they are also called "wrinkle ridges," are attributed to compressional tectonics. These distinctive features form ridges up to 1 km high, several kilometers wide and hundreds of kilometers long (Figure 4.19a). The narrow crenulated ridge often lies near the summit of a much broader upwarp. On the Moon mare ridges are confined to the mare lava plains. They have now been observed on lava plains on all of the terrestrial planets, with the apparent exception of the Earth. Even on the Earth, an unusual type of ridge found on the lava plains of the Columbia Plateau in the Yakima Fold Belt may provide local examples of this otherwise ubiquitous tectonic feature. The internal structure of lunar mare ridges was probed by the Apollo 17 electromagnetic sounder experiment, but those images did not lead to a clear understanding of the nature of these ridges. However, they resemble small-scale (meters to tens of meters) compressional structures formed in soil during the 1968 Meckering, Australia, earthquake (Bolt, 1970) and so have been generally attributed to compression. Analysis of Martian mare ridges using the MOLA laser altimeter revealed offsets in the elevation of plains on opposite sides of the ridges, strongly implying a kind of incipient fault-bend folding over deep-seated detachment faults (Mueller and Golombek, 2004). See also Figure 4.6. Earlier studies of lunar mare ridges also indicated elevation changes across the ridge. Nevertheless, the en-echelon structure of the mare ridges continues to suggest a component of strike-slip motion (Tjia, 1970) in the formation of at least the lunar ridges (Figure 4.19b). At the present time, the nature of mare ridges is not fully understood.

Detachment faults. One of the principal types of fault not described by Anderson's theory is the detachment fault, also called a low-angle thrust fault. First mapped in the Alps, detachment faults are characteristic of compressive mountain belts. As described above, they are the foundation of the theory of fault-bend folding. Detachment faults elude Anderson's fault classification because they are nearly horizontal, parallel to the free surface. None of the faults shown in Figure 4.15 have this orientation. Either a major reorientation of the principal stress axes must take place between the free surface and the fault plane (Melosh, 1990b), or the detachment surface must be a plane of extreme weakness.

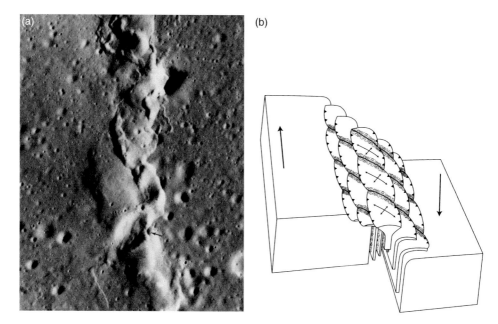

Figure 4.19 Panel (a) illustrates a prominent mare ridge in Mare Serenitatis on the Moon. North is toward the top and the frame is 3.5 km across. NASA Apollo 17 image AS17 2313 (P). Panel (b) is a schematic drawing of the "flower structure" that develops along faults that are simultaneously compressed and sheared (transpression). After Figure 9 of Lowell (1972).

The first geologists to map detachment faults could scarcely believe their eyes: Many require displacements of kilometer-thick plates of rock over horizontal distances up to 100 km. It was quickly realized that such faults must be exceedingly weak: If normal frictional forces act across the fault plane, the overlying plates of rock would crumble long before they could be pushed along the detachment surface. In some cases fluid materials, such as salt or evaporites, underlie the detachment and it is plausible that the overlying plate moved on this viscous substrate. However, in many others the detachment cuts through hard rocks that cannot be reasonably treated as viscous.

Strength paradox. The apparent low strength of detachment faults has long been attributed to high fluid pressures in the deforming rocks, beginning with a seminal paper by Hubbert and Rubey (1959). The empirically determined yield strength of rock was discussed in Section 3.4.2 and is well described by equations similar to (3.23) and (3.24). A widely used low-pressure limit of this type of equation that explicitly shows the effect of fluid pore pressure p_f is:

$$|\sigma_s| = Y_0 + f_F \, (p - p_f) \qquad (4.30)$$

where Y_0 is the cohesion, f_F the "coefficient of internal friction" and p is the total overburden pressure. It is clear from (4.30) that if the pore pressure becomes comparable to the overburden pressure the strength of rock may become very small: On a pre-existing fault, for which Y_0 is already zero, the strength may vanish. High water pressures are indeed found in many terrestrial subduction zones where thrust faulting is active. However, there are some areas (such as the Heart Mountain detachment in Wyoming, USA) where a case for high fluid pressure is difficult to make. And if Venusian fold belts or lunar mare ridges are related to detachment faults at depth, it is hard to make a case for subsurface fluids of any kind.

Strike-slip faults have their own strength problems. Since the late 1960s it has been suspected that the San Andreas Fault in California is sliding with an apparent coefficient of friction at least a factor of 10 smaller than expected from rock friction experiments (Zoback *et al.*, 1987). Although this is not the place to air the ongoing controversy about the strength of active faults, it is important to indicate that a full understanding of tectonic processes has not yet been achieved, even on the well-studied and relatively accessible Earth. This situation, however, adds extra interest and urgency to comparative tectonic studies of the planets. Because other planets offer examples of tectonic processes in vastly different settings of gravity, pressure, and the abundances of subsurface fluids, comparative planetary studies of tectonics may be of great importance in resolving the fundamental problem of what determines the strength of faults.

4.7 Tectonic associations

4.7.1 Planetary grid systems

Geologic mapping of the Moon began well before the spaceflight era. Mare ridges, normal faults, and grabens were all recognized before about 1960. In addition, several "selenologists," a group that included mining geologists familiar with lineament analysis, proposed a global system of faults that became known as the "lunar grid" (Baldwin, 1963). Deduced from straight segments of crater walls (Figure 4.20a), bona fide faults, and, in some cases, confused by troughs created by impact ejecta from the Imbrium basin, the lunar grid appears to delineate an ancient tectonic fabric buried deep within the lunar crust. Although the status of this feature is still somewhat contentious, the prominent NW–NE peaks in both the number and lengths of linear features (Figure 4.20b), with an acute E–W opening angle, strongly suggest a network of small-offset conjugate strike-slip faults that formed very early in the Moon's history (Strom, 1964).

The possible reality and significance of global grid systems got a major boost in 1974, when Mariner 10's images of Mercury also revealed a Moon-like planetary grid. In the case of Mercury, it was quickly suggested that the fault pattern is consistent with strike-slip faulting expected from a relaxing rotational bulge, perhaps a remnant of a time early in Solar System history when Mercury spun faster than its current 59-day period (Melosh and Dzurisin, 1978). In addition, Mercury is covered with a possibly north–south oriented system of lobate thrust fault scarps that indicate planetary contraction, perhaps also influenced by the despinning stress field. Mars and Venus are so complex tectonically that no global grid system has been found.

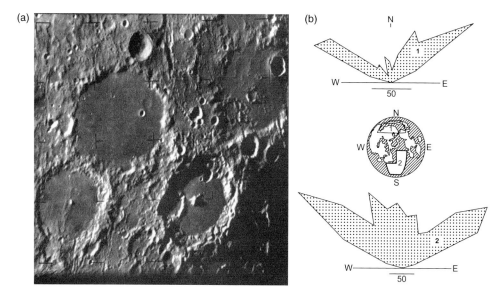

Figure 4.20 Panel (a) is an image returned by the Ranger 9 B-camera 18.5 minutes before it impacted the Moon on March 24, 1965. The large, flat-floored hexagonal crater in the center is Ptolemaeus, 164 km in diameter. Its straight wall segments are probably controlled by pre-existing joints in the Moon's crust. Crater Alphonsus (108 km diameter) is at the lower left and Albategnius (114 km diameter) is at the right. (NASA Ranger 9, B001). Panel (b) shows "rose diagrams" plotting the frequency of occurrence of lineaments with different orientations on the Moon. Top shows lineaments in Region 1 of the index map in the center, bottom is lineaments in Region 2. Orientations are plotted only in the upper half circle because a full circle is redundant. After Strom (1964).

In addition to a decrease in spin rate, tidal distortion (plus early despinning?) and reorientation of the lithosphere with respect to the rotation axis may generate global stress patterns. Such a system has been suggested for Enceladus (Matsuyama and Nimmo, 2008) and evidence for other global systems is currently being sought on other moons. The major requirements for the production of a coherent global tectonic pattern are a one-plate planet and a globally coherent source of stress.

4.7.2 Flexural domes and basins

Roughly circular upwarps and downwarps produce distinctive tectonic associations. Kilometer-scale domelike upwarps over salt domes or igneous intrusions (laccoliths) were the first to be recognized. As shown in Figure 4.21, stretching over an uplifted rock layer produces brittle fracture that results in gaping tensile fractures on a very small scale, or more usually normal faults and grabens. As uplift begins, failure is most likely along spoke-like radial faults that are later connected by more irregular circular normal faults. A good planetary example of this tectonic pattern is seen on the floor of the lunar crater Humboldt.

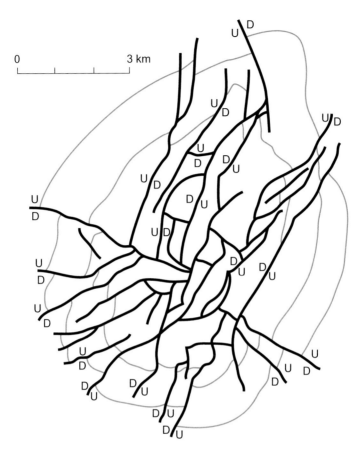

Figure 4.21 The pattern of faults and contours in strata uplifted and domed by the Hawkins salt dome in Texas. Drawn on the top of the Woodbine sand horizon, the outermost contour (light line) is 1400 m below the surface and the inner contour is 1300 m below the surface. The dark lines are faults, where U stands for up and D for down, giving the relative displacement across the fault. Redrawn from Suppe (1985), Figure 8–20.

On a much larger scale, the entire Tharsis region on Mars is underlain by a broad rise up to 9 km high and approximately 5000 km broad. Created by a combination of intrusive and extrusive volcanism, the rise is traversed by numerous approximately radial extensional faults or dikes that transition into a compressional aureole more than 2000 km away from the center of the uplift (Banerdt *et al.*, 1992). The prominent radial grabens cutting the oldest terrain near the crest of the Tharsis Rise strongly suggest extensional doming and led some of the first workers to propose that the rise originated as a broad, isostatically supported uplift (Wise *et al.*, 1979). However, further analysis and gravity data reveals a more complex picture: Beyond the dome itself, most tectonic features indicate deformation of the lithosphere by loads that produced a circumferential girdle of compressional tectonic features. The compressional terrain coincides with a topographic low that is best explained as a thick-lithosphere flexural response to the load of Tharsis (Phillips *et al.*, 2001). The load itself is attested by a large (500 mGal) positive free-air gravity anomaly. The only simple way to reconcile these disparate observations is to suppose that most of the elevation of the Tharsis dome is due to magma intruded into the crust at depths below most of the grabens. Horizontal intruded magma bodies, therefore, loaded the lithosphere while the ancient surface above the intrusions was uplifted and stretched contemporaneously with the intrusion.

Mascon loads provide a cleaner example of tectonics created by subsidence. Large impact basins on the Moon's nearside were flooded by several kilometers of basalt, the last outpourings ending about 1000 Myr after basin formation. This age difference provided ample time for a coherent lithosphere to form before the lava loads were imposed. The heavy basalt loads visibly depressed the underlying terrain and fractured the surface rocks into a characteristic tectonic pattern. Radial and concentric mare ridges occupy the interiors of the mascon basins, while the basins themselves are surrounded by concentric sets of grabens. Theoretical computations of the stress pattern expected at the surface of a flat, floating elastic plate subjected to mascon-like loads showed, as observed, an inner zone of compressional faulting surrounded by an annulus of concentric normal faults (Melosh, 1978). However, the theory indicates that these two zones are separated by an unobserved annulus of strike-slip faults (Figure 4.22). The absence of the expected strike-slip faults became so troublesome that it was known as the "strike-slip fault paradox" until more detailed modeling finally showed that on the Moon the lithospheric shell's curvature, among other things, suppresses strike-slip faulting (Freed *et al.*, 2001). The diameter of the transition between compressional and extensional tectonics, in relation to the diameter of the load (measured from gravity anomalies) serves to determine the thickness of the lithosphere and has been widely used for this purpose, on the Moon and elsewhere (Comer *et al.*, 1979). The Moon's elastic lithosphere varied between 25 and 75 km in thickness during the time of the mare basalt emplacement.

4.7.3 Stress interactions: refraction of grabens by loads

Some of the most striking tectonic patterns reflect interference between different stress systems. Figure 4.23a illustrates the broad loop created by extensional grabens on Mars, as they seem to sidestep the Alba Patera volcanic center on their otherwise gently curving course from the south to north (Cailleau *et al.*, 2005). Similar patterns are observed on Venus (Cyr and Melosh, 1993), where grabens approaching corona centers either avoid the center of the corona load, shown in Figure 4.23b, or approach the extensional stress system of a nascent corona or "nova" dome in an "arachnid"-style corona. These patterns can be understood as the superposition of two stress systems: A broad regional system of extension that creates grabens overprints an existing stress field due to flexure of the lithosphere beneath an existing load. The sum of these two stress systems produces a tectonic response that clearly indicates its multiple origin. Interference patterns of this kind can be

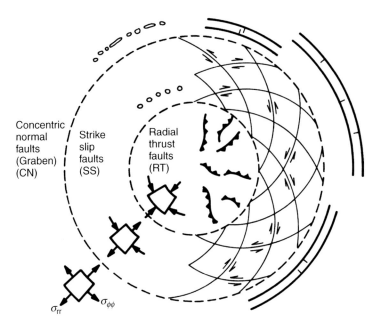

Figure 4.22 Pattern of stresses and faults around a circular mascon-like load on a flat floating elastic plate. The inner region is characterized by compressional tectonics, predicted to be radial thrust faults in this approximation. This region is surrounded by an annulus of strike-slip faults where the circumferential stress is compressive while the radial stress is extensional. This is, in turn, surrounded by an outer annulus of concentric normal faults, where both horizontal stresses are extensional, but the radial stress is more extensional than the circumferential stress. Volcanic dikes reaching the surface, represented here by concentric arcs of small circles, occur wherever the radial stress is extensional. On the Moon, the annulus of strike-slip faults is conspicuously absent, a fact explained mainly by the curvature of the Moon's lithospheric shell. After Figure 2 of Melosh (1978).

quantitatively useful for estimating the sizes of stress acting in the lithosphere if the size of one of the stress systems is known (Turtle and Melosh, 1997). The stresses created by loads can often be computed if the gravity anomalies associated with the load are known and the thickness of the lithosphere estimated. When this is possible, the angles of deflection of tectonic structures then determine the size of the other stress system. In this way the stresses acting in a planetary lithospheric shell may be mapped both in direction and absolute magnitude.

4.7.4 Io's sinking lithosphere

Like the legendary Sisyphus, fated to eternally roll a rock to the top of a hill only to have it roll back again, the crust of Jupiter's moon Io is eternally renewed by volcanism, only to sink once again. Io's internal heat comes from powerful tidal dissipation, its mantle flexing every 1.8 days as the eccentricity of its orbit is maintained by resonances with Jupiter's

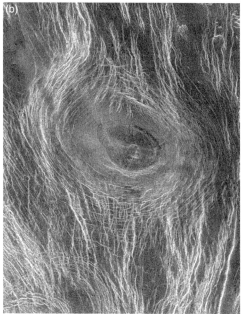

Figure 4.23 Panel (a) Mosaic of Alba Patera on Mars showing the divergence of the trends of graben around the volcanic load. NASA Viking Orbiter frames 783A11 to 16. Panel (b) Similar pattern around 85 km diameter Kvasha Patera on Venus: Portion of NASA Magellan image FL-09S069 centered on 9.5° S, 69° E. North is to the right to facilitate comparison with panel (a).

other large moons. Io's heat seems to keep its mantle in a semi-molten state, from which the excess heat is advected by magmas that flow onto Io's surface. The resurfacing rate is estimated to be 1.5 cm per year on average. Io seems to function as a kind of planetary lava lamp, in which the crust is renewed at the surface even as deeper portions are simultaneously heated and recycled back onto the surface.

The tectonic consequences of all this recycling now seem inevitable, but were so surprising that it took many years before they were fully appreciated. Schenk and Bulmer (1998) first attributed Io's mesa-like mountain masses to thrust faulting and realized that the sinking crust must lead to compressive stresses. More detailed exploration of this idea (McKinnon *et al.*, 2001) showed that, while some of the compressive stress is generated by the decrease in radius of the sinking crustal layers, by far the largest portion is produced by the heating and expansion of the layers as they sink, warm, and expand. Io's crust is thus in compression, which leads to thrust faulting and tilted mountain blocks that rise up to 20 km above its sulfur-coated surface. The visible mountains do not form any apparent global pattern, although an anticorrelation between mountain location and volcanic centers makes some sense from the viewpoint of lithosphere thickness. If Io did not possess a sinking lithosphere that advects heat downward, its lithosphere would be much thinner than it actually is and the observed high mountains could not be supported.

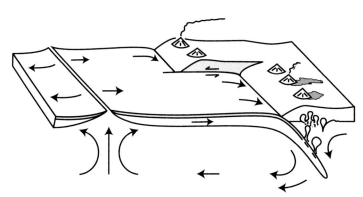

Figure 4.24 Plate tectonics, Earth style. Tectonic plates are created at oceanic spreading centers, travel horizontally across the surface of the Earth, thickening and cooling as they go, then plunge back into the mantle in subduction zones where they are partly reabsorbed. Plates may also slide past one another along strike-slip faults, of which the San Andreas Fault is the most famous example.

The lesson we learn from the bizarre-looking moon Io is that, although established tectonic principles seem to apply to all of the terrestrial planets and moons, we can still be surprised and initially baffled by the circumstances in which they occur.

4.7.5 *Terrestrial plate tectonics*

The Earth possesses a unique tectonic style now known as plate tectonics (Figure 24). Whether or not this tectonic style is an inevitable consequence of Earth's highly vigorous convective interior is presently unresolved. Venus, Earth's "twin" planet, certainly does not operate in the same manner as the Earth, but perhaps this is a consequence of Venus' high surface temperature, which may render the existence of large, effectively rigid plates impossible. Or, on the contrary, perhaps the presence of water on the Earth makes plate boundaries so weak that large expanses of the lithosphere can move as rigid blocks until they meet another plate, where they deform with little resistance.

Whatever the ultimate cause of plate tectonics, the surface of the Earth can best be described as consisting of a dozen or so large lithospheric plates that glide over the surface of the planet, interacting mainly at their edges (Turcotte and Schubert, 2002). There are two basic types of lithosphere composing the plates: Oceanic lithosphere is created at divergent, extensional mid-ocean ridges. It scrapes past adjacent plates at strike-slip transform plate boundaries and is consumed in compressional subduction zones where the oceanic lithosphere sinks back into the mantle and is eventually recycled by mantle convection. Continental lithosphere is chemically different from oceanic lithosphere. It is topped by more silica-rich, less dense and thus more buoyant rocks, which often ride over a deep keel of depleted upper mantle material. Because of its buoyancy, continental lithosphere cannot be subducted, at least not very deeply, and so preserves a record of events much older than the oceanic lithosphere (whose oldest portions date back only about 180 Myr). Because of its high silica content, the deeper portions of the continental crust are more fluid than oceanic lithosphere and may, therefore, deform more readily.

How subduction starts is still controversial: Both types of lithosphere resist small downward deflections. Even oceanic lithosphere probably needs to be forced downward into the mantle by some amount before it becomes unstable. Once started, however, the negative buoyancy of subducted oceanic lithosphere drives the plate tectonic engine. Extensional stresses are transmitted from subduction zones throughout the plates as they are pulled inexorably into the oceanic trenches. Plate velocity correlates closely with the negative buoyancy of attached subduction zones. Increasingly sophisticated computer simulations of terrestrial convection patterns suggest that plate tectonics develops only when the plate boundaries are extremely weak: Stresses of only a few tens of megapascals are typically required for the lithosphere of a planet to organize into independently moving plates.

Plates exhibit the gamut of possible tectonic styles: Extensional faulting at spreading centers where oceanic lithosphere is born, strike-slip transform faulting as they move past other plates, and compressional thrust faulting where they plunge back into the mantle in subduction zones. In addition, folds and detachments often develop where continental crust is compressed, and nearly flat-lying normal faults may form where it is extended. Chains of volcanoes develop on the overriding limb of subduction zones, where deep melting is enhanced by the infusion of subducted water and carbonate. Mountain belts arise where continental lithosphere meets continental lithosphere in collapsing subduction zones. Rift zones tear continental lithosphere apart when a new spreading center develops in the midst of previously intact continents.

Earth and its complex tectonic system is certainly the best understood of all the planets in the Solar System. However, even terrestrial tectonics is not fully understood and remains the subject of vigorous ongoing research. Outstanding questions focus on the strength of the lithosphere and its role in plate tectonic processes. The deformation modes of rock materials at very low strain rates, and how creep is affected by volatiles such as water still engage the attention of large numbers of Earth scientists. Nevertheless, careful studies of the tectonics of other planets is certainly not premature, because tectonic activity in very different settings than those on Earth illuminates these processes under conditions not accessible on our own planet. Another puzzle in its own right is that, for reasons that are not presently understood, compression is almost completely absent on the icy satellites, while extension dominates their tectonics.

Further reading

The prescient writings of James Hutton on Earth's tectonic and sedimentary cycles are so obscurely written that few even of his contemporaries read them. Fortunately, his friend John Playfair presented them in a clear, appealing style that can be read with profit even today (Playfair, 1964). The fundamental reference on geodynamics and tectonics for this chapter, as for the last, is the famous textbook by Turcotte and Schubert (2002). A good, more intuitive introduction to mantle convection is provided by G. Davies (1999), which concentrates on understanding the problem using simple geometries, while the massive tome by Schubert *et al.* (2001) covers all aspects of the convection problem. A classic introduction to fracture mechanics of brittle materials at a high mathematical level is provided by Lawn and Wilshaw (1975). A clear and appealing discussion of the buckling theory of folding and boudinage, among other topics, is in Arvid Johnson's book (Johnson, 1970). The best and most comprehensible introduction to structural geology and tectonics without excessive dumbing-down is Suppe (1985). Suppe's book also contains a nice exposition of the fault-bend model of large-scale folding, which he largely created. A recent summary of tectonics from a planetary perspective has just appeared, with individual chapters by many authors (Watters and Schultz, 2010).

Exercises

4.1 Hot and cold asteroids

The steady-state heat conduction equation for a spherically symmetric planet is:

$$q(r) = -k\frac{dT}{dr}$$

where $q(r)$ is the heat flux passing through a spherical shell of radius r in units of J/m²-s, T is the temperature in K and k is the thermal conductivity (about 2 W/m-K for "typical" rock). Use this equation to show that the internal temperature of a solid, spherical, asteroid with uniform heat production H W/m³ is:

$$T(r) = T_S + \frac{H}{6k}(a^2 - r^2)$$

where T_S is the asteroid's surface temperature and a is its radius. If $H\rho = 5.23 \times 10^{-12}$ W/kg (be careful of units!!), the average value for carbonaceous chondrites, compute the present-day difference between the surface and central temperatures of:

Asteroid Name	Radius (km)
1 Ceres	512
3 Juno	134
4 Vesta	263
243 Ida	~15 (not actually very spherical!)
433 Eros	~5 (not actually very spherical!)

4.2 The towering inferno

Images returned by the Galileo orbiter reveal that Io hosts some of the tallest mountains in the Solar System: The highest peaks rise some 17 km above the surrounding plains. Although Io's surface is topped with a colorful frosting of sulfur compounds, the mountains are believed to be composed of silicates, as are the extremely hot lavas that are extruded at the Ionian paterae. Supposing that Io has a basaltic crust (density about 3000 kg/m³) of unknown thickness overlying a peridotitic mantle (density about 3300 kg/m³), what is the *minimum* depth of an Airy-isostasy root that could underlie the mountains?

The average heat flux on Io lies between 2 and 2.5 W/m². Most of this is lost by volcanism at the paterae, so that the conductive heat flux through the balance of the crust may be much smaller. Suppose that it is as low as the Earth's, 80 mW/m². Further supposing that the thermal conductivity of Io's crust is typical of that of silicates, $k = 2$ W/K-m, what is the temperature at the base of this root? It may be useful to know that the surface temperature of Io is about 100 K.

What does this tell you about the support of high topography on Io?

4.3 The fuzzy lithosphere

Assume that the temperature in the upper mantle of the Earth is 0°C at the surface and rises *linearly* to 1200°C at 100 km depth (roughly the source depth of basaltic rocks that erupt at this temperature). Still deeper the temperature, controlled by convection, is approximately constant.

Use the wet olivine creep law of Chopra and Paterson (1984) for Anita Bay dunite,

$$\dot{\varepsilon}^{Y}(s^{-1}) = 10^{4.0} \, \sigma^{3.4} \, (\text{MPa}) \exp\left(-444\text{kJ mol}^{-1} / RT\right)$$

with $R = 8.314$ J/K, to estimate the Maxwell time as a function of depth for a typical stress $\sigma \approx 30$ MPa (produced by loads ca. 1 km thick). The upper mantle shear modulus $\mu = 65$ GPa.

How thick is the "lithosphere" for loads applied for: 1 Myr, 10 Myr, 100 Myr, and 1 Gyr? How sensitive is this thickness to the assumed stress?

The bottom line: Is the floating elastic plate approximation any good for loads of this duration?

Beware of unit conversions!

4.4 Hot, flexed lithospheres

Jupiter's satellite Europa is squeezed and flexed by tides every 3.55 days as it orbits the giant planet. Stresses in its icy surface shell vary sinusoidally with an amplitude of about 10^5 Pa. Using the Maxwell viscoelastic model, compute the strain as a function of time in a cubic meter of the ice shell, assuming a shear modulus of 10^{10} Pa and an effective viscosity (at 10^5 Pa) of 10^{13} Pa-s. How does the power dissipated depend on frequency? (Remember that this is stress times strain rate – you might want to integrate this over some standard time interval if it varies throughout the tidal cycle!) Under present Europan conditions, estimate the surface heat flux from the flexing ice shell if it is 10 km thick.

For this exercise it is enough to consider only one stress component. The energy dissipated per cycle per unit volume is then:

$$\Delta E = \oint_{\text{cycle}} \dot{\varepsilon}_{xz} \sigma_{xz} \mathrm{d}t.$$

4.5 A primary Europan concern

As you learned in the previous homework problem, the ice shell of Europa is continually flexed by tides over the 3.55 day orbital cycle. Stresses in the shell oscillate sinusoidally with amplitudes of about 10^5 Pa. In Problem 4.4 we assumed that the viscous part of the deformation (the only portion that dissipates heat) is the same as the steady-state creep rate. However, in real life, the creep curve has two major parts (shown schematically in Figure 4.25) in addition to the elastic deformation. In particular, the primary, or transient, portion of the creep curve has been ignored in all publications thus far on the dissipation of tidal heat in Europa. Your task is to *qualitatively* discuss how primary creep may contribute to the budget of heat dissipation due to tidal flexing. Try to couch your analysis in terms of the Maxwell time associated with steady-state creep, $\tau_M = $ (elastic strain)/(steady creep strain rate) $= \eta/\mu$, and that due to primary creep relaxation, τ_R, versus the period P of tidal flexing. Remember that the rate of heat dissipation (power per unit volume) due to stress and strain variations is given by:

$$W = \frac{1}{P} \sum_{i,j} \oint_{\text{cycle}} \dot{\varepsilon}_{ij}(t)\, \sigma_{ij}(t)\, \mathrm{d}t$$

where ε_{ij} is the strain tensor and σ_{ij} is the stress tensor.

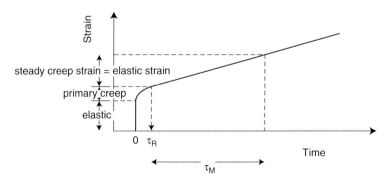

Figure 4.25

4.6 Radiant Earth

Within hours of the impact of a Mars-sized body that struck the ancient Earth, internal readjustments in the Earth and fallback of debris from the projectile distributed the projectile's initial kinetic energy into heat that raised the mean temperature of the entire mantle of the planet to about 3000 K, well beyond the melting point of olivine (which is about 2000 K to sufficient accuracy for our purposes). Not surprisingly, the liquid silicate mantle began to cool from the top, which induced vigorous convection throughout its 3000 km depth. The surface temperature of this magma ocean rose to about 900 K, limited largely by the rate that heat could be radiated from the surface.

a) Using your knowledge of thermal convection, estimate the Rayleigh number of the convecting magma ocean, then compute the surface heat flux and the thickness of the thermal boundary layer. Does it seem likely that chilled surface material will form a crust that may inhibit convection? Compare the convected heat flux to the rate at which radiation can carry it away from the surface.

b) Neglecting, for now, the temperature dependence of viscosity (which is much smaller for a liquid silicate than for crystalline silicates), estimate how long the magma ocean took to cool to the melting point of olivine. If you were to visit the early Earth sometime during its 100 Myr long history of accretion, how likely is it that you would find it in a totally molten state?

Some numerical data you may need: Thermal expansion coefficient of melt: 2×10^{-5} K^{-1}, Thermal diffusivity of melt: 10^{-6} m^2/s, Thermal conductivity: 2 W/(mK), heat capacity: 3.5×10^6 J/m^{3}K, kinematic viscosity (=η/ρ): 3 m^2/s, Stefan–Boltzmann constant, $\sigma = 5.67 \times 10^{-8}$ W/(m^2 K^4).

4.7 Planetary warps

Consider a disk-shaped region of planetary lithosphere of diameter L and thickness H (Figure 4.26). Suppose it is bowed up in the center by an amount U. A central plane, the *neutral sheet* (dashed in Figure 4.27), is unchanged in length as upbowing occurs. The top surface is stretched as shown.

a) Show that the increase in *circumference* of a circle originally located half way between the disk's center and its rim, using purely geometrical considerations, is given by $\delta = HU/L$. From this result, deduce the circumferential strain $\varepsilon_c = \Delta l/l$, where l is the circumference. Compute the elastic stress $\sigma = E\varepsilon_c$ produced by this upwarping if Young's modulus, $E = 6 \times 10^{10}$ Pa, typical of basalt, $L = 500$ km, and $U = 10$ km. Compare this to the "strength" of rock.

b) Suppose this strain is taken up by three radial *graben* that form as upwarping proceeds. Each graben is composed of two normal faults, dipping 60° toward one another (Fig. 4.28). If the upwarped region is 500 km in diameter and is uplifted 10 km, how deep will the graben be? Do you think they could be detected from spacecraft? What if more than three graben form?

c) Suppose the disk is downwarped by the same amount it was upwarped in (b). Suppose now that three radial thrust faults form, each dipping at 30°. How high will the thrust scarps become if they accommodate all circumferential strain?

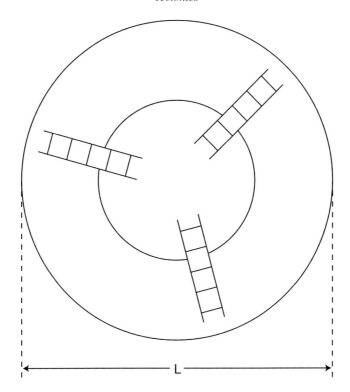

Figure 4.26

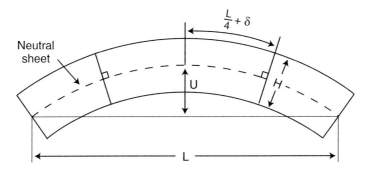

Figure 4.27

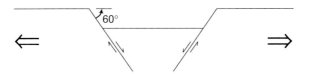

Figure 4.28

NB: This problem is oversimplified. If you were to attack these problems professionally, you would use elastic-plastic *flexure theory* to evaluate stresses and strains more exactly. The curvature of planetary surfaces also cannot be neglected for broad upwarps. However, the geometrical approach in this problem works reasonably well. For more information see Turcotte's review (Turcotte, 1979) or Turcotte and Schubert (2002).

4.8 Boiled cantaloupe lands on Triton

A large percentage of Triton's surface is covered by "cantaloupe terrain," a peculiar land-form dominated by irregular round-to-oval spots 30–50 km in diameter with dark rims that stand 300–800 m higher than their centers. Although this terrain was initially supposed to arise from viscously relaxed impact craters, the rather uniform size distribution suggests an internal origin. It has been proposed that the pattern was created by the convection of a near-surface ice layer that overturned in response to a sudden heating event (perhaps tidal heating when Triton was captured by Neptune). The surface terrain would, thus, be analo-gous to the patterns preserved in a congealed pot of boiled oatmeal. The composition of the upper crust of Triton is unknown, but probably consists of a mixture of H_2O, CO_2, or NH_3 ices. The radius of Triton is 1350 km and its mean density is 2070 kg/m^3, which gives a surface acceleration of gravity of 0.78 m/s^2.

Using your knowledge of convection, estimate:

a) The depth of the ice layer beneath the cantaloupe terrain.
b) Because there is no evidence of melt formation in the cantaloupe terrain, the bottom of the ice layer could at most have reached the melting point of water ice. Triton's surface temperature is about 45 K, so compute the *maximum* temperature difference ΔT that could have developed across the ice layer. Use this number to estimate the effective viscosity of the ice layer for convection to begin. In the light of your knowledge of the rheology of ice and other similar substances, does this seem reasonable? Note that for water ice the coefficient of thermal expansion $\alpha = 3 \times 10^{-6} K^{-1}$, thermal diffusivity $\kappa = 3 \times 10^{-6} m^2/s$.
c) Finally, speculate on whether the convection hypothesis is consistent with the observa-tion that the centers of the spots are lower than the rims.

4.9 Io's crunched crust

Estimates of the volcanic resurfacing rate on Io indicate that new surface material is added at a global average rate of 1.5 cm/yr. As new material is added the lithosphere sinks, warmed by heat conducted from the interior, until it softens and deviatoric stresses relax at a depth of about 25 km.

(a) Estimate the percentage change in area (area strain) of a lithospheric shell formed at the surface, at Io's average radius of 1815 km, as it sinks to the softening depth of 25 km. How is the area strain related to the linear strain of a great circle inscribed on the surface of the sinking crust?

(b) Compute the stress in the plate just before it softens and begins to flow. Compare this to the crushing strength of granite, about 0.1 GPa. The relation between horizontal compressive stress σ and the breadth L of a plate uniformly compressed in its own plane by an amount ΔL is:

$$\sigma = -\frac{E}{(1-\nu)}\frac{\Delta L}{L} = \frac{E}{(1-\nu)}\varepsilon$$

where E is the Young's modulus of the crust, about 65 GPa, ν is its Poisson's ratio, about 0.25, and ε is the horizontal strain.

(c) Compare the compressive strain due to sinking, to the strain caused by thermal expansion as the lithospheric shell warms from Io's surface temperature of 100 K to ½ of the melting point of olivine, $T_m = 2200$ K (presumed to approximate the geologic softening temperature). Which is more important, compression due to sinking or compression due to heating? The volume expansion coefficient α_V of olivine is 2.5×10^{-5} K^{-1} and the linear thermal strain is $\alpha_V \Delta T/3$.

Reference: McKinnon *et al.* (2001).

4.10 Venusian ridge belts

Box 4.1 is required for this problem!

Magellan radar images, confirming hints from the much lower resolution Venera 16 and 17 images, have clearly revealed the presence of fold-like ridges on many parts of the planet. These ridges are hundreds of kilometers long, spaced 15–20 km apart and attain heights of hundreds of meters to several kilometers. Using the theory of buckling of floating elastic plates:

(a) Compute the flexural rigidity and thickness of a single plate that buckles to produce $\lambda = 15$–20 km folds ($g_{venus} = 8.6$ m/s^2, $\rho_c \cong 2.7$ gm/cm^3). Also compute the buckling stress.
(b) If the maximum stress supportable by the surface rocks is 0.1 GPa horizontal compression, what is the mean layer thickness required to obtain the wavelength observed? (Beware! This is not so simple as it seems. You will have to solve coupled algebraic equations to answer this question, which requires you to consider multiple layers.)
    After you have completed this exercise, you may be interested to consult Solomon and Head (1984) and Brown and Grimm (1997) for a more detailed look at this problem.

5

Volcanism

> Burning mountains and volcanoes are only so many spiracles serving
> for the discharge of the subterranean fire … And where there happens
> to be such a structure on conformation of the interior parts of the Earth,
> that the fire may pass freely and without impediment from the caverns
> therein, it assembles unto these spiracles, and then readily and easily gets
> out from time to time …
>
> Bernhard Varenius, 1672, quoted in Sigurdsson (1999), p. 148

5.1 Melting and magmatism

The German geographer Varenius (1622–1650) was one of the first to suggest that volcanic activity is ultimately caused by the escape of hot melted rock from the interior of our planet. Written at a time when most geologists believed that the Earth's interior is filled with molten rock, the source of the melt was not problematic: Any break or fracture would allow molten rock to leak out to the surface, just as puncturing the skin of an animal allows blood to flow out. However, with the study of solid earth tides and the advent of seismology at the end of the nineteenth century, it became plain that the bulk of the Earth is solid and the origin of magma became less obvious.

At the present time we believe that melted rock is a secondary manifestation of the thermal regime of our planet and that heat transport by magma is of slight importance compared to thermal conduction and lithospheric recycling, at least on the Earth. Volcanism and its subsurface accompaniment, igneous intrusion, is, nevertheless, an important process affecting the surface of the terrestrial planets, to the extent that almost no planetary surface seems to have escaped its effects.

Once-molten material from their interiors diversifies the surfaces of many, if not most, planets and satellites. Basalt has even erupted onto the surfaces of large asteroids such as Vesta, although this took place at a time when now-extinct heat-producing elements were active. In the outer Solar System the surfaces of icy satellites exhibit flows of congealed water-rich melts reflecting low-temperature "cryovolcanism." Although the materials differ, the morphology of all these flows is similar, a manifestation of similar physical processes.

Table 5.1 *Heat budget of the Earth*

Source	Heat flow (10^{12} W)	Percent of total
Oceanic plate recycling, excluding crust formation	23.1	55
Intra-crustal radiogenic heat (mostly continental)	6.6	16
Conduction through continental plates	5.0	12
Oceanic crust formation[a]	3.1	7.4
Conduction through oceanic plates	2.9	6.9
Intra-continental advection (erosion, orogeny, magmatism)	1.1	2.6
Volcanic centers	0.2	0.5
Totals	42.	100.

Data from Sclater *et al.* (1980). Estimated accuracy of each entry ca. 10%.

[a] Assumes oceanic crust is generated at 18 km^3/yr, which advects 1.81 MJ/kg.

5.1.1 Why is planetary volcanism so common?

One might suppose that volcanism acts as a kind of relief valve for pent-up heat in the interior of large Solar System bodies. Although there may be some truth to this idea (the planetary volcanic resurfacing event that affected Venus some 700 Myr ago is often supposed to mark an episode of high subsurface temperatures), a detailed inventory of the Earth's heat flow shows that volcanic heat transport accounts only for a small fraction of the total heat lost from our own planet's interior (Table 5.1), which is otherwise dominated by plate recycling (62% of the total) and lithospheric conduction (37%, including heat transported by oceanic crust formation). The amount of heat transported by volcanism, q_{vol}, can be estimated from the volume rate of eruption, Q_E, by:

$$q_{vol} = \rho(c_P\Delta T + \Delta H_f)Q_E \qquad (5.1)$$

where ρ is the density of the magma brought to the surface, c_P its heat capacity, ΔT the temperature difference between the erupted magma and the surface, and ΔH_f is the enthalpy of fusion of the solid magma. The volume rate of eruption is often estimated from the volume of magma observed on the surface and the duration of an eruption, allowing estimates of the heat flux from volcanism alone. When compared to the heat flux conducted through the lithosphere, the volcanic flux is usually found to be small, except in the case of Io, whose heat transport does seem to be dominated by the eruption of magma (probably ultramafic silicate magma, accompanied by large amounts of the volatiles sulfur and SO_2). We do not have good estimates of the eruption rate on Venus during its short volcanic resurfacing event, but there, too, volcanic heat transport may have dominated its planetary heat flow. Venus should serve as a warning against too-simple classifications of planetary heat transfer: Different processes may dominate at different times in a planet's history.

Classifications are, nevertheless, useful, if one keeps their approximate nature in mind. A common and appealing classification of planetary heat transfer is the triangular ternary diagram (Solomon and Head, 1982), borrowed from igneous petrology and illustrated in Figure 5.1a. Dividing planetary heat transfer into lithospheric conduction, plate recycling and volcanism, whose sum must equal 100%, one plots small bodies and one-plate planets close to the conduction vertex, the Earth near the plate-recycling vertex and volcanic bodies like Io close to the volcanic vertex. An alternative classification diagram is shown in Figure 5.1b in which heat transport is plotted versus mean mantle temperature, normalized by the melting temperature. On this one-dimensional diagram we also see the three processes, conduction, convection, and volcanism, but now laid out on a line reflecting the role of increasing internal heat generation. Volcanism, however, is not considered a unique mode of heat transport, but as an accessory phenomenon that can occur in any of the three regimes, although it becomes more important as the mean temperature rises.

The transition between pure conduction and convection occurs where the temperature of the mantle rises to about half its melting temperature, when viscous creep becomes important and the mantle's viscosity drops to about 10^{21} Pa-s. Because of the strong temperature dependence of solid-state creep, discussed in the last chapter, this regime tends to be self-regulating and can accommodate a large range of heat transport. However, once large-scale melting occurs the viscosity drops very rapidly to 10^3 Pa-s or even less, and the rate of heat transport, proportional to the inverse cube root of viscosity – see Equations (4.19) and (4.26) – increases million-fold. Heat transport rates higher than those sustained by such magma oceans are possible, but here we enter the realm of giant planet or even stellar interiors for which the concept of a planetary surface disappears, so we defer this topic to other texts.

Volcanism occurs on planets dominated by each of the major modes of heat transport. This is partly due to the physical nature of melts: no matter where they are produced, melts, once formed, are typically more mobile and less dense than their parent materials and so they may rise to the surface, leaving their parents behind. However, the principal reason for volcanism seems to be the highly variable susceptibility of planetary materials to melting. The ease with which materials melt varies with both position in the planet and composition of the material.

Melting in a convecting planet is most likely to occur just below the conductive thermal boundary layer. The boundary layer itself is, of course, the coldest part of the convecting system and temperatures rise linearly with increasing depth (refer back to Figure 4.1). This steep rise ends at the base of the thermal boundary layer where the temperature gradient becomes approximately adiabatic (Box 5.1). If the melting temperature were constant, melting would begin much deeper still, at the bottom of the convecting region. However, pressure increases the melting point, so the location where melting begins is established by a competition between the rate at which the temperature rises and the rate at which the melting point increases.

Clausius–Clapeyron equation. É. Clapeyron (1799–1864) in 1834 first deduced that the pressure derivative of the melting point of a substance is proportional to its latent heat of

Volcanism

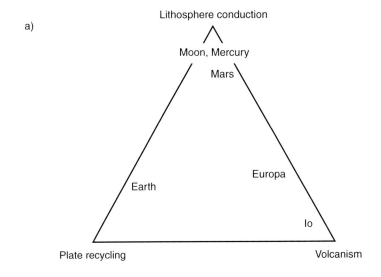

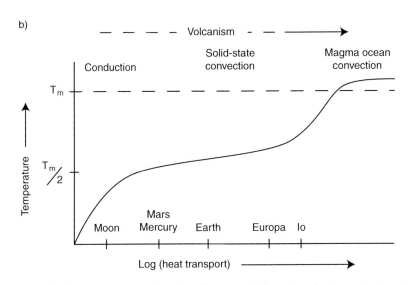

Figure 5.1 Mode of heat transport as a function of process. (a) Location of various planets and moons with respect to the three major heat transport processes of heat conduction through the lithosphere, plate recycling, and volcanism. (b) Dependence of mean mantle temperature and the transport process on net heat transport. As the amount of heat transport increases, the mode of transport switches from conduction to solid-state convection to liquid-state magma ocean convection. Surface volcanism may occur at any stage of this progression, although it is more likely to occur at net heat transport increases.

Box 5.1 **The adiabatic gradient**

The pressure experienced by materials in a planetary interior frequently changes as they move about. Because of the slow conduction of heat into large masses of material, this motion commonly takes place with little or no heat exchange to the surroundings. Nevertheless, compression or expansion of the material does change its internal energy and, therefore, its temperature varies as it ascends or descends. Such temperature changes are called *adiabatic* (equivalently, constant entropy or isentropic, as long as the process is also reversible). Adiabatic temperature changes play an important role in both convection and volcanism. The characteristic adiabatic gradient is a crucial concept in describing these processes.

The adiabatic gradient can be derived from fundamental thermodynamics in various ways. One of the simplest is to note that the entropy, S, is a state variable that is typically a function of only the pressure P and temperature T (in more special circumstances it may also depend on the composition, especially the latent heat during melting, the magnetic field, or other variables that affect the internal energy of the moving material), expressed as $S(P,T)$. Then standard multivariable calculus tells us that a change in the entropy can be expressed in terms of changes in temperature and pressure as:

$$dS = \frac{\partial S}{\partial T} dT + \frac{\partial S}{\partial P} dP. \tag{B5.1.1}$$

Recall the definitions of the heat capacity at constant pressure, c_P, and volume thermal expansion coefficient α_V:

$$c_P \equiv T \frac{\partial S}{\partial T}$$

$$\alpha_V \equiv \frac{1}{V} \frac{\partial V}{\partial T} = \rho \frac{\partial V}{\partial T} \tag{B5.1.2}$$

where ρ is the density of the material. To these equations add the thermodynamic Maxwell relation between derivatives:

$$\frac{\partial S}{\partial P} = -\frac{\partial V}{\partial T}. \tag{B5.1.3}$$

Then, inserting (B5.1.3) into (B5.1.1) and using the definitions (B5.1.2), we obtain an expression for the entropy change in terms of material properties:

$$dS = \frac{c_P}{T} dT - \frac{\alpha_V}{\rho} dP. \tag{B5.1.4}$$

The adiabatic (reversible, no heat exchange) condition assures that $dS = 0$, so that the temperature changes with pressure according to:

$$\left. \frac{dT}{dP} \right|_{\text{adiabatic}} = \frac{\alpha_V T}{\rho c_P}. \tag{B5.1.5}$$

Box 5.1 (cont.)

Finally, applying the hydrostatic relation between pressure and depth z and local gravitational acceleration g, $dP = \rho g dz$, we obtain the standard expression:

$$\left.\frac{dT}{dz}\right|_{\text{adiabatic}} = \frac{\alpha_V g T}{c_P}. \tag{B5.1.6}$$

For the Earth's mantle, typical values of these constants are $\alpha_V = 3 \times 10^{-5}\ \text{K}^{-1}$, $T = 1600$ K, $c_P = 1000$ J/(kg K) and $g = 10$ m/s^2, resulting in a gradient of about 0.5 K/km. This is far smaller than the conductive temperature gradient near the Earth's surface, about 30 K/km. Nevertheless, across the entire 3000 km thickness of the Earth's mantle, if the gradient were independent of depth, then a temperature increase of about 1500 K would be implied, exclusive of the temperature jumps across the hot and cold conductive boundary layers at the bottom and top of the mantle. The adiabatic gradients in the other planets and satellites are smaller than that of the Earth, but the gradient, nevertheless, plays an important role in planetary-scale bodies, especially in relation to the gradient in the melting temperature of planetary materials.

---

melting. Stated in modern terms as the Clausius–Clapeyron equation, the pressure derivative of the melting temperature T_m is:

$$\frac{dT_m}{dP} = \frac{\Delta V_m}{\Delta S_m} \tag{5.2}$$

where $\Delta V_m = V_{\text{liq}} - V_{\text{solid}}$ is the volume change upon melting and $\Delta S_m = S_{\text{liq}} - S_{\text{solid}}$ is the entropy change upon melting, often expressed as the latent heat L divided by the melt temperature, $\Delta S_m = L/T_m$. The volume change upon melting is typically about 10% of the specific volume of a substance and ΔS_m is typically a few times the gas constant, based on Boltzmann's relation $S = R \ln W$, where R is the gas constant and W the probability of a given state (Pauling, 1988, p. 387). Table 5.2 lists the slope of the melting curve for several pure substances of geologic interest.

Decompression melting. Comparing the slopes of these melting curves with the adiabatic gradient, about 16 K/GPa for silicates, makes it clear that the adiabatic gradient is generally less steep than the melting curve of common minerals, at least at low pressures. We will see later that this must be modified at high pressures where mineral phase transformations take place, but at least near the surface of silicate planets it is clear that melting is most likely to occur just below the lithosphere, while at greater depths the planet may remain solid.

When hot, deep-seated solid material rises toward the surface by solid-state creep, its temperature gradually approaches the melting curve and, if the melting curve is reached, magma forms. This is one of the principal causes of melting in the Earth and probably the other silicate planets. It is called pressure-release or decompression melting. The other principal cause of melting, at least in the Earth, is the reduction of the melting point of silicates through the addition of volatiles, principally water. This is known as flux melting.

Table 5.2 *Dependence of melting point upon temperature for various minerals*

Mineral	Volume change upon melting, ΔV_m^a (cm³/mol)	Volume change as a fraction of molar volume, $\Delta V_m/V$	Entropy change upon melting, ΔS_m (J/mol-K)	Slope of melting curve at 1 bar, dT_m/dP (K/GPa)
Ice Ih (0 to 0.2075 GPa)	−1.634	−0.083	22.0	−74.3
Ice VI (0.6 to 2.2 GPa)	1.65	0.120	16.2	102.0
Quartz	1.96	0.086	5.53	355.0
Forsterite	3.4	0.072	70.0	48.0
Fayalite	4.6	0.094	60.9	75.0
Pyrope	8.9	0.077	162.0	55.0
Enstatite	5.3	0.157	41.1	128.0

Silicate data from Poirier (1991), water data from Eisenberg and Kauzmann (1969).
^{a} ΔV_m adjusted to agree with melting curve slope.

The peculiar melting behavior of water in icy bodies complicates this picture: Increasing pressure *decreases* the melting point of ice to a minimum of 251 K at a pressure of 0.208 GPa, after which the melting point increases again at an average rate of about 55 K/GPa. This permits the existence of stable subsurface liquid oceans in bodies such as Europa and possibly others in the outer Solar System, but makes it difficult for pure water to reach the surface.

Although the effect of pressure on the melting curve is of great importance in planetary interiors, the effect of composition, and of mixed compositions in particular, is probably even greater. It is not possible to understand volcanism on the Earth and other silicate planets without understanding the melting of heterogeneous mixtures of different minerals. Even on the icy satellites the melting behavior of mixtures of ices probably dominates their cryovolcanic behavior.

5.1.2 Melting real planets

Planetary composition: rocky planets. The composition of planetary bodies is ultimately determined by the mix of chemical elements that they inherited when they condensed from an interstellar cloud of gas and dust. Most of this material is too volatile to condense into rocky or icy planets: 98% of the mass of the Solar System is H and He, augmented by about 0.2% of "permanent" noble gases such as Ne and Ar. Of the remaining 1.8% of the mass, about 3/4 comprises "ices" and only 1/4 forms the high-temperature *rocky* material from which the terrestrial planets are built. Table 5.3 lists oxides of the major elements present in the "rocky" component of Solar System material. More than 90% of this rocky material is composed of only four elements: O, Fe, Si, and Mg.

The listing of elements as oxides in Table 5.3 is purely conventional: These elements are actually found in more or less complex minerals that are combinations of the simple

Table 5.3 *Cosmic abundances of metal oxides*

Metal in oxide combination	Abundance by mass (%)
FeO	38.6
SiO_2	30.6
MgO	21.7
Al_2O_3	2.2
CaO	2.1
Na_2O	1.9
All others	2.9

Data from Table VI.2 of Lewis (1995).

oxides. The four most abundant elements typically produce a mixture of the minerals olivine (Mg_2SiO_4 or Fe_2SiO_4) and pyroxene ($MgSiO_3$ or $FeSiO_3$) plus metallic Fe. Doubly charged iron and magnesium ions are rather similar in size and readily substitute for one another in the crystal lattices of olivine and pyroxene, which, thus, commonly occur as solid-solution mixtures of both elements. These minerals, along with their high-pressure equivalents, comprise the bulk of the Earth and other terrestrial planets, moons, and asteroids. Geologists are more familiar with rocks that contain a higher proportion of the less abundant elements Al, Ca, Na, and others. These elements generally form minerals of lower density than olivine and pyroxene and so have become concentrated in the surface crusts of differentiated planets.

Planetary composition: icy bodies. "Icy" materials, listed in Table 5.4, are more volatile than those forming the terrestrial planets and are, thus, mostly confined to the outer Solar System. The format in Table 5.4 is also conventional, listing the elements O, C, N, and S as chemical species that condense from a slowly cooling gas of average solar composition that is dominated by H. Water ice is by far the most abundant species, but carbon may occur as methane, as listed here, or as the more oxidized CO or CO_2. In comets, carbon seems to be present mainly as CO and CO_2 ices, often loosely bound in water as clathrates, as well as in more complex hydrocarbon "tars." Nitrogen may occur as N_2 rather than as ammonia, and sulfur may form compounds with other elements.

The main lesson from this brief discussion of planetary composition is that all planets and satellites are heterogeneous mixtures of a variety of different chemical species. Although the proportions may vary depending upon their location in the Solar System and the chemical and physical accidents of their assembly, planets are anything but pure chemical species. This fact is the root cause of volcanic phenomena, and it requires a careful inquiry into the complex process of melting.

Melting rocks. Our most common experience with the melting of a solid is also one of the most misleading: Everyone is familiar with the conversion of solid ice into liquid water as heat is added. Probably everyone also knows that this takes place at a fixed temperature,

Table 5.4 *Cosmic abundances of ices*

Chemical species	Abundance by mass (%)
Water, H_2O (6.4% bound to silicates)	53.8
Methane, CH_4	33.0
Ammonia, NH_3	11.0
Sulfur, S	2.2

Data from Table VI.2 of Lewis (1995).

$0°C$ at 1 bar, and that this temperature remains constant until all of the ice is converted into water. This behavior is typical of pure and nearly pure materials, such as fresh water. However, planetary materials are usually far from pure and so the melting behavior of heterogeneous mixtures of materials is most relevant in a planetary context.

Studies of the melting and reactions of complex mixtures of materials is as old as the "science" of alchemy, but a clear understanding of its basic principles only emerged in 1875 when Yale physicist J. W. Gibbs (1839–1903) published his masterwork "On the Equilibrium of Heterogeneous Substances." Austerely written and published only in the *Transactions of the Connecticut Academy*, it took many years for the scientific community to absorb the principles that he set forth. In addition to mechanical work and internal energy, Gibbs associated energy, which he called the "chemical potential," with each different chemical species and phase in a mixture of materials. Using the tendency of the entropy of an isolated system to increase, he defined the conditions under which reactions and phase changes occur in thermodynamic equilibrium and showed how to perform quantitative computations of the abundances of each species in a chemical system, once the chemical potential of each reactant is known. Much of the research in petrology and physical chemistry over the subsequent century has centered about measuring these chemical potentials (now known as the Gibbs' free energy) of a wide variety of substances at different temperatures and pressures. The detailed application of these methods to the melting of ices and minerals is discussed in standard texts, such as that of McSween *et al.* (2003).

Melting of solid solutions. For a basic understanding of volcanic processes, it is enough to recognize that there are two fundamental types of melting in mixed systems. The first occurs in systems in which the different components dissolve in one another in both the liquid and the solid phase. Illustrated in Figure 5.2a for a mixture of iron and magnesium olivine at atmospheric pressure, the pure end member forsterite (Mg_2SiO_4) melts at a single temperature of 2163 K, while fayalite (Fe_2SiO_4) melts at the lower temperature of 1478 K. Mixtures of the two components, however, do not have a single melting temperature but melt over an interval that may be larger than 200 K, depending on the mixing ratio. The pair of curves connecting the pure endpoints on Figure 5.2a indicates this melting range. The lower curve, the solidus, marks the temperature at which the first melt appears at the given composition. The upper curve, the liquidus, marks the temperature at which the last solid

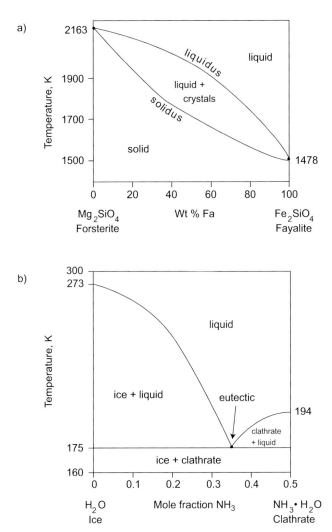

Figure 5.2 Melting relations in heterogeneous mixtures of substances. (a) Melting of a mixture of two materials that dissolve in one another, forming a solid solution. This example is for the pair of silicate minerals forsterite and fayalite. Note that melting of the solution occurs over a considerable interval of temperature. (b) Melting of a mixture of materials that do not form a solid solution but crystallize as distinct phases. This example is for water ice and water–ammonia clathrate, simplified after Figure 1 of Durham *et al.* (1993). Melting also takes place over a range of temperatures except at a single point, the eutectic.

crystal disappears. In between, a mixture of crystals and liquid is present. The crystals and melt do not, however, have the same composition: The crystals are always richer in magnesium (the higher melting point end member) than the melt. The equilibrium compositions of crystals and melt at any given temperature can be read off the diagram at the intersection of a horizontal line drawn at the given temperature and initial composition on the plot with the liquidus and solidus lines.

Eutectic melts. The second type of melting behavior is illustrated in Figure 5.2b for a simplified version of the melting of a mixture of pure water and water–ammonia clathrate, $NH_3 \cdot H_2O$. Water and ammonia clathrate do not mix in arbitrary proportions, but form nearly pure compounds that separately melt at 273 K and 194 K, respectively. When crystals of the two species are mechanically mixed together, however, they react to form a liquid solution at about 175 K, less than the melting temperature of either pure end member. This first melt, called the eutectic composition, is composed of about 0.36 NH_3 by mole fraction. As the temperature continues to rise, what happens depends upon the mixture of crystals. The crystals always remain pure, but the melt composition changes as more crystals melt and mix into the liquid. If the overall composition is richer in ice than the eutectic melt, then all of the ammonia clathrate reacts with the water to form the liquid and only ice crystals are present in contact with the melt, whose composition is indicated by the line to the left of the eutectic point. As the temperature continues to rise, more and more of the water reacts with the melt until an upper temperature is reached at which all of the ice crystals disappear, also indicated by the line to the left of the eutectic point. If the original mixture of crystals is richer in ammonia clathrate than the eutectic composition, then all of the water crystals disappear at the eutectic temperature and a mixture of liquid plus ammonia clathrate crystals persists until the temperature reaches the line to the right of the eutectic point.

Complex melting. Real materials may exhibit either type of behavior in different ranges of composition due to partial solubility of one phase within another or more complex relations due to thermal decomposition of one phase before it finally melts. Despite the complexity of such behavior, it is all governed by Gibbs' rules and, with sufficient experimental data, the melting relations of any mixture of materials can be understood. The number of components that must be added to understand real rocks, however, is distressingly large. Ternary and quaternary diagrams have been devised to represent the mixtures of three or four components (Ehlers, 1987), but in the Earth as well as the other planets, many more than four elements, in addition to volatiles such as water and CO_2, are present and the situation commonly exceeds the ability of any graphical method to illustrate the outcome. At the present time one of the frontiers in this field is gathering all of the data that has been collected by several generations of petrologists and physical chemists into computer programs that use Gibbs's thermodynamics and various models of mixing to predict the outcome of any natural melting event.

Pyrolite and basalt. A simple generalization, however, is possible for the terrestrial planets. As described above, most of their mass is composed of a mixture of olivine and pyroxene. Adding to this the "second tier" elements such as Al, Ca, Na, and K in the form of either feldspar (at low pressure) or garnet (at high pressure), geochemist A. E. Ringwood (1930–1993) concocted a hypothetical material he called "pyrolite" that forms a fair approximation of the bulk composition of any rocky body in our Solar System. When this material melts the first liquid appears at a eutectic temperature of about 1500 K and has a composition generally known as "basaltic." Basalt is a name applied to a suite of dark-colored rocks rich in Fe, Mg, and Ca, among others, with about 50% of SiO_2. It is the typical rock produced on Earth at mid-ocean ridges, forms the dark mare of the Moon, and underlies extensive plains on Venus and Mars. Eucrite meteorites, believed to originate from the large asteroid Vesta, are basaltic in composition. Basalt is the quintessential volcanic melt from rocky bodies in the Solar System. Even on the Earth's continents, where more SiO_2-rich volcanic rocks are common, volcanologists have been known to describe volcanic activity as "basically basalt" because basalts from the Earth's mantle are now believed to provide most of the heat for even silica-rich volcanism.

Role of pressure in melting. Pressure affects the melting behavior of rock both by changing the properties (volume, entropy) of a given mineral phase and by changing the stable phases of the minerals themselves. As olivine is compressed it undergoes a series of transformations to denser phases, assuming the structure of the mineral spinel beginning at about 15 GPa (depending on composition and temperature; there are actually two spinel-like phases), then transforming to a still denser perovskite phase at 23 GPa. These phase changes are clearly seen seismically as wave velocity jumps in the Earth's interior. The phase changes are reflected in the melting curve, as shown in Figure 5.3a, which illustrates the behavior of the solidus and liquidus of the olivine–pyroxene rock known as peridotite (equivalent to basalt-depleted pyrolite, and taken to represent the Earth's mantle) as a function of pressure up to 25 GPa (equivalent to that at a depth of 700 km in the Earth or the core-mantle boundary in Mars). It is clear that the steep melting gradients computed for individual minerals in Table 5.2 cannot be extrapolated to great depths, although the average melting temperature does continue to rise as pressure increases. Figure 5.3b expands the low-pressure region to show several adiabats along with the solidus and liquidus. Note the low slope of the adiabats in comparison with the solidus, as well as the change in slope of the adiabats when melt is present. A hot plume rising along one of the adiabats begins to melt when the adiabat intersects the solidus and continues melting as it rises further, although in reality the melt separates from the solid when more than a few percent of liquid is present, causing a compositional change in the residual material that must be modeled in more detail than is possible from Figure 5.3b.

Flux melting. The properties of silicate magmas are strongly controlled by small quantities of volatile species, particularly water, which has a high affinity for the silica molecule. The dramatic effect of water in lowering the melting point of rock of basaltic composition is illustrated in Figure 5.4, where water saturation is seen to lower the melting point by as much as 500 K at a pressure of a few gigapascals. This strong dependence of melting temperature on water content leads to the possibility of melting caused by addition of volatiles – flux melting. The name comes from the practice of adding "fluxes" like limestone to iron smelters to produce a low melting point slag. On Earth, fluxing by water is particularly important in subduction zones, where hydrated minerals formed in the oceanic crust sink into the mantle. As these minerals heat up, they lose much of their water, which then invades the overlying crustal wedge and lowers the melting point of these rocks. The result is the massive volcanism associated with the overriding plate in subduction zones. As a measure of the importance of water fluxing, note that the average eruption temperature

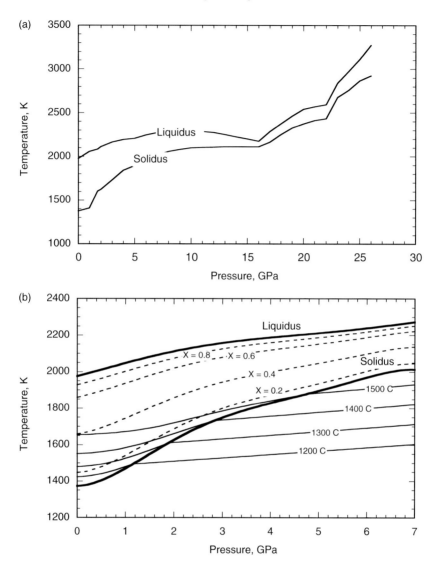

Figure 5.3 Phase diagram of peridotite representative of the Earth's mantle. (a) The solidus and liquidus temperatures of peridotite to a pressure of 25 GPa. The inflections in these curves are due to phase transformations of the constituent minerals as the pressure is increased. After Ito and Takahashi (1987). (b) Detail of the peridotite phase curve (heavy solid lines) to 7 GPa showing adiabats for various temperatures (light solid lines; temperatures indicated in centigrade to facilitate comparison with petrologic data) and light dashed lines showing different degrees of partial melting. Data from Ito and Takahashi (1987), computational method after McKenzie and Bickle (1988).

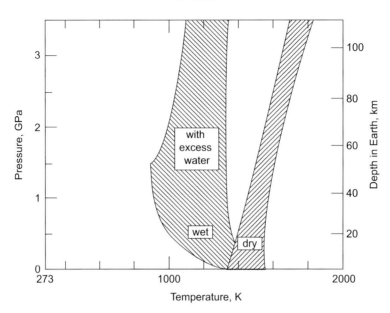

Figure 5.4 The effect of water on the melting temperature for magma of gabbro composition. The shaded regions indicate the interval between the solidus and liquidus. Note that the gabbro–eclogite transition takes place just below the solidus temperature of dry gabbro. The presence of excess water in the magma dramatically lowers the melting temperature at high pressure, by as much as 500 K for pressures near 1 GPa. After Figure 6–12 of Wyllie (1971).

of silica-rich island arc magmas is only about 1200 K, whereas the eruption temperature of basaltic lava is typically about 1500 K. Because silica-rich rocks are more susceptible to the fluxing effects of water than silica-poor rocks, the overall effect of flux melting in subduction zones is to enhance the abundance of silica in melts produced in this environment. The high-silica granitic rocks of Earth's continents may thus be a direct consequence of subduction plus water. It is presently unknown whether this mechanism also plays a role on planets that lack plate recycling, but this makes the discovery of silica-rich granitic rocks on other terrestrial planets a question of great interest. To date, there is no clear evidence for such rocks on Mars, in spite of a brief flurry of excitement over the possible discovery of an andesitic composition rock by the Pathfinder mission – since retracted by the discovery team.

Carbon dioxide also plays a large role in magmas as its presence reduces the solubility of water – water and CO_2 are the dominant gases released in volcanic eruptions. On the Moon, where water is scarce and the lunar mantle is more reducing, CO seems to have been the major gas released during eruptions of basaltic magma. The important role of these volatiles in driving explosive volcanic eruptions will be discussed below in more detail.

Cryovolcanism. Cryovolcanism on the icy satellites presents a still-unsolved problem. We do not yet have samples of the material that flowed out on their surfaces, so the exact composition of the cold "lavas" on these bodies is still conjectural, except that spectral reflectance studies reveal an abundance of water ice. But pure water has a relatively high melting temperature for these cold worlds and, moreover, it is denser than the icy crusts through which these "cryomagmas" apparently ascend, making it difficult to understand how pure liquid water could reach the surface. So either the crusts are mixed with some denser phase (silicate dust or maybe CO_2 ice), or the liquid water is impure, mixed perhaps with ammonia or bubbles of some more volatile phase that lower its average density.

5.1.3 Physical properties of magma

The mechanics of volcanic eruptions and flows are largely dependent upon the physical properties of magma as it separates from its source rock, rises from depth, and spills out onto a planetary surface. Eruptions and flows generally occur so rapidly that chemical equilibrium is not attained and so it is physics, not equilibrium chemistry, which governs the final stages of magma evolution. The most important property in eruptions is the viscosity of the melt, which is a strong function of temperature, composition, pressure, and crystal content.

The viscosity of silicate magmas depends strongly upon their silica content. The small tetravalent silicon ion bonds strongly with four oxygen ions to form very stable tetrahedra. In silica-poor minerals, such as olivine, the tetrahedra are isolated and their charge is balanced by adjacent metal ions. However, as the silica content increases the tetrahedra share corners, forming linear chains in pyroxenes, then sheets and ultimately create a space-filling lattice in pure SiO_2. Silica tetrahedra can, thus, form long polymers in silica-rich melts and, similar to the polymers in carbon-based compounds, the viscosity increases rapidly as the length and abundance of the polymer chains increase. At the same temperature, granitic melts with SiO_2 abundances in the range of 65–70% typically have viscosities about 10^5 times larger than basaltic melts with SiO_2 near 50%. Moreover, as temperature increases and breaks up the polymers, the viscosity of silica-rich melts changes much faster than that of basaltic melt. These relations are shown in Figure 5.5.

Water and silicate magma. The tendency of silica tetrahedra to polymerize also explains the strong dependence of the viscosity of silica-rich magmas on water content. Water does not dissolve in a silicate melt as the triatomic molecule H_2O. Instead, one of its hydrogen atoms bonds with an oxygen ion in the melt to form a pair of OH– radicals that bond with the silica tetrahedra. In the process, as shown in Figure 5.6, silica polymers are broken into shorter pieces and the viscosity consequently drops. This effect is important only in very silica-rich melts: The viscosity of basaltic melts is hardly affected by dissolved water, as shown in Table 5.5, which compares the viscosity of wet and dry silica magmas at typical terrestrial eruption temperatures. Thus, water both lowers the melting temperature of silica-rich melts, such as granites, and decreases their viscosity. Granitic melts, thus, typically arrive at the surface with high water contents and low temperatures, whereas basaltic magmas are hot and relatively dry. This circumstance has major implications for the processes

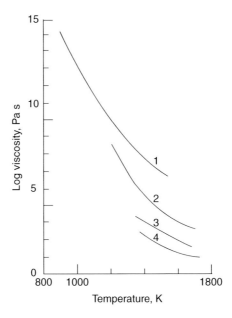

Figure 5.5 The viscosity of silicate melts depends strongly upon both temperature and silica content. The four curves are shown for dry but increasingly silica-poor rocks: (1) rhyolite, (2) andesite, (3) tholeiitic basalt, (4) alkali basalt. After Figure 2–4 of Williams and McBirney (1979).

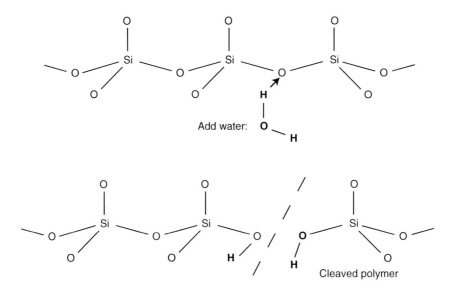

Figure 5.6 Water has a strong affinity for silica polymers. When a water molecule reacts with a long silica polymer chain in the upper half of this figure, it easily breaks the polymer into pieces, while inserting OH groups into the resulting silicate chains. Simple counting of O–H and Si–O bonds shows the same number before and after the insertion, so this process is almost energy-neutral. Breakdown of the silica polymers lowers the viscosity of silicate magma.

Table 5.5 *Effects of water on the viscosities of silicate melts*

Melt composition	Approximate SiO_2 content (Wt %)	Temperature (K)	Viscosity dry (Pa s)	Viscosity with dissolved H_2O (Pa s)
Granite	72	1058	10^{11}	10^4 (5% H_2O)
Andesite	60	1423	10^3	$10^{2.5}$ (4% H_2O)
Tholeiitic basalt	50	1423	$10^{2.2}$	10^2 (4% H_2O)
Olivine basalt	48	1523	10^1	10^1 (4% H_2O)

Data from Williams and McBirney (1979).

that occur in an eruption and lend silica-rich magmas a dangerous tendency to explode due to exsolution of their water.

Viscosity and crystal content. Magmas are typically not entirely liquid. Upon eruption, most magmas contain a heterogeneous assemblage of liquid, crystals, and bubbles of gas. The presence of both crystals and bubbles strongly affects the viscosity and flow properties of the mixture. If the magma contains more than about 55% by volume of solids it may not be able to flow at all: The solid crystals interlock with one another and the flow of the bulk magma is controlled by solid-state creep of the crystal framework rather than the liquid matrix. For crystal contents ϕ up to about 30% by volume, the Einstein–Roscoe formula gives the average viscosity of the liquid mass η in terms of the viscosity η_0 of the liquid alone (McBirney and Murase, 1984):

$$\eta = \frac{\eta_0}{\left(1 - \phi\right)^{2.5}}. \tag{5.3}$$

The viscosity of natural magmas varies with crystal content even more than this equation suggests, because the crystals that form in a cooling magma are typically less rich in silica than the magma itself. Thus, as crystallization proceeds the melt becomes progressively enriched in silica and, because of polymerization, the viscosity rises dramatically. Because of its importance in terrestrial volcanology, the rheology of silicate melts has received a great deal of attention. Unfortunately, the same cannot yet be said of the water-rich melts present in the icy satellites of the outer Solar System, for which only a few rheological measurements exist.

Bingham rheology. In addition to increasing the viscosity of the melt, the presence of a dense mass of crystals alters the flow properties of magma still more profoundly. In 1919 the chemist E. C. Bingham (1878–1945) discovered the peculiar behavior of a dense emulsion of solids while trying to measure the viscosity of paint (Bingham and Green, 1919). Paint, like magma, is a dispersion of small solid particles in a liquid, although in the case of paint it is usually compounded from a ceramic powder like titanium oxide mixed into organic oil. Bingham was trying to measure its viscosity by forcing paint under pressure through narrow capillary tubes and using a well-established equation for the flow of a viscous fluid through a tube to determine its viscosity. Much to his surprise, he discovered that the paint would not flow at all until the pressure reached some finite threshold, after which the rate of flow depended linearly on the pressure, as he had expected. Bingham quickly realized that he had made an important discovery and could, for the first time, explain why wet paint does not immediately flow off a vertical wall: Paint has a finite yield stress, now called the "Bingham yield stress," that must be exceeded before it can flow. The thickness of a layer of fresh paint is proportional to this yield strength, and does not depend on the viscosity. Previous to Bingham's work, viscosity alone was used to determine the quality of paint until the American Society for Testing Materials compared 240 samples of paint at its Arlington, VA, laboratory. In this test many samples, prepared to have the same viscosity but, unbeknownst to the testing staff, having different yield stresses, ran off the boards of a fence and left gaping, unsightly, bare spots. Following his success in explaining this fiasco, Bingham went on to a distinguished career during which he coined the term "rheology," introduced the "poise" as a unit of viscosity, and founded the Society of Rheology.

Although it might seem that the rheology of paint has little in common with volcanic phenomena, it has been abundantly shown that lava is also a Bingham material and that the Bingham yield stress is a crucial parameter for computing the length and thickness of lava flows and domes. Indeed, almost any dense mixture of solid and liquid is likely to behave as a Bingham material: Even kitchen staples such as mashed potatoes (surely you have noticed that mashed potatoes can only be piled so high, after which the pile collapses – their Bingham yield stress has been exceeded!), apple sauce, and pudding are properly described as Bingham materials, as are basaltic magma, mudflows, and rock glaciers. Although the Bingham yield stress is thus a central parameter in many applications, it unfortunately cannot be computed from first principles for nearly any mixture. There are literally hundreds of empirical equations relating the Bingham yield stress to solid volume fraction, particle shape, size and liquid composition, but the ability of these formulas to predict the yield stress of previously unmeasured materials is practically nil. The reason for this failure is that the Bingham stress depends on the surface energy of contact between the solids in the mixture. This surface energy depends on so many presently unknown factors that prediction is nearly impossible: All current work is empirical.

The relation between shear strain rate and shear stress for a Bingham material is given by:

$$\begin{aligned} \dot{\varepsilon} &= 0 & \text{for} \quad \sigma < Y_B \\ \dot{\varepsilon} &= (\sigma - Y_B)/\eta_B & \text{for} \quad \sigma \geq Y_B \end{aligned} \tag{5.4}$$

where Y_B is the Bingham yield stress and η_B is the Bingham viscosity. This equation will be used below to describe the behavior of lava flows on a planetary surface. First, however, we need to understand how magma gets up to the surface of a planet in the first place.

5.1.4 Segregation and ascent of magma

When solid material from deep within a planetary body begins to melt, small pockets of melt first form at high-energy locations such as grain boundary intersections and where different crystals can react to produce eutectic liquids. At first, these tiny melt pockets have no tendency to join together and remain trapped in the rock. At this stage an often-overlooked phenomenon controls the fate of these small particles of melt. If the surface contact energies of the melt and crystals surrounding them permits the melt to wet the crystal faces and run along the grain boundaries, melt will begin to accumulate into larger volumes. On the other hand, if the contact angle between the melt and the solid crystals is greater than about 60°, the melt beads up and much larger volumes of melt must form before the melt can separate from its parental rock (Watson, 1982). In both silicate rocks and water–ammonia mixtures the contact angle is small and melt readily percolates out of the matrix. However, some combinations of materials, such as liquid iron and silicate, have larger contact angles and percolation is strongly inhibited.

Magma percolation flow. In silicate rocks, when the melt fraction exceeds a few percent, the melt begins to percolate along grain boundaries and flows out of the source rock. The process of "Rayleigh distillation" then controls the chemical evolution of the melt, in which the continuously extracted melt carries away the elements that enter the liquid and, thus, the composition of the source material gradually changes. The contrasting process of "batch melting" occurs when the melt remains in chemical communication with the parent rock.

If the matrix through which the melt percolates can be treated as a rigid structure, the flow of the melt is described by the Darcy equation (which was originally devised to describe the percolation of water through porous rock: See Section 10.2.2). In one dimension, this equation relates the volume discharge of fluid (magma in this case) per unit area Q to the gradient of pressure driving the flow and the viscosity η of the liquid,

$$Q = -\frac{k}{\eta}\frac{dP}{dz}.$$
(5.5)

The permeability k has dimensions of $(\text{length})^2$ and depends upon the size and spacing of the pores through which the magma percolates (see Turcotte and Schubert, 2002 for more on permeability and how to calculate it). The most uncertain part of this equation is the permeability, but this equation has, nevertheless, often been used to estimate the length of time necessary to, say, differentiate a basaltic crust on a heated asteroid or to produce enough magma to feed an observed surface flow. In this case the vertical pressure gradient is equal to the difference in density between the liquid and solid matrix times the gravitational acceleration, $dP/dz \simeq \Delta\rho\, g$. Taking the permeability very roughly to equal the square of the grain size d, the timescale for magma to flow out of a layer of thickness h is given by:

$$t_{\text{percolate}} = \frac{\phi\,\eta\,h}{d^2\Delta\rho\,g}.$$
(5.6)

For an asteroid like Vesta, the magma percolation time is only about 60 yr for a 125 km thick mantle (half of Vesta's radius), assuming a density difference of 300 kg/m^3, a grain size of 1 mm, total melt fraction of 10%, and a basaltic magma viscosity of 10 Pa–s. This timescale is very short compared to the thermal heating timescale, and indicates that the rate-determining step in Vesta crust formation is not melt percolation but the rate at which its mantle heats up. This seems to be the case in many circumstances: In general, melt percolation is so fast that melt leaves its parental rock as fast as it is formed.

Diapirs vs. dikes. Studies of depleted source rocks on the Earth suggest that the simple rigid percolation model is quite inadequate. It appears that melt in hot rocks, especially if they are deforming, quickly collects into pockets and veins that are much larger than the grain size. These melt rivulets join to form larger veins that drain the mass of source rock more efficiently than uniform percolation. As the magma accumulates in ever larger bodies, the difference in density between the melt and matrix becomes more important and buoyant bodies of melt may begin to slowly rise through the high-viscosity source rock. Many book illustrations depict magma, especially highly viscous silica-rich magma, in this stage as rising in mushroom-shaped diapirs, similar to those depicted in Figure 4.9. However, petrologists are currently in doubt about the validity of this picture.

A low-density fluid, such as magma, enclosed in a higher density but deformable matrix, has two means of ascending through the matrix. One is a diapir, discussed above. The other is a dike, a vertical fluid-filled crack that pierces directly through the matrix and permits much more rapid ascent of the fluid. Whereas diapirs exploit the viscous property of a fluid, developing when a low-density fluid displaces an overlying higher density viscous fluid, dikes exploit the elastic property of the enclosing material. Hot rocks, however, exhibit both viscosity and elasticity, depending on the timescale: They are best described as Maxwell solids, as described in Section 3.4.3. Whether the rise of magma is governed by the viscous or elastic response of the surrounding rocks depends on timescale, and, thus, on the ratio between the viscosity of the magma and that of the host rocks. The precise conditions for the dominance of one process or the other are still somewhat uncertain: It is presently an area of active research (Rubin, 1993). However, near the surface it seems clear that most basaltic magmas ascend via dikes. This is also true for at least some granitic magmas (Petford, 1996), so the following discussion focuses on the mechanics of dike ascent. An older, and now-discredited model of volcanic eruption is discussed in Box 5.2. Whereas this "standpipe" model has some apparent successes, in the light of the discussion below it cannot possibly be correct, but it is still of interest because the elastic dike model cannot, as yet, reproduce the major success of the old, impossible model!

Dikes differ from simple cracks for two major reasons: They are filled with a viscous liquid and, because of gravity, the pressure that the fluid exerts on the walls of the vertical crack differs greatly between its top and bottom. Much of our present understanding of dikes rests on the summer research of material scientist Johannes Weertman. Weertman, who has made fundamental contributions to the study of dislocations and creep in solids, is also a highly regarded glaciologist. He may have become interested in dikes while watching a stream flowing over the surface of a glacier disappear into a crevasse and wondered

Box 5.2 **The standpipe model of magma ascent**

A striking observation made by pilots flying along the volcanic range of the Andes or the Aleutian volcanic chain is that the summits of all the volcanoes seem to be at nearly the same elevation (Ben-Avraham and Nur, 1980). This holds even though the bases of the volcanic constructs often differ greatly in elevation: Extruded lavas build their ultimate cone up to an apparently fixed elevation. A good planetary example of the same relation is the summit elevations of the four major volcanic centers on Mars. The summit calderas of the three major volcanoes on the Tharsis Rise – Arsia Mons, Pavonis Mons, and Ascraeus Mons – all rise to an elevation of almost 20 km above the Martian datum, each about 12 km above their bases near the summit of the Tharsis Rise (Table B5.2.1). Olympus Mons, however, rising from the lowlands of Amazonis Planitia far to the west of the Tharsis Rise, *also* rises to about 20 km, even though its base lies 1 km below the Martian datum.

The accordance of these summits was stunningly discovered near the beginning of the Mariner 9 orbital mission. Arriving at Mars during the height of a planetary dust storm in 1972, the first images showed only a uniform haze of red dust. However, as the dust began to settle, four dark spots – the tops of the major volcanoes – began to emerge almost simultaneously. Planetary geologist Hal Masursky, seeing these spots, immediately proposed that they were the summits of giant volcanoes, much to the astonishment of the planetary geologic community. Their simultaneous appearance out of the haze was an indication of the similarity of their summit elevations.

Even though the concordances of height of most volcanoes are approximate, their consistency does not seem accidental. This concordance has been "explained" for many years by a model that might be called the "standpipe" model, because it posits that the magma rises up a rigid open channel (pipe) and settles at the level of hydrostatic equilibrium.

Illustrated in Figure B5.2.1, the model supposes that the magma rises from its source depth through a rigid lithosphere of density ρ_l. The pressure in the magma, of density ρ_m, is equal to the pressure of the lithosphere in the magma source region at the base of the lithosphere. Because the magma is less dense than the lithosphere, the column of magma must be taller than the thickness of the lithosphere, of thickness t, by an amount h, equal to the height of the volcano's summit above the planetary datum, independent of the actual height of the volcano's base. Balancing pressures,

$$(t + h)\, \rho_m g = t\, \rho_l\, g. \tag{B5.2.1}$$

Table B5.2.1 *Elevations of the major Martian volcanoes*

Volcano	Edifice height (km)	Base elevation (km)	Summit elevation (km)
Olympus Mons	22	−1	21.229
Ascraeus Mons	15	3	18.225
Pavonis Mons	10	4	14.058
Arsia Mons	12	6	17.761

Data from USGS (2003).

Box 5.2 **(cont.)**

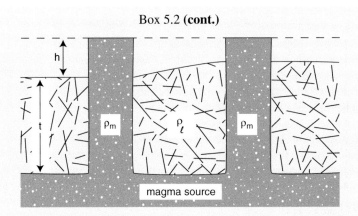

Figure B5.2.1 The standpipe model of volcanic summit heights. According to this model, the summit heights of volcanoes are all the same, independent of the elevation of the top of the crust (here represented as supported by Pratt isostasy, although Airy isostatic support gives the same result) because the magma originates at a common depth and connects to the surface through a rigid standpipe.

The gravitational acceleration at the planet's surface cancels out of this equation and we obtain a linear relation between the summit height and lithosphere thickness:

$$h = t \left(\frac{\rho_l - \rho_m}{\rho_m} \right). \tag{B5.2.2}$$

This model neatly explains the concordant summits of chains of volcanoes, supposing only that the magma originates at the same depth. It has been widely used to estimate lithosphere thickness or source depth and gives very reasonable results. On the other hand, it cannot possibly be correct, because the lithosphere thickness derived, typically of the order of 100 km, is so large that the assumption of a rigid open channel makes no sense: The difference in pressure between the magma and the walls along the channel is so large that the walls would deform elastically, expanding near the top and closing off the channel near the bottom, as described in the text.

The standpipe model is a beautiful example of a simple, clear model that neatly explains the data, but cannot possibly be right. Unfortunately, the more detailed, "correct" model of dike intrusion cannot easily explain the observational fact of accordant summits or the heights themselves. Obviously, this is a problem that needs more work.

whether the stream would ever be able to fill the crevasse. These ruminations led to a fundamental paper (Weertman, 1971) that elucidated the role of gravity and elasticity of the wall rock in determining the shape of a vertical dike (from a mechanical point of view, a dike is just a water-filled crevasse turned upside-down). One of his most important discoveries is that dikes cannot be arbitrarily deep: They are strictly limited in their vertical extent by the

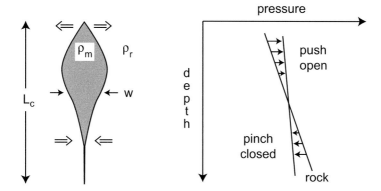

Figure 5.7 Pressure as a function of depth in a vertical dike. Because the magma in the dike is less dense than the surrounding rock, the pressure in the dike falls less rapidly (here shown in highly exaggerated form as a vertical line in the right half of the figure). If the average pressure in the dike and the surrounding rock are equal, the pressure in the dike exceeds that of the enclosing rock at the top of the dike, whereas the pressure in the rock exceeds that in the dike at its bottom. The magma thus pushes the crack open at its head, while it is squeezed closed at its tail, forcing its way upward through the rock. The actual length of the crack L_c is a function of the elastic properties of the surrounding rock as well as the pressure drop in the crack and is computed in the text.

elastic deformation of the surrounding matrix. This fact implies that the quantity of magma that can rise in a dike is quantized into a fixed volume.

Dike mechanics. The full analysis of the length and ascent velocity of a fluid-filled crack is complex (Rubin, 1995), but some simple order-of-magnitude estimates can illustrate the main outlines of the theory. As magma rises into a vertical, slowly moving crack, the low density of the magma compared to the wall rock means that the pressure in the crack falls less rapidly than in the surrounding denser rock (Figure 5.7). If the magma and rock are at the same pressure in the source region, the pressure at the head of the crack will, thus, exceed that in the adjacent rock and the top of the crack will tend to balloon outward. This extra stress at the crack tip, if large enough, may rip apart the rock ahead of the crack and permit the mass of magma to ascend. However, as the tip of the crack balloons, the tail of the crack grows narrower because of the elastic reaction of the surrounding medium (this is similar to the bulge that forms adjacent to a loaded area on an elastic half space – or the bulge next to a person who sits down on a springy sofa). As the crack lengthens the tail nearly pinches off, although the fluid in the crack prevents it from closing completely. At this stage the rising crack can accept no more magma and it continues to ascend toward the surface as an independent pod of hot magma.

A simple relation between the length L and width w of such a crack is obtained by equating the average stress generated by the crack in the elastic medium to the pressure drop between the head and the tail of the crack. The elastic stress is computed from the strain, $\varepsilon = w/L$, times Young's modulus E: $\sigma = Ew/L$, according to the definition (3.9). The pressure drop through the crack is just $(\rho_l - \rho_m)gL = \Delta\rho gL$. Equating these stresses, we obtain a relation between the critical crack length and width that is nearly identical to the relation derived from the detailed theory of crack-tip dynamics (Rubin, 1995):

$$L_c = \sqrt{\frac{E\,w}{\Delta\rho\,g}}. \tag{5.7}$$

For parameters appropriate for the Earth, this tells us that a 1 m wide vertical dike would have a height of about 2 km, in reasonably good accord with geologic observations. This result, however, does not give us an independent way of estimating the crack width and length. Weertman solved this difficulty by balancing the elastic stress of the crack against a regional extensional stress T, which he considered necessary to permit the crack to ascend. Although dikes do generally ascend perpendicular to regional extensional stress, experiments on the injection of dyed liquids into gelatin matrices suggest that extension is not an essential factor in dike ascent. What has been neglected so far is the flow of the fluid included in the dike. Magma-filled dikes ascend at rates limited by the viscosity of the fluid and the width of the dike. A magma-filled dike cannot ascend so fast that the pressure drop in the viscous magma exceeds the pressure gradient in a static crack (otherwise the pressure gradient would reverse and the magma would decelerate), so the pressure drop due to the fluid flow provides another equation to determine w in terms of the rate at which magma flows in the dike. Using the definition of viscosity, Equation (3.12), it is easy to show that the mean velocity $\bar{v}$ of a viscous fluid flowing through a channel of width w is, up to factors of order 2, given by:

$$\bar{v} \approx \frac{w^2}{2\eta}\frac{dP}{dz} = \frac{w^2}{2\eta}\Delta\rho\,g. \tag{5.8}$$

The volume discharge of a planar dike Q_d per unit length is equal to $\bar{v}w$. Solving for w in terms of the discharge, and inserting it into Equation (5.7), one can show that the length of a dike carrying a fixed discharge of magma Q_d is given by:

$$L_c = \frac{E^{1/2}(2Q_d\eta)^{1/6}}{(\Delta\rho\,g)^{2/3}}. \tag{5.9}$$

The most notable feature of this equation is its dependence on $1/g$. This has the surprising and important implication that the volume of a magma "quantum" ascending from a magma source, which is roughly equal to $L_c^2 w$, actually *increases* as the gravitational acceleration decreases. This makes a good deal of sense: In order to break through to the surface of a small body, the magma's buoyancy force must overcome the resistance of the elastic medium through which it ascends. On a low-gravity body this means that the volume of magma must increase. On Earth these magma quanta are relatively small, a few 100 m³, and may account for the almost regular pulsing activity seen in erupting volcanoes: Each pulse represents the arrival of a new package of magma traveling up a dike connecting the surface with the magma reservoir below.

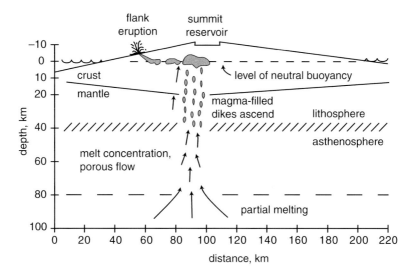

Figure 5.8 Ascent and eruption of magma beneath the Hawaiian volcanoes. Magma originates in the hot mantle below the volcanoes, concentrates into pods and pockets, and finally ascends to the surface in dikes. As it reaches the level of neutral buoyancy it stalls, collecting in magma chambers and moving laterally to emerge in flank eruptions. Summit eruptions occur when the driving pressure from new material becomes high enough to push magma from the level of neutral buoyancy up to the summit of the edifice. Figure simplified after Tilling and Dvorak (1993).

Eruption volume and gravity. This analysis shows that volcanic eruptions should become both more voluminous and more rare as the size of the body decreases. Lunar volcanic eruptions should be much larger and more catastrophic than terrestrial eruptions, a deduction that seems to agree well with the large volume of lunar lava flows. Going to still smaller bodies, many of the small icy satellites seem to have been volcanically resurfaced only once in their history, by an eruption so large that it covered nearly their entire surface.

Level of neutral buoyancy. The behavior of magma near the surface depends largely upon the density contrast between the magma and the surrounding rocks. Although the height to which magma can ascend depends upon the driving pressure at depth, this factor often seems to be eclipsed by the level at which magma achieves neutral buoyancy: That is, where the density of the magma equals that of the surrounding rock. First expressed by G. K. Gilbert in his famous monograph on the Geology of the Henry Mountains (Gilbert, 1880), this concept has found abundant support from detailed studies of the Hawaiian volcanoes (Ryan, 1987). Magma beneath Kilauea caldera rises until it encounters rocks of similar density, then spreads out laterally beneath the surface in the form of vertical dikes (Figure 5.8). The depth to the upper portion of these dikes fluctuates as the magmatic pressure fluctuates. When this pressure becomes especially high the top of the dike reaches the surface and magma pours out in a fissure eruption.

As magma-filled dikes approach the surface dissolved volatiles may come out of solution and form a pocket of gas that leads the liquid toward the surface, as discussed in more detail in the next section. When this occurs the average density of the fluid filling the dike decreases and it may be possible for the dike to rise higher than the buoyancy of the liquid magma itself might suggest. Interactions between the liquid, gas and surrounding solid rocks then become complex and their consequences have yet to be fully understood.

Intrusion vs. extrusion. Magma reaching its level of neutral buoyancy has no further tendency to ascend and may become "stalled" underground, creating magma chambers and reservoirs. It may also spread out in the form of either vertical dikes or horizontal sills, depending on whether the local minimum principal stress is horizontal (dikes) or vertical (sills). When sills extend to a critical size that is controlled by the elastic properties of the overlying rock, they may bodily lift the overlying rocks and create a turtle-shaped intrusive mass named a "laccolith" by Gilbert. Laccoliths on Earth may reach several kilometers in diameter and uplift the overlying rocks by up to 1 km.

An important statistic for volcanism on any planet is the ratio between the volume of magma extruded on the surface and that intruded below the surface. Estimates for the Earth suggest that far more is injected below ground than ever reaches the surface, perhaps by as much as a factor of 40. The oceanic crust typically consists of about 0.5 km of extrusive pillow basalts that overlie a total of about 5–10 km of vertical dikes and plutonic gabbro (the intrusive equivalent of basalt), giving a ratio of intrusion to extrusion of 10:1 to 20:1. In continental rifts, basaltic lava has more difficulty reaching the surface through the low-density continental crust and this ratio may be still larger: A large fraction of the magma of the well-studied Central Atlantic Magmatic Province that formed as Africa rifted away from North America appears to be intrusive.

5.2 Mechanics of eruption and volcanic constructs

5.2.1 Central versus fissure eruptions

A common observation is that volcanic materials on the surface of a planet may either pile up in a heap, recognized as a volcanic center or mountain, or they may spread over a broad, low-lying area. In the latter case the source of the volcanic flow is often obscure, but when it can be located it frequently turns out to be a long fissure from which lava poured over a relatively short interval to feed the surface flows. Because such feeder fissures tend to cover themselves, they can be difficult to find without detailed topographic maps supported by high-resolution images.

What determines the pattern of central vs. plains volcanism is unclear: Every planet appears to possess both volcanic mountains and extensive volcanic plains, although the proportion of central to fissure eruptions varies greatly from one planet to another, or even from one location to another on a single planet. On Earth, we recognize volcanic centers that create steep-sided volcanoes such as Japan's Mount Fuji, and low domes such as the islands of Hawaii. Earth also possesses broad volcanic plains such as the US's Columbia Plateau or India's Deccan Plateau. Venus similarly exhibits large central volcanoes, such as Sif Mons, that contrast with the broad volcanic plains that underlie most of the planet's surface. Even the Moon, which seems to lack steep-sided volcanic mountains in favor of broad mare basalt plains, possesses a volcanic center on the Aristarchus plateau. Recent images of Mercury from the MESSENGER spacecraft revealed a volcanic center southwest of the Caloris Basin, which contrasts with the otherwise planet-wide volcanic plains.

Mantle plumes and hot spots. Regional concentrations in the intensity of volcanic activity are often related to the activity of hot, buoyant plumes that rise through the planet's mantle (see Figure 4.9), carrying heat from depth as part of the normal convective heat engine that cools the planet's interior (Ernst and Buchan, 2003). As a new teardrop-shaped plume head nears the surface and its pressure drops, its temperature may cross the solidus. The melts that, thus, form quickly separate and rise further, either intruding the crustal rocks just below the surface or erupting onto the surface. Plumes eventually spread out in the mantle beneath the crust, generating melts at a lower rate than the initial spurt. Hot mantle material may continue to rise for a long time along the warm trail (the "plume tail") left in the wake of the original plume head. This extended flow of hot material becomes the source of a long-lived volcanic center. Earth's Hawaiian island chain is believed to originate as the Pacific plate drifts over such a long-lived source of magma rising from deep in the mantle. The Tharsis Rise on Mars may similarly be located over a long-lived plume in the Martian mantle. In the case of Mars, however, there are no moving tectonic plates and the large volume of Tharsis is attributed to the long-term accumulation of volcanic melts at one location. The evolution of Venusian coronae is likewise linked to the activity of plumes rising from its deeper mantle (Squyres *et al.*, 1992).

Although plumes can explain regional concentrations of volcanic activity, they do not determine whether the eruption will be either of the central or fissure type: On Earth, plume sources are invoked both for volcanic centers such as Hawaii or Yellowstone and for plateau basalts such as the Deccan or Siberian flows. Does magma composition play a role in determining whether activity is centralized or diffused into regional fissure systems? What about the persistence of the heat source – do long-lived sources of magma (plume tails) favor central activity, while short, hot pulses (plume heads) favor fissure-fed plains? These questions are presently unresolved.

The Earth is unique among the terrestrial planets in its possession of plate tectonics. Subduction-zone magmas are highly enriched in silica as well as volatiles from subducted oceanic plates bearing sediments and hydrated minerals. Subduction zones also tend to persist for geologic periods. The result is long, linear chains of central volcanoes such as the Andes or Aleutians. Hot, plume-generated, basaltic magmas may also intrude the base of silica-rich continental crust, melting the overlying rocks and creating local pockets of silica-rich melts that underlie volcanic centers such as Yellowstone,which are not related to subduction zones.

5.2.2 Physics of quiescent versus explosive eruptions

Volcanic eruptions can, in principle, be simple outpourings of liquid magma onto the surface of a planet. In practice, however, they usually involve a complex mixture of solid, liquid, and gas components whose behavior upon reaching the surface is anything but simple. One of the most consistent observations of terrestrial volcanic eruptions is that

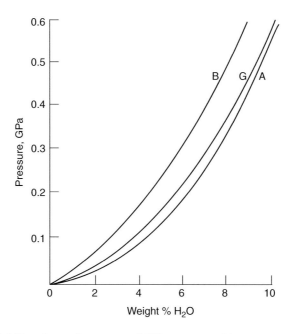

Figure 5.9 The solubility of water in magmas of different composition at a common temperature of 1373 K: B is basalt, G granite, and A is andesite magma. Decreasing the pressure greatly decreases the solubility for all compositions. Although all three materials can dissolve similar amounts of water at a given temperature and pressure, the eruption temperature of these three kinds of magma varies greatly on Earth. After Figure 2–13 in Williams and McBirney (1979).

they progress from gas-rich initial phases into the eruption of successively gas-depleted magmas, although the course of any one eruption may involve a complex alternation of gas-rich to gas-depleted pulses (Williams and McBirney, 1979). This progression is readily explained by the separation of gas from liquid magma as it rises toward the surface: The upper tip of rising dikes or the near-surface zones of masses of magma become enriched in gas that is vented as the dike or magma column breaches the surface. Gas-depleted liquid magma (often mixed with crystals) then follows.

The tendency for dissolved gases to separate from their parent magmas near the surface is a simple consequence of the pressure and temperature dependence of gas solubility. Gas solubility in a liquid generally increases with increasing pressure. This is especially true for water in silicate melts because of the strong affinity of water for silica. Measurements of the solubility of water in magma as a function of pressure, Figure 5.9, show that deep in planetary crusts silica-rich magmas may hold up to 10% water by weight, but at surface pressures this drops by almost two orders of magnitude. This tendency for depressurized liquids to exsolve gas is familiar to anyone who has quickly opened a sealed container of a carbonated drink: Upon opening, the pressure suddenly drops and bubbles of carbon dioxide gas appear throughout the liquid.

The consequences of gas exsolution near the surface depend strongly upon the viscosity of the magma and the rate at which the pressure drops. Fluid magmas, such as basalt, tend to erupt relatively quietly. Their viscosities are low (Figure 5.5) so that bubbles of exsolved water or carbon dioxide readily escape the magma, collecting in large pockets of gas at the tip of rising dikes. The presence of such gas pockets in moving dikes contributes to the seismically observed "harmonic tremor" that often precedes Hawaiian eruptions. When such a dike breaches the surface, the first material to erupt is mostly gas, driving the dramatic (but localized) fire-fountain activity that ushers in the main flow of magma onto the surface. Fire-fountaining may be renewed during a prolonged eruption as new dikes arrive to discharge their own gas pockets, then add their magma to the overall flow. Basaltic eruptions on the sea floor may not possess a gas-rich phase because the pressure beneath 4 km of seawater (about 0.04 GPa) is too large for much gas to exsolve. Such deep-sea eruptions are much more quiescent than eruptions onto the Earth's surface. Venus' surface pressure may likewise be large enough to suppress intense exsolution of volatiles and, thus, preclude explosive eruptive activity (presuming, of course, that Venus' interior possesses Earth-like quantities of water or carbon dioxide).

Silica-rich magmas, such as andesites or rhyolites, have such high viscosities that volatiles have great difficulty separating from them. As pressures drop in rising magmas of this type, the volatiles form bubbles that remain trapped in the viscous melt. This fact, along with the strong pressure dependence of solubility, leads to a dangerous tendency for silica-rich volcanoes to catastrophically "explode." Note that, although the word "explosion" is commonly used to described catastrophic volcanic eruptions, these events are not actually explosions in the sense that a rapid conversion of solid to gas results in a sudden increase of pressure: In volcanic eruptions the pressure always *decreases*. This behavior strongly differentiates volcanic eruptions from impacts, in which large pressure increases do occur. The shocked minerals that characterize impacts have never been reliably associated with volcanic eruptions. The presence of such minerals, thus, serves to discriminate the two types of event.

Catastrophic, silica-rich eruptions can proceed from either fissures or central volcanoes, as demonstrated by deposits in Earth's geologic record. However, no fissure eruptions have been observed during recorded history, so the following discussion will focus mainly on the eruptions of central volcanoes.

Silica-rich magmas are seldom observed reaching the surface directly. Instead, they accumulate for some time beneath the surface, cooling by conduction and interaction with surrounding groundwater and thus partially crystallizing as they lose their initial heat. Fluid pressure in the magma increases as more of the melt solidifies into crystals because fluids are excluded from the regular crystal lattice. Eventually this fluid pressure is released, either as a gas-rich eruption or in response to some unrelated event, such as the landslide that preceded the 1980 eruption of Mount St. Helens and suddenly uncapped its magma chamber. A sudden pressure release, from whatever cause, initiates a rapid chain reaction that we recognize as an explosive eruption.

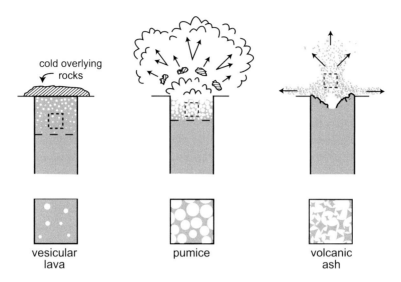

Figure 5.10 The upper row illustrates the sequence of events that occurs during an explosive volcanic eruption. At the beginning, pressure builds up below a strong cap sealing the volcanic vent. Bubbles exsolve from the magma as it cools, which create vesicular lava when it erupts on the surface (lower row). In the middle panel, the increasing pressure removes the cap overlying the vent and the sudden pressure release causes more gas to exsolve. The gases expand explosively, accelerating volcanic debris and magma to high speed. Frothy magma erupted at this stage is called pumice. In the right panel, so much material has been removed from the vent that it collapses, sealing in the magma below. The expanding gases disrupt frothy magma into small glass shards that mix with varying amounts of ambient air and produce volcanic "ash."

Any sudden pressure release causes vapor to exsolve from the liquid magma. In a highly viscous, silica-rich magma the bubbles formed by this vapor cannot easily escape from the magma. Near the surface, where pressure is low, the bubbles grow large and the volume of the magma suddenly increases by a large amount. This bubbly froth spills out of the magma chamber onto the surface, continuing to expand as the pressure drops. It expands upward as well as outward and may accelerate to velocities of hundreds of meters per second. The magma may be completely dispersed by this large expansion, forming an emulsion of gas and magma fragments that is commonly called "volcanic ash," even though no actual combustion takes place. The vapor cools rapidly as it expands, chilling the magma fragments, which often form tiny glass shards whose shapes are recognizably portions of the walls of former liquid bubbles. In more fluid magmas the bubble walls may have time to reform into spherical liquid droplets, as is observed for the silica-poor glassy products of lunar fire fountains.

The ultimate fate of such erupting gas/liquid emulsions depends strongly upon the ratio of gas to liquid (glass). Gas-poor magmas erupt as bubbly liquids in which the bubbles are widely separated from one another (see Figure 5.10, lower row). These magmas chill to form lavas containing small, often roughly spherical cavities known as vesicles. This material is thus called vesicular lava. Even lunar basalts contain vesicles whose gas phase is now believed to have been carbon monoxide, CO. Magmas containing more gas cool to form a rock whose bubbles are nearly in contact, giving it an average density that may be less than that of water. Lavas of this type are called pumice. Volcanic ash results when the bubbles coalesce and the emulsion's volume is dominated by gas.

Gas-rich emulsions may expand at high speed. The ultimate velocity of such an expanding mixture is determined by the thermodynamic properties of the gas, in particular by its molecular weight and initial temperature. The expanding emulsion often incorporates other material from the vent walls, clots of gas-depleted magma or even cold rocks. These accidental inclusions are accelerated with the gas and may be thrown substantial distances from the vent. Such fragments are called volcanic bombs and pose a major hazard to volcanologists trying to approach the site of an active eruption. The impact of large volcanic bombs sometimes forms craters a few to ten meters in diameter many kilometers from the vent. Although volcanic debris may be thus accelerated to high speed, it is unlikely to exceed escape velocity of even moon-sized bodies. Box 5.3 explains how the maximum ejection speed is determined. Note that the size of the volcano itself is not a factor in determining the ejection speed – only the eruption temperature and nature of the gas are important.

As a gas-rich emulsion of hot gas and melt fragments expands above the surface it may reach substantial heights before falling back. Its initial velocity at the surface, v_{ej}, gives the maximum height that this mixture can reach ballistically. Equating its kinetic and gravitational potential energies, this height is given by $v_{ej}^2 / 2g$, where g is the surface acceleration of gravity (Figure 5.11a). On an airless body the mixture of gas and fragments may rise to considerable heights – the Prometheus plume on Io rises about 75 km above its surface – then spreads laterally, driven by the entrained gas, into a wide umbrella that rains back onto the surface over a broad area (Figure 5.11b).

When an atmosphere is present, the erupting emulsion may incorporate some of the surrounding atmosphere. The average density of this mixture may become lower than that of the ambient atmosphere as the incorporated gas is heated. In this case the erupted material rises still further as a buoyant plume (Figure 5.11c). On the Earth, such buoyant eruption columns commonly rise into the stratosphere, tens of kilometers above the surface. The incorporated glassy particles then drift with the local winds and rain out later, at a rate depending on their size. Fine volcanic ash may, thus, spread globally over the Earth, although most of the coarser volcanic ash falls closer to the vent. Many terrestrial volcanic ash deposits can be recognized thousands of kilometers from their sources.

Even when an eruption plume does not become buoyant, it spreads rapidly away from the vent. The emulsion of gas and glassy particles is then denser than the surrounding atmosphere, but the mass is still fluidized by the gas phase, much as a dry snow avalanche is fluidized by air incorporated with the snow particles. It, thus, spreads as a density current, often overrunning topographic obstacles near the vent. Such hot, mobile density currents are known as pyroclastic flows. They can be devastating to human life and buildings

Box 5.3 A speed limit for volcanic ejecta

When magma rises to the surface and dissolved gases come out of solution, the hot, high-pressure gases expand and lower the pressure. As the gas expands it accelerates both itself and any entrained solid or liquid droplets to high velocity. However, thermodynamics imposes a strict limit on the maximum velocity to which this material can expand. This maximum is mainly a function of the initial temperature and composition of the gas phase.

The gas itself attains the highest velocity: Any burden of entrained material lowers the ultimate velocity of the mixture. We thus focus on computing the maximum expansion velocity of the gas alone, with the understanding that any admixture of solid or liquid material slows the final velocity in proportion to the square root of its additional mass loading. We assume that, for the time period under consideration, the flow is approximately steady, with a reservoir of hot, high-pressure gas at pressure P_1 expanding though some complex but energy-conserving process to a final pressure P_2.

Referring to Figure B5.3.1, consider two successive times during the steady expansion, t_A and t_B. At time t_A there is a mass m of gas on the left with a total energy equal to the sum of its specific internal energy e_1 and kinetic energy $1/2\ u_1^2$ per unit mass times m:

$$m\left(e_1 + \frac{1}{2}u_1^2\right) + (\text{energy of hatched region})\,. \text{ At time } t_B \text{ the same mass of gas has emerged}$$

on the right (this is the steady-flow approximation) and the total energy is now given by:

$$m\left(e_2 + \frac{1}{2}u_2^2\right) + (\text{energy of hatched region})\,. \text{ Furthermore, in steady flow the energy in the}$$

hatched region has not changed. Between the times t_A and t_B, the work done by the pressure P_1 on the mass on the left is $P_1 \Delta V_1$, where the volume change $\Delta V_1 = m/\rho_1$, where ρ_1 is the density

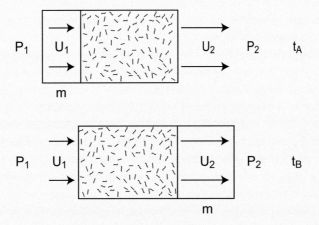

Figure B5.3.1 Schematic illustration of volcanic gases expanding steadily through a complex vent, represented by a hatched block. So long as the expansion does not add or subtract energy, the details of the expansion do not matter. The high-pressure gas on the left at time t_A expands so that its increase in kinetic energy per unit mass at t_B is equal to its decrease in enthalpy per unit mass, as described in the text.

Box 5.3 (cont.)

of the fluid on the left. Similarly, the work done by P_2 on the right is $P_2 \Delta V_2 = m P_2 / \rho_2$. Equating the sum of the energy and the work done at t_A and t_B then gives:

$$m\left(e_1 + \frac{1}{2}u_1^2\right) + m\frac{P_1}{\rho_1} = m\left(e_2 + \frac{1}{2}u_2^2\right) + m\frac{P_2}{\rho_2} \tag{B5.3.1}$$

where the energy of the hatched region cancels out, because it does not change from t_A to t_B: This is the crucial assumption that no energy is added or lost during the expansion. We can cancel the common factor m and note that the definition of the specific enthalpy h is:

$$h \equiv e + \frac{P}{\rho} \tag{B5.3.2}$$

so Equation (B5.3.1) becomes:

$$h_1 + \frac{1}{2}u_1^2 = h_2 + \frac{1}{2}u_2^2 \tag{B5.3.3}$$

or, collecting like terms:

$$u_2^2 - u_1^2 = 2(h_1 - h_2). \tag{B5.3.4}$$

If the initial velocity u_1 is either zero or much smaller than u_2, as is usually the case in a volcanic eruption, we can neglect it and set u_2 equal to the maximum expansion velocity u_{max}, which from Equation (B5.3.4) we find:

$$u_{max} = \sqrt{2(h_1 - h_2)}. \tag{B5.3.5}$$

For a perfect gas at temperature T the specific enthalpy is:

$$h = c_p T \tag{B5.3.6}$$

where c_P is the specific heat at constant pressure (SI units are J/kg-K) and T is the temperature in K. Note that, although the molar heat capacity for many substances is similar, the heat capacity per unit mass depends strongly on the molecular weight of the material, so that low-molecular-weight gases typically have a much higher specific heat than high-molecular-weight gases and consequently expand faster. Hydrogen-powered volcanoes thus eject material much faster than volcanoes erupting water vapor or CO_2.

If h_2 is much less than h_1 because of the strong cooling during adiabatic expansion, then we can write simply:

$$u_{max} \simeq \sqrt{2 c_P T_{magma}} \tag{B5.3.7}$$

where T_{magma} is the pre-eruption temperature of the magmatic gases. This equation provides a convenient and easily evaluated estimate of the maximum expansion velocity possible in volcanic eruptions. It gives a good estimate of the maximum observed velocity of volcanic bombs from terrestrial volcanoes. Some typical results are given in Table B5.3.1.

Box 5.3 **(cont.)**

Table B5.3.1 *Maximum expansion velocity in volcanic eruptions, Equation (B5.3.7)*

Working gas	Heat capacity, c_P (kJ/kg-K)	Magmatic temperature (K)	u_{max} (km/s)
H_2O	2.0	1473	2.4
CO	1.02	1473	1.7
CO_2	0.84	1473	1.6
H_2	14.3	1473	6.5

It is possible to incorporate modifications of this simple formula for the presence of solids or liquids loading the gas, but there are no simple formulas because the results are sensitive to the details of how heat is exchanged between the gases and solid. As a start, one notes that the solids or liquids make only a negligible contribution to the pressure. Defining $s = m_{solids}/m_{gas}$, we can write for the density $\rho = (1+s)\rho_{gas}$, so using the perfect gas law the enthalpy of the gas itself is given by:

$$h_{gas} = \left(c_P - \frac{s}{1+s} R \right) T \qquad (B5.3.8)$$

where R is the gas constant. Values of s greater than zero clearly lower the enthalpy of the gas, but this neglects heat transfer between the entrained solids and liquids and gas during expansion. Further treatment requires modeling beyond the level of this book.

tens to hundreds of kilometers from a volcanic vent. The phenomenology of such flows is both complex and of great interest from many points of view. For more information the reader is referred to the recent monograph of Branney and Kokelaar (2002).

The ultimate deposits of explosive volcanic eruptions range from welded tuffs, which are deposited from density currents that are so hot that the glass particles weld together once they come to rest, to airfall tuffs in which the chilled, glassy magma particles fall relatively gently to the surface through either air or water. Less is known of deposits on airless bodies such as the moon. There, the chilled droplets of magma must be emplaced ballistically, with less ability to flow far from their original vent. An important characteristic of all such deposits is that they tend to blanket pre-existing terrain and, unlike water-laid sediments, may be deposited with initial slopes that follow the terrain rather than lying in initially horizontal beds.

Back at the vent from which the emulsion of gas and liquid magma erupted, changes occur as more and more material is ejected. The pressure on the deeper-lying magma is relieved as material from the top of the magma chamber is erupted (Figure 5.10). Volatiles dissolved in this deeper-seated magma then exsolve, creating more bubbles and increasing

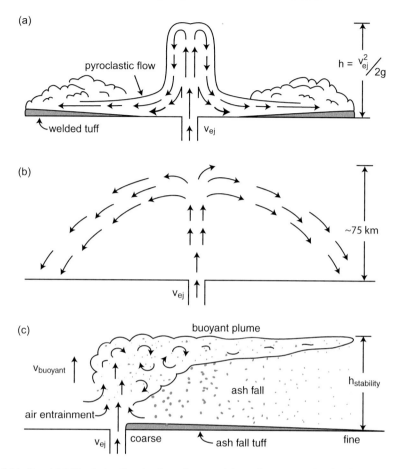

Figure 5.11 Panel (a) illustrates the ejection of an emulsion of magma and volcanic gas from a vent at high speed. The plume rises ballistically to a height h determined by the conversion of its kinetic energy to gravitational energy and then falls back. As it descends, it forms a hot density current known as a pyroclastic flow. When this material finally vents its gas and settles into a deposit on the surface it creates an ignimbrite or welded tuff. In Panel (b) the eruption occurs on an airless body. In this case the plume rises so high above the surface that it cools (by thermal radiation) before it descends to the surface. It spreads out as it rises, forming an umbrella-shaped plume such as those on Io. Panel (c) illustrates an eruption on a body with a dense atmosphere, such as the Earth or Venus. In this case the plume incorporates atmospheric gases and becomes buoyant, rising until its density equals that of the surrounding atmosphere at $h_{stability}$, after which it drifts laterally, responding to local winds. As it drifts, cooled volcanic ash falls out of the cloud and is deposited as loose volcanic tuff on the surface. The largest particles fall out first, the finer material later, producing a deposit in which the particle size grades laterally from coarse near the vent to fine farther away.

the magma's volume. This material then follows the first-erupted material out of the vent, further expanding and accelerating as it reaches surface pressures. This initiates a chain reaction in which progressively deeper material expands and flows out onto the surface at progressively higher velocities. This chain reaction ceases only when rock surrounding the magma chamber loses its lateral support and collapses into the void left by the magma that has now erupted onto the surface. Overlying rocks, if not initially removed by the violence of the eruption, also subside to fill the void. The final result is a depression, a volcanic caldera, surrounded and perhaps partially buried by the magma that has so recently left its original location. The size of the depression reflects the size of the original magma chamber and its included volume of unstable magma. The vents producing small eruptions, such as that of the 1980 eruption of Mount St. Helens, may not totally collapse because the surrounding rock is strong enough to support the evacuated cavity. Much larger magma bodies, however, leave caldera depressions that reach dimensions of hundreds of kilometers. Such collapse calderas are observed on most volcanically active bodies, including Venus, Mars, and Io as well as the Earth.

5.2.3 Volcanic surface features

Although nearly every volcanic eruption involves gas and liquid phases, the relative contributions of these two major phases vary widely from one eruption to another. At one extreme, some eruptions are driven almost entirely by gas, in which little or no magma may appear at the surface. At the other extreme, enormous volumes of liquid magma pour rapidly onto the surface with little or no accompanying gas. The morphology of the final deposits depends strongly upon the relative volumes of gas and liquid.

Maar craters. The structures most likely to be confused with impact craters are produced by gas-dominated eruptions. Eruptive vents are often nearly circular, are surrounded by a raised rim of ejecta, and are depressed below the pre-eruption surface, similar to impact craters. Such vents are called maar craters or tuff rings. On Earth, they frequently contain lakes (hence the name maar, which derives from the German word for a small lake). They form as volcanically heated gases breach the surface and eject a gas-fluidized mass of rock debris that frequently contains no magma. The gas may either be deep-seated, rising from depths of a few hundred kilometers in the case of diamond-bearing kimberlite pipes, or near-surface water that is explosively vaporized by contact with hot, rising magma. In either case the gas/debris expulsion velocity may be very high, several hundred meters per second, and the vent is often surrounded by a bedded deposit containing debris from great depth displaying bedforms that indicate high-velocity radial outflows of gravity currents initiated by the eruption. The volume of this ejecta (the tuff ring surrounding the vent) is often much smaller than the volume of the crater itself, indicating withdrawal or subsidence of the magma following the eruptive phase. Craters of this type on Earth include many examples in the Franconian Lake District, Germany, MacDougal and Sykes craters, among others, in the Pinacate Mountains of Sonora, Mexico, and Ubehebe crater in California's Death Valley. The latter crater is often described as a phreatic explosion crater because the source of the water in this case appears to be near-surface groundwater. Maar craters are typically only a few kilometers in diameter.

It is difficult to recognize maar-type volcanic craters on other planets, largely because of their close resemblance to impact craters. In this case one often must appeal to the departure of volcanic vents from perfect circularity, their tendency to form linear clusters over subsurface dikes, and their small-volume rim deposits compared to impact craters. However, in some cases there is no ambiguity because gas-rich eruptions have been caught in the act. The prominent SO_2-rich plumes discovered on Io, the water geysers on Enceladus, and nitrogen geysers on Triton are all spectacular examples. Cometary jets and outbursts may be further examples of gas-fluidized eruptions on the surfaces of small extraterrestrial bodies.

Cinder cones. Cinder cones are steep-sided conical mounds of pumice and volcanic bombs that build up near the sources of more magma-rich volcanic flows. Driven by gas bubbling out of silica-poor magmas, these landforms often dot the traces of dikes feeding large flows, or stand in isolation over the sources of smaller flows. They may be built, destroyed, and rebuilt many times during any given eruption on Earth. They form because of the relatively small range to which pumice and volcanic bombs can be ejected from the vent and so pile up until landslides transport loose debris farther away from the vent. They are typically only a few kilometers across and 1 km high, with sides standing near the angle of repose for loose rock debris, around 30° from the horizontal. Cinder cones have been identified on Venus and Mars as well as Earth. On the Moon and other low-gravity bodies, the range of volcanic debris may be so large that cinder cones are much more spread out and thus do not achieve steep sides (McGetchin and Head, 1973). In this case cinder cones grade insensibly into pyroclastic deposits and the name cinder cone might not be appropriate.

Shield volcanoes. Silica-poor magmas, when they persistently erupt from a single center, produce low, broad volcanic mountains known as shield volcanoes. This name comes from the profile of an ancient Greek soldier's shield, placed concave-side down on a flat surface. The gases dissolved in silica-poor magmas readily escape near the vent and eruptions are characterized mainly by outpouring lava accompanied by only minor fire-fountaining. These eruptions are not explosive and mainly feed low-viscosity flows that carry liquid lava away from the vent. Shield volcanoes may spread hundreds of kilometers horizontally while achieving elevations of a few to a few tens of kilometers. The largest known example in the Solar System is Olympus Mons on Mars, with a basal diameter of 550 km and an elevation of 21 km above its base. Its summit is crowned by a complex caldera depression about 80 km across and 3 km deep. The slopes of shield volcanoes are low, much less than the angle of repose, and they are built up from many thousands of individual lava flows that either proceed from a summit caldera or are erupted from dikes breaching the flanks of the mound. The shield often contains a central magma chamber where rising magma accumulates before eruption. Eruption events that partially drain the magma chamber produce collapse calderas near the summit that alternately fill and reform during the life of the volcano. Shield volcanoes have been identified on Venus

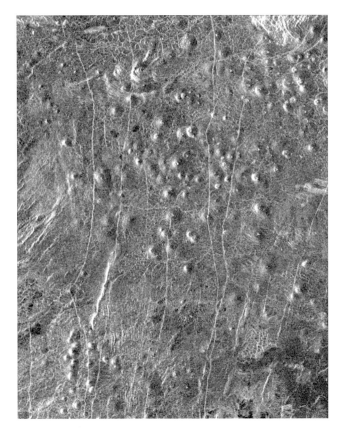

Figure 5.12 An extensive field of shield volcanoes on Venus. This field contains approximately 200 small volcanoes ranging from 2 to 12 km in diameter, many of which possess summit calderas. These are identified as shield volcanoes, although some cinder cones may occur among them. NASA Magellan image, left portion of PIA00465. Located at 110°E and 64° N. North is to the top.

(Figure 5.12), Earth (the Hawaiian volcanoes are the iconic examples), Mars, and perhaps on Mercury. Shield-like constructs also form the Moon's Aristarchus plateau and Io's Prometheus patera.

Long, fissure eruptions produce broad, nearly level volcanic plains that will be discussed in more detail in the next section. Near the fissure, exsolution of small quantities of gas may build cinder cones or, if less gas is present, small, typically elongated spatter ramparts around the dike where magma wells out of the ground.

Composite cones. Silica-rich magmas are highly viscous and volcanic material tends to build up around their eruptive centers because viscous flows are typically thick and short. As described above, their eruptions tend to be much more violent than silica-poor magmas because this type of magma is likely to be internally ruptured by bubble formation. Volcanic ash is created abundantly by such eruptions and pyroclastic flows spread it far and wide, blanketing pre-existing terrain with smooth-surfaced deposits that may reach hundreds of meters in thickness. This type of magma is more likely than any other to form recognizable volcanic mountains. These constructs are often very steep-sided and are frequent sites of catastrophic rock avalanches that limit the size and extent of silica-rich volcanic mountains. Called "composite cones," these piles of volcanic materials consist of alternating lava flows and volcanic ash deposits. All known examples occur on Earth and are the direct product of plate tectonics, their parent magmas being created by flux melting of silica-rich rocks in subduction zones, although it has recently been suggested that Ascraeus Mons on Mars might be a composite cone.

Laccoliths. In addition to lavas that reach the surface, a few volcanic features are created by lava that is intruded beneath the surface. Horizontal sills may extend tens of kilometers from their source, but are difficult to recognize because they uplift the surface only a few meters to perhaps 100 m over a broad area. However, if the overburden above a sill is thin enough, it may become elastically unstable, flexing upward into a broad dome known as a laccolith. First recognized and named by G. K. Gilbert in the Henry Mountains of Utah (Gilbert, 1880), laccoliths have been found in many locations on Earth and the mechanics of their formation has been analyzed in detail (Johnson, 1970). Updoming results in characteristic radial fractures that may themselves be the source of surface lava flows. Laccoliths are typically a few kilometers to tens of kilometersin diameter. It has been suggested that several features on Mars are laccoliths, but definitive proof of a laccolithic structure is difficult on the basis of remote sensing alone.

Calderas. Calderas are one of the few negative-relief volcanic surface features. When large volumes of magma ascend over the same route from a persistent source, the liquid magma may accumulate beneath the surface at the level of neutral buoyancy and grow into a compact, long-lived, hot mass by displacing or partially assimilating the pre-existing cold rock. Called a magma chamber, slow cooling of the magma results in gradual crystallization and compositional changes. Rocks formed from large magma masses that slowly cool below the surface are called plutonic rocks and are characterized by large crystal sizes, typically millimeters to centimeters.

Eruptions may rapidly deplete the volume of liquid magma in a magma chamber. When this occurs the sudden loss of volume undermines the overlying rocks, which then collapse into the space formerly occupied by the magma, forming a depression. Such volcanic depressions are called calderas and their size reflects the diameter of the underlying magma chamber. Calderas on the terrestrial planets range from a few kilometers up to 100 km in diameter and range from a few 100 m in depth to many kilometers. They form at the sites of both silica-poor and silica-rich volcanic eruptions.

Not all volcanic centers display calderas: If the rate of magmatic replenishment to the chamber can keep up with the eruption rate, wholesale foundering of the surface into the magma chamber does not occur and a caldera never forms. Nevertheless, erupted magma piles up on the surface and a large volcanic edifice may be constructed from the products of many small eruptive events.

5.3 Lava flows, domes, and plateaus

Lava flows form when predominantly liquid magma reaches the surface and flows away from the vent or fissure, eventually to cool into a solid surface deposit. Magmas are not simple liquids: They almost universally contain bubbles formed by exsolved volatiles and crystals created during cooling before eruption. Rapidly erupted magmas may also be loaded with solid xenoliths, inclusions of rock from either a deep source region or the walls of the rock surrounding the magma's route of ascent. For this reason they exhibit complex rheologies, which are reflected by their deposits.

5.3.1 Lava flow morphology

Magma that erupts onto a planetary surface is called lava. When this hot, complex liquid flows out onto the surface of a planet it suddenly enters a much cooler environment. The response of the lava's surface to this rapid cooling creates a variety of textures that depend strongly on the magma composition and cooling rate. The interior of the flow cools more slowly and behaves very differently from the lava at its surface.

Highly viscous, silica-rich lavas form short, stubby flows that usually contain a large proportion of glass upon cooling. Differential thermal stresses fracture the surface of the hot magma into glassy fragments that bury and then insulate the interior of the flow. Slow, viscous flows usually exhibit a wavy, strongly textured surface on the scale of tens to hundreds of meters that is later cut by deep cooling cracks.

Pillow lava. More fluid lavas develop surface textures that are described as being of four general types. Submarine flows cool very quickly upon contact with water. The glassy surface is splintered into fine glassy fragments (called palagonite), while the mass of the magma collects into lava-filled sacks that often detach from the flow front and pile up in front of it, forming structures known as pillows. Individual pillows range from a few tens of centimeters to meters in diameter and their presence in a cooled lava flow is diagnostic of an underwater eruption. Subaerial flows cool more slowly and produce surface textures denoted pahoehoe, aa, or block types.

The peculiar surface textures of silica-based lavas derive in part from the tendency of silica tetrahedra to polymerize. At high temperatures (between about 800°C and 1070°C) highly polymerized silica glass behaves like rubber (a high-polymer substance based on carbon). Capable of sustaining large strains without breaking, the rubbery surface of cooling silicate lavas can inflate with liquid lava like a balloon or, under compression, collapse into folds resembling drapery on a scale of tens of centimeters. Incautious Hawaiian volcanologists once impressed onlookers by jumping up and down on fresh lava flows, which responded like a giant waterbed (this practice is now forbidden due to several unfortunate accidents).

Aa lava. Subaerial lava that is depleted of volatiles or moving rapidly develops a clinkery, fractal surface known by the Hawaiian name aa. Its surface disaggregates into porous, decimeter-sized fragments as the flow moves. Glassy spines pull out of the hot, separating fragments like the sugar-rich spines of pulled taffy. Aa surfaces are almost impossible to walk across and tend to shred shoes or boots. The fronts of aa flows are typically a few to tens of meters high and advance like a caterpillar tractor tread, with cooled chunks of aa falling from the top of the steep-fronted flow as hot lava in its interior overruns the previously fallen debris. A section of a cooled aa flow, thus, shows both a clinkery basal zone and a rubbly top, separated by a more massive zone of vesicular lava. Although the surfaces of aa flows are unforgettable for those who have experienced them, they are not voluminous on the Earth's surface.

Block lava. Block lavas, as their name suggests, are composed of meter-scale polyhedral blocks of lava. Formed from more viscous lava than either aa or pahoehoe, on Earth they form thick, short flows and seem to advance in a mode similar to aa. They are also not very voluminous on Earth.

Pahoehoe lava. Pahoehoe is, by far, the most abundant type of lava on Earth and probably on the other terrestrial planets. It forms both small flows and the giant sheets that blanket Earth's most extensive flood basalt provinces. Its name is derived from a Hawaiian word that indicates "smooth going." The surface of fresh pahoehoe is smooth, with a thin glassy rind that weathers rapidly on Earth. Its surface is marked by broad billows and swales and is locally puckered into drapery-like folds. Pahoehoe advances by a unique mechanism that has only recently been understood. Called "inflation," this mode of motion permits it to travel large distances with only modest eruption rates (Self *et al.*, 1998). Inflation was discovered during observations of active pahoehoe lava flows in Hawaii (Hon *et al.*, 1994). Flows first advance as a thin sheet that rapidly covers the terrain. The upper surface of the sheet cools rapidly, forming an insulating blanket under which hot lava continues to flow. On a timescale of hours, hot, fluid lava intrudes beneath the chilled surface, uplifting it and creating a nearly planar surface. Deep cracks form near the margins of inflated flows as the upper surface is lifted above the original base and slabs of chilled crust tilt away from the main mass of the flow. Uninflated areas form irregular depressions in the overall, nearly uniform, lava surface. Individual pahoehoe flows achieve thicknesses of tens of meters in flood basalt provinces and may continue to inflate and spread for more than a year. Over time, the lava feeding an inflated flow organizes into distinct streams, or "tubes," beneath the thickening crust. The flow velocity in an individual tube thus exceeds the rate of advance of the lava flow as a whole. When the supply of fresh lava from the vent declines, these tubes may drain and remain open as lava caves, or in the case of a thinner cover, the empty lava tubes may collapse to leave sinuous channels on the flow surface. The surface of large pahoehoe flows is locally marked by pits where inflation failed to occur, mounds called "tumuli" where lava locally broke through the surface when lava channels became blocked and small mounds accumulated, and "rootless cones" where the lava flowed over wet ground, creating small phreatic explosions that threw up blocky mounds (often called "hornitos") or larger rimmed craters up to a few hundred meters in diameter.

Columnar jointing. As lava flows cool and crystallize the lava shrinks and cracks open throughout the solidified flow. Because lava flows cool from outside inward, cracks initiate on their surfaces and propagate toward their interiors. Uniform contraction of a thin surface layer typically produces a polygonal fracture pattern, familiar from the surface of drying mud puddles. As cooling cracks propagate into the flow perpendicular to the cooling surface, they elongate, creating a pattern of polygonal columns called "columnar jointing." Columnar joints are often prominent at the eroded edges of lava flows, forming spectacular outcrops at Devils Postpile in California and the Giant's Causeway in Ireland. Columnar joints form in both silica-poor and silica-rich lava flows as well as pyroclastic deposits. Columns are typically tens of centimeters to meters in width. They have recently been observed on Mars, in lava flows outcropping on the rims of the great canyons. The formation of columnar joints is a universal process that occurs for any cooling, shrinking mass and can be analyzed in terms of fundamental principles (Goehring *et al.*, 2009).

5.3.2 The mechanics of lava flows

Spurred by the hazards posed by advancing lava flows, volcanologists have made many efforts to compute the length and width of lava flows from their basic properties. Planetary scientists, observing lava flows on other planets, have inverted these efforts to compute the properties of lava from the morphology of the final flow. Although early efforts treated lava as a Newtonian fluid, it is now agreed that lava behaves as a Bingham fluid and modern flow models are based on this rheology. The Bingham model provides a number of simple relations for the dimensions of a lava flow (Moore *et al.*, 1978). One of the goals of planetary volcanology is to relate estimates of the physical properties of the lava to its composition and, thus, learn about the chemistry of planetary crusts and mantles without having to directly sample them.

The thickness H of an extensive sheet of lava resting on a uniform slope standing at angle α to the horizontal is given in terms of the Bingham yield stress as simply:

$$H = \frac{Y_B}{\rho g \sin \alpha} \qquad (5.10)$$

where g is the acceleration of gravity and ρ is the density of the lava. This formula strictly applies only to an infinitely wide sheet of lava and is derived by setting the shear stress at the base of the lava flow equal to the Bingham yield stress (Figure 5.13a). If the thickness and other parameters of a lava flow can be estimated, this equation can be inverted to determine the Bingham yield stress. This model makes the uncomfortable prediction that a lava flow on a level surface ($\alpha = 0$) is infinitely thick. A more sophisticated model, still applying the Bingham rheology, examines the force equilibrium of a lava sheet of variable thickness resting on a level surface (Figure 5.13b). If $y(x)$ is the thickness of the lava flow and x is the distance to the edge of the flow, by balancing the horizontal thrust of the lava toward its edge at x, given by $\frac{1}{2} \rho g y(x)^2$, by the basal force at the yield stress, $x Y_B$, for a thin slice of the flow, an equation for the thickness of the lava flow at any distance from its edge is found:

$$y(x) = \sqrt{\frac{2 x Y_B}{\rho g}}. \qquad (5.11)$$

(a)

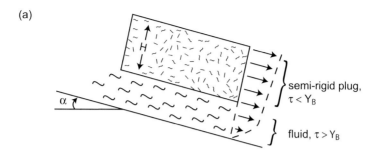

(b)

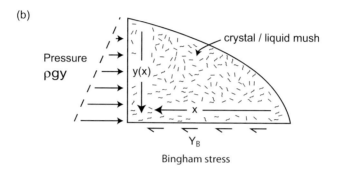

Figure 5.13 Panel (a) shows the shear stress acting on the base of an infinitely wide sheet of lava resting on a surface sloping at a uniform angle α. The shear stress acting on its base is $\rho g H \sin \alpha$. The Bingham rheology implies that, so long as the shear stress is less than the Bingham yield stress Y_B, the lava does not deform. At the higher shear stresses found beneath the uppermost semi-rigid plug, the lava flows like a viscous fluid. Panel (b) shows the stresses near the edge of a lava flow, where the horizontal thrust of the pressure in the lava flow to the left is balanced against the shear resistance of the Bingham material to the right. The profile of the lava flow margin is a parabola, as described in the text.

This is the equation of a parabola and, indeed, measurements of the shape of the edges of many lava flows give approximately parabolic profiles from which the Bingham yield stress can be computed. This equation is often applied to estimate the Bingham yield stress from the width of a lava flow. So long as the cross-profile of the flow is crudely parabolic (not flat-topped: If the flow has a flat top, this indicates that the lava flow thickness is controlled by the slope, through Equation (5.10)), then the width of the flow $W = 2x$ and its centerline thickness H determines the Bingham yield stress:

$$Y_B = \frac{\rho g H^2}{W}. \tag{5.12}$$

Bingham yield stress. Equation (5.11) makes the prediction that the thickness of a lava flow on a level surface increases as its width increases. This might be a sensible prediction, but planetary surfaces are seldom level over very long distances. A better model results from combining Equations (5.10) and (5.11) to model the edge of a wide lava flow on a

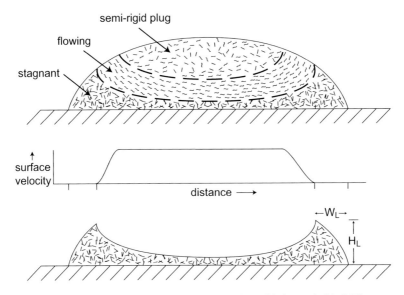

Figure 5.14 Cross section of a lava flow and the leveed channel it leaves behind. The upper section shows the profile of the lava flow at its most active. The chilled bottom is frozen to the bed, but the hotter lava above it flows as a viscous fluid so long as the shear stress exceeds the Bingham yield stress. In low-stress regions the material moves as a semi-rigid plug, similar to the infinite flow in Figure 5.13a. The surface velocity of the flow is shown in the middle panel. The lower section shows the levees that remain after the lava drains away. The width and height of the levees can be related to the Bingham yield stress.

gently sloping surface. In this case, the thickness of the center of a wide lava flow is given by Equation (5.10) while its edges are described by (5.11). The combined equations also give a nice description of the formation of the levees sometimes observed flanking lava flows. So long as lava is erupting at a high rate, the profile of the lava flow bulges up in the middle. However, as the eruption rate declines, the lava runs out of the channel, leaving the material in the cooler levees behind (see Figure 5.14). The levee width W_L and height H_L are easily measured on images of the flow so, applying Equation (5.11), the Bingham yield stress can be computed from these observables:

$$Y_B = \frac{\rho g H_L^2}{2W_L}. \tag{5.13}$$

An alternative formula can be derived by combining this equation with (5.10) to eliminate the flow thickness H_L:

$$Y_B = 2\,\rho g W_L \sin^2 \alpha. \tag{5.14}$$

The Bingham yield stress measured in this way suffers from the problem that the yield stress applies to the stagnant levees, not the hotter active portion of the flow, but it does give an order-of-magnitude estimate useful for comparisons between lava flows.

Lava viscosity. Once the Bingham yield stress is exceeded, a Bingham material flows as a Newtonian viscous fluid. The Newtonian rheology has also been widely used to estimate the viscosity η of actively flowing lava by measuring the surface velocity U_{surf} and estimating the flow depth H. In the case of laminar flow, the velocity profile is parabolic and the surface velocity is:

$$U_{surf} = \frac{\rho g H^2 \sin \alpha}{2\eta}. \tag{5.15}$$

This velocity is approximately related to the total lava discharge Q_E (volume/unit time) of the channel of width W_c by:

$$Q_E = \frac{2}{3} U_{surf} H W_c. \tag{5.16}$$

Similar equations can be derived for turbulent flow, but because most known lava flows seem to have moved in the laminar regime, with low Reynolds number, they are not tabulated here.

Equations (5.15) and (5.16) are the basis of a widely used viscosity estimate (Nichols, 1939) called the Jeffreys' equation (after a 1925 paper by Harold Jeffreys):

$$\eta = \frac{\rho g H^3 W_c \sin \alpha}{n Q_E} \tag{5.17}$$

where n is a numerical parameter equal to 3 for broad flows and 4 for narrow flows. This equation assumes that lava is a Newtonian fluid and so its predictions for planetary lava flows are open to question.

For a Bingham fluid, the flow depth H should properly be the depth of the fluid excluding the semi-rigid plug of lava on top of the flow (the flowing portion must be subject to a shear stress exceeding the Bingham yield stress, and so requires some overburden before this stress is reached) and the overall discharge should include the volume of this plug. In practice these distinctions are usually ignored, as there is no simple way of separating them from image data. Thus, planetary estimates of the viscosity and Bingham yield stress are only rough estimates, not precise measurements.

Lava effusion rate. The effusion rate of lava from a feeder vent provides information about how magma is transported beneath the surface of a planet and is, thus, important for comparing volcanism on different bodies. It also plays a role in determining the maximum length of a lava flow. The total volume V_L of a lava flow can be estimated from its area A and thickness H: $V_L = A H$. If the duration of the eruption t_e is known, then the average effusion rate is simply $Q_E = V_L / t_e = AH/t_e$. However, the duration of past eruptions cannot be measured directly and so other ways of estimating eruption rates must be found. One current approach is to use a dimensionless measure of heat transport, the Grätz number, Gz, which is the ratio between the heat advected in a flow to the heat conducted in a flow. The Grätz number for a lava flow is defined as:

$$Gz = \frac{Q_E \, H}{\kappa A} \tag{5.18}$$

where κ is the thermal diffusivity of the lava composing the flow. It has been observed that this number is typically about 300 for basaltic lava flowing in channels (Gregg and Fink, 1999). Using this number, Equation (5.18) can readily be inverted to determine the effusion rate. However, tube-fed pahoehoe flows probably cool much more slowly and this may greatly overestimate the effusion rate for such flows. Unfortunately, no systematic estimates yet exist for the Grätz number of pahoehoe flows, although one estimate of the cooling time of the crust (Hon *et al.*, 1994) suggests that it may be of order 10. Another estimate from the 1300 km^2 Roza flow of the Columbia River Basalt Province gives $Gz \sim 32$. Eruption rates for inflated flows based on the Grätz number for channel flows may thus be about 10 times too large.

An estimate of the effusion rate for a single lava flow on a surface of average slope α and Bingham yield stress Y_B can be obtained by combining Equations (5.10) and (5.18):

$$Q_E = \frac{\kappa A \, Gz}{Y_B} \rho g \sin \alpha. \tag{5.19}$$

Equations of this sort can be used to make models of lava flow formation if the material parameters of the flow can be estimated. Rearrangement of this equation indicates that the effusion rate is probably the major factor in controlling the final area of a lava flow. Table 5.6 collects a number of estimates of the parameters describing lava flows on the terrestrial planets and moons.

Lava composition and rheology. It is widely believed that the Bingham yield stress is correlated with silica content, as is the viscosity. Data compilations suggest that yield stresses in the range of 100 Pa indicate basalt with silica contents of about 40 wt%, whereas values near 10^5 Pa indicate andesitic lavas with silica contents around 65 wt%. However, this relationship does not seem to be monotonic at higher silica contents, as rhyolites also have yield stresses in the range of 10^5 Pa (Moore *et al.*, 1978). It is clear from Table 5.6 that the properties of silicate lavas are generally similar on all of the terrestrial planets. Silica-rich lavas tend to have higher Bingham strengths and viscosities than silica-poor lavas. The major exception is Io, where very high eruption temperatures produce weak, highly fluid lavas. Eruption rates are highly variable and it seems that the eruption rate, more than any other factor, is the primary determinant of the size of an individual lava flow.

5.3.3 Lava domes, channels, and plateaus

Lava flows are highly complex landforms whose surface features are too varied to treat in a chapter of this length, so the interested reader who wants to go further is urged to consult some of the specialized works listed at the end of this chapter. However, a few features that are prominent in images of planetary surfaces deserve a brief mention here.

Table 5.6 *Rheologic properties of lava flows in the Solar System*

Location (type)	Bingham yield strength (Pa)	Viscosity (Pa-s)	Effusion rate (m³/s)
Earth			
Mauna Loa, HI (basalt)	$(3.5–72) \times 10^2$	$1.4 \times 10^2 –5.6 \times 10^6$	417–556
Makaopuhi, HI (basalt)	$(8–70) \times 10^3$	$(7–45) \times 10^2$	–
Mount Etna, Italy (basalt)	9.4×10^3	9.4×10^3	0.3–0.5
Sabancaya, Peru (andesite)	$5 \times 10^4 – 1.6 \times 10^6$	$7.3 \times 10^9 – 1.6 \times 10^{13}$	1–13
Mono Craters, CA (rhyolite)	$(1.2–3) \times 10^5$	–	–
Moon			
Mare Imbrium	$(1.5–4.2) \times 10^2$	–	–
Gruithuisen domes	$(7.7–14) \times 10^4$	$(3.2–14) \times 10^8$	5.5–120
Venus			
Artemis festoons	$(4.1–13) \times 10^4$	$7 \times 10^6 – 7.3 \times 10^9$	$(2.5–10) \times 10^3$
Atalanta Festoon	1.2×10^5	2.3×10^9	950
Mars			
Arsia Mons	$(2.5–3.9) \times 10^3$	9.7×10^5	$(5.6–43) \times 10^3$
Ascraeus Mons	$(3.3–83) \times 10^3$	$(2.1–640) \times 10^3$	18–60
Tharsis plains[a]	$(1.2–2.4) \times 10^2$	$(8–58) \times 10^2$	$(2–25) \times 10^2$
Io			
Shield volcanoes[b]	$(1–10) \times 10^1$	$10^3–10^5$	~3000
Ariel cryovolcanism			
Flows[c]	$(6.7–37) \times 10^3$	$(9–45) \times 10^{14}$	–

Data from Hiesinger *et al.* (2007), except:
[a] Hauber *et al.* (2010)
[b] Schenk *et al.* (2004)
[c] Melosh and Janes (1989)

Pancake domes. Figure 5.15 illustrates a cluster of "pancake domes" on Venus. These domes resemble silica-rich rhyolite domes on the Earth (an example is the Mono Craters in California), although the 25 km-wide Venusian domes are larger than most terrestrial occurrences. These constructs are believed to form from the extrusion of highly viscous lava that flows only a short distance from an underlying vent. When lava of this type erupts onto steep slopes the lava oozes downhill to freeze into thick, stubby, elongated lobes.

Sinuous rilles. First named by German astronomer Johann Schröter (1745–1816) in 1787, sinuous rilles immediately caught the attention of some of the first observers of the Moon, who thought that they had discovered river valleys. They often head in circular to irregular pits and tend to decrease in width as they meander downslope in wide curves (Figure 5.16). Typically several kilometers wide and hundreds of meters deep, they may

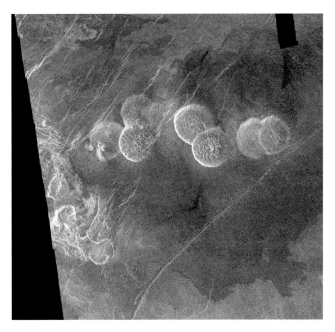

Figure 5.15 Pancake domes on Venus. These seven domes in eastern Alpha Regio are each about 25 km in diameter and 750 m high. They are interpreted as extrusions of very viscous lava from a central source. NASA Magellan radar image PIA00215. North is up.

Figure 5.16 Apollo 15 image of the Aristarchus plateau of the Moon, showing flooded craters and sinuous rilles of this highly volcanic region. The flooded crater in the foreground is Prinz, 46 km in diameter. Aristarchus crater, 40 km diameter, is in the right background. NASA image A15_m_2606.

stretch hundreds of kilometers in length. Some appear to stop for a short distance and then begin again as if whatever flowed through them went underground for a while. Sinuous rilles occasionally degenerate downslope into a line of rimless pit craters, again suggesting collapse into an underlying cavity. A small rille on the floor of the much larger Schröter's Rille seems to wind enigmatically in and out of its side walls. They are invariably associated with lava plains. The Apollo 15 astronauts famously landed on the margin of Hadley Rille, but were unable to demonstrate its origin other than to show that the rocks outcropping on its rim are basalts.

Sinuous rilles are observed on the Earth, Mars, and Venus as well as the Moon. All of these characteristics suggest that sinuous rilles represent channels through which lava once flowed. However, much controversy still centers about how completely they were filled, when they were active, and whether they are collapsed lava tubes or were once open to the sky. Some of them cut deep into the surrounding surface, suggesting that the flowing lava somehow deepened its channel. Much discussion has focused on the process of thermal erosion, by which the hot, flowing lava heats and softens the underlying rocks, melting them away and, thus, slowly deepening the channel. On the other hand, flowing lava is quite capable of quarrying away jointed blocks in its bed through the hydrodynamic "plucking" process by which water carves channels into bedrock. While thermal erosion is theoretically capable of deepening channels, its action has never been unambiguously demonstrated on the Earth, whereas there are good Hawaiian examples of plucking.

Lava plateaus. Lava plateaus are stacks of many individual lava flows. When a rising, hot, mantle plume arrives near a planet's surface it may deliver both heat and differentiated melts to the base of the crust for a long period of time. In this case magma spills out onto the surface in many, perhaps hundreds to thousands, of individual flows. So long as magma can continue to penetrate the lava flows already congealed on the surface, more material builds up, creating a thick pile of individual flows with a combined volume that may rival the volume of the crust itself.

On the Moon, lava began to erupt through the thin nearside crust about 500 Myr after the large basins were themselves created by impacts. Because of the long time interval between basin formation and the lava flows, there does not seem to be any genetic connection: The impacts did not initiate the volcanism. Instead, the impacts created topographically low basins and thin crust through which the lava extruded. Although individual lunar flows are hundreds of meters thick, owing to the low lunar gravity, it required the accumulation of many individual flows to build up the multi-kilometer thick piles of lava that we now recognize as the lunar mare.

Similar stacks of individual lava flows from long-continued volcanism created the tens of kilometers thick Tharsis plateau on Mars. Thinner lava plains cover much of the rest of the surface of Mars, with individual thin flows extending thousands of kilometers. Most of the surface of Venus was covered by extensive lava flows around 700 Myr ago, obliterating most of whatever surface Venus originally possessed. A single lava channel, Baltis Vallis, on Venus stretches 6800 kilometers across its surface. The Earth possesses dozens of broad, large igneous provinces that have erupted hundreds of thousands of cubic kilometers of basalt throughout the geologic history of our planet, each eruptive episode lasting only a few million years. Recent images returned from the MESSENGER spacecraft show that Mercury's extensive intercrater plains are volcanic, again indicating the importance of repeated volcanic eruptions.

Io is the volcanic body *par excellence*, whose crust seems to be entirely created by repeated volcanic flows, which renew its surface at the average rate of 1.5 cm/yr. Although its visible surface is largely coated with volatile sulfur and SO_2, its overall density and the high temperatures of its lavas measured by the Galileo probe indicate that its volcanism is predominantly silicic, perhaps dominated by ultramafic lavas.

Less is known of the water-mediated cryovolcanism on the icy satellites, but it is clear that flowing liquids have covered many of the icy satellites' surfaces. Volcanism has been active on nearly all of the solid bodies in the Solar System, so understanding this process is a vital part of the study of planetary surfaces in general.

Further reading

The history of investigation of volcanoes and magma on the Earth is engagingly told in the history by Sigurdsson (1999). The book by Williams and McBirney (1979) provides excellent quantitative coverage of the basic ideas of volcanic phenomena, but is now becoming somewhat dated. A more up-to-date reference with similar coverage is Schmincke (2003), but the best and most readable reference, although not as quantitative as Williams and McBirney is Francis and Oppenheimer (2003). A good general reference to the chemistry of rocks and melting, as well as much more, is McSween *et al.* (2003). Io is the volcanic moon *par excellence* and an entire book is now devoted to it Davies (2007).

Exercises

Note: As for all problems of this kind, your best guide to a correct answer is to make sure that the dimensional units of all results are correct!

5.1 Squeezing magma sponges

Derive Equation (5.6) for the timescale over which magma percolates out of a magma-saturated layer of thickness h by equating the volume discharge per unit area Q times the percolation time $t_{percolate}$ to the volume of melt ϕh in the layer.

Use this equation to estimate the timescale over which the 100 km thick asthenosphere underlying oceanic plates on Earth would lose 10% of its melt if the grain size in the mantle is about 1 mm and the viscosity of hot basaltic melt is 10^4 Pa-s.

Now shift to the outer Solar System and *estimate* the rate at which water "magma" segregated from the body of Uranus' 1160 km diameter satellite Ariel (Voyager imaged lava flows on its surface). Ariel's mean density is 1660 kg/m^3 and the viscosity of water at its melting point is about 1.5×10^{-3} Pa-s. What does this tell you about the ability of planets to retain melt in their interiors?

5.2 Feeding Ariel's volcanoes

If pure liquid water ascends from the interior of Ariel by means of dikes of width about 1 cm, use Equation (5.8) to estimate the mean velocity of the water in the dike. Next, use Equation (5.7) to estimate the vertical depth L_c of the dike as it rises toward the surface. Finally, assuming that the horizontal extent of a dike segment is approximately the same as its vertical depth, compute the volume of water carried by this dike. Combining the mean velocity and area of the dike as it breaches the surface, estimate the eruption rate of water onto the surface. At this rate, how long would it take to build up a cryovolcano 10 km in diameter and 1 km high? How many dikes must discharge their contents to build the volcano?

5.3 Plumbing Enceladus' geysers

The NASA–ESA Cassini Mission discovered that water-dominated geysers erupt continuously from the south pole of Enceladus, a Saturnian satellite about 500 km in diameter. The geysers spew out of four long fissures called "tiger stripes" whose hottest portions extend about 50 km horizontally. The CISR Infrared Spectrometer observed an excess heat flow from the entire region of about 6 GW. If the liquid water freezes and cools to the ambient temperature of 69 K, use Equation (5.1) to estimate the volume eruption rate, per unit length of the fissures, of water necessary to supply this heat flux. Supposing that the geysers are fed by pure liquid water, apply Equation (5.8) (plus a little creative thinking) to estimate the width of the dikes feeding water to the surface and the mean velocity of the water moving up the dike. Finally, use either Equation (5.7) or (5.9) to estimate the depth of the fissures feeding the eruption. If these eruptions are fed by "quanta" of water trapped in rising dikes of these dimensions, what is the duration of a single eruptive pulse as an individual dike breaches the surface? Note that there are multiple valid ways to get the correct answer.

Useful data: The latent heat of freezing for water is 334 kJ/kg and its heat capacity is 4.2 kJ/kg-K. The viscosity of water is about 1.5×10^{-3} Pa-s. Young's modulus of Enceladus' ice crust is about 10^{10} Pa. The density contrast between the erupting fluid and the surrounding ice is almost completely unknown. A crude estimate is to suppose that it is about 10% of the density of ice. The surface acceleration of gravity on Enceladus is 0.11 m/s^2.

5.4 Volcanic bombs in orbit: a natural answer to StarWars

Ronnie, an inquisitive sixth-grade visitor to the planetarium, wants to know if big, noisy volcanoes on Earth can eject rocks into space. Johnnie, another budding intellect, thinks that lunar volcanoes eject tektites. Use your knowledge of the thermodynamics of expansion to compute the *maximum* expansion velocity $v_\infty = \sqrt{2h}$, where h is the specific enthalpy of the expanding gas for the volcanic gases CO, H_2O, and H_2, at a typical eruption temperature of 1200°C. Compare these velocities to the escape velocities of the Earth, Mars, and the Moon.

If water vapor is the primary volcanic gas on Mars, how high an eruption temperature would be required for Martian volcanoes to eject volcanic bombs from the planet? How hot a volcano is needed for it to eject material from Earth? Do you think that Ronnie's and Johnnie's ideas make any sense?

Useful data: The specific enthalpy h of a gas is approximately $h = c_p T$, where T is the absolute temperature and cP is the specific heat at constant pressure, $cP = 7/2\ R$, where R is the gas constant, 8.317 J/mol-K.

5.5 Go with the flow

(a) Steep flow fronts are observed at the edges of broad, extensive lava flows on the lunar Mare. These lava flows are considerably thicker than terrestrial lava flows, often reaching 100 m in height. The average slope of one such flow is about 0.5°. Use Equation (5.10) to compute the Bingham yield stress of this lava flow and compare it to terrestrial lava flows. Is lunar magma substantially stronger than terrestrial magma?

(b) Viking images of Olympus Mons on Mars reveal numerous leveed lava flows running down its flanks, which slope at about 7° to the horizontal. The widths of the levees are easily measured from orbit. One large flow is observed to possess levees about 1 km wide. Estimate the Bingham yield stress of this flow and compare it to that measured on terrestrial lava flows. What can you deduce (if anything) about the magma that created this flow? This leveed channel is observed to have fed a small lava flow on the flanks of Olympus Mons that expanded to about 8 km wide and ran 50 km down the slope. Estimate the effusion rate of this flow. What additional information would you need to estimate the viscosity of the lava? Can you envisage obtaining this information from orbit?

5.6 Big volcanoes on little planets

Use the theory of lava flow lengths derived in Section 5.3.2 to relate the radius of a volcanic edifice, L, with a height H to the eruption rate Q_E. Assume that central eruptions last long enough that the length of each flow that builds up the edifice is limited by its solidification time. That is, if L is of order $Q_E t_e/h$, where t_e is the duration of the flow and h is its thickness, then t_e is of order h^2/κ, where the thermal diffusivity of rock, κ, is about 10^{-6} m²/s. Assume that the thickness of the flow is given by the Bingham yield stress Y_B,

$$h = Y_B/(\rho g \sin \alpha)$$

where α is the mean angle of the volcano's slope, $\tan \alpha = H/L$. Derive an expression relating volcano radius (that is, maximum lava flow length, L) to eruption rate Q_E and height H. Note that you will have to make some approximations to account for the fact that the volcano is circular in plan while the estimates above for L and h are on a per unit width basis.

Use this relation to derive an eruption rate and mean flow thickness for Olympus Mons on Mars, a 400 km diameter central volcano made predominantly of basaltic lava, where Y_B = 10^3 Pa. The average surface slope of this 24 km high volcanic edifice is about 7°. Compare the derived eruption rate to the typical eruption rate of terrestrial basaltic volcanoes, ca. 3×10^7 m³/day. What does this mean?

Note: There is no single "right answer" to this problem, which requires you to make a number of "reasonable" approximations. This problem is a thinking exercise in how simple theories are concocted.

6

Impact cratering

> The dominant surface features of the Moon are approximately circular
> depressions, which may be designated by the general term craters …
> Solution of the origin of the lunar craters is fundamental to the unravel-
> ing of the history of the Moon and may shed much light on the history of
> the terrestrial planets as well.
>
> E. M. Shoemaker (1962)

Impact craters are the dominant landform on the surface of the Moon, Mercury, and
many satellites of the giant planets in the outer Solar System. The southern hemisphere
of Mars is heavily affected by impact cratering. From a planetary perspective, the *rarity
or absence* of impact craters on a planet's surface is the exceptional state, one that needs
further explanation, such as on the Earth, Io, or Europa. The process of impact cratering
has touched every aspect of planetary evolution, from planetary accretion out of dust or
planetesimals, to the course of biological evolution.

The importance of impact cratering has been recognized only recently. E. M. Shoemaker
(1928–1997), a geologist, was one of the first to recognize the importance of this process
and a major contributor to its elucidation. A few older geologists still resist the notion that
important changes in the Earth's structure and history are the consequences of extraterrestrial impact events. The decades of lunar and planetary exploration since 1970 have, however, brought a new perspective into view, one in which it is clear that high-velocity impacts
have, at one time or another, affected nearly every atom that is part of our planetary system.
Impact cratering is crucially important for the accumulation of the planets in the first place
and has played major roles from the formation of the most ancient planetary landscapes
to the creation and maintenance of the modern regolith of airless bodies. In an important
sense, impact cratering is the most fundamental geologic process in the Solar System.

6.1 History of impact crater studies

Craters were discovered in 1610 when Galileo pointed his first crude telescope at the Moon.
Galileo recognized the raised rims and central peaks of these features, but described them
only as circular "spots" on the Moon. Although Galileo himself did not record an opinion
on how they formed, astronomers argued about their origin for the next three centuries.

Astronomer J. H. Schröter first used the word "crater" in a non-genetic sense in 1791. Until the 1930s most astronomers believed the Moon's craters were giant extinct volcanoes: the impact hypothesis, proposed sporadically over the centuries, did not gain a foothold until improving knowledge of impact physics showed that even a moderately oblique high-speed impact produces a circular crater, consistent with the observed circularity of nearly all of the Moon's craters. Even so, many astronomers clung to the volcanic theory until high-resolution imagery and *in situ* investigation of the Apollo program in the early 1970s firmly settled the issue in favor of an impact origin for nearly every lunar crater. In the current era spacecraft have initiated the remote study of impact craters on other planets, beginning with Mariner 4's unexpected discovery of craters on Mars on July 15, 1965. Since then craters have been found on almost every other solid body in the Solar System.

Meteor Crater, Arizona, was the first terrestrial structure shown unambiguously to be of impact origin. D. M. Barringer (1860–1929) investigated this 1 km diameter crater and its associated meteoritic iron in detail from 1906 until his death in 1929. After Barringer's work a large number of small impact structures resembling Meteor Crater have been found. Impact structures larger than about 5 km in diameter were first described as "cryptovolcanic" because they showed signs of violent upheaval but were not associated with the eruption of volcanic materials. J. D. Boon and C. C. Albritton in 1937 proposed that these structures were really caused by impacts, although final proof had to wait until the 1960s when the shock-metamorphic minerals coesite and stishovite proved that the Ries Kessel in Germany and subsequently many other cryptovolcanic structures are the result of large meteor impacts.

Finally, theoretical and experimental work on the mechanics of cratering began during World War II and was extensively developed in later years. This work was spurred partly by the need to understand the craters produced by nuclear weapons and partly by the fear that the "meteoroid hazard" to space vehicles would be a major barrier to space exploration. Computer studies of impact craters were begun in the early 1960s. A vigorous and highly successful experimental program to study the physics of impact was initiated by D. E. Gault (1923–1999) at NASA's Ames facility in 1965.

These three traditional areas of astronomical crater studies, geological investigation of terrestrial craters, and the physics of cratering have blended together in the post-Apollo era. Traditional boundaries have become blurred as extraterrestrial craters are subjected to direct geologic investigation, the Earth's surface is scanned for craters using satellite images, and increasingly powerful computers are used to simulate the formation of both terrestrial and planetary craters on all size scales. The recent proposals that the Moon was created by the impact of a Mars-sized protoplanet with the proto-Earth 4.5 Gyr ago and that the Cretaceous era was ended by the impact of a 15 km diameter asteroid or comet indicate that the study of impact craters is far from exhausted and that new results may be expected in the future.

6.2 Impact crater morphology

Fresh impact craters can be grossly characterized as "circular rimmed depressions." Although this description can be applied to all craters, independent of size, the detailed form of craters varies with size, substrate material, planet, and age. Craters have been observed over a range of sizes varying from 0.1 μm (microcraters first observed on lunar rocks brought back by the Apollo astronauts) to the more than 2000 km diameter Hellas Basin on Mars. Within this range a common progression of morphologic features with increasing size has been established, although exceptions and special cases are common.

6.2.1 Simple craters

The classic type of crater is the elegant bowl-shaped form known as a "simple crater" (Figure 6.1a). This type of crater is common at sizes less than about 15 km diameter on the Moon and 3 to about 6 km on the Earth, depending on the substrate rock type. The interior of a simple crater has a smoothly sloping parabolic profile and its rim-to-floor depth is about 1/5 of its rim-to-rim diameter. The sharp-crested rim stands about 4% of the crater diameter above the surrounding plain, which is blanketed with a mixture of ejecta and debris scoured from the pre-existing surface for a distance of about one crater diameter from the rim. The thickness of the ejecta falls off as roughly the inverse cube of distance from the rim. The surface of the ejecta blanket is characteristically hummocky, with mounds and hollows alternating in no discernible pattern. Fields of small secondary craters and bright rays of highly pulverized ejecta that extend many crater diameters away from the primary may surround particularly fresh simple craters. Meteor Crater, Arizona, is a slightly eroded representative of this class of relatively small craters. The floor of simple craters is underlain by a lens of broken rock, "breccia," which slid down the inner walls of the crater shortly following excavation. This breccia typically includes representatives from all the formations intersected by the crater and may contain horizons of melted or highly shocked rock. The thickness of this breccia lens is typically 1/2 to 1/3 of the rim-to-floor depth.

6.2.2 Complex craters

Lunar craters larger than about 20 km diameter and terrestrial craters larger than about 3 km have terraced walls, central peaks, and at larger sizes may have flat interior floors or internal rings instead of central peaks. These craters are believed to have formed by the collapse of an initially bowl-shaped "transient crater," and because of this these more complicated structures are known as "complex craters" (Figure 6.1b). The transition between simple and complex craters has now been observed on the Moon, Mars, Mercury, and the Earth, as well as on some of the icy satellites in the outer Solar System. In general the transition diameter scales as g^{-1}, where g is the acceleration of gravity at the planet's surface, although the constant in the scaling rule is not the same for icy and rocky bodies. This is consistent with the idea that complex craters form by collapse, with icy bodies having only about 1/3 the strength of rocky ones. The floors of complex craters are covered by melted and highly shocked debris, and melt pools are sometimes seen in depressions in the surrounding ejecta blanket. The surfaces of the terrace blocks tilt outward into the crater walls, and melt pools are also common in the depressions thus formed. The most notable

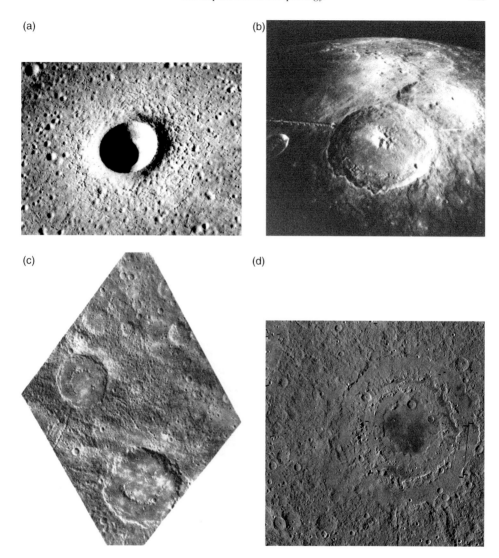

Figure 6.1 Impact crater morphology as a function of increasing size. (a) Simple crater: 2.5 km diameter crater Linné on the Moon (Apollo 15 Panometric Photo strip 9353). (b) Complex crater with central peak: 102 km diameter crater Theophilus on the Moon (Apollo 16 Hasselblad photo 0692). (c) Complex crater with internal ring: Mercurian craters Strindberg (165 km diameter) to the lower right and Ahmad Baba (115 km) to the upper left (Mariner 10 FDS 150, rectified). (d) Multiring basin: 620 km diameter (of most prominent ring) Orientale basin on the Moon (LROC WAC mosaic. Full width of image PIA13225 is 1350 km. NASA/GSFC/ASU).

structural feature of complex craters is the uplift beneath their centers. The central peaks contain material that is pushed upward from the deepest levels excavated by the crater. Study of terrestrial craters has shown that the amount of structural uplift h_{su} is related to the final crater diameter D by:

$$h_{su} = 0.06 \, D^{1.1} \tag{6.1}$$

where all distances are in kilometers. The diameter of the central peak complex is roughly 22% of the final rim-to-rim crater diameter in craters on all the terrestrial planets.

Complex craters are shallower than simple craters of equal size and their depth increases slowly with increasing crater diameter. On the Moon, the depth of complex craters increases from about 3 km to only 6 km while crater diameters range from 20 to 400 km. Rim height also increases rather slowly with increasing diameter because much of the original rim slides into the crater bowl as the wall collapses. Complex craters are thus considerably larger than the transient crater from which they form: estimates suggest that the crater diameter may increase as much as 60% during collapse. A useful scaling relation suggests that the rim-to-rim diameter of a complex crater is related to the transient crater D_t diameter by:

$$D = 1.17 \frac{D_t^{1.13}}{D_{s-c}^{0.13}} \tag{6.2}$$

where D_{s-c} is the diameter at the simple to complex transition, about 3.2 km on the Earth and 15 km on the Moon.

As crater size increases, the central peaks characteristic of smaller complex craters give way to a ring of mountains (Figure 6.1c). This transition takes place at about 140 km diameter on the Moon and about 20 km diameter on the Earth, again following a g^{-1} rule. Known as "peak-ring craters," the central ring is generally about 0.5 of the rim-to-rim diameter of the crater on all the terrestrial planets.

The ejecta blankets of complex craters are generally similar to those of simple craters, although radial troughs and ridges replace the "hummocky" texture characteristic of simple craters as size increases. Fresh complex craters also have well-developed fields of secondary craters, including frequent clusters and "herringbone" chains of closely associated, irregular, secondary craters. Very fresh craters, such as Copernicus and Tycho on the Moon, have far-flung bright ray systems.

6.2.3 Multiring basins

The very largest impact structures are characterized by multiple concentric circular scarps, and are, hence, known as "multiring basins." The most famous such structure is the 930 km diameter Orientale basin on the Moon (Figure 6.1d), which has at least four nearly complete rings of inward-facing scarps. Although opinion on the origin of the rings still varies, most investigators feel that the scarps represent circular faults that slipped shortly after the crater was excavated. There is little doubt that multiring basins are caused by impacts: most of them have recognizable ejecta blankets characterized by a radial ridge-and-trough pattern. The ring diameter ratios are often tantalizingly close to multiples of $\sqrt{2}$, although no one has yet suggested a convincing reason for this relationship.

In contrast to the simple/complex and central peak/internal ring transitions discussed above, the transition from complex craters to multiring basins is not a simple function of

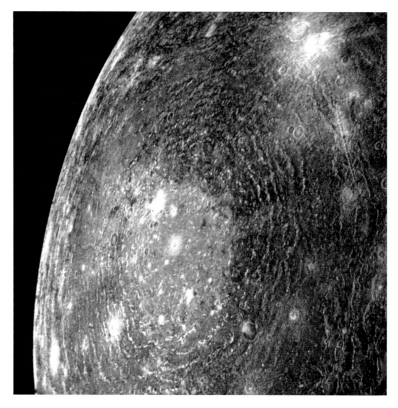

Figure 6.2 The Valhalla basin on Callisto. The original impact was within the central bright patch, which is 300 km in diameter and may represent ejecta from a still smaller (now unrecognizable) crater. This central zone is surrounded by an annulus of sinuous ridges, which in turn is surrounded by an annulus of trough-like grabens, which can be recognized up to 2000 km from the basin center. Voyager 1 mosaic PIA02277. NASA/JPL.

g^{-1}. Although multiring basins are common on the Moon, where the smallest has a diameter of 410 km, none at all has been recognized on Mercury, with its two times larger gravity, even though the largest crater, Caloris Basin, is 1540 km in diameter. The situation on Mars has been confused by erosion, but it is difficult to make a case that even the 1200 km diameter Argyre Basin is a multiring structure. A very different type of multiring basin is found on Jupiter's satellite Callisto, where the 4000 km diameter Valhalla basin (Figure 6.2) has dozens of closely spaced rings that appear to face outward from the basin center. Another satellite of Jupiter, Ganymede, has both Valhalla-type and Orientale-type multiring structures. Since gravity evidently does not play a simple role in the complex crater/multiring basin transition, some other factor, such as the internal structure of the planet, may have to be invoked to explain the occurrence of multiring basins. The formation of such basins is currently a topic of active research.

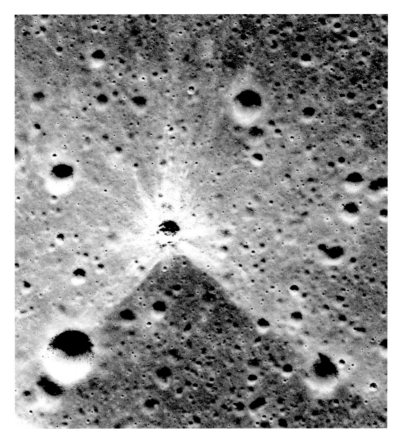

Figure 6.3 Distinctive asymmetrical ejecta surrounds a 370 m diameter crater on the lunar mare near the Linné crater 25° N, 17° W. This pattern is typical of ejecta surrounding a crater formed by an impact at an angle between 20° and 45° from the horizontal. The crater itself is still circular, but there is an uprange wedge in which very little ejecta is present. Portion of Apollo 15 Panometric image AS15-P-9337.

6.2.4 Aberrant crater types

On any planetary surface a few craters can always be found that do not fit the simple size–morphology relation described above. These are generally believed to be the result of unusual conditions of formation in either the impacting body or the planet struck. Circular craters with asymmetric ejecta blankets (Figure 6.3) or elliptical craters with "butterfly-wing" ejecta patterns are the result of very low impact angles. Although moderately oblique impacts yield circular craters, at impact angles less than about 6° from the horizontal the final crater becomes elongated in the direction of flight. Small, apparently concentric, craters or craters with central dimples or mounds on their floors are the result of impacts into a weak layer underlain by a stronger one. The ejecta blankets of some Martian craters show petal-like flow lobes that are believed to indicate the presence of liquid water in the excavated material. Craters on Ganymede and Callisto have central pits at a diameter where internal rings would be expected on other bodies. The explanation for these pits is still unknown. In spite of these complications, however, the simple size–morphology relation described above provides a simple organizing principle into which most impact craters can be grouped.

6.2.5 Degraded crater morphology

All crater morphologies are observed in either "fresh," pristine landforms or as *erosionally degraded* forms. The dominant degradational process determines the detailed changes in crater morphology. Other impact craters are often the exclusive agent of degradation on airless bodies like the Moon (although burial by floods of lava can be locally important there as well). This form of erosion fuels a variety of surface creep (see Section 8.1 for more detail). Sharp terrain features such as crater rims are rounded and battered out of line by smaller impacts, crater bowls are gradually filled and slopes become gentler. At the extreme limit, craters may fade into invisibility as their place is occupied by large numbers of overlapping craters. Moon mappers have established "degradation classes" for lunar craters that range from fresh to nearly invisible and depend upon the initial crater size. Used in conjunction with crater density data, the numbers of craters in different degradation classes can be used to infer the age and cratering history of a given site for different populations of impactors.

On ancient Mars, fluvial processes dissected impact craters by gullying and channel formation. Old craters there were filled with sediment and lava. Wind-blown sand and dust fill small craters on Mars today and erode their rims into crenulated yardang ridges.

Erosion on Earth is so active that craters are among the most rare landforms. Fluvial deposition fills in closed depressions, such as crater bowls, and fluvial erosion gullies rims and quarries away ejecta blankets. Many of the craters that are fortunately preserved were once completely buried, preserving them, and are only now being exhumed: The Ries crater in Germany is an example of this fortunate circumstance. Differential erosion of the various rock units etches out the present morphology of the crater to create its modern landscape.

The varieties of degraded crater morphologies are as diverse as the different agencies of erosion or deposition. Recognition of degraded crater forms must, thus, take the behavior of each process into account as observers attempt to reconstruct the original structure of an impact crater.

6.3 Cratering mechanics

The impact of an object moving at many kilometers per second with the surface of a planet initiates an orderly sequence of events that eventually produces an impact crater. Although this is really a continuous process, it is convenient to break it up into distinct stages that are each dominated by different physical processes. This division clarifies the description of the overall cratering process, but it should not be forgotten that the different stages really grade into one another and that a perfectly clean separation is not possible. The most commonly used division of the impact cratering process is into contact and compression, excavation, and modification.

6.3.1 Contact and compression

Contact and compression is the briefest of the three stages, lasting only a few times longer than the time required for the impacting object (referred to hereafter as the "projectile") to traverse its own diameter, $t_{cc} \approx L/v_i$, where t_{cc} is the duration of contact and compression, L is the projectile diameter, and v_i is the impact velocity. During this stage the projectile first contacts the planet's surface (hereafter, "target") and transfers its energy and momentum to the underlying rocks. The specific kinetic energy (energy per unit mass, $\frac{1}{2} v_i^2$) possessed by a projectile traveling at even a few kilometers per second is surprisingly large. A. C. Gifford, in 1924, first realized that the energy per unit mass of a body traveling at 3 km/s is comparable to that of TNT. Gifford proposed the "impact-explosion analogy," which draws a close parallel between a high-speed impact and an explosion. During contact and compression the projectile plunges into the target, generating strong shock waves as the material of both objects is compressed. The strength of these shock waves can be computed from the Hugoniot equations, first derived by P. H. Hugoniot in his 1887 thesis, that relate quantities in front of the shock (subscript 0) to quantities behind the shock (no subscript):

$$
\begin{aligned}
\rho(U - u_p) &= \rho_0 U \\
P - P_0 &= \rho_0 u_p U \\
E - E_0 &= \frac{(P + P_0)}{2}\left(\frac{1}{\rho_0} - \frac{1}{\rho}\right).
\end{aligned}
\tag{6.3}
$$

In these equations P is pressure, ρ is density, u_p is particle velocity behind the shock (the unshocked material is assumed to be at rest), U is the shock velocity, and E is energy per unit mass. These three equations are equivalent to the conservation of mass, momentum, and energy, respectively, across the shock front. They hold for all materials, but do not provide enough information to specify the outcome of an impact by themselves. The Hugoniot equations must be supplemented by a fourth equation, the equation of state, that relates the pressure to the density and internal energy in each material, $P = P(\rho, E)$. Alternatively, a relation between shock velocity and particle velocity may be specified, $U = U(u_p)$. Since this relation is frequently linear, it often provides the most convenient equation of state in impact processes. Thus,

$$
U = c + S u_p
\tag{6.4}
$$

where c and S are empirical constants (c is the bulk sound speed and S is a dimensionless slope). Table 6.1 lists the measured values of c and S for a variety of materials. These

Table 6.1. *Linear shock-particle velocity equation of state parameters*

Material	ρ_0 (kg/m^3)	c (km/s)	S
Aluminum	2750	5.30	1.37
Basalt	2860	2.6	1.62
Calcite (carbonate)	2670	3.80	1.42
Coconino sandstone	2000	1.5	1.43
Diabase	3000	4.48	1.19
Dry sand	1600	1.7	1.31
Granite	2630	3.68	1.24
Iron	7680	3.80	1.58
Permafrost (water saturated)	1960	2.51	1.29
Serpentinite	2800	2.73	1.76
Water (25°C)	998	2.393	1.333
Water Ice (−15°C)	915	1.317	1.526

Data from Melosh (1989).

equations can be used to compute the maximum pressure, particle velocity, shock velocity, etc. in an impact.

Planar impact approximation. A rough estimate of the parameters describing the highest pressure portion of the contact and compression stage is obtained from the planar impact approximation (sometimes called the impedance matching solution), which is valid so long as the lateral dimensions of the projectile are small compared with the distance the shock has propagated. This approximation is, thus, valid through most of the contact and compression stage. A simultaneous solution to the Hugoniot jump equations is obtained in both the target and projectile by noting that, at the interface between the two, both the particle velocity and pressure must be the same in both bodies. Unfortunately, there is no simple formula for this approximation. The simplest expression is for the particle velocity in the target, u_t (the particle velocity in the projectile is v_i-u_t by the velocity matching condition), which is the solution of a simple quadratic equation:

$$u_t = \frac{-B + \sqrt{B^2 - 4AC}}{2A} \tag{6.5}$$

where *A, B,* and *C* are defined as:

$$\begin{aligned} A &= \rho_{0t} S_t - \rho_{0p} S_p \\ B &= \rho_{0t} c_t + \rho_{0p} c_p + 2\rho_{0p} S_p v_i. \\ C &= -\rho_{0p} v_i (c_p + S_p v_i) \end{aligned} \tag{6.6}$$

The subscripts *p* and *t* refer to the projectile and target respectively. The above equation can be used in conjunction with the Hugoniot equations and equation of state to obtain

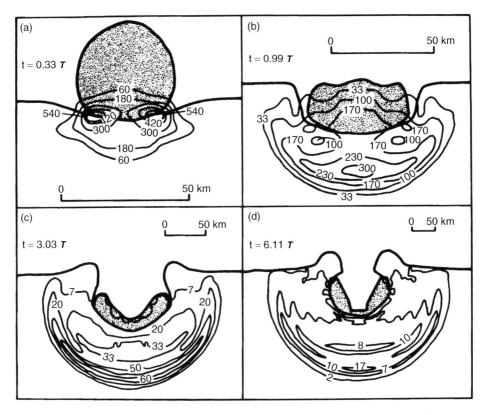

Figure 6.4 The first three frames (a to c) illustrate the evolution of shock waves in the contact and compression stages of the vertical impact of a 46.4 km diameter iron projectile on a gabbroic anorthosite target at 15 km/s. The last frame (d) is a very early phase of the excavation stage. Pressure contours are labeled in GPa, and the times given are in multiples of the time that the projectile takes to pass through its own diameter, about 3 s in this case. Note the change in length scale from frame to frame.

any other quantities of interest. Thus, the pressure behind the shock is given by $P = \rho_{ot}\, u_t$ $(c_t + S_t\, u_t)$. The pressures in both the target and projectile are the same by construction of the solution.

As the projectile plunges into the target, shock waves propagate both into the projectile, compressing and slowing it, and into the target, compressing and accelerating it downward and outward (Figure 6.4). At the interface between target and projectile the material of each body moves at the same velocity. This equals 1/2 the impact velocity if they are composed of the same materials (note that in the above equation, $A = 0$ in this case, but the numerator also vanishes and the right-hand side of the equation approaches $-C/B$, which equals $v_i/2$). The shock wave in the projectile eventually reaches its back (or top) surface. When this happens, the pressure is released as the surface of the compressed projectile expands upward and a wave of pressure relief propagates back downward toward the projectile–target interface. The contact and compression stage is considered to end when this relief wave reaches the projectile–target interface. At this time the projectile has been compressed to high pressure, often reaching hundreds of gigapascals, and upon decompression it may be in the liquid or gaseous state due to heat deposited in it during the irreversible compression process. The projectile generally carries off 50% or less of the total initial energy if the density and compressibility of the projectile and target material do not differ too much, while the balance of the energy moves into the target. This energy will eventually be expended in opening the crater as well as heating the target. The projectile–target interface at the end of contact and compression is generally less than one projectile diameter L below the original surface.

Contact and compression are accompanied by the formation of very high-velocity "jets" of highly shocked material. These jets form where strongly compressed material is close to a free surface, for example near the circle where a spherical projectile contacts a planar target. The jet velocity depends on the angle between the converging surface of the projectile and target, but may exceed the impact velocity by factors as great as 5. Jetting was initially regarded as a spectacular but not quantitatively important phenomenon in early impact experiments, where the incandescent streaks of jetted material only amounted to about 10% of the projectile's mass in vertical impacts. However, recent work on oblique impacts indicates that in this case jetting is much more important and that the entire projectile may participate in a downrange stream of debris that carries much of the original energy and momentum. Oblique impacts are still not well understood and more work needs to be done to clarify the role of jetting early in this process.

6.3.2 Excavation

During the excavation stage the shock wave created during contact and compression expands and eventually weakens into an elastic wave, while the crater itself is opened by the much slower "excavation flow." The duration of this stage is roughly given by the period of a gravity wave with wavelength equal to the crater diameter D, equal to $t_{EX} \sim (D/g)^{1/2}$, for craters whose excavation is dominated by gravity g (this includes craters larger than a few kilometers in diameter, even when excavated in hard rock). Thus, Meteor Crater, Arizona, was excavated in about 10 s, while the 1000 km diameter Imbrium Basin on the Moon took about 13 minutes to open. Shock wave expansion and crater excavation, while intimately linked, occur at very different rates and may be usefully considered separately.

The high pressure attained during contact and compression is almost uniform over a volume roughly comparable to the initial dimensions of the projectile, a region called the "isobaric core." However, as the shock wave expands away from the impact site the shock pressure declines as the initial impact energy spreads over an increasingly large volume of rock and loses energy to heating the target. The pressure P in the shock wave as a function of distance r from the impact site is given roughly by:

Impact cratering

Figure 6.5 Shatter cones from the Spider Structure, Western Australia, formed in mid-Proterozoic orthoquartzite. This cone-in-cone fracture is characteristic of shattering by impact-generated shock waves. The scale bar is 15 cm long (courtesy of George Williams).

$$P = P_0 \left(\frac{a}{r} \right)^n \qquad (6.7)$$

where a ($= L/2$) is the radius of the projectile, P_0 is the pressure established during contact and compression, and the power n is between 2 and 4, depending on the strength of the shock wave (n is larger at higher pressures – a value $n = 3$ is a good general average).

Shock metamorphism. The shock wave, with a release wave immediately following, quickly attains the shape of a hemisphere expanding through the target rocks. The high-shock pressures are confined to the surface of the hemisphere: the interior has already decompressed. The shock wave moves very quickly, as fast or faster than the speed of sound, between about 6 and 10 km/s in most rocks. As rocks in the target are overrun by the shock waves, then released to low pressures, mineralogical changes take place in the component minerals. At the highest pressures the rocks may melt or even vaporize upon release. As the shock wave weakens high-pressure minerals such as coesite or stishovite arise from quartz in the target rocks, diamonds may be produced from graphite, or maskelynite from plagioclase. Somewhat lower pressures cause pervasive fracturing and "planar elements" in individual crystals. Still lower pressures create a characteristic cone-in-cone fracture called "shatter cones" (Figure 6.5),which are readily recognized in the vicinity of impact structures. Indeed, many terrestrial impact structures were first recognized from the occurrence of shatter cones. Table 6.2 lists a number of well-established shock metamorphic changes and the pressures at which they occur.

Table 6.2. *Petrographic shock indicators*

Material	Indicator	Pressure (GPa)
Tonalite (igneous rock)	Shatter cones	2–6
Quartz	Planar elements and fractures	5–35
	Stishovite	15–40
	Coesite	30–50
	Melting	50–65(?)
Plagioclase	Planar elements	13–30
	Maskelynite	30–45
	Melting	45–65(?)
Olivine	Planar elements and fractures	5–45
	Ringwoodite	45
	Recrystallization	45(?)–65(?)
	Melting	>70
Clinopyroxene	Mechanical twinning	5–40(?)
	Majorite	13.5
	Planar elements	30(?)–45
	Melting	45(?)–65(?)
Graphite	Cubic diamond	13
	Hexagonal diamond	70–140

Data from Melosh (1989).

Spallation. The expanding shock wave encounters a special condition near the free surface. The pressure at the surface must be zero at all times. Nevertheless, a short distance below the surface the pressure is essentially equal to P, defined above. This situation results in a thin layer of surface rocks being thrown upward at very high velocity (the theoretical maximum velocity approaches the impact speed v_i). Since the surface rocks are not compressed to high pressure, this results in the ejection of a small quantity of unshocked or lightly shocked rocks at speeds that may exceed the target planet's escape velocity. Although the total quantity of material ejected by this "spall" mechanism is probably only 1–3% of the total mass excavated from the crater, it is particularly important scientifically as this is probably the origin of the recently discovered meteorites from the Moon, and of the SNC (shergottite, nakhlite, and chassignite) meteorites, which are widely believed to have been ejected from Mars.

Seismic shaking. The weakening shock wave eventually degrades into elastic waves. These elastic waves are similar in many respects to the seismic waves produced by an earthquake, although impact-generated waves contain less of the destructive shear-wave energy than earthquake waves. The seismic waves produced by a large impact may have significant effects on the target planet, creating jumbled terrains at the antipode of the impact site if they are focused by internal planetary structures, such as a low-velocity core.

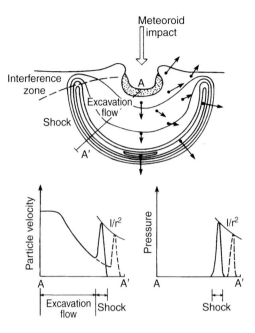

Figure 6.6 Illustration of the expanding shock wave and excavation flow following a meteorite impact. The contours in the upper part of the figure represent the pressure at some particular time after the impact. The region of high-shock pressure is isolated or "detached" on an expanding hemispherical shell. The lower graphs show profiles of particle velocity and pressure along the section AA'. The dashed lines on these graphs show the particle velocity and pressure some time later than those shown by the solid lines, and the solid curves connecting the peaks are portions of the "envelopes" of peak particle velocity and peak pressure.

This effect has been observed opposite Caloris Basinon Mercury and opposite Imbrium and Orientale on the Moon. The equivalent Richter magnitude M caused by an impact of energy E (= $1/2\, m_p v_i^2$) is given approximately by:

$$M = 0.67 \log_{10} E - 5.87. \tag{6.8}$$

Excavation mechanics. Target material engulfed by the shock wave is released a short time later. Upon release the material has a velocity that is only about 1/5 of the particle velocity in the shock wave. This "residual velocity" is due to thermodynamic irreversibility in the shock compression. It is this velocity field that eventually excavates the crater (Figure 6.6). The excavation velocity field has a characteristic downward-outward-then upward pattern that moves target material out of the crater, ejecting it at angles close to 45° at the rim. The streamlines of this flow cut across the contours of maximum shock pressure, so that material ejected at any time may contain material with a wide range of shock levels (Figure 6.7). Nevertheless, the early, fast ejecta generally contain a higher proportion of highly shocked material than the later, slower ejecta. Throughout its growth the crater is lined with highly shocked, often melted, target material.

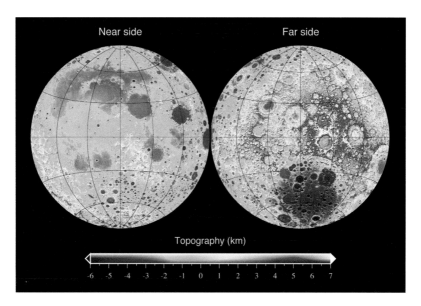

Figure 1.4 The topography of the Moon referenced to a sphere with a radius of 1737.4 km. Data were obtained from the Lunar Orbiter Laser Altimeter (LOLA) that was flown on the mission Lunar Reconnaissance Orbiter (LRO). The color-coded topography is displayed in two Lambert equal-area images projected on the near and farside hemispheres. Courtesy Mark Wieczorek, 7 August 2010.

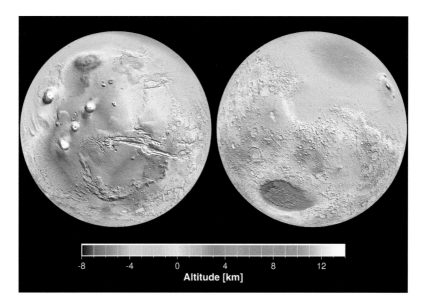

Figure 1.7 False color topography of Mars from the MOLA instrument aboard the Mars Global Surveyor spacecraft. The left hemisphere is dominated by the Tharsis Rise with its enormous volcanoes. Olympus Mons rises to the upper left. The gigantic trough of Valles Marineris extends to the right center. The northern lowlands and the Borealis plains dominate the upper half of the right hemisphere. The deep circular basin to the lower left is Hellas and the smaller basin near the center is Utopia. NASA/JPL image PIA02820.

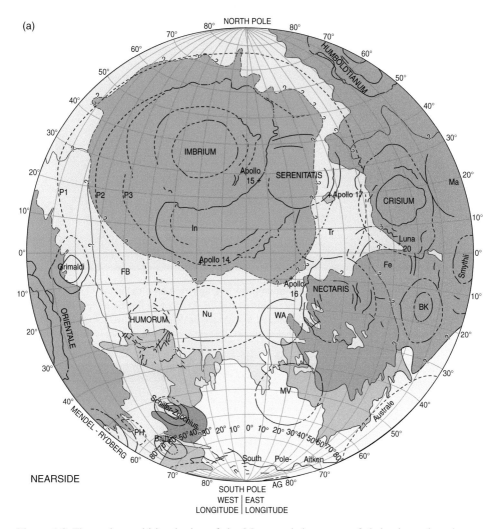

Figure 1.3 The major multiring basins of the Moon and the extent of their ejecta deposits are indicated. Curved lines indicate major rings. Panel (a) is the Moon's nearside and (b) is its farside. Blue indicates the deposits of the youngest (Imbrian) basins, yellow-orange Nectarian, dark brown Pre-Nectarian. After Plates 3A and 3B in Wilhelms (1987).

(b)

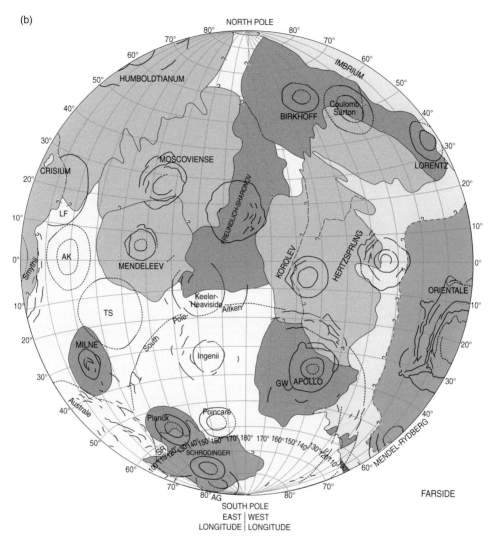

Figure 1.3 (cont.)

(a)

Planetary Radius (km)

6048 6050 6052 6054 6056 6058 6060 6062

Figure 1.6 Topographic elevations from the Magellan radar altimeter. Panel (a) is centered on 0°
Longitude, panel (b) is centered on 180°. The surface of Venus is occupied by seemingly randomly
spaced rises and plains with a few highlands such as the Lakshmi Plateau in panel (a), near Venus'
north pole. Maxwell Montes, the highest point on Venus, rises above the plateau. Note the extensive
chain of circular coronae extending across the lower half of panel (b). This chain ends in the large
incomplete circle of Artemis Chasma. Very few impact craters are visible. Panel (a) is NASA/JPL/
USGS PIA00157 and panel (b) is PIA00159

(b)

Planetary Radius (km)
6048 6050 6052 6054 6056 6058 6060 6062

Figure 1.6 (*cont.*)

Figure 1.8 The Galilean satellites of Jupiter as imaged by the Galileo spacecraft. In order from left to right are Io, Europa, Ganymede, and Callisto. Volcanoes dominate Io, Europa is covered with an ice shell, Ganymede's surface is a patchwork of bright young and dark old terrain, and Callisto is an undifferentiated mixture of ice and rock. NASA/JPL/DLR image PIA01400.

Figure 1.9 Global view of Titan's surface from the VIMS spectrometer aboard the Cassini spacecraft. This false color composite is constructed from three wavelengths in the infrared that penetrate Titan's hazy atmosphere (1.3 μm is shown in blue, 2 in green and 5 in red). The dark region in the center of the image is named Xanadu and may be the site of a large ancient impact. NASA/JPL/University of Arizona image PIA09034.

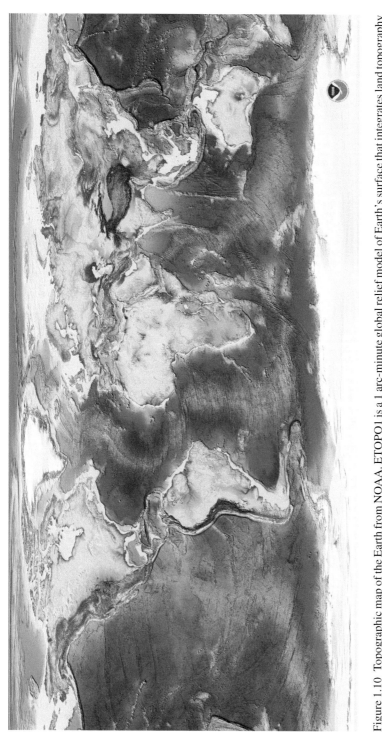

Figure 1.10 Topographic map of the Earth from NOAA. ETOPO1 is a 1 arc-minute global relief model of Earth's surface that integrates land topography and ocean bathymetry. It was built from numerous global and regional data sets (Amante and Eakins, 2009).

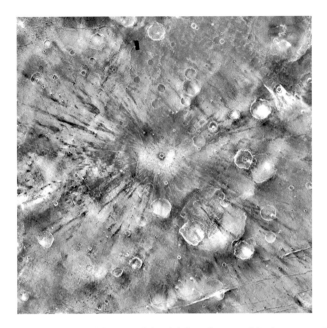

Figure 6.17 Night-time IR thermal image of the 6.9 km diameter Martian crater Gratteri, located at 17.8° S, 202° E. The dark streaks are created by secondary impact craters that extend up to 500 km from the crater center. In images of this type, dark regions are cold and emit little IR radiation because they have low thermal inertia, indicating that the streaks are composed of fine-grained material compared to their surroundings. The overall image measures 545 x 533 km across. THEMIS image courtesy of Phil Christensen. NASA/JPL/ASU.

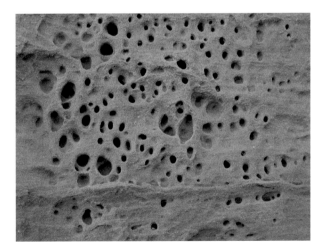

Figure 7.14 Cavernous weathering surface in sandstone. Image is about 1 m across, showing deep pits and columns that have become detached from the mass of the rock behind them. This variety is often called *honeycomb* weathering from its appearance. Canyonlands National Park, Utah, USA. Photo courtesy of Ingrid Daubar-Spitale, 2010.

Figure 7.16 Boulder-strewn surfaces on Venus, Earth, Mars, and Titan. Panel (a) is from the Soviet lander, Venera 13, image VG00261. Panel (b) near Yuma, AZ, looking north toward the Cargo Muchacho Mountains from Indian Pass Road (photo courtesy of Mark A. Dimmitt). (c) is a panorama from the Viking 2 lander. (d) The surface of Titan as viewed by the Huygens lander. The lander evidently set down in a former riverbed, only the "rocks" in the foreground are water ice and the liquid that transported them was liquid methane. The two "rocks" just below the middle of the image are 15 cm and 4 cm in diameter, respectively, and lie about 85 cm from the Huygens probe. The dark fine material on which the "rocks" lie is probably a mixture of water and hydrocarbon ice. Image PIA07232. ESA/NASA/JPL/University of Arizona.

Figure 8.17 (a) a landslide triggered by the 2002 Denali earthquake in Alaska, looking west toward the divide of the Black Rapids and Susitna glaciers. Image courtesy Dennis Trabant and Rod Marsh, USGS. For panel (b), see p. 342 in text.

Figure 9.14 Yardangs on Mars. For panel (a), see p.376 in text. (b) Panoramic view of a soft rock unit dissected into yardangs south of Olympus Mons. The three flat regions in the fore-, middle-, and background measure about 17 x 9 km in this oblique view. Mars Express HRSC image, orbit 143, ESA/DLR/FU Berlin.

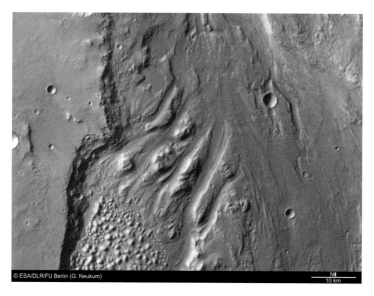

Figure 10.9 Ares Vallis is one of the large outflow channels on Mars. This image shows the transition between Iani Chaos region to the lower left and the plains of Xanthe Terra to the top (north). The spurs between the individual channels have been shaped into crude streamlined shapes by massive floods of water. 10 km scale bar is at lower right. Mars Express images by ESA/DLR/FU Berlin.

Figure 10.18 Liquid methane lakes near the North Pole of Titan imaged by the Cassini synthetic aperture radar. Dark regions are smooth lake surfaces and brighter regions are the surface. Intermediate brightness levels near the lake shores indicate some radar return from the lake bottoms. Image is centered near 80° N and 35° W and the strip is 140 km wide. Smallest details are about 500 m across. The radar strip was foreshortened to simulate an oblique view from the west. Image PIA09102. NASA/JPL/USGS.

Figure 11.12 Ice-wedge polygons on Mars. For panel (a), see p. 461 in text. (b) Troughs are spaced 1.5 to 2.5 m apart near the Phoenix landing site at 68°N and 26° W. They are believed to represent ice-wedge activity. On Earth, ice wedges may also be manifested by either ridges or troughs. NASA/ JPL/University of Arizona.

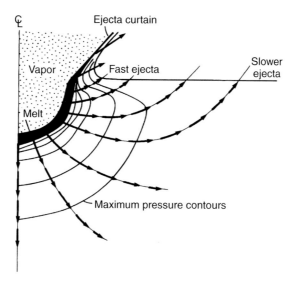

Figure 6.7 Geometry of the excavation flow field that develops behind the rapidly expanding shock front, which has moved beyond the boundaries of this illustration. The lines with arrows indicate streamtubes along which material flows downward and outward from the crater. The streamtubes cut across the contours of maximum shock pressure, showing that material ejected at any given range from the impact is shocked to a variety of different maximum pressures. When material flowing through a streamtube crosses the initial surface it forms part of the ejecta curtain. Ejecta emerging near the impact site travel at high speed, whereas ejecta emerging at larger distances travel at slower velocities.

Inside the growing crater, vaporized projectile and target may expand rapidly out of the crater, forming a vapor plume that, if massive enough, may blow aside any surrounding atmosphere and accelerate to high speed. In the impacts of sufficiently large and fast projectiles some of this vapor plume material may even reach escape velocity and leave the planet, incidentally also removing some of the planet's atmosphere. Such "impact erosion" may have played a role in the early history of the Martian atmosphere. Even in smaller impacts the vapor plume may temporarily blow aside the atmosphere, opening the way for widespread ballistic dispersal of melt droplets (tektites) above the atmosphere and perhaps permitting the formation of lunar-like ejection blankets even on planets with dense atmospheres, as has been observed on the Soviet Venera 15/16 images of Venus.

Crater growth rate. The growing crater is at first hemispherical in shape. Its depth $H(t)$ and diameter $D(t)$ both grow approximately as $t^{0.4}$, where t is time after the impact. Hemispherical growth ceases after a time of about $(2H_t/g)^{1/2}$, where H_t is the final depth of the transient crater. At this time the crater depth stops increasing (it may even begin to decrease as collapse begins), but its diameter continues to increase. The crater shape, thus, becomes a shallow bowl, finally attaining a diameter roughly three to four times its depth. At this stage, before collapse modifies it, the crater is known as a "transient" crater. Even simple craters experience some collapse (which produces the breccia lens), so

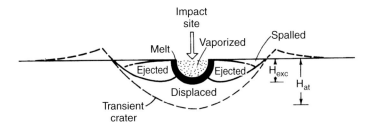

Figure 6.8 Provenance of material expelled from an impact crater. Vaporized material expands outward in a vapor plume. Of the remaining material, some is ejected and some is displaced out of the crater and deforms the adjacent rocks, uplifting the surface near the rim and downwarping rocks beneath the crater floor. The ejected material is excavated from a maximum depth H_{exc} that is only about one-third of the transient crater depth or one-tenth of the transient crater diameter. The dashed lines show the profile of the transient crater.

that the transient crater is always a brief intermediate stage in geological crater formation. However, since most laboratory craters are "frozen" transient craters, much of our know-ledge about crater dimensions refers to the transient stage only, and must be modified for application to geological craters.

Maximum depth of excavation. Laboratory, field, and computer studies of impact craters have all confirmed that only material lying above about 1/3 of the transient crater depth (or about 1/10 of the diameter) is thrown out of the crater. Material deeper than this is simply pushed downward into the target, where its volume is accommodated by deformation of the surrounding rocks (Figure 6. 8). Thus, in sharp contrast to ejecta from volcanic craters, material in the ejecta blankets of impact craters does not sample the full depth of rock inter-sected by the crater, a surprising fact that has led many geologists astray in their estimation of the nature of the ejected debris.

The form of the transient crater produced during the excavation stage may be affected by such factors as obliquity of the impact (although the impact angle must be less than about 6° for a noticeably elliptical crater to form at impact velocities in excess of about 4 km/s), the presence of a water table or layers of different strength, rock structure, joints, or initial topography in the target. Each of these factors produces its own characteristic changes in the simple bowl-shaped transient crater form.

6.3.3 Modification

Shortly after the excavation flow has opened the transient crater and the ejecta has been launched onto ballistic trajectories, a major change takes place in the motion of debris within and beneath the crater. Instead of flowing upward and away from the crater center, the debris comes to a momentary halt, then begins to move downward and back toward the center whence it came. This collapse is generally attributed to gravity, although elastic rebound of the underlying, compressed rock layers may also play a role. The effects of collapse range from mere debris sliding and drainback in small craters to wholesale alteration of the form of larger craters in which the floors rise, central peaks appear, and the rims sink down into wide zones of stepped terraces. Great mountain rings or wide central pits may appear in still larger craters (Figure 6.9).

These different forms of crater collapse begin almost immediately after formation of the transient crater. The timescale of collapse is similar to that of excavation, occupying an interval of a few times $(D/g)^{1/2}$. Crater collapse and modification, thus, take place on timescales very much shorter than most geologic processes. The crater resulting from this collapse is then subject to the normal geologic processes of gradation, isostatic adjustment, infilling by lavas, etc. on geologic timescales. Such processes may eventually result in the obscuration or even total obliteration of the crater.

The effects of collapse depend on the size of the crater. For transient craters smaller than about 15 km diameter on the Moon, or about 3 km on the Earth, modification entails only collapse of the relatively steep rim of the crater onto its floor. The resulting "simple crater" (see Figure 6.1a) is a shallow bowl-shaped depression with a rim-to-rim diameter D about five times its depth below the rim H. In fresh craters the inner rim stands near the angle of repose, about 30°. Drilling in terrestrial craters (Figure 6.10) shows that the crater floor is underlain by a lens of broken rock (mixed breccia) derived from all of the rock units intersected by the crater. The thickness of this breccia lens is typically 1/2 the depth of the crater H. Volume conservation suggests that this collapse increases the original diameter of the crater by about 15%. The breccia lens often includes layers and lenses of highly shocked material mixed with much less-shocked country rock. A small volume of shocked or melted rock is often found at the bottom of the breccia lens.

Complex craters (Figure 6.1b,c) collapse more spectacularly. Walls slump, the floor is stratigraphically uplifted, central peaks or peak rings rise in the center, and the floor is overlain by a thick layer of highly shocked impact melt (Figure 6.11). The detailed mechanism of collapse is still not fully understood because straightforward use of standard rock mechanics models does not predict the type of collapse observed (see Box 8.1). The current best description of complex crater collapse utilizes a phenomenological strength model in which the material around the crater is approximated as a Bingham fluid, a material that responds elastically up to differential stresses of about 3 MPa, independent of overburden pressure, and then flows as a viscous fluid with viscosity of the order of 1 GPa-s at larger stresses. In a large collapsing crater the walls slump along discrete faults, forming terraces whose widths are controlled by the Bingham strength, and the floor rises, controlled by the viscosity, until the differential stresses fall below the 3 MPa strength limit. A central peak may rise, and then collapse again in large craters, forming the observed internal ring (or rings). Figure 6.9 illustrates this process schematically. The rock in the vicinity of a large impact may display such an unusual flow law because of the locally strong shaking driven by the large amount of seismic energy deposited by the impact.

The mechanics of the collapse that produces multiring basins (Figure 6.1d) is even less well understood. Figure 6.12 illustrates the structure of the Orientale basin on the Moon with a highly vertically exaggerated cross section derived from both geological and

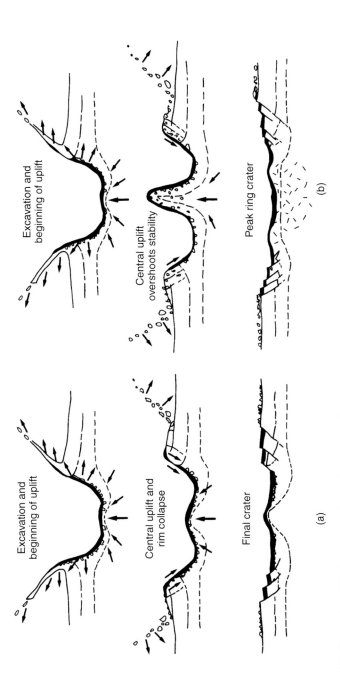

Figure 6.9 Illustration of the formation of complex craters of either (a) central peak morphology or (b) peak ring morphology. Uplift of the crater floor begins even before the rim is fully formed. As the floor rises further, rim collapse creates a wreath of terraces surrounding the crater. In smaller craters the central uplift "freezes" to form a central peak. In larger craters the central peak collapses and creates a peak ring before motion ceases.

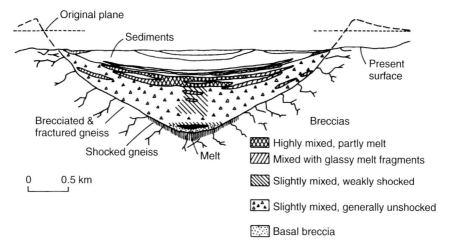

Figure 6.10 Geologic cross section of the 3.4 km diameter Brent Crater in Ontario, Canada. Although the rim has been eroded away, Brent is a typical, simple crater that forms in crystalline rocks. A small melt pool occurs at the bottom of the breccia lens and more highly shocked rocks occur near its top.

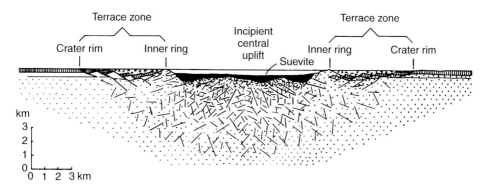

Figure 6.11 Geologic cross section of the 22 km diameter Ries Crater in Germany. Drilling and geophysical data suggest that this is a peak ring crater. Its central basin is filled with suevite, a mixture of highly shocked and melted rocks and cold clasts.

geophysical data. Note the ring scarps are interpreted as inward-dipping faults above a pronounced mantle uplift beneath the basin's center. One idea that is currently gaining ground is that the ring scarps are normal faults that develop as the crust surrounding a large crater is pulled inward by the flow of underlying viscous mantle material toward the crater cavity (Figure 6.13). An important aspect of this flow is that it must be confined in a low-viscosity channel by more viscous material below, otherwise the flow simply uplifts the crater floor, and radial faults instead of ring scarps are the result. Special structural conditions are, thus, needed in the planet for multiring basins to form on its surface, so that a g^{-1} dependence for the transition from complex craters to multiring basins is not expected (or observed). This

---

Box 6.1 **Maxwell's Z model of crater excavation**

The details of the excavation flow can be determined only by experiment or by elaborate numerical computations. Even such numerical work may have difficulty in correctly computing the final dimensions of the transient crater. However, in 1973 D. Maxwell and K. Seifert proposed a simple analytical model of excavation flow (Maxwell, 1977). This model gives a useful kinematic, although not dynamic, description of the cratering flow field. Like all approximate models, it should not be used to determine fine details.

Maxwell and Seifert noted that in explosion cratering computations the radial component of the excavation flow velocity u_r usually falls as a simple inverse power of distance r from the explosive charge.

$$u_r = \frac{\alpha(t)}{r^Z} \tag{B6.1.1}$$

where $\alpha(t)$ is a function of time and describes the strength of the flow, while Z is a dimensionless power.

The incompressibility of the excavation flow, $\nabla \cdot \mathbf{u} = 0$, requires that the angular component of the flow velocity u_θ in polar coordinates (r, θ), is:

$$u_\theta = (Z-2)\, u_r\, \frac{\sin\theta}{1+\cos\theta}. \tag{B6.1.2}$$

The geometry of the velocity field defined by this model, seen in Figure 6B.1, is remarkably similar to that computed in both explosion and impact cratering events $Z \approx 3$. The equation of streamlines in polar coordinates is:

$$r = r_0 (1 - \cos\theta)^{\frac{1}{Z-2}} \tag{B6.1.3}$$

where r_0 is a constant that is different for each streamline. It is equal to the radius at which the streamline emerges from the surface ($\theta = 90°$). Taking $r_0 = D_{at}/2$, the radius of the transient crater, the maximum depth of excavation H_{exc} is:

$$H_{exc} = \frac{D_{at}}{2}(Z-2)(Z-1)^{\frac{1-Z}{Z-2}}. \tag{B6.1.4}$$

For $Z = 3$ the maximum depth of excavation $H_{exc} = D_{at}/8$, or about one-third of the final transient crater depth.

The total mass ejected from a crater described by the Z model, M_{ej}, is a fraction of the total mass displaced from the transient crater M_e:

$$M_{ej} = \frac{Z-2}{Z-1} M_e. \tag{B6.1.5}$$

The Z model also predicts that the vertical and horizontal velocity components u_V and u_H of the ejecta launched at a distance s along the surface from the impact point are:

$$u_V = \alpha / s^Z$$
$$u_H = (Z-2)\, u_V \tag{B6.1.6}$$

Box 6.1 **(cont.)**

or that the angle of ejection is $\phi = \tan^{-1} (Z - 2)$, equal to 45° for $Z = 3$.

The Z-model presented thus far is a kinematical model useful for describing the form of the excavation flow. Maxwell and Seifert attempted to give it more dynamical content by computing the function $\alpha(t)$ in Equation (B6.1.1). This function gives the strength of the flow at any particular time. Its value is different for each streamtube in the flow. It is estimated by using energy conservation in each of these streamtubes, neglecting interactions with adjacent tubes. Thus, the sum of the kinetic, gravitational, and distortional energies is found in each streamtube at some initial time. The total energy in each streamtube is conserved as the flow progresses. However, the kinetic energy declines at the expense of the gravitational and distortional energies, so that the net flow velocity declines. Unfortunately, this aspect of the Z model has not worked out well in practice: The actual course of the excavation flow is best determined through detailed dynamical models.

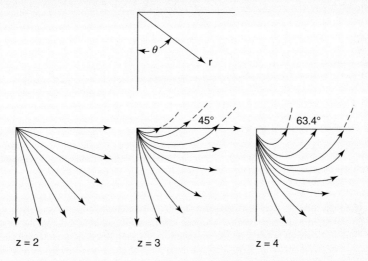

Figure B6.1

The Z-model described here can be (and has been) improved and extended in several ways. One of the most straightforward is to move the source of the flow, $r = 0$, from the surface to some depth below the surface, taking into account the depth of the effective center of the shock wave (Croft, 1980). Other workers have attempted to refine Maxwell and Siefert's methods of estimating energies in the streamtubes. The Z-model, however, is fundamentally limited by its neglect of interactions between the streamtubes. For this reason, it can never become an exact description of the cratering flow, however accurately the dynamics within a single streamtube is represented.

In spite of all its faults, the Z-model gives a reasonably accurate representation of the gross geometric features of the cratering flow and can even be used to predict some first-order dynamical properties. It has the unfortunate feature of not being a truly dynamical model, so that further refinements are not necessarily closer approximations to the full dynamical equations of motion. Nevertheless, the excellent properties of this model are probably still far from being fully exploited.

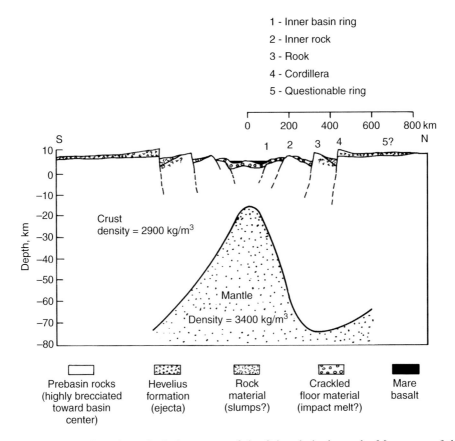

Figure 6.12 Geologic and geophysical structure of the Orientale basin on the Moon, one of the freshest and best-studied multiring basins. A dense mantle uplift underlies the center of the basin. The crustal thinning above the uplift is due to the ejection of about 40 km of crustal material from the crater that formed the basin. The great ring scarps shown in cross-section formed during collapse of the crater. Note the 10X vertical exaggeration necessary to show the ring scarps.

theory is capable of explaining both the lunar-type and Valhalla-type multiring basins as expressions of different lithosphere thicknesses.

6.4 Ejecta deposits

A deposit of debris ejected from the crater interior surrounds essentially all impact craters. The only exceptions are craters on steeply sloping surfaces or on satellites with too little gravity to retain the ejecta or too much porosity to produce it. This ejecta deposit is thickest at the crater rim and thins with increasing distance outwards. Where this deposit is recognizably continuous near the crater it is called an "ejecta blanket." Ejecta beyond the edge of the continuous deposit are thin and patchy. Secondary craters occur in this zone and beyond it. Figure 6.14 shows the ejecta blanket of the 30-km diameter lunar crater Timocharis. The pre-existing

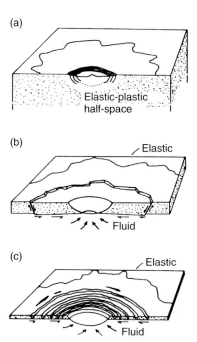

Figure 6.13 The ring tectonic theory of multiring basin formation: (a) shows the formation of a normal complex crater on a planet with uniform rheology; (b) shows the inward-directed flow in a more fluid asthenosphere underlying a lithosphere of thickness comparable to the crater depth and the resulting scarps; (c) shows a Valhalla-type basin developing around a crater formed in a very thin lithosphere.

terrain within about one crater radius of the rim is buried and mostly obliterated by the continuous ejecta blanket. Light patches show where thinner deposits overlie the mare surface.

The thickness of the ejecta deposit varies greatly with direction away from the rim. Azimuthal thickness variations, at the same radius, can be as large as ten to one. The ejecta is concentrated into rays that often are observed to form at angles of about 30° to one another. However, it is also true that, at a given azimuth, the thickness falls rapidly and systematically with distance away from the crater center. Compilation of many data sets from both impact and explosion craters shows that the thickness $\delta(r)$ as a function of radius r away from the crater center is given by an approximate inverse cube relation:

$$\delta(r) = f(R)\left(\frac{r}{R}\right)^{-3\pm0.5} \tag{6.9}$$

where R is the radius of the crater rim. Integration of this relation indicates that most of the mass of the ejecta is located near the crater rim. According to the approximate "Schröter's rule," the volume of the ejecta is approximately equal to the volume of the crater bowl. Although this rule seems to make a great deal of sense (however, it does ignore the increase in volume of the ejected material due to fragmentation), it is unverifiable in practice because the original ground surface can seldom be located with adequate precision.

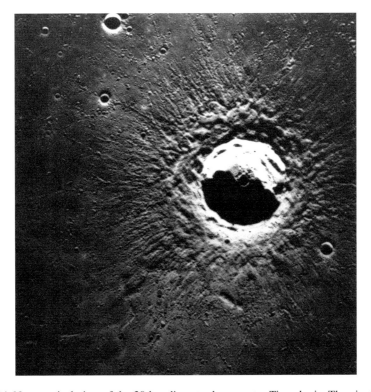

Figure 6.14 Near-vertical view of the 30 km diameter lunar crater Timocharis. The ejecta deposits are dunelike near the crater rim but grade into a subradial facies beyond about 2 R from the crater's center. Secondary craters occur at greater distances. A pattern of bright material surrounding the crater indicates the presence of ejecta too thin to greatly modify the pre-existing terrain at this scale. Apollo 15 photo AS15–1005.

The rims of many fresh craters are littered with blocks of rock ejected from beneath the crater. Compilations of data show that block size generally decreases as a function of distance from the rim. The maximum size of block observed on the rim of a crater is related to the size of the crater itself and an empirical relation that holds over a wide range of crater sizes relates the mass of the largest ejected fragment, m_f, to the total mass ejected from the crater, M_e. If both masses are expressed in kilograms, this relation is:

$$m_f = 0.8 M_e^{0.8}. \tag{6.10}$$

6.4.1 Ballistic sedimentation

The ejecta deposited around an impact crater on an airless planet are emplaced ballistically, that is, ejecta are thrown from the crater with some initial velocity, follow a nearly parabolic trajectory above the planet's surface, then fall back to the surface, striking with the

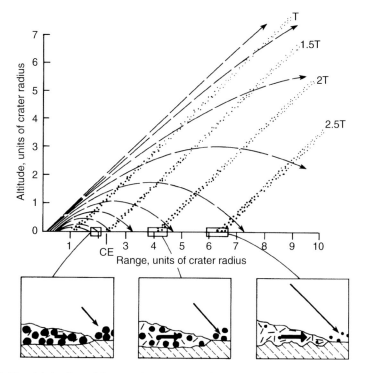

Figure 6.15 The debris ejected from an impact crater follows ballistic trajectories from its launch position within the final crater (the rim of the final crater is located at a range equal to one crater radius). The innermost ejecta are launched first and travel fastest, following the steepest trajectories shown in the figure. Ejecta originating farther from the center are launched later and move more slowly, falling nearer the crater rim. Because of the relationship between position, time, and velocity of ejection, the debris forms an inverted cone that sweeps outward across the target. This debris curtain is shown at four separate times during its flight, at 1, 1.5, 2, and 2.5 T_f, where T_f is the crater formation time, $(D/g)^{1/2}$. Coarser, less-shocked debris travels near the base of the curtain, whereas the fast, highly shocked ejecta fraction tends to travel near the top. The three lower figures show details of the pre-existing ground surface when the ejecta curtain arrives. As the range increases the ejecta strike with progressively larger velocities, incorporating larger amounts of surface material and imparting a larger net horizontal velocity.

same velocity as on ejection. Some interaction may occur between ejecta fragments in the denser parts of the ejecta curtain, but the general motion is dominated by ballistics alone. The ground surface around the crater is profoundly affected by the ejecta as it lands, and its interaction with this falling debris determines the character of the ejecta deposits of large craters.

The debris ejected from an impact crater travels together in the form of an "ejecta curtain." Although each fragment follows a parabolic trajectory, the times and velocities of ejection from the crater are organized so that most of the debris lies on the surface of an expanding inverted cone. Figure 6.15 illustrates the parabolic trajectories of a number of fragments from an impact crater. The positions of these fragments at several times are indicated. At any one time the debris lies on the surface of a cone that makes an angle of about 45° with the ground surface. This cone sweeps rapidly outward from the crater rim. Debris from the curtain strikes the ground at its base, impacting first near the crater rim, then at greater distances as time progresses. The size of the ejecta fragments near the base of the ejecta curtain is expected to be larger than the fragments higher in the curtain, and the proportion of highly shocked fragments and glass increases with height in the curtain.

Ballistically emplaced debris falling near the crater rim strikes with a low velocity because it travels only a short distance. At the rim itself this velocity is so low that rock units may retain some coherence and produce an overturned flap with inverted stratigraphy. Eugene Shoemaker first recognized such overturned beds at Meteor Crater, Arizona. At greater distances from the crater rim, the debris strikes with a higher velocity. When this velocity is large enough, surface material is eroded and mixes with the debris. The falling ejecta also possess a radially outward velocity component. Although the vertical velocity is cancelled when the debris strikes the surface, the mixture of debris and surface material retains its outward momentum (see the lower inserts in Figure 6.15). This mixture moves rapidly outward as a ground-hugging flow of rock debris, similar in many ways to the flow of a large dry-rock avalanche. Depositional features such as dunes, ridges, and radial troughs indicative of high-speed flow may result from this motion. The deposit itself consists of an intimate mixture of primary crater ejecta and of secondary material scoured from the pre-existing ground surface.

6.4.2 Fluidized ejecta blankets

The ejecta blankets of impact craters on Mars are dramatically different from those on the Moon or Mercury. Martian craters smaller than about 5 km in diameter closely resemble their counterparts on the Moon. However, craters between 5 and 15 km in diameter have a single ejecta sheet that extends about one crater radius from the rim and ends in a low concentric ridge or outward-facing escarpment. These are called "rampart craters," from the continuous ridge surrounding the ejecta deposit. The ejecta of most craters larger than 15 km in diameter are divided into petal-like lobes that extend two or more radii from the rim (Figure 6.16), approximately twice as far as the continuous ejecta deposits of lunar or Mercurian craters. A few large craters with lunar-type ejecta blankets are known, but they are rare. Many Martian craters have abnormally large central peaks and other internal collapse structures compared with lunar or Mercurian craters, also suggesting the presence of some fluidizing agent peculiar to Mars.

The fluidized ejecta blankets of Martian craters appear to have been emplaced as thin, ground-hugging flows. When impeded by topographic obstacles that could not be overridden, the flows are deflected and either spread out elsewhere or pond against the obstacle. Ejecta lobes fail to overtop low hills and mesas that are only a few times

Figure 6.16 The 19 km diameter crater Yuty on Mars is surrounded by thin, petaloid flow lobes that extend approximately twice as far from the crater as the continuous ejecta deposits of lunar or Mercurian craters. Viking Orbiter frame 3A07.

higher than the flow thickness itself, suggesting that the lobe could not have traveled as dispersed clouds of the base-surge type, nor were they emplaced by ballistic sedimentation, because the ejecta curtain should have fallen on the topographic obstacle from above.

The peculiar form of Martian ejecta blankets is generally attributed to the presence of liquid water in the substrate. Ejected along with subsurface material, liquid water mixed into the ejecta would greatly enhance the mobility of the debris, converting the dry, fragmental ejecta flows characteristic of lunar craters to fluid debris flows similar to terrestrial mudflows. Nevertheless, not all Mars researchers agree with this interpretation and an alternative viewpoint attributes at least part of the mobility of these flows to interaction with the thin atmosphere of Mars.

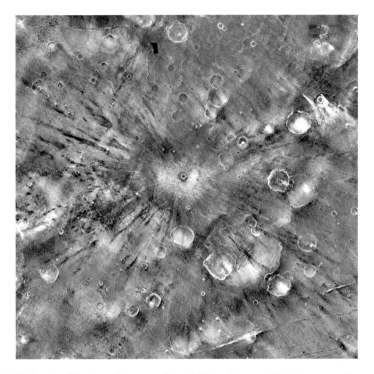

Figure 6.17 Night-time IR thermal image of the 6.9 km diameter Martian crater Gratteri, located at 17.8° S, 202° E. The dark streaks are created by secondary impact craters that extend up to 500 km from the crater center. In images of this type, dark regions are cold and emit little IR radiation because they have low thermal inertia, indicating that the streaks are composed of fine-grained material compared to their surroundings. The overall image measures 545 x 533 km across. THEMIS image courtesy of Phil Christensen. NASA/JPL/ASU. See also color plate section.

6.4.3 Secondary craters

Numerous secondary impact craters, variously occurring either singly or in loops, clusters and lines, surround large impact craters. Figure 6.17 shows the secondary crater field around the crater Gratteri on Mars, as revealed by the THEMIS thermal mapper. Recognizable secondary craters extend from just beyond the continuous ejecta blanket out to distances of up to thousands of kilometers from their source crater. Close to the primary crater, secondary craters are produced by relatively low-velocity impacts and are, thus, irregular in shape, shallow, and obviously clustered, and are often separated by V-shaped dunes known as the "herringbone pattern." Farther from the primary impact, velocities are larger and secondary craters are more dispersed, which makes them difficult to discriminate from small primary craters.

   An important controversy is presently raging about the importance of secondary craters in masking the primary flux. If a majority of the small craters (less than a few 100 m diameter) on a planet's surface are secondary, then ages assigned to cratered surfaces based on the assumption that the craters are primary will be too great. On the other hand, many experienced crater counters claim that they can exclude secondary craters because they are clustered, a claim that is disputed by other experts. At the moment there is no consensus on this problem.

The maximum size of secondary craters is approximately 4% of the primary diameter on the Moon, Mars, and elsewhere. However, on Mercury obvious secondary craters are apparently several times larger. This observation is at odds with the larger impact velocity on Mercury, which is expected to result in smaller ejected fragments, not larger ones. Is the crust of Mercury somehow stronger than that of the Moon or Mars? At the moment, the solution to this conundrum is unknown.

6.4.4 Oblique impact

Although high-velocity impact craters are circular down to very low angles of approach, the pattern of the ejecta may betray impact obliquity at angles as large as 45°. The first sign of an oblique impact is an asymmetric, but still bilaterally symmetrical, ejecta blanket. The ejecta in the uprange direction are thinner and less extensive than those in the downrange direction at impact angles near 45°. At impact angles near 30° an uprange wedge free of ejecta develops, an example of which is shown in Figure 6.3. As the angle decreases still farther, to 10°, ejecta-free regions appear in both downrange and uprange directions, although bright streaks may extend downrange in very fresh craters. At such low angles the crater itself becomes elliptical, with its long axis parallel to the flight direction of the projectile. At these highly oblique angles the projectile essentially plows a furrow into the target surface, throwing ejecta out to both sides to form a "butterfly-wing" pattern.

6.5 Scaling of crater dimensions

One of the most frequently asked questions about an impact crater is, "How big was the meteorite that made the crater?" Like many simple questions this has no simple answer. It should be obvious that the crater size depends upon the meteorite's speed, size, and angle of entry. It also depends on such factors as the meteorite's composition, the material and composition of the target, surface gravity, presence or absence of an atmosphere, etc. The question of the original size of the meteorite is usually unanswerable, because the speed and angle of impact are seldom known. The inverse question, of how large a crater will be produced by a given-sized meteorite with known speed and incidence angle is in principle much simpler to answer. However, even this prediction is uncertain because there is no observational or experimental data on the formative conditions of impact craters larger than a few tens of meters in diameter, while the impact structures of geologic interest range up to 1000 km in diameter. The traditional escape from this difficulty is to extrapolate beyond experimental knowledge by means of scaling laws.

C. W. Lampson, who studied the craters produced by TNT explosions of different sizes, introduced the first scaling law in 1950. Lampson found that the craters were similar to one another if all dimensions (depth, diameter, depth of charge placement) were divided by the cube root of the explosive energy W. Thus, if the diameter D of a crater produced by an explosive energy W is wanted, it can be computed from the diameter D_0 of a crater produced by a smaller explosive energy W_0 using the proportion:

$$\frac{D}{D_0} = \left(\frac{W}{W_0} \right)^{1/3}. \tag{6.11}$$

An exactly similar proportion may be written for the crater depth, H. This means that the ratio of depth to diameter, H/D, is independent of yield, a prediction that agrees reasonably well with observation. In more recent work on large explosions the exponent 1/3 in this equation has been modified to 1/3.4 to account for the effects of gravity on crater formation.

Although impacts and explosions have many similarities, a number of factors make them difficult to compare in detail. Thus, explosion craters are very sensitive to the charge's depth of burial. Although this quantity is well defined for explosions, there is no simple analog for impact craters. Similarly, the angle of impact has no analog for explosions. Nevertheless, energy-based scaling laws were very popular in the older impact literature, perhaps partly because nothing better existed, and many empirical schemes were devised to adapt the well-established explosion scaling laws to impacts.

6.5.1 Crater diameter scaling

This situation has changed rapidly in the last few decades, however, thanks to more impact cratering experiments specifically designed to test scaling laws. It has been shown that the great expansion of the crater during excavation tends to decouple the parameters describing the final crater from the parameters describing the projectile. If these sets of parameters are related by a single, dimensional "coupling parameter" (as seems to be the case), then it can be shown that crater parameters and projectile parameters are related by power-law scaling expressions with constant coefficients and exponents. Although this is a somewhat complex and rapidly changing subject, the best current scaling relation for impact craters forming in competent rock (low-porosity) targets whose growth is limited by gravity rather than target strength (i.e. all craters larger than a few kilometers in diameter) is given by:

$$D_{tc} = 1.161 \left(\frac{\rho_p}{\rho_t} \right)^{1/3} L^{0.78} \, v_i^{0.44} \, g^{-0.22} \sin^{1/3} \theta \tag{6.12}$$

where D_{tc} is the diameter of the transient crater at the level of the original ground surface, ρ_p and ρ_t are densities of the projectile and target, respectively, g is surface gravity, L is projectile diameter, v_i is impact velocity and θ is the angle of impact from the horizontal. All quantities are in SI units.

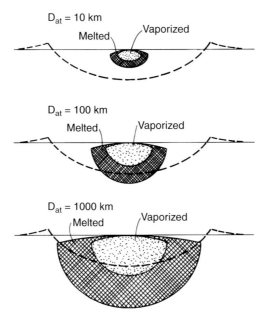

Figure 6.18 The different scaling laws for crater diameter and melt or vapor volume imply that as the crater diameter increases, the volume of melted or vaporized material may approach the volume of the crater itself. This figure is constructed for impacts at 35 km/s on the Earth.

The transient crater depth, H_{tc}, appears to be a constant times the diameter D_{tc}. Although a few investigations have reported a weak velocity dependence for this ratio, the experimental situation is not yet clear.

6.5.2 Impact melt mass

The amount of material melted or vaporized by an impact is a strong function of the impact velocity itself. The melt mass depends principally upon the velocity and mass of the projectile, but does not depend upon the gravitational acceleration of the target body. Very little melt is produced until a threshold velocity of about 12 km/s is reached. Once this threshold is exceeded, the mass of melt M_m is given in terms of the mass of the projectile M_p as:

$$\frac{M_m}{M_p} = 0.25 \frac{v_i^2}{\varepsilon_m} \sin\theta, \qquad v_i \geq 12 \text{ km/s} \qquad (6.13)$$

where ε_m is the specific internal energy of the Rankine–Hugoniot state from which isentropic decompression ends at the 1 bar point on the liquidus. It is equal to 5.2 MJ/kg for granite, which can be taken to be representative of crustal rocks.

Although the mass of melt does not depend on the gravitational acceleration of the planet, the crater size does, through Equation (6.12). The volume of the melt relative to the volume of the crater is relatively small for small craters, but as the crater size increases it becomes a progressively larger fraction of the total crater volume, as shown in Figure 6.18. At some sufficiently large diameter (about 1000 km on the Earth), the volume of the melt equals the volume of the crater itself and substantial changes in the morphology of the crater can be expected, although these changes are not well understood at the present time.

6.6 Atmospheric interactions

As fast-moving meteoroids enter the atmosphere of a planet, they are slowed by friction with the atmospheric gases and compressed by the deceleration. Small meteoroids are often vaporized by frictional heating and never reach the surface of the planet. Larger meteoroids are decelerated to terminal velocity and fall relatively gently to the surface of the planet. The diameter of a meteorite that loses 90% of its initial velocity in the atmosphere is typically about 1 m for Earth and 60 m for Venus. However, this assumes that the projectile reaches the surface intact, whereas in fact aerodynamic stresses may crush all but the strongest meteorites. Once fractured, the fragments of an incoming meteorite travel slightly separate paths and strike the surface some distance apart from one another. This phenomenon gives rise to the widely observed strewn fields of meteorites or craters on the Earth, which average roughly a kilometer or two in diameter. On Venus, clusters of small craters attributed to atmospheric breakup are spread over areas roughly 20 km in diameter. Rather surprisingly, clusters of small craters are also observed on Mars, where the spread of small craters in a cluster averages only a few tens of meters to hundreds of meters across.

The aerodynamic crushing stress experienced by an incoming meteoroid is of the order of the stagnation pressure, given by

$$P_{\text{stagnation}} \approx \rho_a v^2 \qquad (6.14)$$

where ρ_a is the density of the atmosphere through which the meteoroid is traveling and v is the relative velocity of the meteoroid and atmosphere. Evidently, even the thin Martian atmosphere is enough to fracture and partially disperse weak incoming meteoroids.

It can be shown that the dispersion of a cluster of fragments is a maximum when breakup occurs at twice the atmospheric scale height, H_s. In this case the expected dispersion ΔY is given by

$$\Delta Y \approx \frac{2H_s}{\sin\theta}\sqrt{\frac{\rho_a}{\rho_p}} \qquad (6.15)$$

where θ is the angle of entry of the meteoroid with respect to the horizontal and ρ_p is its density. Because the scale heights of the atmospheres of the Earth, Venus, and Mars are all similar, the dispersion of clusters of fragments is expected to be roughly a factor of ten different among these bodies, increasing from Mars to Earth to Venus, as observed.

The atmospheric blast wave and thermal radiation produced by an entering meteorite may also affect the surface: The 1908 explosion at Tunguska River, Siberia, was probably

Figure 6.19 Oblique view of a heavily cratered landscape on the Moon. This area is to the northeast of crater Tsiolkovskiy on the Moon's farside. The large crater near the center is about 75 km in diameter, but craters as small as a few tens of meters in diameter can be discerned in the foreground. Apollo 17 photo 155–23702 (H).

produced by the entry and dispersion of a 30 to 50 m diameter stony meteorite that leveled and scorched about 2000 km^2 of meter-diameter trees. Radar-dark "splotches" up to 50 km in diameter on the surface of Venus are attributed to pulverization of surface rocks by strong blast waves from meteorites that were fragmented and dispersed in the dense atmosphere.

6.7 Cratered landscapes

Impact craters have been treated as individual entities in the preceding sections. However, as spacecraft images abundantly illustrate, the surfaces of most planets and satellites are scarred by vast numbers of impact craters that range in size from the limit of resolution to a substantial fraction of the planet's or satellite's radius. In some places impact craters are the dominant landform: little of the observed topography can be ascribed to any other process (see Figure 6.19). Craters on such a surface exhibit degrees of preservation ranging from fresh craters with crisp rims and bright rays to heavily battered or buried craters that may only betray their presence by a broken rim segment or a ragged ring of peaks.

The present crater population on surfaces such as the lunar highlands or the more lightly cratered lunar mare is the outcome of a long history of impact cratering events. Analysis of the existing crater population in conjunction with some assumptions about the rate of crater formation may reveal a great deal about the geologic history of a surface. A typical population is composed of craters with a wide range of sizes, some of which are relatively fresh, with sharp rims, extensive rays, and crisp fields of small secondary craters, whereas others are progressively more degraded. On parts of such airless bodies as the Moon, the principal agent of degradation is other impacts, producing a surface that appears to be crowded with craters. In other regions volcanism has created plains that are more or less sparsely cratered. Elsewhere in the Solar System the activities of wind, water, or tectonic processes such as subduction erase craters within a short time of their formation, leading to landscapes like the Earth's, where impact craters are among the rarest of landforms, or like that of Venus, where the low abundance of craters may be due to an ancient era of resurfacing that obliterated most pre-existing craters.

Study of crater populations is, thus, a powerful tool for geologic investigation of the surfaces of other planets and satellites. If the flux and size distribution of the impacting bodies were known, studies of crater populations could yield absolute ages of the surface and some of its features. Although the original flux is often unknown, relative ages can usually be obtained. Before geologic inferences can be drawn from crater populations, however, we must have an effective means of describing and comparing them. Unfortunately, a large number of descriptions have evolved over the years as each group of scientists studying a particular problem created their own specialized means of presenting population data, making it difficult to compare the results of different groups. In this section I adopt the major recommendations of a NASA panel convened in 1978 (NASA, 1978) to standardize the presentation of crater population data.

6.7.1 *Description of crater populations*

The first step in an investigation of the crater population on a given surface is to select an area that is believed to have had a homogeneous geologic history. It would make little sense, for example, to combine the crater population of a sparsely cratered lava plain with that of a densely cratered upland. Once such an area is selected, the craters that lie within it are counted. Most crater population studies include all recognizable craters, regardless of their state of degradation. Where a large enough population exists, more specialized studies may be performed in which the numbers of fresh craters, slightly degraded craters, degraded craters, etc., are counted separately. Although these studies leave some room for interpretation as to what a "recognizable" or "fresh" crater is, intercomparison of results between different groups of crater counters has generally shown good agreement.

Craters occur in a wide variety of sizes, so that the principal information about a crater population is the number of craters per unit area as a function of crater diameter. It is presumed that impact cratering is a random process and that there is no significance to the particular location of craters within the selected area, so that only data on the number and diameter of craters is kept.

Incremental distribution. Numerous ways of representing the number of craters as a function of diameter have been developed. One very simple method is to list the number N of craters per unit area with diameters between two limits, say between D_a and D_b. The problem with this method is that the resulting number of craters depends upon the interval $\Delta D = D_b - D_a$, and different crater counters may choose different intervals. Furthermore, if the interval ΔD is fixed at, say, 1 km, this might be convenient for craters with diameters between 5 and 20 km, but would be too large for craters with diameters less than 1 km and too small for craters larger than 100 km. A simple way to overcome this problem is to let the interval depend on crater size. Thus, the number of craters may be tabulated between D and $2D$, where the intervals increase in octaves. Actually, this binning has been found to be too coarse in practice, so that most such *incremental* size-frequency distributions use an interval of D to $\sqrt{2}D$. The incremental distribution still suffers, however, from the arbitrary choice of a starting diameter D. It is now recommended that the bins be chosen so that one bin boundary is at $D = 1$ km.

Cumulative distribution. Although the incremental size-frequency distribution could be successful if the same bin sizes and boundaries are universally adopted, it lacks fundamental simplicity. Another distribution has long been used that is independent of bin size: this is the *cumulative* size-frequency distribution. In this distribution the number N_{cum} of craters per unit area with diameters greater than or equal to a given diameter D is tabulated. Not only is the resulting distribution $N_{\text{cum}}(D)$ independent of bin size, but any desired incremental distribution can be easily generated from it, since the number of craters N per unit area in the interval between D_a and D_b is simply:

$$N(D_a, D_b) = N_{\text{cum}}(D_a) - N_{\text{cum}}(D_b) \qquad (6.16)$$

where $D_b > D_a$. $N(D_a, D_b)$ is necessarily positive or zero by the definition of the cumulative number distribution. The only disadvantage of the cumulative distribution is that the cumulative number of craters at some given diameter depends upon the number of craters at all larger diameters. Although this is rarely a major problem, cumulative distributions in limited-diameter intervals (often controlled by the size of the region being analyzed) have to be adjusted in overall value to join with the distributions from other diameter ranges. The slope of the cumulative plot is not, of course, affected by such adjustments.

It has been found in practice that the cumulative number distribution closely approximates a power function of diameter:

$$N_{\text{cum}} = c\,D^{-b} \qquad (6.17)$$

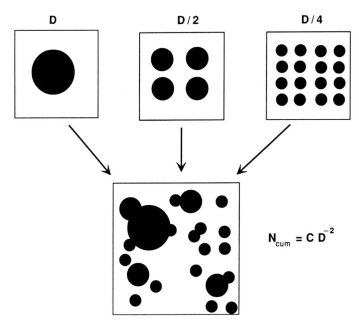

Figure 6.20 A population of craters with a slope of $b = 2$ is structured such that for each crater of diameter D, there are four times as many craters of diameter $D/2$ and 16 times as many as for diameter $D/4$. When mixed together, the population looks very much like that seen on the heavily cratered areas of the Moon and other planets. Note that the total area in each size class of this distribution is the same, a characteristic that may explain why such distributions are so common when formed by either coagulation or impact fragmentation – processes that both depend on surface area.

where $b = 1.8$ for post-mare craters on the Moon between 4 and several hundred kilometers in diameter (and, within the limited data, also seems to hold for impact craters on the Earth).

It is intriguing that the power b is close to 2. A power 2 in Equation (6.17) is a kind of "magic" number because when $b = 2$ the coefficient c is dimensionless (remember N_{cum} is number per unit area). There is no fundamental length or size scale in a crater population with this power law, so that such a population looks the same at all resolutions (see Figure 6.20 for an illustration of such a distribution). It is impossible to tell from a photograph of such a cratered landscape whether the scale of the photograph is 100 km or 1 m (of course, other clues than crater population alone may give a hint about the actual scale). A population of craters described by a power near 2 might arise either from a simple formation process in which there is truly no fundamental length scale or from a series of independent processes that are so complex and chaotic that no one scale dominates.

Because the cumulative number distribution of Equation (6.17) falls as a power of D, it is conventional to graph such distributions on a log-log plot on which a power law is a straight line with slope equal to $-b$ (Figure 6.21). Unfortunately, cumulative number distributions plotted in this way have a tendency to look all the same, apparently differing only in the absolute number density of craters. Although this may be adequate or even desirable for some purposes,

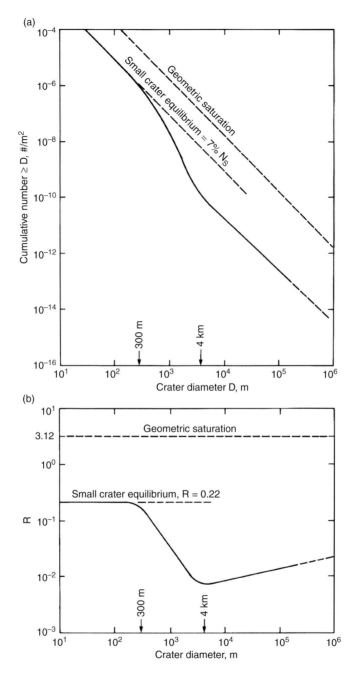

Figure 6.21 The post-mare crater population on the Moon in both a cumulative plot (a) and R-plot (b) format. The population shows three distinct segments. At small diameters ($D < 300$ m) the population is in equilibrium and the cumulative number is proportional to D^{-2}. R is constant for this population. At intermediate diameters (300 m $< D <$ 4 km) the cumulative number is proportional to $D^{-3.4}$, whereas at larger diameters ($D > 4$ km) it falls as $D^{-1.8}$. The curves are dashed at the largest diameters ($D > 200$ km) because no craters of this size have yet formed on the mare. The changes in slope of the crater population are especially evident in the R-plot format. The dashed line labeled "geometric saturation" is an upper limit to the crater density on any surface. After Melosh (1989, Figure 10.2).

such a mode of presentation obscures slight but significant differences in crater populations. Since these differences can be useful in deciphering the source of the crater population or other geologic processes that acted on it, another type of plot is in common use.

The R plot. This type of plot exploits the close approach of b to 2 in Equation (6.17) by graphing essentially the ratio between the actual crater distribution and a distribution with slope –2. A crater population in which the actual slope is −2 would, thus, plot as a horizontal line. The conventional plot of this type is called an R plot (R stands for "Relative"). It is based on an incremental distribution with $\sqrt{2}$ intervals between diameter bins. Note that the slope of this type of incremental plot is the same as the cumulative distribution of Equation (6.17), although the coefficient is different. It is easy to show that if b is constant over the interval D to $\sqrt{2}D$, then the incremental number density is:

$$N(D, \sqrt{2}D) = c\,(1 - 2^{-b/2})\,D^{-b}. \tag{6.18}$$

The definition of the R plot includes several numerical factors. In terms of the cumulative number distribution it is given by:

$$R(D) \equiv \frac{2^{3/4}}{\sqrt{2}-1} D^2 \left[N_{\mathrm{cum}}(D) - N_{\mathrm{cum}}(\sqrt{2}D) \right]. \tag{6.19}$$

Figure 6.21 compares the cumulative and R plots for pose-mare craters on the Moon. A useful interpretation of R is to note that, up to a factor of 3.65, R is equal to the fraction $f_c(D)$ of the total area covered by craters in the diameter interval D to $\sqrt{2}D$:

$$R(D) = 3.65\, f_c(D) \tag{6.20}$$

where both R and f_c are dimensionless numbers. With this interpretation it is easy to see that in a crater population with $b = 2$, for which R and f_c are constant, craters in every size interval occupy the same fraction of the total area. If $b < 2$, as it is for post-mare craters on the moon, $R(D)$ increases as D increases so that large craters occupy a larger fraction of the surface than small craters. If $b > 2$, small craters occupy a larger fraction of the surface than do large ones.

The production population. The impact of the primary meteoroid flux on a planetary surface results in some definite rate of crater production as a function of diameter. As the surface, initially taken to be craterless, ages, more and more craters accumulate on it. The integral of the crater-production rate over the age of the surface is a special, theoretical, crater population called the production population. The production population is the size-frequency distribution of all the craters, excluding secondary craters, that have ever formed since craters began to accumulate on the surface. The population is theoretical in the sense that it neglects all crater-obliteration processes and is, hence, formally unobservable, although the crater population on lightly cratered surfaces may approach the production population closely enough for practical applications.

The production population is a useful concept for the study of the evolution of crater populations. Such studies usually begin with an assumed or inferred production population and

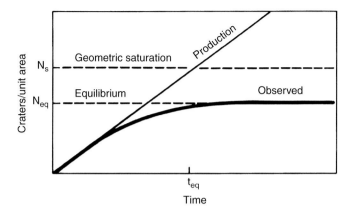

Figure 6.22 Evolution of a crater population in which all craters are the same size. The graph shows how the crater density increases as a function of time. Although the production population rises linearly with time, the number of craters that can be counted on the surface eventually reaches a limit well below the geometric saturation limit. Once the population has reached equilibrium, each additional crater obliterates, on average, one old crater.

then postulate a model of crater obliteration that will, it is hoped, result in a predicted crater population that matches the observed population. The crater-obliteration model is a function of the process being modeled. In the next section we shall consider only the process of crater obliteration by other craters. As we shall see, this can be a surprisingly complex process.

6.7.2 Evolution of crater populations

Given the fact that craters accumulate randomly on the surface of a planet at a rate that is, on average, constant, it should be possible to begin with a knowledge of the rate at which craters form as a function of their size and then predict the crater population at any future time. Furthermore, it should be possible, within certain limits, to invert an evolved crater population and deduce both how long it has been accumulating (that is, the age of the surface) and the rate of crater formation as a function of size.

Simple as these propositions may seem, there has been great difficulty in actually implementing them. The main problem is the interaction between craters of different sizes: the formation of a single large crater on a surface may obliterate many smaller craters, while it takes many small craters to batter a large crater beyond recognition. It has taken many years to fully understand the effects of this interaction on crater populations, and some aspects of it are not completely understood today.

The conceptually simplest population is one in which all craters are the same diameter. Although no natural examples of such a population are known, the study of its evolution introduces several important concepts. Moreover, there are crater populations, such as that on Mimas, in which many of the craters fall within a relatively narrow size range and that may, therefore, be approximated by a population of craters all of the same size.

The equilibrium population. The evolution of this crater population is illustrated in Figure 6.22 as a function of time. The plot shows that the observed crater population and the production population initially increase at the same rate. However, as the density of craters on the surface increases a few older craters are either overlapped by new ones or are buried by their ejecta. As this process continues, some older craters are completely obliterated by younger craters and the observed crater density falls below the production line. Eventually, the crater density becomes so high that each new crater that forms obliterates, on average, one older crater. At this stage the crater population has reached *equilibrium*: no further increase in crater density is possible, although new craters continue to form and the production population increases steadily in number.

The attainment of equilibrium places severe constraints on attempts to date planetary surfaces by crater counting. Up until equilibrium is attained at time t_{eq} in Figure 6.22 the crater density increases with the age of the surface, so that knowledge of the crater production rate permits computation of an absolute age from the crater density. Even if the production rate is not known, the relative ages of two surfaces may be obtained by comparing their crater densities. However, once equilibrium is attained the crater density becomes constant and only a lower limit on the age can be obtained. The relative ages of two surfaces in equilibrium are also completely unconstrained, since their crater densities are identical even though the surfaces may be widely different in age.

Geometric saturation. A useful concept introduced by Don Gault in 1981 (Project, 1981) is that of *geometric saturation*. The idea is to define a crater density that serves as an upper limit to the number of craters that can possibly be recognized on a heavily cratered surface. For a population of craters all of the same size this limit is simply the number of craters per unit area in a hexagonally closest packed configuration, neglecting any possibility of obliteration by overlapping ejecta blankets. In this case, $N_s = 1.15 \, D^{-2}$. The definition of a limiting crater density when craters are of different sizes is less objective. Gault proposed the limit:

$$N_{cs} = 1.54 \, D^{-2}. \tag{6.21}$$

This crater density corresponds to R = 3.12 or a fractional area coverage $f_c = 0.85$. These limits are shown in Figure 6.21: It is clear that lunar crater populations fall well short of this limiting density, although the mare surface is apparently in equilibrium for crater diameters less than about 300 m. The crater density on even the most heavily cratered surfaces seldom reaches more than about 3–5% of the geometric saturation limit, Equation (6.21).

6.8 Dating planetary surfaces with impact craters

It has long been recognized that the number of impact craters per unit area could date planetary surfaces. Shoemaker *et al.* (1963), who wrote the classic paper on this subject, realized that relative dates on the same planet could always be attained, but absolute dates require knowledge of the flux of impacting bodies.

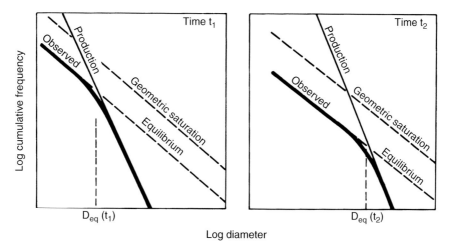

Figure 6.23 Evolution of a crater population with slope $b > 2$. The production population exceeds the equilibrium line at small crater diameters. Small craters are, thus, in equilibrium up to some diameter D_{eq}, above which the observed population follows the production population. The left panel illustrates the population at a relatively early time t_1 and the right panel shows how the population has changed at a later time t_2. The equilibrium diameter D_{eq} increases as a function of time, although this increase is generally not linear.

A full understanding of how to interpret crater populations as a function of size has been long in coming. The attainment of equilibrium by crater populations divides into two distinct cases (and a trivial intermediate). The first, and simplest, case occurs when the production population has a slope b steeper than 2. This case was studied by Gault (1970) and offers the fewest conceptual difficulties. Unfortunately, this distribution is only appropriate for small (≤ 300 m) craters on the lunar mare (see Figure 6.21), although at the time Gault performed his analysis it was believed to be valid for all crater sizes. The size-frequency distribution of larger lunar craters follows a power law with a slope $b = 1.8$. This distribution fits the second case of an evolving population with slope b smaller than 2, and will be treated shortly.

6.8.1 $b > 2$ population evolution

Figure 6.23 illustrates the evolution of a production population with slope b steeper than 2. The left frame depicts the population at an early time t_1 and the right frame is at a later time t_2. Because the production population is steeper than the geometric saturation line, mutual obliteration must occur for sufficiently small craters (craters smaller than $D_{eq}(t_1)$ in Figure 6.23) no matter how early the time, unless, of course, the population has had so little time to evolve that the statistics of small numbers of craters begins to play a role. Because the actual crater density cannot reach the geometric saturation limit, the observable crater density reaches equilibrium somewhere below this line. The difference in the

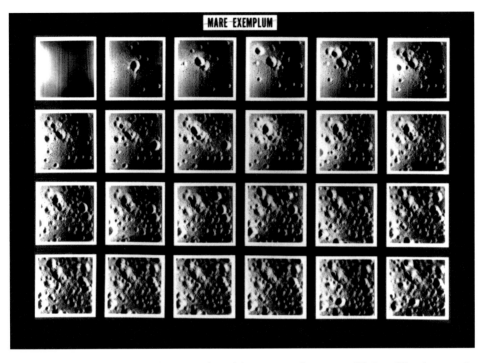

Figure 6.24 A laboratory-scale demonstration of the concept of crater equilibrium. The photographs are of a box 2.5 m square filled 30 cm deep with quartz sand. The sand is topped with 2 cm of carborundum powder to provide a color contrast. Six sizes of projectile were fired into the box at random locations, simulating a production population with a slope index $b = 3.3$, similar to that of small craters on the Moon. Time increases from the upper left horizontally to lower right. Equilibrium is attained about halfway through the simulation: Although individual surface details vary from frame to frame, the crater population in the later frames remains the same. From Gault (1970); photo courtesy of R. Greeley.

number of craters between the projected production population and the observed population is equal to the number of craters obliterated by later impacts. It seems intuitively reasonable that the equilibrium population should follow a line parallel to, but below, the geometric saturation distribution. Gault showed empirically in small-scale impact experiments (Figure 6.24) that this is the case, and subsequent work has confirmed this result both analytically (Soderblom, 1970) and by Monte Carlo computer simulations (Woronow, 1977). In Gault's experiments the crater equilibrium density depends on the slope b of the production population, with steeper slopes giving a lower equilibrium crater density. Similar results were also obtained from the theoretical studies.

The observed population at any one time in Figure 6.23 is, thus, composed of two branches. Small craters follow an equilibrium line with slope $b = 2$. Larger craters follow the steeper production population curve. The inflection point between these two curves is at diameter $D_{eq}(t_1)$ where the production curve crosses the equilibrium line. The right-hand

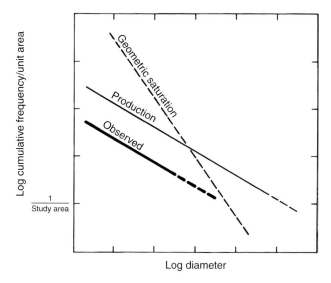

Figure 6.25 Evolution of a crater population with slope $b < 2$. The production population exceeds the geometric saturation line at large diameters. Under these circumstances the evolution of the population is dominated by large impacts that allow a number of small craters to accumulate before wiping the surface clean. The observed population line is always parallel to the production population, but may lie considerably below it. The production population line is dashed at the large-crater end because there must be at least one crater within the study area for the line to be meaningful. The observed population may approach geometric saturation if the entire study area is the site of one large impact, but the population will be dominated by fluctuations because of the statistics of small numbers.

frame in Figure 6.23 shows the crater density at a later time. The crater population is qualitatively similar to that at the earlier time, except that the transition diameter $D_{eq}(t_2)$ is larger. If the rate at which craters accumulate is constant (that is, c in Equation (6.17) is a linear function of time), it is easy to show that the transition diameter $D_{eq}(t)$ grows as $(time)^{1/(b-2)}$. Conversely, if the crater production rate, slope b, and the present D_{eq} are known, the age of the surface can be computed.

6.8.2 b < 2 population evolution

Figure 6.25 illustrates the evolution of a crater population with slope b less than 2. Note that in the unlikely event that b exactly equals 2, the observed crater density simply maintains the slope of 2 until the density reaches equilibrium (which occurs simultaneously at all diameters) after which the observed crater density remains constant and the slope remains 2. When b is less than 2 the situation is more complex. In this case the production curve exceeds the geometric saturation line at the large-diameter end of the scale. Gault's (1970) model could not deal with this situation and it took nearly 15 years before the implications of this large-diameter crossing were understood, in spite of the developing knowledge that $b \sim 1.8$ for craters more than about 4 km in diameter on the lunar mare. The situation was resolved by C. R. Chapman (Chapman and McKinnon, 1986) who performed a Monte Carlo simulation of crater population evolution that included a wider range of diameters than had previously been possible.

Chapman realized that, first of all, the large-diameter end of the cratering curve is dominated by the statistics of small numbers. Even though the production curve apparently crosses the geometric saturation line at all times, enough time must pass after the formation of the initial uncratered surface that at least one large crater has formed on it. The crossing of the two curves makes little sense unless the cumulative number of craters larger than D_s is at least 1 in the finite area under study. Since the probability of a small impact is larger than the probability of a large impact for any $b > 0$, a population of small craters initially develops that follows the production curve closely. Eventually, however, a large impact occurs. With $b < 2$, there is a high probability that this large crater obliterates all or a significant fraction of the study area, wiping out nearly all previous smaller craters. The number of observed craters thus suddenly drops below the production curve. As time passes, small craters again accumulate on the surface. The slope of this new population is equal to that of the production curve but, as shown in Figure 6.25, the cumulative number of such craters is smaller than the production population. The number of small craters continues to grow until the next large impact wipes the slate clean once more.

Under these circumstances there is no "equilibrium" population, although the observed crater density is always well below the geometric saturation limit. The observed crater density fluctuates widely and irregularly, controlled by the large, rare catastrophic-impact events. In spite of these wide variations in density, the slope of the observed population at any given time is roughly equal to that of the production population. The crater densities on a surface of this type are spatially patchy, being low at the sites of recent large impacts and high in areas that have not been struck for a long time by one of the large impacts. Dating such a surface is nearly impossible after the first large impact, unless some area can be found that has escaped all large impacts. In practice, all that can be determined from the crater density on such a surface is the date since the last large impact that affected the particular study region.

6.8.3 Leading/trailing asymmetry

An important assumption in the relative dating of planetary surfaces is that the cratering rate is uniform over the entire body. Although this is true to a high degree of approximation for most bodies in the Solar System, it may be badly violated for synchronously locked planets and moons. Our Moon and the Galilean satellites of Jupiter circulate about their primary with the same face always leading. Just as a car driving into a rainstorm encounters more raindrops on its front windscreen then on the back window, the impact flux on synchronous satellites is higher on the leading side than the trailing side, leading to an asymmetry in cratering rate that may be as large as a factor of 20 or more, depending on the cratering population (Zahnle *et al.*, 2001). The degree of asymmetry depends on the orbits of the impactors. The asymmetry is largest for impactors in orbit about the same primary or moving at a low relative velocity. In this case the impact velocity on the leading hemisphere is the sum of the orbital velocities of the satellite and impactor, while the velocity on the trailing hemisphere is the difference, so that not only is the flux of impactors different between the leading and trailing hemispheres, but the different impact velocities mean that the same size of impactor yields a different size of final crater. The leading/trailing asymmetry is smaller for heliocentric populations of impactors: Comets, for example, produce much more uniform crater populations than co-orbiting objects on synchronous satellites.

Neptune's moon Triton has a large leading/trailing crater asymmetry, but the Galilean satellites of Jupiter show much less crater asymmetry than expected on theoretical grounds. Some process must, therefore, occasionally reorient their surfaces relative to Jupiter: Either an exchange of their primary-facing and opposite hemispheres due to exchange of their A and B moments of inertia by internal or external (cratering?) mass transfers, or perhaps the momentum impulse of a large cratering event.

6.9 Impact cratering and planetary evolution

Over the last few decades it has become increasingly clear that impact cratering has played a major role in the formation and subsequent history of the planets and their satellites. Aside from their scientific interest, impact craters have also achieved a modest economic importance as it has become recognized that the fabulously rich Sudbury nickel deposit in Ontario, Canada, is a tectonically distorted 140-km diameter impact crater. Similarly, the Vredefort structure in South Africa is a large, old, impact crater. Oil production has been achieved from a number of buried impact craters, such as the 10-km diameter Red Wing Creek crater in the Williston Basin and the 3.2-km diameter Newporte structure in North Dakota. On a more homely level the 60-km diameter ring-shaped depression in the Manicouagan crater in eastern Quebec is currently used as a reservoir supplying water to New York City.

6.9.1 Planetary accretion

Modern theories of planetary origin suggest that the planets and the Sun formed simultaneously 4.6×10^9 yr ago from a dusty, hydrogen-rich nebula. Nebular condensation and hydrodynamic interactions were probably only capable of producing ca. 10-km diameter "planetesimals" that accreted into planetary-scale objects by means of collisions. The timescale for accretion of the inner planets by mutual collisions is currently believed to be between a few tens and one hundred million years. Initially rather gentle, these collisions became more violent as the random velocities of the smaller planetesimals were increased during close approaches to the larger bodies. The mean random velocity of a swarm of planetesimals is comparable to the escape velocity of the largest object, so as the growing planetary embryos reached lunar size, collisions began to occur at several kilometers per second. At such speeds impacts among the smaller objects were disruptive, whereas the larger objects had sufficient gravitational binding energy to accrete most of the material

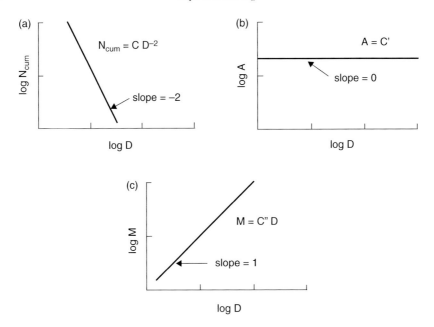

Figure 6.26 The numerical foundation of impact catastrophism. The size-frequency distribution of a typical crater population with slope $b = 2$ is shown in panel (a). In any such distribution there are many more small craters than large ones. Panel (b) shows the area of craters as a function of diameter. This distribution is flat: The total area of small craters and large craters is equal (refer back to Figure 6.20). Panel (c) shows, however, that the mass is concentrated at the large-size scale, so that despite the overwhelming numerical superiority of small craters, the few large ones are the most important in terms of either mass or energy delivered.

that struck them. Infalling planetesimals bring not only mass, but also heat to the growing planets. In the past it was believed that the temperature inside a growing planet increased in a regular way from near zero at the center to large values at the outside, reflecting the increase in collision velocity as the planet became more massive. However, it now seems probable that the size distribution of the planetesimal population was more evenly graded between large and small objects, and that each growing planetary embryo was subjected to many collisions with bodies comparable in size to the embryo itself. Such catastrophic collisions would deposit heat deep within the core of the impacted body, wiping out any regular law for temperature increase with increasing radius and making the thermal evolution of growing planets rather stochastic.

6.9.2 Impact catastrophism

Although small craters (and the impactors that created them) are overwhelmingly more abundant than large ones in terms of numbers, the characteristic $b \approx 2$ slope of most such distributions in the Solar System implies that most of the mass actually resides at the large end of the size spectrum (see Figure 6.26). This circumstance gives rise to the idea of impact catastrophism: Although small impacts are relatively common, all of the small impacts combined do less geologic work than a few large impacts. This is the formal definition of a catastrophic process and it is a good description of the way that impacts affect the geological evolution of the planets. It also implies that stochastic processes cannot be neglected where impacts play a role, except in situations where b is substantially larger than 2, such as for the Shoemaker–Gault theory of regolith evolution on the Moon.

Note that the usual statistical mechanical idea of energy equipartition does not play a role in the Solar System: The dynamics of the Solar System are such that the time to relax to a state of thermal equilibrium is far, far longer than the age of the Solar System, so that the biggest, most massive objects also carry the most energy.

6.9.3 Origin of the Moon

The origin of the Moon is now attributed to a collision between the proto-Earth and a Mars-size protoplanet near the end of accretion 4.5×10^9 yr ago. This theory has supplanted the three classic theories of lunar origin (capture, fission, and co-accretion) because only the giant impact theory provides a simple explanation for the Moon's chemistry, as revealed in the lunar rocks returned by Apollo. One view of this process is that a grazing collision vaporized a large quantity of the proto-Earth's mantle, along with a comparable quantity of the projectile. While most of the mass of the projectile merged with the Earth (incidentally strongly heating the Earth: If the Earth was not molten before this impact it almost certainly was afterward), one or two lunar masses of vapor condensed into dust in stable Keplerian orbits about the Earth and then later accumulated together to form the Moon.

6.9.4 Late Heavy Bombardment

Sometime after the Moon formed, and before about 3.8×10^9 yr ago, the inner planets and their satellites were subjected to the "Late Heavy Bombardment," an era during which the impact fluxes were orders of magnitude larger than at present. The crater scars of this period are preserved in the lunar highlands and the most ancient terrains of Mars and Mercury. A fit to the lunar crater densities using age data from Apollo samples gives a cumulative crater density through geologic time of:

$$N_{cum}\,(D > 4 \text{ km}) = 2.68 \times 10^{-5}\,[T + 4.57 \times 10^{-7}\,(e^{\lambda T} - 1)] \qquad (6.22)$$

where $N_{cum}(D > 4 \text{ km})$ is the cumulative crater density (craters/km^2) of craters larger than 4 km diameter, T is the age of the surface in Gyr ($T=0$ is the present) and $\lambda = 4.53$ Gyr^{-1}. The current cratering rate on the moon is about 2.7×10^{-14} craters with $D > 4$ km/km^2/yr. On the Earth the cratering rate has been estimated to be about 1.8×10^{-15} craters with $D > 22.6$ km/km^2/yr, which is comparable to the lunar flux taking into account the different minimum sizes, since the cumulative number of craters $N_{cum}(D) \sim D^{-1.8}$. There is currently much debate about these cratering rates, which might be uncertain by as much as a factor of 2.

Although Equation (6.22) describes an impact flux that decreases monotonically with time, there is currently much debate about the possible reality of an era of "heavy bombardment" between about 4.2 and 3.9 Gyr ago in which the cratering rate reached a local peak (Strom *et al.*, 2005). It is now supposed that eccentric asteroids from the main asteroid belt caused this peak in flux. These asteroids were mobilized as destabilizing orbital resonances swept through the asteroid belt when Jupiter and Saturn underwent an episode of planetary migration. The heavily cratered surfaces of the Moon and terrestrial planets are supposed to have formed at this time. The intense flux obliterated any evidence of an earlier surface on these bodies.

The cratering rates on Mars and Venus are believed to be comparable to that on the Earth and Moon; however, the exact fraction of the Earth/Moon rate is presently uncertain and a subject of controversial discussion. The exact rates will probably remain unknown until radiometric dates on cratered surfaces are determined by means of sample returns from these bodies. Cratering rates in the outer Solar System are even more uncertain and controversial: Most of the craters on the satellites of Jupiter and Saturn are probably formed by comets, whose flux is very uncertain at the small sizes represented by most observed craters (Dones *et al.*, 2009).

The high cratering rates in the past indicate that the ancient Earth must have been heavily scarred by large impacts. Based on the lunar record it is estimated that more than 100 impact craters with diameters greater than 1000 km should have formed on the Earth. Although little evidence of these early craters has yet been found, it is gratifying to note the recent discovery of thick impact ejecta deposits in 3.2 to 3.5 Gyr Archean greenstone belts in both South Africa and Western Australia. Since rocks have recently been found dating back to 4.2 Gyr, well into the era of heavy bombardment, it is to be hoped that more evidence for early large craters will be eventually discovered. Heavy bombardment also seems to have overlapped the origin of life on Earth. It is possible that impacts may have had an influence on the origin of life, although whether they suppressed it by creating global climatic catastrophes (up to evaporation of part or all of the seas by large impacts), or facilitated it by bringing in needed organic precursor molecules, is unclear at present. The relation between impacts and the origin of life is currently an area of vigorous speculation.

6.9.5 Impact-induced volcanism?

The idea that large impacts can induce major volcanic eruptions is one of the recurring themes in the older geologic literature. This idea probably derives from the observation that all of the large impact basins on the Moon's nearside are flooded with basalt. However, radiometric dates on Apollo samples made it clear that the lava infillings of the lunar basins are nearly 1 Gyr younger than the basins themselves. Furthermore, the farside lunar basins generally lack any lava infilling at all. The nearside basins are apparently flooded merely because they were topographic lows in a region of thin crust at the time that mare basalts were produced in the Moon's upper mantle. Simple estimates of the pressure release caused by stratigraphic uplift beneath large impact craters make it clear that pressure release melting cannot be important in impacts unless the underlying mantle is near the melting point before the impact (Ivanov and Melosh, 2003). Thus, it is probably safe to say that, to date, there is no firm evidence that impacts can induce volcanic activity. Impact craters may create fractures along which pre-existing magma may escape, but themselves are probably not capable of producing much melt. Nevertheless, there were massive igneous intrusions associated with both the Sudbury and Vredefort structures whose genesis is sometimes attributed to the impact, although in this case they may have been triggered by the uplift of hot lower crust. Further study of these issues is needed.

6.9.6 *Biological extinctions*

The most recent major impact event on Earth was the collision between the Earth and a 10 to 15 km diameter asteroid 65 Myr ago that ended the Cretaceous era and caused the most massive biological extinction in recent geologic history. Evidence for this impact has been gathered from many sites over the last decade, and is now incontrovertible (Schulte *et al.*, 2010). First detected as an enrichment of the siderophile element iridium in the ca. 3 mm thick K/Pg (Cretaceous/Paleogene) boundary layer in Gubbio, Italy, the iridium signature has now been found in more than 100 locations worldwide, in both marine and terrestrial deposits. Accompanying this iridium are other siderophile elements in chondritic ratios, shocked quartz grains, coesite, stishovite and small (100–500 μm) spherules resembling microtektites. All these point to the occurrence of a major impact at the K/Pg boundary. The impact crater is located beneath about 1 km of sedimentary cover on the Yucatan Peninsula of Mexico. Known as the Chicxulub crater, it is about 170 km in diameter and is presently the subject of intensive study.

Further reading

A general survey of all aspects of impact cratering from which several sections of this chapter were abstracted was published by Melosh (1989). This book is presently out of print and often difficult to obtain, but many university libraries possess a copy. A more popular but generally clear and accurate description of terrestrial impact craters can be found in Mark (1987). A good description of the three largest craters on Earth can be found in Grieve and Therriault (2000). Although deeply eroded or otherwise obscured, these craters provide the "ground truth" about impacts that images of extraterrestrial craters cannot give. The surface of our Moon is dominated by impact craters and so the account of its geology by Wilhelms (1987) is necessarily an account of the process of impact cratering. Don Wilhelms also wrote an insightful and very readable historical account of lunar exploration that emphasizes the growing appreciation of impact cratering though the course of the Apollo missions (Wilhelms, 1989). The importance and mechanics of impact crater collapse and the simple-to-complex transition is treated by Melosh and Ivanov (1999), while the special features of oblique impacts are the subject of a review by Pierazzo and Melosh (2000). The best current summary of the scaling of impact crater ejecta was stimulated by analysis of the ejecta from the Deep Impact Mission cratering event (Richardson *et al.*, 2007). A fine review of the details of crater counting and the application of crater counts to the dating of surfaces appears in a rather unlikely-sounding book, which was the subject of a collaborative project called the *Basaltic Volcanism Study Project* (Project, 1981). The near-field effects of an impact on the Earth are reported by an on-line computer program, whose basic algorithms are described in detail by Collins *et al.* (2005).

Exercises

6.1 Crater dimensions

a) Suppose that the transient crater that collapsed to form the Imbrium Basin on the Moon was $D_{tc} = 800$ km in diameter. Assuming that it struck vertically at $v_i = 22$ km/s, what was the diameter of the asteroid that produced Imbrium?

Use the revised Schmidt–Holsapple scaling law (this form implicitly assumes that the target and projectile have the same density):

$$d = 0.671\, D_{tc}^{1.28} \left(\frac{g}{v_i^2} \right)^{0.277}. \qquad \text{(MKS units)}$$

b) If this object struck at 45° to the vertical, what is the period of the Moon's rotation thus imparted, assuming the Moon was free in space (unlikely, but easy to analyze)? Compare this to the Moon's current rotation rate. What can you conclude about the ability of large impacts to unlock the Moon's present synchronous rotation state?

6.2 Crater collapse

a) The empirical relation for the thickness T of an impact crater's ejecta blanket indicates that the average thickness declines as $(R_c/r)^3$, where R_c is the crater radius and

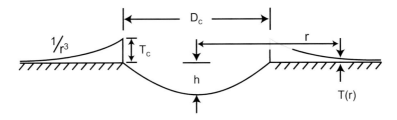

Figure 6.27

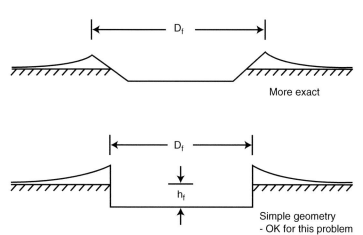

Figure 6.28

r is distance from the crater's center. Assuming that the interior shape of a simple (or transient) crater is described by a paraboloid of revolution with diameter/depth ratio $D_c/h \approx 5$ (use the rim-to-rim diameter D_c and depth h below the original ground surface), and assuming that the volume of the ejecta blanket and crater bowl are equal, derive an equation for the ejecta blanket thickness T_c at the rim. (Hint: It is permissible at this level of accuracy to treat the inner edge of the ejecta blanket as if it were a vertical cliff. An exact result, however, *can* be obtained with a bit of effort, and anyone who can legitimately get the result $T_c = h/5$ should get extra credit for this assignment!)

b) Now suppose the crater collapses to a uniform, constant depth h_f (3 km for the Moon) over its entire interior. Again using volume conservation, how much does the rim-to-rim diameter increase for a 100 km (final diameter) crater?

6.3 Swedish rock rain

In addition to the 5 proven and 33 possible impact crater scars that adorn the country of Sweden, Lilljequist and Henkel (1996) proposed the existence of a truly world-class crater, the Uppland structure, which is supposed to be 320 km in diameter. It encompasses the area around Stockholm up to Uppsala. Using Shoemaker's estimated cratering rate on Earth, N_{cum} ($D > 22.6$ km) $= 1.8 \times 10^{-15}$ craters/km^2/yr, and the post-mare size exponent $b = 1.8$ ($N_{cum} \sim D^{-1.8}$), compute the largest crater likely to have formed in Sweden (area $\sim$ 450 000 km^2) since the formation of its surface rocks (1600 to 2300 Myr; use 2000 Myr as a reasonable average).

How does this compare to the size of the putative Uppland structure? Put another way, what is the probability of the formation of a crater as large as Uppland in the past 2000

Myr in Sweden? What does this tell you about the Uppland "crater"? Note that the largest *confirmed* crater in Sweden is the 55 km-diameter Siljan structure.

Using the same cratering flux, compute the maximum size of crater likely to be found in the UK (now it is the student's task to look up the area and mean geologic age of that fair but damp archipelago). Note that a single, probably 4 km diameter, impact crater was discovered in the North Sea by Stewart and Allen (2002) – the Silverpit crater!

6.4 Those blasted Martian rocks!

F. Hörz *et al.* (1999) argued that dish-shaped depressions on large rocks at the Pathfinder landing site on Mars, along with split boulders, might be due to the impact of millimeter to centimeter diameter meteorites at speeds of 300 m/s or more (corresponding to the muzzle velocity of a high-speed rifle).

Evaluate the ability of the Martian atmosphere (surface pressure $P_S = 600$ Pa) to stop small meteoroids (average pre-atmospheric speed 7 km/s) by comparing the mass of the projectile to the mass of atmospheric gases swept out of the cylindrical volume defined by the meteoroid's path (assume a straight trajectory for the purposes of this estimate).

The total mass of atmospheric gas above a unit area on the surface of a planet is P_S/g, where g is the acceleration of gravity at the surface. Perhaps surprisingly, you do not need to know the density of the atmosphere as a function of height to do this problem (convince yourselves of this!).

Using momentum conservation, estimate how much the meteoroid is slowed upon arriving at the surface as a function of its mass and density. Finally describe how reasonable the Horz–Cintala proposal really is for both stony and iron meteoroids. Could this process be effective on high mountaintops on Earth? Why or why not?

Extra credit

As the meteoroid traverses the atmosphere it encounters a ram pressure of $\rho_a v^2$ on its leading face (ρ_a is the density of the atmosphere, 0.0104 kg/m³ at the surface of Mars), while its rear is nearly a vacuum for supersonic flight. Estimate the maximum stresses thus acting to crush the meteoroid and compare them to the typical strength of rock or iron for the meteoroids in the problem above. (NB, you *do* have to know the height distribution of density for this problem – you may want to know that the scale height of the Martian atmosphere is about 10 km.)

6.5 Titan gets its kicks!

One of the mysteries about Saturn's large moon Titan is its (slightly) eccentric orbit, $e = 0.028$. If Titan could dissipate tidal energy as Io or Europa does, its orbit would have circularized long ago. It has been proposed that Titan's Xanadu region is a gigantic impact structure, 700 km in diameter (with an ejecta blanket bringing the diameter of the entire feature to 1800 km). Using the impact crater scaling relation in Problem 6.1, estimate the size of the object that could have made this crater, assuming a cometary impact velocity of 30 km/s (do not forget the difference between the transient and final crater!). If this object struck the equator of Titan at the optimum place and angle to transform its linear momentum to orbital angular momentum, could this impact have imparted enough momentum to account for Titan's present eccentricity? You may need to know that the orbital angular momentum H of a planet of mass m circling a body of mass M with a semimajor axis of R and eccentricity e is:

$$H = m\sqrt{GMR(1-e^2)}.$$

I will leave you with the task of looking up the values of the necessary parameters. Don't forget that you are looking for a *change* of orbit from circular ($e = 0$) to elliptical. Note that this problem is akin to the oft-asked question of whether the K/Pg impact that killed the dinosaurs could also have knocked the Earth "out of its orbit." What do you think about this possibility?

7

Regoliths, weathering, and surface texture

> It is generally accepted that the dynamic nature of the [lunar] regolith is primarily the result of meteoritic bombardment. Although other contributing mechanisms have been suggested (e.g. electrostatic levitation …) the formation of agglutinates, breccias, diaplectic glasses, and the presence of significant contamination of the regolith with meteoritic material are collectively unambiguous evidence in support of the dominant role meteoritic impact has played in regolith evolution.
>
> Gault *et al.* (1974)

7.1 Lunar and asteroid regoliths: soil on airless bodies

Impact pioneer Gene Shoemaker and his colleagues introduced the modern concept of a planetary regolith in 1967 (Shoemaker *et al.*, 1967). Following the first soft landing on the Moon's surface by the Russian probe Luna 9 in 1966, Surveyors 1, 3, and 5 were the first successful landers of the American Surveyor program. Detailed analysis of their images by Shoemaker's team was the first step toward clarifying the nature of an airless planetary body's surface. Geologist R. B. Merrill had coined the word "regolith" in 1897 to designate the fragmental layer of rock debris that mantles the Earth's surface. Compounded from the Greek words for "blanket" (*rhegos*) and "stone" (*lithos*), the word had become obsolete by 1967, allowing Shoemaker to revive it in the context of the Moon's surface.

Although the lunar surface layer is sometimes referred to as "lunar soil," it is utterly different from the familiar agricultural soil of the Earth. Terrestrial soils typically contain large amounts of organic carbon and weathered rock material that has been transported by wind or water. Living organisms play a major role in creating Earth's soil layers. In contrast, the lunar surface is blanketed by a loose breccia composed of broken, angular rock fragments that range in size from fine, submicron dust to meter-sized blocks. The lunar regolith has never been disturbed by wind, water, or organic life. Instead, it is created and maintained by the steady hail of meteoritic debris and radiation from open space (Figure 7.1).

The lunar regolith is still, by far, the best understood surface of an airless body thanks to the many *in situ* investigations and sample returns by both manned and unmanned space missions (Heiken *et al.*, 1991). Although the ubiquitous regolith prevented the Apollo

276

Figure 7.1 Apollo 17 panorama showing details of the regolith at Station 1 on EVA 1. Note the chaotic mixture of dust and rock fragments outcropping on the surface, and impact craters of all sizes in the foreground. The hills in the background display the "elephant hide" texture typical of lunar slopes. Wessex Cleft and the Sculptured Hills are visible in the background. Hasselblad photo AS17–134-2048.

astronauts from directly sampling any bedrock exposures, it was quickly realized that 90% of the rocks lying on and in the regolith are of local origin: The fact that the mare basalt/ highland contacts are still visible after 4 Gyr of meteoritic bombardment makes it clear that lateral transport of regolith is very inefficient. Nevertheless, the remaining 10% of the rego-lith is composed of material exotic to the collection site and represents debris thrown great distances by large meteorite impacts. Up to about 2% is composed of meteoritic material foreign to the Moon. A random regolith sample is, thus, much more informative about both local and global geology than a comparable grab-bag sample of the Earth's soil.

By terrestrial standards, lunar regolith is a most peculiar deposit. Its density is very low at the surface, about 1400 kg/m^3, but it rapidly compacts to about 2000 kg/m^3 only a meter below the surface (Figure 7.2). It is a gray, gritty, and extremely abrasive powder, full of poorly sorted glass shards and angular rock fragments. Most of its particles range in size from around 40 μm up to a few hundred microns, with rare, larger rock clasts. Its electrical conductivity is low and it easily acquires electric charges so that it sticks to everything: The astronauts found that it coated any surface that it came in contact with and rapidly fouled the interior of their spacecraft and the seals of their space suits. Even the loose surface layers have a small cohesive strength, so that the astronaut's bootprints retained vertical sides a few centimeters high. The regolith has a large coefficient of internal friction, so the lunar surface has a considerable bear-ing strength, allaying initial fears that landers would sink into a deep powdery surface layer.

On the Moon, the regolith ranges from 4 to 5 m deep on the mare to perhaps 20 m deep in the highlands. Its depth is highly irregular even over short horizontal distances

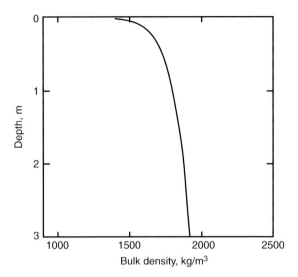

Figure 7.2 Calculated *in situ* bulk density of the lunar regolith as a function of depth below the surface. The curve shown is a power-law fit, $\rho = 1799 \, z^{0.056}$, where depth z is in meters. Simplified after Figure 9.16 of Heiken *et al.* (1991).

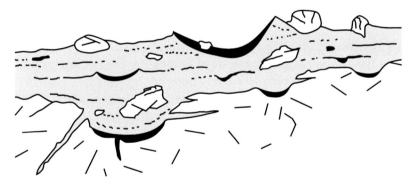

Figure 7.3 Schematic cross section of the regolith (gray), indicating its irregular character. The lunar regolith was created by repeated impacts of all sizes that generated a heterogeneous deposit containing abundant impact melt (black) and broken fragments of bedrock (white).

(Figure 7.3). Drive tube cores reveal many layers whose thickness varies from about 1 to 10 cm. Spectrally, the lunar regolith is much darker than fresh rock of the same composition and reflects a larger fraction of long-wavelength (red) light. Spectral absorption bands of minerals composing the bedrock are muted (Figure 7.4). Mineral fragments in the regolith are riddled with tracks produced by energetic cosmic rays, and solar wind gases are implanted below the surfaces of many grains. Although the Moon's interior seems to be nearly devoid of water, recent remote spectral observations have detected OH and H_2O molecules produced as solar wind protons bond chemically with oxygen in silicate minerals on the surface (Sunshine *et al.*, 2009). Neutrons liberated by primary cosmic rays

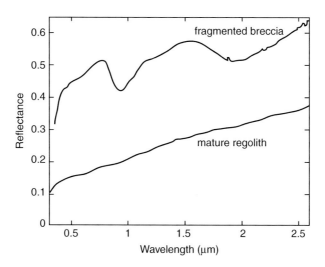

Figure 7.4 Spectral reflectance of mature regolith and fragmented breccia. As regolith matures it becomes darker (lower reflectance) and the prominent absorption bands at 0.9 and about 2 μm in the fresh breccia disappear. Bidirectional reflectance spectra of noritic breccia 67455 showing pyroxene absorption bands and mature regolith 68501, both collected by the Apollo 16 mission. After Figure 4 of Clark *et al.* (2002).

penetrate many meters into the surface, creating exotic isotopes whose presence records the relentless irradiation and permits the total duration of exposure to be determined.

The regoliths of asteroids and other airless bodies are much less well understood than that of the Moon. It was once believed that asteroids could not retain loose surface material because meteorite impacts would eject it from their weak gravity fields. Surprisingly, regoliths on the few asteroids for which thicknesses have been estimated are, if anything, deeper than the lunar regolith. The regoliths of Ida and Eros are at least tens of meters thick, while Phobos' regolith may be as deep as 200 m. Although it is now clear that most asteroids are, in fact, mantled with loose surface material, asteroid regoliths seem to be much less "mature" than that of the Moon, containing less glassy material. Nevertheless, asteroid reflectance spectra clearly show the effect of "space weathering" that alters the way that they reflect light and makes it difficult to connect astronomically observed asteroids with meteorites in the laboratory (Clark *et al.*, 2002). The space environment also affects the spectra of icy bodies in still unknown ways: Controversies presently rage over whether radiation or hydrated salts control the spectra of icy satellites such as Europa. Organic "tars" on the surfaces of comets and objects in the outer Solar System play similarly obscure roles.

7.1.1 Impact comminution and gardening

The major process creating and shaping the regolith is meteorite impact. When a fresh rock is exposed on the surface of the Moon it is immediately pockmarked by small meteorites. Indeed, the faceplates of the Apollo astronauts' space suits recorded an extensive sample of micrometeorite impacts after only a few hours' exposure to the lunar environment. As time passes, more and more impacts accumulate on exposed rock surfaces. Surface rocks are occasionally flipped over, so that small craters are found on their undersides. Small impacts occur much more frequently than large ones (see the discussion of crater populations in Section 6.7) so that a surface is typically covered over with small impacts, before it becomes more deeply excavated by large impacts. Boulders lying on the Moon's surface are eventually destroyed by an impact large enough to burst it into small fragments.

Crater populations are best described by the cumulative size-frequency distribution $N_{cum}(D)$, which is the number of craters per unit area of diameter equal to or greater than D. Observation shows that this number density usually approximates a power-law (fractal) relationship:

$$N_{cum}(D) = c\,D^{-b} \qquad (7.1)$$

where c is equal to the number density of craters of diameter $D = 1$ km and the exponent b indicates how sharply the density of craters falls as their diameter increases. If $b = 2$, craters occupy the same fraction of the surface's area, independent of their diameter (crater area is proportional to D^2, so when $b = 2$ the total area occupied by craters of diameter equal to or larger than D is proportional to $N_{cum}(D)\,D^2$, which is independent of D only when $b = 2$). However, impact crater distributions at the sizes important for regolith formation typically show $b > 2$: It ranges between 2.9 and 3.4 for lunar craters smaller than 4 km in diameter. For distributions of this type, the surface is completely covered over by small craters before it is covered with large craters. The case $b < 2$ does occur, but it is much more complicated and the reader is referred to Section 6.8.2 for more information on this interesting situation. Fortunately, this case seems important only for large impact craters.

To better understand regolith formation, pretend, for a moment, that all impact craters are the same size. As these impact craters begin to accumulate on a freshly exposed surface, their number grows steadily with time. Each new crater blasts out virgin rock and throws it radially away from its rim, heaping it into an ejecta blanket outside the crater. As craters become denser, their ejecta blankets begin to overlap and the ejecta from one crater partially fills earlier craters. New craters then form in both virgin rock and in previously excavated ejecta. As still more craters accumulate, less and less virgin rock remains and more craters form in pre-existing ejecta. Eventually, the surface is entirely covered over with craters and, on average, each new crater obliterates an older crater. When this occurs, the number of observable craters ceases to grow and the population reaches "equilibrium." New craters continue to form, but the material they excavate is the ejecta from former craters. Because craters of a given size fracture and excavate the rock to a depth of about ¼ of their diameter, the bottom of this continually overturned layer is also about ¼ of the crater diameter and its depth does not increase with time: The broken debris covering the surface is just recycled as impact after impact accumulates. The overturn time of this layer is roughly the time needed for the area of the accumulated craters to equal the total surface area.

Now recognize that craters of all sizes are actually forming. In a population with $b > 2$, however, small craters cover the area faster than big craters, so the surface reaches

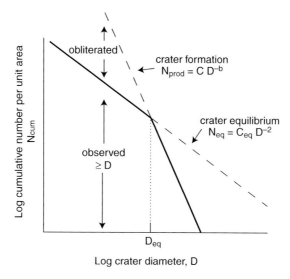

Figure 7.5 The population of impact craters observed on the lunar regolith is in equilibrium at small sizes (below D_{eq}) and reflects the production population at larger sizes for populations with slopes $b > 2$. The heavy curve shows the observed population, while the extension of the production population indicates the total number of craters that have struck the surface. The difference in these two curves indicates how many craters have been obliterated.

equilibrium for small craters long before it equilibrates for big craters (Figure 7.5). While the big craters are still blasting out virgin rock, small impacts simply recycle the shallower surface layers (with an occasional addition of deeper-seated material from a larger impact). The regolith layer now gradually deepens as it comes into equilibrium at progressively larger sizes, while the upper layers are recycled at rates that depend on their depth. This rapid overturn at shallow depths and slow overturn for deeper material is known as "gardening," a term that aptly describes this process. It is analogous to a garden whose owner digs it up once a year, hoes it weekly, and lightly rakes it every day.

Mathematical models of regolith growth were developed and successfully applied to the Moon by D. Gault, E. M. Shoemaker, and their colleagues in the late 1960s (e.g. Gault, 1970). In these models the base of the regolith is the depth excavated by craters (of diameter D_{eq}), whose number is just sufficient to reach equilibrium. Of course the exact location of each of these craters is determined by chance, so the base of the regolith is very irregular, with some areas cratered more than once by impacts that land on top of one another, adjacent to areas that have, for the time being, entirely escaped cratering at this diameter. One can show that the depth of such a regolith grows at the rate of $(\text{time})^{1/(b-2)}$, assuming a constant rate of impact (see Box 7.1). This rate is not linear unless $b = 3$, even for a constant impact rate. However, given that the observed values of b are close to 3, it is approximately correct to say that the lunar regolith has accumulated at the rate of about 1.2 m/Gyr since the era of heavy bombardment. This is a very slow rate and hardly suggests that lunar regolith is a renewable resource for future lunar colonists.

Box 7.1 **Growth of the lunar regolith**

The lunar regolith grows by the gradual accumulation of meteorite impacts. The earliest theoretical studies of the growth of the regolith by E. M. Shoemaker *et al.* (1969) and all subsequent studies are based on the evolution of crater populations with slopes $b > 2$. Although b is generally less than 2 for large craters on the lunar mare, it typically ranges between 2.9 and 3.4 for craters smaller than 4 km in diameter, so the assumption that $b > 2$ is valid for the post-mare lunar regolith and we shall consider only this case.

As shown in Figure 6.21, at any given time a $b > 2$ crater population is in equilibrium up to some maximum crater diameter $D_e(t)$, where $D_e(t)$ increases as (time)$^{1/(b-2)}$ for a constant rate of meteoritic bombardment. The crater population is in equilibrium with the production rate at diameters $D < D_e$ and at the smaller diameters many more craters have been obliterated than presently can be counted on the surface. Shoemaker *et al.* (1969) realized that an initially craterless surface subjected to this type of bombardment would rapidly develop a layer of debris with a wide variation of depths. They supposed that the maximum depth likely to occur is some fraction of the depth of the largest crater that has reached equilibrium. Although the first crater of this diameter in a given region might leave an open cavity, the ejecta blankets of subsequent craters would eventually fill it, leaving a layer of fragmental debris whose thickness is locally equal to the crater depth. This maximum likely thickness h_{eq} is taken to be:

$$h_{eq} = \frac{D_{eq}}{4}. \tag{B7.1.1}$$

The equilibrium crater diameter may be determined either directly, if it can be observed as a kink in the crater population curve, or indirectly, by equating the cumulative number of craters in the production population, $N_{cum} = c\, D^{-b}$, to the number of craters in the equilibrium population, $N_{ceq} = c_{eq}\, D^{-2}$. In this case:

$$D_{eq} = \left(\frac{c}{c_{eq}} \right)^{\frac{1}{b-2}} \tag{B7.1.2}$$

where c_{eq} is dimensionless and independent of diameter, although it may depend somewhat on b. If it is supposed that equilibrium takes place at a crater density of about 4% of geometric saturation, Equation (6.2.1), then $c_{eq} = 0.056$. Conversely, the coefficient c describing the production population may be expressed in terms of D_{eq} and c_{eq}:

$$c = c_{eq} D_{eq}^{b-2}. \tag{B7.1.3}$$

I will use this representation of c in the rest of this section because it leads to more compact formulas without fractional powers of units.

In addition to the maximum likely regolith thickness, Shoemaker and his colleagues introduced the concept of a minimum regolith thickness. This is perhaps the least well-founded concept of the model and does not agree well with the data. Unfortunately, it is also an integral part of the model. The definition of the minimum thickness begins with a straightforward integral: the fraction of the total area covered by craters with diameters between D and D_{eq}, defined as $f_c(D, D_{eq})$, is given by:

Box 7.1 (cont.)

$$f_c(D, D_{eq}) = -\frac{\pi}{4} \int_D^{D_{eq}} D^2 \frac{dN_{com}}{dD} dD = \frac{\pi b c_{eq}}{4(b-2)}\left[\left(\frac{D_{eq}}{D}\right)^{b-2} - 1\right] \qquad (B7.1.4)$$

where the minus sign in front of the integral is a consequence of the negative slope of the cumulative number distribution: $dN = -(dN_{cum}/dD)\, dD$.

When $f_c(D, D_{eq})$ exceeds some fixed fraction f_{min} at diameter D_{min} the bottoms of all craters with this diameter or less are presumed to be connected and the regolith is at least as thick as this crater's depth, $h_{min} = D_{min}/4$. Shoemaker *et al.* argued that this occurs when the target area is covered twice over by craters between D_{min} and D_{eq}, so that $f_{min} = 2$. Solving Equation (B7.1.4) for $f_c(D_{min}, D_{eq}) = f_{min} = 2$, they obtained:

$$h_{min} = h_{eq}\left[\frac{4(b-2)f_{min}}{\pi b c_{eq}} + 1\right]^{-\frac{1}{b-2}}. \qquad (B7.1.5)$$

The probability $P(h)$ of finding a patch of regolith of depth h, where h must lie between h_{min} and h_{eq}, is given by the ratio of the fractional area covered by craters between diameters $D = 4h$ and D_{eq}, to f_{min}:

$$P(h) = \frac{f_c(4h, D_{eq})}{f_{min}} \qquad (B7.1.6)$$

After some algebraic manipulation $P(h)$ can be written in the convenient form:

$$P(h) = \left[\frac{(h/h_{eq})^{b-2} - 1}{(h_{min}/h_{eq})^{b-2} - 1}\right] \quad \text{for } h_{min} < h < h_{eq}. \qquad (B7.1.7)$$

The median regolith depth $<h>$ occurs when $P(<h>) = 0.5$. Solving the above equation, it is easy to show that:

$$<h> = 2h_{eq}\left[\left(\frac{h_{eq}}{h_{min}}\right)^{b-2} + 1\right]^{-\frac{1}{b-2}}. \qquad (B7.1.8)$$

Table B7.1.1 compares the predictions of this model to the observed regolith thickness distributions for four areas on the moon. These thicknesses were obtained from the morphology of small craters. This table shows that the model tends to overestimate the regolith thickness by about 40%. Although some adjustments could be made to bring theory and observation into better agreement ($f_{min} = 3$ would improve the fit greatly), such adjustments are probably unjustified given the other assumptions of the theory. On this level, 40% agreement is quite good. Although Equation (B7.1.7) is a good fit at the larger depths, it fits poorly at shallow depths. This is mainly a consequence of the artificial choice of a minimum depth, although it may also be due to neglect of the previously formed regolith in blanketing the surface (Gault, 1970).

Box 7.1 **(cont.)**

Table B7.1.1 *Predicted vs. observed regolith thickness from Equation (B7.1.8)*

Location[a]	D_{eq}^{b} (m)	h_{eq} (m)	h_{min} (m)	<h> (m)	Observed[c] < h > (m)
LO III P-11	80	20	2.3	4.5	3.3
LO III P-13b	120	29	3.4	6.6	4.6
LO II P-7b	180	45	5.3	10.0	7.5
LO V 24	410	100	12.0	23.	16.

[a] Locations are designated by Lunar Orbiter mission and frame. For more detail see Oberbeck and Quaide (1968).
[b] Computed from Equation (B7.1.2) using $c_{eq} = 0.056$. $b = 3.4$ for all sites and $c = 25$, 43, 79, and 250, respectively, for the four sites, where c is in units of $m^{1.4}$.
[c] Oberbeck and Quaide (1968)

   Another important result of this model is that it predicts that the regolith thickness may increase non-linearly with time even if the cratering rate is constant. Both h_{eq} and h_{min} are related to D_{eq} by constant factors for constant b, so the median depth <h> is also proportional to D_{eq}. Because D_{eq} increases as $(time)^{1/(b-2)}$ for constant cratering rate, so do all these measures of the regolith thickness. Only if $b = 3$ does the regolith accumulate at a constant rate. Data on the crater population, regolith thickness, and ages of the underlying mare at the Apollo 11 landing site indicates $b = 2.9$, so that the lunar regolith there accumulated at a nearly constant rate of about $1.2 / 10^9$ m/yr.
   On the whole, the regolith evolution model of Shoemaker *et al.* (1969) gives an adequate, first-order description of the growth of the regolith. It has the great advantage of being analytic, so that the equations may be easily applied to many different situations, and it relies on the observed crater population for its input parameters. Although it might be improved by a better statistical treatment of the smaller regolith thicknesses, no one has yet published a revised model.

   These mathematical models predict regolith overturn times of about 10^5 yr for the topmost centimeter, 3×10^6 yr for the upper 10 cm and 10^8 yr for the upper meter. These rates are in good agreement with radiometric measurements of cosmic-ray ages of layers in the Apollo and Luna core samples. This process predicts that many grains and rocks currently buried in the regolith once resided at the surface, a prediction amply verified by the recovery of grains containing gases implanted from the solar wind. The lunar regolith has served as a model for the formation of gas-rich meteorites, which are now recognized as compacted regolith from their asteroidal parent bodies.
   In addition to vertical mixing of the regolith, impacts also necessarily mix the regolith laterally. The various lunar landers and Apollo astronauts noted gently sloping "fillets" of regolith piled up around the base of larger rocks lying on the surface. Such accumulations of debris are the result of horizontal motion of the regolith under the influence of small

impacts. Most debris ejected by an impact crater falls within a distance of one or two crater diameters from the rim, so lateral transport is not very efficient. Nevertheless, as shown by the moon-spanning rays of fresh craters such as Tycho, a small amount of material is transported very large distances. Indeed, the current ages of the large craters Tycho and Copernicus are based on the premise that the Apollo 17, 15, and 12 astronauts collected distal ejecta from these craters, whose rays cross the various landing sites.

7.1.2 Regolith maturity

The regolith is formed principally from the overlapping ejecta blankets of innumerable impact craters, large and small. The upper layers are frequently overturned by small impacts, whereas deeper layers are overturned less frequently by large impacts. This mode of formation explains the frequent observation of layers of widely varying thickness observed in the cores returned from the Moon. On the Moon most crater ejecta travels only a few times the diameter of the crater that created it and it nearly all falls back to the Moon somewhere. On smaller bodies such as asteroids, the lower escape velocity means that more material may be ejected into space and so growth rates may be smaller for a given impact flux.

Because the regolith is composed of multiply recycled crater ejecta, material in the regolith is strongly affected by the cratering process. On the Moon the average impact velocity is high, approximately 17 km/s, so that melting and vaporization are common. The lunar regolith is filled with numerous fractured rock fragments, impact-generated glass and vapor-deposited coatings. A common component is "agglutinate" consisting of angular rock fragments welded together by irregular masses of glass. On average, agglutinates account for about 30% of the lunar regolith. Agglutinates are typically smaller than about 1 mm in size. Agglutinate glass itself is riddled with 1 to 10 nm blobs of metallic iron. This "nanophase" iron is largely responsible for the optical darkening and reddening characteristic of the lunar surface. Its origin has been much debated: Nanophase iron was originally attributed to chemical reduction of ferrous iron Fe^{2+} in silicate minerals by solar-wind-implanted hydrogen. More recently it appears that impacts (and impact-analog laser-evaporation experiments) can reduce iron directly, so that no reducing agent is necessary, although the recent observation of OH and H_2O produced by the solar-wind plasma argues that at least some chemical reduction must occur.

Mature lunar regolith is in equilibrium between impacts that simultaneously break down bedrock and build up agglutinate from the broken fragments. The regolith's grain size distribution is apparently in a stable balance between the processes that break the regolith into smaller fragments and agglutinates that weld them together into larger particles. The average grain size of a mature lunar soil is about 60 μm. Less mature soils have a larger average grain size and contain fewer agglutinate particles. They are spectrally bluer and mineral absorption bands are more distinct.

Soil maturity is often gauged remotely by brightness and color: Dark, spectrally red regolith is mature, whereas bright, blue surfaces are immature. Images of the lunar surface typically reveal dark, red intercrater areas interspersed with the bright, blue rays surrounding fresh craters that have penetrated the regolith to fresh bedrock. This same pattern is revealed on asteroids, although with some puzzling differences: Images of the main-belt asteroid 243 Ida by the Galileo probe revealed a color contrast between old surfaces and fresh crater ejecta, but albedo contrasts are low. On the other hand, NEAR images of the near-Earth asteroid 433 Eros show large albedo differences but low color contrasts. Carbonaceous asteroid 253 Mathilde seems to be uniformly dark and does not exhibit either albedo or color contrasts, perhaps due to its large carbon content. At the moment, asteroid regoliths are not well understood. It is presumed that some combination of low gravity and lower impact velocities in the asteroid belt (typically about 5 km/s) results in a very different evolution of the regolith on these bodies, but a clear understanding may have to await sample returns from asteroid surfaces.

7.1.3 Radiation effects on airless bodies

In addition to the impacts of solid bodies, the surfaces of airless bodies are exposed to ionizing radiation from several sources. The Sun emits X-rays and ultraviolet light that can ionize atoms and induce chemical reactions among surface materials. Although this radiation does not penetrate deeply, it can cause radiolysis of ices and carbonaceous compounds. While such changes have been much studied in the context of planetary atmospheres, their effects on surface chemistry are less well known, although the red color of Martian soil has been attributed to oxidation by ultraviolet light that penetrates its thin atmosphere. The positive indications of life by the Viking GCMS experiment are now attributed to peroxides created on surface minerals by the intense ultraviolet radiation bathing Mars' surface. The present surface environment of Mars is extremely hostile to life because of this radiation and the highly oxidized chemical species it generates.

In addition to X-rays and ultraviolet light, the surfaces of airless, or nearly airless, bodies are affected by a variety of ionizing particle radiations. In order of increasing particle energy, there is solar-wind plasma, solar-flare radiation, and galactic cosmic rays. Most of this particle flux is composed of hydrogen nuclei (protons) and electrons, but an important fraction is nuclei of heavier elements. Its depth of penetration depends critically on particle energy and varies from microns to meters. In addition to the primary radiation flux, energetic particles also generate neutrons below the surface that can induce further nuclear reactions.

The solar wind consists of ionized gas ejected from the Sun's corona that impinges on the surfaces of exposed bodies with energies between 0.3 and 3 keV/nucleon. It penetrates only microns into the surfaces of exposed mineral grains, but can induce sputtering and implant gases below their surfaces. Large fluxes of the solar wind may destroy the crystalline surfaces of minerals and create an amorphous rim. The solar wind is the main source of the elements H, C, N, and the noble gases that are otherwise very rare in the lunar regolith. Solar flares occur sporadically, but an individual event may emit large fluxes (up to 10^6 protons/cm^2-s) of 1–100 MeV/nucleon nuclei from the Sun's atmosphere. These particles

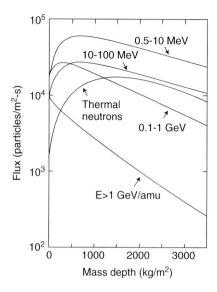

Figure 7.6 Fluxes of galactic cosmic rays as a function of depth (here measured in terms of mass above the given depth per unit area, kg/m²: 2 m of lunar regolith is about 3000 kg/m²). This plot indicates both primary and secondary radiation. Thermal neutrons, in particular, are only created by collisions of the primary radiation with nuclei in the regolith. After Fig. 3.20 of Herken *et al.* (1991).

penetrate about 1 cm into the lunar surface and produce visible tracks of displaced atoms through mineral crystals, but do not create much secondary radiation.

Galactic cosmic rays are the most energetic and penetrating radiation that impinges on the surface of an airless body. Energies range from about 0.1 GeV/nucleon to above 10 GeV/nucleon (rarely, much more than this) and penetration depths are measured in meters for the lighter nuclei. Heavy nuclei in the cosmic-ray flux penetrate only centimeters, but because they cause a great deal of ionization they severely damage mineral structures, not to mention electronic components of spacecraft. These energetic particles produce abundant secondary radiation when they strike nuclei below the surface, creating cascades of neutrons and other exotic particles. A consequence of this cascade is that the total radiation exposure at some depth below the surface is larger than that at the surface itself. The radiation dose peaks at a depth of about 1000 kg/m² and declines at greater depths (Figure 7.6). Galactic cosmic rays produce visible tracks in minerals and their secondary neutrons induce nuclear reactions that create a suite of exotic isotopic species. These isotopes are useful for dating regolith exposure ages and overturn rates.

The direct impingement of ionizing radiation on airless bodies has proved very useful for remote sensing of their surface composition. X-ray fluorescence has been used to map elemental abundances on the lunar and asteroid surfaces. X-rays emitted from atoms excited by particle radiation are similarly used as a kind of natural microprobe to determine the identity and abundance of different elements. The neutrons and gamma rays emitted by secondary nuclear reactions have revealed the presence and distribution of many elements in the regolith from orbit around the Moon, Mars, and the asteroid Eros. Slow neutrons thermalized by hydrogen in the surface of Mars and the Moon permit mapping the distribution of water in the upper few decimeters beneath their surfaces. Similar instruments aboard the Mercury MESSENGER mission will shortly be used to map elemental abundances on that planet. This project was delayed during the initial flybys of Mercury due to an unexpected lack of solar activity that reduced the Sun's X-ray flux below the level needed for clear detection by the fluorescence experiment.

The Galilean satellites of Jupiter are immersed in the most intense radiation environment in the Solar System, with surface radiation doses 100 to 1000 times more intense than experienced by our Moon (Johnson *et al.*, 2004). Plasma and energetic particles trapped in Jupiter's magnetosphere, along with solar ultraviolet radiation, strongly affect the surfaces of these satellites down to depths of about a centimeter. Impact gardening then churns this material into their meters-deep regoliths. Radiation damage of ice grains creates thin layers of amorphous ice that are detected spectrally, while the radiation itself breaks chemical bonds producing exotic chemical species. Considerable research has gone into understanding the effects of radiation on water ice. The predominant process is radiolysis of the water molecules; ultimately producing oxidized species such as H_2O_2, HO_2, and O_2 after H is lost from the surface by diffusion into space. Carbon, either implanted from the radiation itself or created by decomposition of organics present in the surface, yields CO_2, while sulfur is oxidized to SO_2. Intense radiation seems to brighten the surfaces of the icy satellites, perhaps by surface condensation of water molecules sputtered from ice grains. An intriguing suggestion is that Ganymede's bright "polar cap" may be created by radiation funneled to its surface by its internal magnetic field. This would explain both the coincidence of the bright cap with the location of open magnetic field lines and the otherwise puzzling fact that Callisto, which lacks an internal magnetic field, also lacks a bright polar cap.

Io's volcanoes eject considerable amounts of sulfur, some of which is incorporated into the plasma of Jupiter's magnetosphere. This sulfur is then swept out to Europa and beyond, where it preferentially impacts the trailing hemisphere of the outer Galilean satellites and is implanted in their surfaces. The result is a clear "bull's-eye" ultraviolet spectral anomaly on Europa that is also observed on Ganymede and Callisto. In addition to the trailing hemisphere sulfur implanted from the magnetosphere, there are other spectral indications of hydrated sulfur compounds, especially on Europa. It is debated whether the origin of this sulfur signature is due to radiation-induced sulfuric acid or endogenic sulfate salts from Europa's interior ocean. At the moment the exogenic/endogenic origin debate is unresolved.

7.2 Temperatures beneath planetary surfaces

The surface temperature of a planet is one of the most important facts about its surface environment. It is one of the first things that most people want to know about another planet. Temperature sets the stage for which elements and chemical compounds are present, in what form (solid, liquid, or gas), and the rates of chemical and physical changes that may occur. Planetary habitability largely depends on the occurrence of surface temperatures in the range where liquid water is possible.

Table 7.1 *Effective temperatures of the planets*

Planet	Distance from Sun (AU)	Albedo	Effective temperature (K)
Mercury	0.39	0.058	442
Venus	0.72	0.71	244
Earth	1.00	0.33	253
Mars	1.52	0.17	216
Ceres	2.77	0.09	164
Jupiter	5.20	0.73	87
Saturn	9.54	0.76	63
Uranus	19.19	0.93	33
Neptune	30.06	0.84	32
Pluto	39.53	0.58 (average)	32
Eris	67.67	0.86	21

The approximate surface temperature of a planet is set by the balance between solar radiation received and thermal radiation emitted. A standard calculation (Houghton, 2001) leads to the "effective" temperature estimate:

$$T_{\text{eff}} = \left[\frac{(1-A)\,L_\odot}{16\,\pi\sigma R^2} \right]^{1/4}$$ (7.2)

where $L_\odot$ is the solar luminosity, the planetary albedo A measures the fraction of solar radiation absorbed, R is the distance from the Sun and σ is the Stefan–Boltzmann constant. This crude estimate assumes that the planet's surface temperature is uniform and ignores the effect of an atmosphere. Atmospheric modifications may be profound and are the subject of a huge literature of their own, but are outside the scope of this book. A good, concise, introduction to this literature is Houghton (2001).

Planetary effective temperatures range from a high of 442 K on Mercury down to 32 K on Pluto (Table 7.1), falling roughly as the inverse square root of distance from the Sun. Albedo plays a large role: Even though the effective temperature of Mars is a chilly 216 K, the nearly black surface of comet Tempel 1 registered a sweltering 300 K well outside the orbit of Mars when it was visited by the Deep Impact spacecraft in 2005. Earth itself is pretty cold in terms of effective temperature: Its current habitable state is largely due to atmospheric greenhouse warming.

7.2.1 Diurnal and seasonal temperature cycles

The effective temperature estimate is a global average that ignores diurnal, seasonal, and equator-to-pole temperature differences. These differences can be profound: In spite of Mercury's very high effective temperature, polar temperatures are estimated to fall as low as 167 K, making it likely that ice is stable in high-latitude craters (Paige *et al.*, 1992). Mercury's 59-day rotational period and the peculiar 2:3 resonance between its rotational and orbital periods yield further extremes between different longitudes along its equator. Daily maximum temperature estimates range from about 570 K at an equatorial "warm pole" to 700 K at the "hot poles" located 90° in longitude from the warm poles (Morrison, 1970). Minimum temperatures drop to about 100 K at all longitudes. On the Moon, diurnal (27-day) variations in temperature range from a high of 374 K down to 92 K at the Apollo 15 site (Heiken *et al.*, 1991). Polar temperatures in shadowed lunar craters may fall as low as 40 K.

Surface temperature cycles on planetary surfaces vary enormously in duration. On Earth we are accustomed to the 24-hour day/night temperature fluctuation, but because of the Earth's obliquity we also experience large yearly seasonal cycles. Climatic oscillations vary on timescales that range from tens of thousands of years (the waning of the last ice age) to millions of years. Other planets experience a similar wide range of timescales: The Martian day is very similar in length to the Earth's and its seasons are approximately twice as long, but its layered polar deposits suggest climatic cycles due to obliquity variations with periods of millions of years or even longer when chaotic orbital variations are taken into account (Carr, 2006). Seasonal cycles on Saturn's large moon Titan last 29 yr, 165 yr on Neptune's moon Triton, and 249 yr on Pluto's Charon. The high obliquities of Uranus and Pluto suggest that seasonal temperature swings are larger than diurnal alternations.

7.2.2 Heat transfer in regoliths

Computation of subsurface temperatures from surface temperatures is a straightforward process using Fourier's heat conduction Equation (4.13). The major practical difficulty is knowing the thermal conductivity of the surface layers, which can be very complex in a realistic regolith, because it depends on composition, grain size, grain packing, the presence of atmospheric gases or liquids, and the temperature (Presley and Christensen, 1997). In the idealized case of a harmonic surface temperature variation at the surface of a half-space of uniform thermal conductivity, the temperature as a function of depth is given by (Carslaw and Jaeger, 1959, p. 64ff.):

$$T(z,t) = T_m + A\, e^{-\frac{2\pi z}{\lambda}} \cos\left[2\pi\left(\frac{t}{P} - \frac{z}{\lambda}\right)\right] \qquad (7.3)$$

where the thermal "skin depth" λ is given by:

$$\lambda = \sqrt{4\pi\kappa P} \qquad (7.4)$$

where κ is the thermal diffusivity. The surface temperature fluctuates about the mean temperature T_m with amplitude A and period P. The origin of time t is chosen so that the surface temperature is given by $T(0) = T_m + A \cos(2\pi t/P)$. This equation ignores the slow increase of temperature with depth due to internal heat flow.

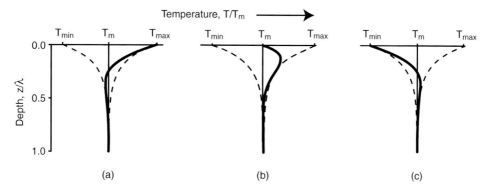

Figure 7.7 Temperature in the lunar regolith as a function of time, computed from the linear diffusion equation. The temperatures in the left panel are plotted at noon, those in the middle panel at dawn or dusk, and those in the right panel are plotted at midnight. The dashed lines show the envelope of the temperature extremes as a function of depth. Note that the mean temperature does not vary with depth.

Equation (7.3) indicates that the amplitude of the temperature fluctuation decreases exponentially with increasing depth below the surface (Figure 7.7). The surface temperature variation not only decreases in amplitude going deeper, but the time at which it reaches maximum is progressively delayed. The depth to which the thermal wave penetrates depends strongly on the period P of the thermal fluctuations: Rapid temperature variations remain in the shallow subsurface, while long-period variations penetrate deeply. Thus, diurnal temperature variations on the Earth may penetrate only centimeters into the soil, while seasonal variations are appreciable at depths of meters. It is interesting to note that terrestrial heat-flow measurements in boreholes hundreds of meters deep must take the cold temperatures of the last ice age into account, or errors will result: The 10 000-year thermal wave is just now reaching this depth. Rapid erosion or sedimentation must likewise be considered in many such measurements (an example is the USGS borehole in Cajon Pass, CA, which initially gave erroneously high estimates of the heat flow because rapid erosion at that location was neglected).

Handbook tabulations of thermal conductivity $k = \rho c_p \kappa$ may suggest that it is approximately constant; however, this is far from the truth. Thermal conductivity depends upon the mineral composition of the surface, porosity, grain size and sorting, presence of included gases, and temperature (Presley and Christensen, 1997). Temperature dependence is especially evident in the lunar regolith where, at high temperatures, heat is transferred through its pores mainly by radiation, not grain-to-grain conduction. The thermal conductivity of the lunar regolith is usually fit by an equation of form (Langseth *et al.*, 1973):

$$k(T) = k_c + k_r T^3 \tag{7.5}$$

where k_c is (approximately) temperature-independent and k_r measures radiative transfer within the regolith. The strong temperature dependence of Equation (7.5) leads to a phenomenon known as "thermal rectification." When thermal radiation can be ignored, the

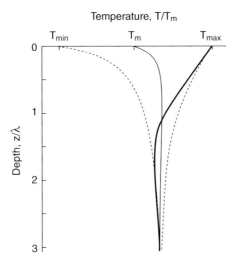

Figure 7.8 Temperature in the lunar regolith for temperature-dependent thermal conductivity that is dominated by radiation transfer. The heavy line indicates the temperature at noon. Note that the mean temperature rises sharply with increasing depth, reflecting the process of "thermal rectification." The temperature extremes (dashed) are also strongly asymmetric, in contrast to Figure 7.7.

heat conduction Equation (4.13) is linear and the average temperature below the surface is the same as the average temperature at the surface itself. However, the T^3 term in Equation (7.5) makes the conduction equation non-linear and the solution (7.3) no longer applies. Thermal conductivity is high when the temperature is high, allowing heat during the day to penetrate into the surface, but drops to low values at night, turning the regolith into an insulator that retains much of the day's heat (Figure 7.8). The net result is to raise the average temperature below the loose surface layer by as much as 40–45 K. This shift in average temperature complicated measurement of the lunar heat flow from the shallow Apollo 15 and 17 heat flow probes. Geophysicists and any future lunar colonists must take this shift in average temperature into account when computing the temperature below the surface of the Moon. At temperatures near absolute zero, thermal conductivity also depends strongly on temperature, so that care must be taken in computing subsurface temperatures of very cold bodies far from the Sun.

A phenomenon closely related to thermal rectification is the "solid-state greenhouse." This concept was invented by Bob Brown and Dennis Matson (1987; Matson and Brown, 1989), who realized that on bright, high-albedo bodies like Europa, sunlight is not absorbed right at the surface but penetrates some distance into it. The Sun's heat is thus absorbed a few centimeters below the surface, where it must reach the surface by conduction before being radiated back into space. It is estimated that this effect may account for about a 10 K rise in temperature over the case where sunlight is absorbed right at the surface. This effect is important for any bright surface that does not readily absorb sunlight, so it may also affect the near-surface temperatures of snowpacks or icy polar deposits.

Heat and material may also be transported by advection in the regoliths of bodies with substantial atmospheres. On the Earth, fresh snow is rapidly metamorphosed by the advection of water vapor within a snowpack. Water vapor evaporates from sharp-edged snowflakes whose small radii of curvature destabilize molecules on their surface. This water recondenses into larger crystals within the mass of snow. It also migrates readily from warmer to colder regions of the porous mass of snow, transferring heat and accumulating into crystalline masses known as hoar frost. Known as "firnification," this process is an important step in the conversion of snow into ice in snowpacks and glaciers (Shumskii, 1964). On Mars, the seasonal cycling of atmospheric pressure drives atmospheric gases into and out of the soil, strongly affecting the distribution of volatile species, such as water vapor, in the near-surface regolith (Haberle *et al.*, 2008). Such vapor-phase advection may play a role in the distribution of ground ice on Mars, as well as having a major effect on near-surface heat transport.

7.2.3 Thermal inertia

The response of different materials to temperature changes depends strongly upon their thermal conductivity, density, and heat capacity. Thus, solid rocks lying on a planetary surface are slow to warm up after sunrise, while loose, powdery soil heats up rapidly. Conversely, after sunset rocks retain their heat for a long time compared to fine-grained powders. This observation has led to the development of thermal emission spectroscopy, which enables orbiting spacecraft (or Earthbound observers) to estimate grain sizes and dust/rock ratios of planetary surfaces remotely, utilizing thermal infrared (IR) emissions (Mellon *et al.*, 2008).

The principal parameter that determines the rate at which a surface warms (or cools) is its *thermal inertia*, I, which is defined as:

$$I = \sqrt{k \rho c_p}. \tag{7.6}$$

This parameter controls how much heat energy is gained or lost from a surface during changes of temperature. Its units are rather complex: $J/m^2\text{-}K\text{-}s^{1/2}$. Table 7.2 lists some representative values of thermal inertia for different materials and surfaces.

Spacecraft observations of temperature changes in response to insolation variations are most readily interpreted in terms of thermal inertia itself. Conversion of these observations to composition, grain size, or surface density is highly model-dependent. Nevertheless, such measurements have proved invaluable in sorting out the nature of different terranes on Mars as well as other bodies, such as the Galilean satellites of Jupiter, where the thermal response of their surfaces to eclipses has long been used to infer their nature.

7.3 Weathering: processes at the surface/atmosphere interface

Planetary surfaces are geological battlefields: They are the frontier between a planet's interior and its atmosphere or, in the case of airless bodies, open space. They are assaulted

Table 7.2 *Thermal inertia of selected materials*

Material	Density (kg/m³)	Heat capacity, c_P (J/kg-K)	Thermal conductivity, k (W/m-K)	Thermal inertia, I (J/m²-K-s$^{1/2}$)
Water ice (273 K)	915	1960	2.3–3.6	2030–2540
Water ice[a] (173 K)	926	1389	3.48	2116
Dense snow[b] (273 K)	540	2100	0.46	722
New snow (273 K)	200	1962	0.078	174
CO_2 ice (150 K)	1560	1083	0.6	1006
Basalt	2900	830–900	1.12–2.38	1642–2492
Tuff	1980	850	0.91–3.20	1237–2320
200 µm sand @ 6 mb	1650	850	0.044	248
10 µm sand @ 6 mb	1375	850	0.01	108
1 µm dust @ 6 mb	1100	850	0.004	61

Data based on Table 18.1 of Mellon *et al.* (2008).

[a] Engineering Toolbox (http://www.engineeringtoolbox.com/).

[b] Data from Clark (1966).

by meteor impacts, ionizing radiation, rapidly varying temperatures, and, especially for the Earth, corrosive gases and reactive liquids (mainly oxygen and water). Rocks erupted or exhumed from the deep interior come into contact with the atmosphere, entering a very different chemical environment with which they are seldom in equilibrium.

The events that transpire in this environment are described by the term "weathering." Weathering transforms rocks from structures and compositions that are at home in the planet's interior to new forms that can survive for long periods of time in the surface environment (Figure 7.9). Depending upon one's point of view, these events are either destructive or creative. The destructive aspect was most strongly emphasized early in the history of geology ("The Earth in Decay" was a favorite theme of early geomorphologists), whereas more modern authors look upon weathering as the source of our agricultural soil. The lunar regolith is regarded as a potential resource and the Martian regolith may harbor life. Indeed, the very origin of life probably required an active, kaleidoscopically changing environment that only occurs on planetary surfaces.

Weathering processes are traditionally divided into "chemical" weathering, which emphasizes chemical changes, and "physical" weathering that deals mainly with macroscopic cracks and changes in mechanical properties. Although this traditional division is respected in this book, there are many processes, such as granular disintegration of rocks caused by swelling of minerals undergoing chemical reactions, which cut across the boundaries of this division. Most chemical reactions would not occur at geologically significant rates if reactants could not penetrate rock masses through cracks created by physical weathering. Classifications are useful, but should not obscure the fact that both chemical and physical processes play essential roles in the overall group of processes that constitute weathering.

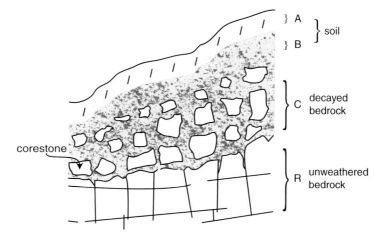

Figure 7.9 Schematic of a weathered soil profile showing the A, B, C, and R (unweathered bedrock) horizons. Slightly weathered corestones are mixed with fine-grained, heavily weathered material in the C horizon.

7.3.1 Chemical weathering

Chemical weathering focuses mainly on the chemical changes that occur in minerals in the near-surface environment. The interiors of all of the terrestrial planets appear to be chemically reduced compared to their atmospheres, so minerals from the interior typically become oxidized in the surface environment. Although there is a long list of elements with variable oxidation states that may be altered by surface exposure (C, P, S, V, Mn, Fe, Co, Ni, and Cu, plus others), the most visible and distinctive changes occur for iron, which may occur in the metallic Fe^0 state, as ferrous Fe^{2+}, or ferric Fe^{3+} ions. The ferrous/ferric transformation is particularly significant, as minerals containing ferrous iron are typically black or green to the human visual system and the ion is highly soluble in water, whereas minerals incorporating the ferric ion are typically red or yellow (the Martian soil is a prime example) and the ion is very insoluble.

We typically think of oxygen O_2 as the eponymous oxidizer; however, H_2O, CO_2, and SO_2 are also significant oxidizers in different planetary environments (as are UV and ionizing radiation). Table 7.3 summarizes the atmospheric composition, surface pressure, and temperature of the terrestrial planets, Titan, and Triton. Titan's nitrogen/methane atmosphere is the only truly reducing environment.

Planetary atmospheres are not passive backgrounds to the chemical drama that unfolds on their surfaces. The reactions that atmospheric gases undergo with surface rocks profoundly affect the composition of the atmosphere itself, often determining the abundance of many atmospheric species. Thus, the long-term abundance of CO_2 in the Earth's atmosphere may be controlled by weathering reactions with rocks erupted from its interior. Earth's oxygen is almost entirely the result of photosynthesis by plants on its surface and many other minor species are in equilibrium with the surface through planetary geochemical cycles that, on Earth, are usually part of the biosphere. The abundance of CO in the atmospheres of Venus

Table 7.3 *Near-surface atmospheres in the Solar System*

Body	Temperature (K)	Pressure (MPa)	Major atmospheric constituents	Chemically active minor constituents
Venus	750	9.7	CO_2 (96.4%)	N_2, SO_2, H_2O, CO
Earth	288	0.1	N_2, O_2 (78.1%, 21%)	H_2O, CO_2
Mars	250	0.0008	CO_2 (95.3%)	N_2, O_2, H_2O, CO
Titan	94	0.15	N_2 (95%)	CH_4
Triton	38	$1.4–1.9 \times 10^{-6}$	N_2	CH_4 (trace)

and Mars may be determined by weathering reactions at the surface, and on Titan ephemeral lakes of liquid methane buffer the methane content of its atmosphere.

The detailed atmospheric composition of a planet is, thus, a reflection of weathering reactions on its surface. The first indication that life is present on distant planets around other stars may be spectroscopic detection of chemical species indicative of biological activity. Closer to home, a recent report of methane detected over the northern hemisphere of Mars has given rise to much excitement over its possible biological origin.

Expositions of chemical weathering unfortunately tend to devolve into rather dull lists of chemical reactions that the student is expected to memorize by rote. The exposition that, in my opinion, best avoids this pitfall is a rather obscure pamphlet published in 1957 (Keller, 1957). A more recent book (Bland and Rolls, 1998) does a good job of explaining the background, but both of these treatments are entirely Earth-focused. There is not yet an extensive description of chemical weathering applicable to the other planets, although individual book chapters (Fegley *et al.*, 1997; Gooding *et al.*, 1992; Wood, 1997) have made a start on this difficult subject (which, admittedly, is not yet constrained by much data). Your author, along with everyone else, has not found a better way to present this topic, but in the following section I attempt to put each set of reactions into its most appropriate planetary context. The classes of reaction are discussed in rough order of their Gibbs energy change and, thus, in some sense, of the ease with which they occur. This is not an exhaustive list of weathering reactions, which would be very long and not very informative, but an exhibition of the most characteristic ones.

The basic weathering reactions are solution, hydration, hydrolysis, and oxidation, along with a few reactions that include carbonation and sulfonation, in rough order of their Gibbs energy change. The values of the reaction Gibbs energy are not, however, listed here because they depend on the pressure, temperature, and, most importantly, on the activities (hence, concentrations) of gaseous or soluble species. These numbers are, thus, highly sensitive to the particular environment in which the reaction occurs. The interested reader is advised to consult standard tables of Gibbs energy (the NIST tables, Chase, 1998, for example) to determine accurate values for reaction energies.

One of the lowest-energy reactions with water is simple solution. Ionic crystals, such as halite, NaCl, dissolve readily in water because the polar water molecules neutralize the ions' electric charges.

$$NaCl \rightleftarrows Na^+(aq) + Cl^-(aq). \tag{7.7}$$

This process allows water to dissolve a large number of other ionic salts, such as $MgCl_2$ or $MgSO_4$, although halite is one of the most soluble. Waters containing large amounts of dissolved salts are known as brines and they play a major role in terrestrial geochemistry. Brines may also be present beneath the surface of Mars and in Europa's subsurface ocean. A slightly more complex process that involves water and CO_2 acting together accounts for the dissolution of limestone:

$$CaCO_3 + H_2O + CO_2 \rightarrow Ca^{2+} + (HCO_3)_2^{2-}. \tag{7.8}$$

Extensive karst terranes on Earth, where limestone rocks have been dissolved by rainwater to form pits, hollows, and underground caves, are the result of this reaction. This apparently simple reaction has many subtleties, depending on the concentration and temperature of the water. A large body of literature exists on karst, which connects the morphologic expression of these terranes to the chemistry of carbonate and water (White, 1988).

Another low-energy reaction with water, not normally listed as a weathering reaction, is clathrate formation. Non-polar molecules of CO_2, CH_4, NH_3, and many others assemble a cage of water molecules about themselves at low temperatures and form distinctive, often stoichiometric, phases in ice. These cold mineral phases may play important roles in the trapping and release of gases in icy bodies far from the Sun. Clathrate densities differ from that of pure water ice and so may also play a role in the tectonics and volcanism on icy satellites.

A slightly more energetic reaction with water results in a stoichiometric association between a mineral and water molecules. This reaction, called hydration, is readily reversible by heat at temperatures of 600–700 K. A common terrestrial reaction with anhydrite (calcium sulfate) produces the evaporite mineral gypsum:

$$CaSO_4 + 2H_2O \rightleftarrows CaSO_4 \cdot 2H_2O. \tag{7.9}$$

Hydrolysis, in which the water molecule dissociates and forms a new -OH bond with the reacting species, is distinct from hydration, in which the water molecule remains intact. Hydrolysis is illustrated here by the reaction of water with minerals common in igneous rocks. We first consider a reaction that, until recently, was thought to be an example of hydration. The iron mineral hematite reacts readily with water to form goethite:

$$Fe_2O_3 + H_2O \rightleftarrows 2FeO(OH). \tag{7.10}$$

A complex mixture of goethite, hematite, and variable adsorbed water forms limonite, which imparts a yellow color to many types of sediment. Limonite is very stable in the Earth's surface environment and is considered the endpoint of weathering of iron-rich rocks.

The second example of hydrolysis focuses on orthoclase, a feldspar common in granitic rocks:

$$KAlSi_3O_8 + H_2O \rightarrow HAlSi_3O_8 + KOH(aq). \tag{7.11}$$

The hydrogenated silicate created by this reaction undergoes further reactions with water to eventually form the clay mineral kaolinite $Al_2Si_2O_5(OH)_4$ while silica, SiO_2, is released into solution. Kaolinite is another of the end products of weathering in the terrestrial environment. The silica and potassium are carried off in water, later to precipitate or add to the ionic content of briny waters.

A third, more complex, example of hydrolysis focuses on the mineral olivine. Olivine is a major component of planetary mantles, chondritic meteorites, and an important constituent of many types of basalt. It typically occurs as a solid solution of the iron and magnesium endmembers, Fe_2SiO_4 and Mg_2SiO_4. The complex suite of reactions with water known as serpentinization begins with the hydrolysis of the iron-rich endmember:

$$3Fe_2SiO_4 + 2H_2O \rightarrow 2Fe_3O_4 + 3SiO_2 + 2H_2. \tag{7.12}$$

This reaction is coupled with a similar reaction of the magnesium-rich endmember that absorbs the silica produced in the first reaction:

$$3Mg_2SiO_4 + SiO_2 + 4H_2O \rightarrow 2Mg_3Si_2O_5(OH)_4. \tag{7.13}$$

The final product, serpentinite, is closely related to kaolinite and is also a terminal weathering product. Note that reaction (7.12) releases hydrogen gas. This hydrogen may react further with carbon in the environment, reducing it to methane, CH_4. This is an abiogenic source of methane, one that has been suggested to account for the observation of methane in the atmosphere of Mars without invoking biological activity.

Oxidation is the quintessential reaction on Earth. One-fifth of our atmosphere is currently oxygen gas, a biological by-product of photosynthesis. Oxygen is not only highly reactive with rocks from Earth's interior, it is also highly toxic to life itself. Early photosynthetic organisms excreted it as a waste product. As these organisms became more abundant and the concentration of oxygen rose in the early Earth's atmosphere, complex biological pathways developed to protect organisms from their own pollutants. Animals now require oxygen as a high-energy chemical propellant, but the cells that use it must still handle it carefully and keep it out of most of their biochemical machinery.

A typical reaction is the oxidation of wüstite, FeO, to hematite. Although wüstite itself is a rare mineral, FeO is a common component of iron-containing igneous rocks and minerals, so that the reaction is representative:

$$4FeO + O_2 \rightarrow 2Fe_2O_3. \tag{7.14}$$

Notice that the iron in this reaction changed from the +2 oxidation state in wüstite to the +3 state in hematite.

Oxygen itself is not the only atmospheric gas capable of oxidizing iron. On Venus the stable form of iron is believed to be the black mineral magnetite, Fe_3O_4, which contains a mixture of iron in the +2 and +3 oxidation states. FeO in minerals erupted from the interior of Venus is oxidized by reaction with CO_2:

$$3FeO + CO_2 \rightleftarrows Fe_3O_4 + CO. \tag{7.15}$$

This reaction is postulated to contribute to the small abundance of CO observed in Venus' atmosphere.

Carbon dioxide can also react with other minerals without changing their oxidation states. Such reactions are known as carbonation, for example the iron-rich olivine fayalite reacts with carbon dioxide to produce siderite plus silica:

$$Fe_2SiO_4 + 2CO_2 \rightarrow 2FeCO_3 + SiO_2. \tag{7.16}$$

Siderite is a dark, dense mineral (density 3800 kg/m^3) that sometimes forms concretions in sedimentary rocks.

Sulfur plays a large role in the surface weathering on Venus. Carbonates do not appear to be stable on Venus, but react with SO_2 in the atmosphere to form anhydrite:

$$SO_2 + CaCO_3 \rightleftarrows CaSO_4 + CO. \tag{7.17}$$

Pyrite, FeS_2, from igneous rocks erupted on the surface of Venus, may play an important role in its sulfur cycle as well, through reactions of type:

$$3FeS_2 + 2CO + 4CO_2 \rightleftarrows 6COS + Fe_3O_4. \tag{7.18}$$

The gas COS has been detected in the lower atmosphere at the level of about 28 ppm. The overall importance of these reactions is currently contentious (Fegley *et al.*, 1997; Wood, 1997), although it seems likely that anhydrite and magnetite are two major weathering products likely to be present on Venus' surface.

A major mystery on Venus is the nature of the "snow" detected on the tops of Venusian mountains. Radar images show that terrain above 4 km elevation is highly reflective, reminiscent of snow-capped mountains on Earth. Snow itself is not a possibility in view of Venus' high surface temperatures (indeed, the best fit to the data was half-seriously identified as the metal, Te). Various suggestions, such as pyrite deposited in the vapor phase by $FeCl_2$ or deposition of volatile metals, have been suggested, but no consensus has yet been reached on this puzzle.

An important reaction that has not yet been mentioned is the so-called "Urey reaction." First proposed by Nobel prize-winning chemist Harold Urey (1893–1981) in his 1952 book *The Planets* (Urey, 1952), this reaction continues to play a major role in discussions of planetary habitability and the evolution of the Earth's atmosphere and climate. Urey noted that the abundance of carbon dioxide might be controlled by reaction with silicate rocks. He illustrated his reaction with the calcium silicate mineral wollastonite, $CaSiO_3$. Wollastonite is not particularly common, but similar reactions apply to many common minerals and we illustrate the reaction here with Urey's own example:

$$CaSiO_3 + CO_2 \rightleftarrows CaCO_3 + SiO_2. \tag{7.19}$$

This reaction occurs at appreciable speed only when liquid water is present, in which the carbon dioxide forms bicarbonate ions and the acid water attacks the silicate.

The vast majority of Earth's carbon dioxide is locked up in limestone, which, at present, is mostly deposited by oceanic organisms. Earth and Venus have similar complements of carbon dioxide, but at Venus' high surface temperatures, reaction (7.19) has proceeded far to the left. At the lower temperatures prevailing on Earth this reaction proceeds to the right and would have removed all of the CO_2 from the Earth's atmosphere but for plate tectonics, which constantly releases the CO_2 from subducted sediments and creates a stable balance over a billion-year timescale (Kasting and Catling, 2003; Kasting and Toon, 1989).

7.3.2 Physical weathering

Physical weathering is mostly about cracks and joints. However, chemical considerations are always present. Chemical reactions cannot occur if the reactants cannot meet, which is why weathering reactions usually involve a liquid or gaseous phase. However, even gases cannot react with a mineral they cannot reach. Cracks at both the macroscopic and microscopic scales provide the means through which weathering agents reach their targets.

Stress corrosion cracking. The growth of cracks themselves, however, often requires chemical assistance. A small initial flaw or microcrack can grow only when the stress at its tip exceeds the strength of the material. Regional stresses are concentrated and, thus, amplified at crack tips, but intact rock or minerals are often too strong for the crack to propagate. However, if chemically reactive fluids are present, they can move through the crack to its tip. Reaching the already-stressed bonds of the host mineral, they preferentially attack these bonds, weakening them and permitting the crack to propagate a little deeper into the intact rock. This process, known as stress corrosion cracking (Atkinson, 1987), proceeds slowly, but often leads to a great reduction in the effective strength of a material. Bridges and buildings that, if kept dry and clean would never collapse, crumble under the onslaught of rain and polluted air.

Rock fracture is a complex process that has its own extensive literature. Rock masses near the surface of a planet are subjected to a variety of crack-producing stresses, of both internal and regional (tectonic) origin. Volcanic rocks are erupted as hot liquids that solidify and cool to ambient temperatures. After they solidify they continue to cool and contract, developing large internal stresses that create cooling joints, like the columnar jointing mentioned in Section 5.3.1. Deep-seated rocks exposed by erosion or tectonic faulting cool as they near the surface, generating large stresses by thermal contraction. Another factor in joint formation is uplift, in which deep-seated rocks are stretched as they move to larger radii on a spherical planet (Suppe, 1985). It is an observed fact that every large mass of rock on Earth is cut by cracks. High-resolution images of the surfaces of Mars and Venus have shown that their rocks, too, are highly fractured, even where meteor strikes have not shattered the rocks.

In addition to macroscopic cracks, all rocks contain microscopic cracks at the scale of their individual mineral grains. The minerals composing igneous rocks expand and contract by different amounts as the temperature changes. Most minerals also expand and contract by different amounts in different directions with respect to their crystal lattices. As an igneous rock cools, stresses develop between adjacent grains and generate microcracks along

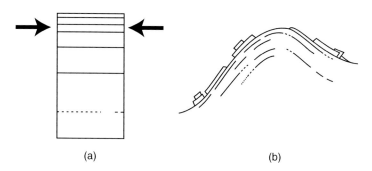

(a)   (b)

Figure 7.10 Sheeting joints form in response to horizontal compression, shown in the left side of the figure. The right panel indicates that sheeting joints follow the profile of the land surface. Their spacing is also typically closer nearer to the free surface due to unloading of the confining stress.

grain boundaries, creating both small amounts of porosity for invading fluids to occupy and incipient flaws from which macroscopic cracks can grow.

Sheeting and exfoliation joints. Besides the ubiquitous, nearly vertical cracks (joints) that cut most large masses of rock, horizontal *sheeting* joints develop near the free surface of homogeneous rocks. Due to the relief of vertical stresses, the rock expands upwards and outwards, separating into stacks of closely spaced slabs that tend to follow the contours of the free surface (Figure 7.10). These cracks grow when the stresses parallel to the surface are compressive, tending to buckle sheets of rock up and out toward the stress-free surface. Spaced only centimeters apart right at the surface, the horizontal joints' spacing increases rapidly with depth, reaching a few meters 20 to 30 m below the surface where vertical and horizontal stresses are more equal. Good examples of this kind of jointing are displayed at Half Dome in Yosemite National Park and Stone Mountain in Georgia, both in the USA.

Spheroidal weathering. A striking landscape may develop where weathering solutions propagate readily along the cracks and disintegrate large jointed blocks from the outside in. Called *spheroidal weathering*, this process creates landscapes that appear to be covered by gigantic (meters to tens of meters) misshapen marbles. Extensive terrains covered by such blocks occur in southern California's Peninsular Ranges and many other places on Earth. They are frequently set aside as tourist attractions, such as at Arizona's Texas Canyon. Below the surface the spherical "marbles" grade into inhomogeneously weathered rock, where intact corestones are imbedded in a matrix of crumbling, disintegrated rock debris (Figure 7.11). It is clear that the rock has been preferentially attacked along the joints, from which a weathering front has moved gradually inward, rounding off the sharp corners and leaving a near-spherical unweathered mass in the center. Recent images from the Spirit Rover confirm that spheroidal weathering occurs on Mars as well as on Earth.

Frost shattering. One of the most widely cited types of physical weathering is frost shattering by freezing water (McGreevy, 1981). A moderate number of people have actually heard the sharp report of a rock or tree suddenly splitting during a cold snap; the physical

Figure 7.11 Spheroidally weathered granite showing incipient corestones in a homogeneous mass. The light bands are aplite dikes, fine-grained dikes that typically cut granite plutons during their final stages of crystallization. These dikes are more resistant to weathering than the coarse-grained granite itself. Outcrop on the north shore of Big Bear Lake, Southern California that was exposed by road construction in the 1970s. Photo by the author. Corestones are about half a meter in diameter.

evidence of fractured rocks in cold climates is clear for anyone to see. This process is widely attributed to the 9% volume expansion of water upon freezing. When freezing water is completely sealed up in a rigid container it can develop truly enormous pressures – 207 MPa if cooled to –22°C – but the problem is, how can water be so completely confined under natural circumstances? If volume expansion were the cause of frost shattering, it would make it a rather special process, occurring only for liquid water, but not for most other liquids, which contract upon freezing. Thus, methane on Titan would not be capable of shattering solid ice "rocks," presuming that it were able to freeze under ambient Titanian conditions, because solid methane does not expand upon freezing. However, it may come as a surprise to many readers of this book (who may even have read that frost shattering is caused by the volume expansion of freezing water in otherwise reputable books on geomorphology) that frost shattering does not, in fact, have anything to do with the peculiar behavior of water and that *any* freezing fluid can shatter a solid in which it crystallizes. The process of frost shattering is closely connected to the similar processes of salt weathering. The downside of our current understanding of frost shattering is that the explanation of what really happens is longer and more complex than simple volume expansion.

Although a large mass of water freezes, by definition, at 0°C (at 1 bar pressure), small volumes of water must be cooled to much lower temperatures before they freeze, a fact first discovered by J. J. Thomson in 1888. This discrepancy is caused by the surface energy of small masses of water (this is also true for any other substance). Similarly, a thin film of water on rock possesses a surface contact energy that keeps it liquid well below 0°C. In fine-grained silts, where water is both dispersed into tiny pores and gains considerable energy

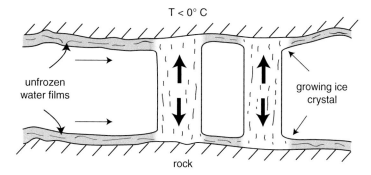

Figure 7.12 Ice freezing in narrow cracks wedges rock apart by the force of crystallization, not by its volume expansion. Thin films of water in immediate contact with the rock do not freeze until the temperature falls well below the normal freezing point of water because of the chemical interaction energy between water and silicate minerals. Hexagonal ice crystals typically grow fastest along their c-axes, shown here as dashed lines, which are, thus, usually perpendicular to the plane of the crack.

from contact with the silicate minerals, some water may remain unfrozen even at $-30°C$. In contrast, water in gravel fills large pores and is entirely frozen just a little below $0°C$.

A thin film of water in a joint or crack in a rock on the surface may, thus, remain liquid at temperatures well below those necessary to freeze a large mass of water. Now suppose that, adjacent to this crack, there is an unusually large pore or widening of the crack. The larger mass of water in this space freezes first, while the water film remains unfrozen (Figure 7.12). Under these circumstances water flows from the film to the frozen mass, which grows larger and begins to wedge open the space in which it is growing. This is also the origin of the ice lenses frequently found in frozen soil, as was first convincingly demonstrated by Taber (1929). Detailed analysis of the thermodynamics of this process shows that the chemical potential of the ice mass is lower than that of the water film, driving this process forward (Dash *et al.*, 2006). The greater the difference between the ambient temperature and the freezing point, the greater is the chemical potential favoring crystal formation. This chemical energy difference translates to a maximum difference in pressure between the crystallizing solid P_s and the liquid, P_l, given by:

$$P_s - P_l = \frac{\rho_s L_m}{T_m}(T_m - T) = C_{fh}(T_m - T).$$

(7.20)

When the pressure difference between the solid and liquid is less than this maximum, water continues to flow to the solid, which grows in volume. If the pressure difference is greater, the water flow reverses and the solid melts. The "frost-heave coefficient" C_{fh} in (7.20) is equal to the density of the solid ρ_s times the latent heat of melting L_m divided by the melt temperature T_m. For water it equals 1.12 MPa/K, which is unusually large for most substances (this coefficient is only 0.32 MPa/K for methane). Thus, ice lenses segregating from a thin film that freezes at $-10°C$ can exert pressures of about 10 MPa – large enough to split most solid rocks. This analysis ignores crystallographic orientations of the ice crystal, which is not actually correct: Hexagonal ice crystals grow preferentially along their c-axes and correspondingly exert their maximum pressure along this axis. Nevertheless, the pressure given in (7.20) gives the correct order of magnitude for frost heaving.

The process of frost shattering is most effective where the temperature makes frequent excursions around the freezing point and ice masses alternately grow and shrink in the interiors of wet rocks. On Earth, this occurs at middle and high latitudes in spring and fall. Where frost shattering is effective the surface may be covered with broken shards of rock that form extensive fields. Because it requires the presence of thin films of liquid water, frost shattering is ineffective where the temperature remains well below freezing or liquid water is absent. Mars, at present, may not be a candidate for this process because its atmospheric pressure is below the triple point of water so that liquid water is unstable at its surface, except for a small region at the bottom of the Hellas crater depression. Temperatures on Titan are presently believed to always remain above the freezing point of methane, so methane-shattering may not be effective there, although it appears that *some* process on Titan shatters its ice "bedrock" and produces ice boulders that can be transported by methane streams.

Salt weathering. Minerals crystallizing from solution exert rock-shattering pressure in a similar manner. In this case it is not the temperature below the nominal freezing point that counts, but the degree of supersaturation of the mineral (often a salt such as halite) in solution. If the concentration of the crystallizing species at saturation is C_s, and the actual concentration is C, then the supersaturation is defined as C/C_s. The mineral remains in solution if $C < C_s$, but C must exceed C_s by some finite amount before precipitation can start. Supersaturations seldom exceed about 10 when nuclei are present to begin the crystallization process. Typically the liquid portion of a solution evaporates, gradually raising the concentration of the mineral solute until it exceeds the saturation limit, at which point crystallization begins. The maximum pressure that can be exerted by a growing crystal is given by (Winkler and Singer, 1972):

$$P_{\text{xtal}} = \frac{R\rho T}{m} \ln \frac{C}{C_s} \qquad (7.21)$$

where R is the gas constant, ρ the density of the crystalline species, T the absolute temperature and m is its molecular weight. Equation (7.21) ignores crystal anisotropy and so can only be trusted for its order of magnitude. Table 7.4 lists the isotropic crystallization pressures of a number of geologically common evaporite minerals at a supersaturation ratio of 2. It is clear that many salts can exert very large pressures as they crystallize and "salt weathering" is an important and effective process where evaporation occurs readily, such as in the hot deserts of Earth. The Phoenix mission to Mars is believed to have detected brines in the immediate subsurface, which suggests the possibility of salt weathering on Mars as well. Hematite, listed in Table 7.4, has been found as concretionary "blueberries" on Mars by the Opportunity rover. Concretions of this type can grow by displacing the enclosing permeable sediment through crystallization pressure.

Table 7.4 *Crystallization pressures of evaporite minerals*

Mineral	Chemical formula	Crystallization pressure at $C/C_S = 2$ and 273 K (MPa)
Halite	NaCl	55.4
Hematite	Fe_2O_3	52.3
Anhydrite	$CaSO_4$	33.5
Gypsum	$CaSO_4 \cdot 2H_2O$	28.2
Kieserite	$MgSO_4 \cdot H_2O$	27.2
Epsomite	$MgSO_4 \cdot 7H_2O$	10.5
Jarosite	$KFe_3(SO_4)_2(OH)_6$	9.7
Natron	$Na_2CO_3 \cdot 10H_2O$	7.8

Data and method for jarosite and hematite from Winkler and Singer (1972).

Frost shattering by freezing pore fluid and salt weathering by crystallizing salts are only two special examples of a general phenomenon. Whenever a chemical alteration, driven forward by a change in chemical potential $\Delta\mu$, results in a volume change ΔV, the reaction will go forward unless the work expended in accommodating the volume change, $P\Delta V$, is greater than the gain in chemical potential. The maximum pressure the reaction can exert is thus given by

$$P_{max} = \frac{\Delta\mu}{\Delta V}. \tag{7.22}$$

Equations (7.20) and (7.21) are merely special cases of (7.22), specialized to the case of a liquid that freezes at a temperature below that of the solid and of a crystallizing solute.

Rock disintegration. Another common realization of this phenomenon is the disintegration of weathered igneous rocks. Initially solid rocks, such as granite, contain a variety of minerals with different susceptibilities to chemical weathering. When a particularly susceptible mineral (biotite or feldspar, for example) becomes hydrated and swells in volume by only a few percent, its expansion generates large tensile stresses that wedge the mineral grains apart and reduce the rock to a pile of coarse, sand-sized debris. This process is so common on Earth that disintegrated granite has its own special name: grus. The formation of grus is of especial importance because the quartz grains thus released from granite are eventually freed of all other igneous minerals. Quartz is uniquely stable in the Earth's surface environment – it is possible to find quartz grains that have been recycled through multiple generations of sandstone for a billion years. This quartz forms the sand that is so abundant in our beaches, streams, seas, and sandy deserts. But because silica-rich granite is an exclusive product of plate tectonics, quartz sand may also be unique to Earth. The process that forms the sand-sized particles in the extensive eolian dune fields on Venus, Mars, and Titan is still a mystery, but these dunes are not made of quartz grains.

Volume changes are caused by many other processes: The reversible swelling and shrinkage of hydrating clays during wet/dry cycles is another such process, one that keeps clay-rich soils slowly seething in perpetual movement. The consequences of this motion will be explored in more detail later in this chapter and the next, but we end this section by examining a much more controversial volume change: Thermal expansion.

Thermal stresses. Diurnal and seasonal temperature changes, as discussed above in Section 7.2, are most pronounced at or near the surface. Rocks exposed on the surface regularly warm up during the day and cool at night. Because the thermal skin depth may be only a few centimeters, rocks larger than the skin depth experience strong gradients in temperature. But because of thermal expansion, different parts of the rock expand or contract by different amounts and this generates internal stresses. It has long been understood that such stresses might shatter the rock (Mabbutt, 1977). This process has therefore, been given the name "insolation weathering." There is little doubt that it actually occurs on Earth: The orientations of cracks in split desert rocks bear clear relationships to the direction of maximum solar heating. The problem is, laboratory experiments have been unable to reproduce this phenomenon and, worse yet, no sign of insolation weathering has been found among lunar rocks, where temperature extremes are much larger than those on Earth.

Much of the doubt about the effectiveness of insolation weathering dates from a series of experiments performed by geologist David Griggs (1936). Griggs cycled dry rocks through an extreme temperature excursion of 110°C, 89 400 times, an equivalent of 244 yr of daily cycles on the Earth. In spite of his very high rates of temperature change as well as extreme temperature excursions, he found no sign of damage to the rock. On the other hand, when small amounts of water were present the rock crumbled after only 1000 cycles. The implication is that temperature variations alone do not cause rock disintegration, but that water is required, perhaps by a chemical attack through hydration or stress corrosion cracking (Moores *et al.*, 2008). "Insolation weathering" is, thus, not a purely mechanical process but requires that diurnal temperature changes be accompanied by some other, probably chemical, weathering process.

It is known that really extreme temperature changes, such as occur in brushfires on the Earth, do cause flakes to spall off rock surfaces. However, even here, temperature changes might not be acting alone, as quartz in silica-rich rocks undergoes a large volume change at 573°C in association with the alpha-beta solid-state phase transition.

7.3.3 Sublimation weathering

Sublimation is not a very important weathering process on Earth. Although it has been often suggested in connection with the evolution of snow and ice surfaces, the latent heat of sublimation of ice, 2.83 MJ/kg (at 0°C), is much higher than its latent heat of melting, 0.34 MJ/kg. Observational investigations of the relative role of melting and sublimation show that melting almost always dominates over sublimation when it is possible.

However, elsewhere in the Solar System melting may not be possible, such as on Mars where liquid water is not stable on the surface, or on Callisto where both the pressure and

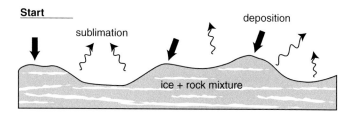

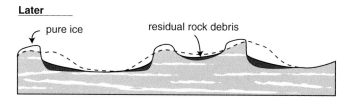

Figure 7.13 Sublimation weathering on Callisto. Water vapor sublimates from exposed warm surfaces and is deposited on cold surfaces, raising its albedo and promoting the separation of the surface into patches of cold, bright ice and ice-free regions mantled by darker silicate dust that absorbs more sunlight and becomes warmer than the bright ice. After Figure 17–24 of Moore *et al.* (2004).

temperature are too low for liquid water. On Mars, CO_2 frosts are common and the triple point of CO_2 is even higher than that of water (its heat of sublimation is much lower than water, 0.571 MJ/kg). Under these circumstances, solid volatiles pass directly into the vapor phase, giving rise to a novel form of weathering.

By itself, sublimation promotes a net transport of material from warm, exposed, areas on a surface to cooler, shady places. On a planetary scale, vapor is transported from equatorial regions toward the poles, and bright polar caps may be the result. However, when a bright volatile ice is mixed with darker, inert material, complex and characteristic landforms may arise. A surface composed of a mixture of dark material such as silicate dust and ice is inherently unstable. The dark material absorbs more sunlight and, thus, remains warmer than the more reflective ice. A thin coating of dark material, thus, enhances the sublimation of the underlying ice, although a sufficiently thick deposit of dark material suppresses sublimation by shielding the underlying ice from the daily high extremes of temperature. Warm, dark areas tend to lose mass and become darker, while cool, bright areas gain more bright ice by vapor deposition (Figure 7.13). The surface becomes dominated by strong positive feedback that produces a landscape of high, bright peaks separated by dark, low-lying plains and hollows. Spires observed on the surface of comet Wild 2 by the Stardust spacecraft and "hoodoos" on comet Hartley 2 observed by the EPOXI mission may have originated by similar processes.

Earth does not have any large-scale analogs of this process, although "suncups" that evolve into spire-like "penitents" on old snowfields provide small-scale examples, and some features in the Antarctic dry valleys have been attributed to sublimation. Mars has many probable examples of this kind of terrain in its deposits of polar volatiles, where irregular to circular pits are etched into ice-rich sediments on a scale of hundreds of meters in the so-called "Swiss-cheese" terrain. Callisto provides the most extensive example of such terrains, where dark, broad plains are dotted with steep-sided knobs and fretted scarps (Moore *et al.*, 2004). Pole-facing slopes are typically brighter than equator-facing slopes, reflecting the tendency of volatiles to deposit in cooler environments.

7.3.4 *Duricrusts and cavernous weathering*

It is common to find that individual boulders and rock surfaces exposed in the Earth's deserts are "case-hardened" by a strong rind or crust a few centimeters thick. Beneath the rind, the rock is often much softer than the intact rock. In some cases the interior has entirely disappeared, leaving an empty shell (sometimes called "turtle" *rocks*, as they look like a giant empty turtle shell). Such rinds, properly called "duricrusts," are another result of chemical weathering coupled with fluid transport.

When an infrequent rain wets a rock in an arid environment, the slightly acid rainwater soaks into the rock only a small distance, typically a few centimeters. Once there, it initiates hydration and hydrolysis reactions and is enriched in some of the soluble weathering products. Later, after the rain stops and the Sun returns, the water evaporates from the rock surface, drawing the now briny water back to the surface by capillary action. The brine becomes concentrated near the surface and deposits its load of solutes as it evaporates completely. In some rocks, this process extracts salts that disintegrate the rock surface by salt weathering. However, in rocks whose minerals are cemented together by $CaCO_3$, SiO_2, or Fe_2O_3, the deposit may enhance the natural cement, hardening the outer layer of the rock while softening the interior by removing its cement. The depth of penetration of the water sets the scale for this process: Weathering rinds are thicker in permeable rocks such as sandstones or tuffs, and thin or absent in very impermeable rocks.

Duricrusts may also form in soils subject to infrequent rainfall followed by evaporation. The process is the same as in rock, with the exception that soluble salts are more likely to harden the soil surface than to disintegrate it. Hard soil caps with names like *gypcrete*, *calcrete*, *silcrete*, or *ferricrete* (for soil duricrusts cemented by gypsum, $CaCO_3$, SiO_2, or Fe_2O_3, respectively) are common in many arid regions on Earth (Mabbutt, 1977). The Viking 2 and Phoenix landers on Mars encountered hard layers in the Martian soil associated with enhanced abundances of SO_4 and Cl that may be duricrusts, although in this case they are more likely to be produced by evaporation of briny soil water rather than rainfall. High thermal inertia measurements over broad regions of Mars suggest that some form of duricrust may be globally distributed.

Duricrust formation is closely related to another striking weathering feature, cavernous weathering. Many rock exposures in the semiarid American West are peppered with shallow holes now termed "niches" or "tafoni" (Figure 7.14). Such holes astounded and perplexed travelers from the humid Eastern states and led to many colorful explanations, ranging from the miners' claim that they were drilled by "rockpeckers," an extinct giant woodpecker that extracted burrowing mammals from solid rock, to an early geologist's assertion that they

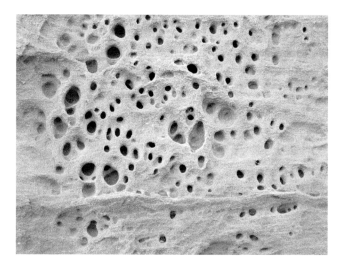

Figure 7.14 Cavernous weathering surface in sandstone. Image is about 1 m across, showing deep pits and columns that have become detached from the mass of the rock behind them. This variety is often called "honeycomb" weathering from its appearance. Canyonlands National Park, Utah, USA. Photo courtesy of Ingrid Daubar-Spitale, 2010. See also color plates section.

are caused by wind (this hypothesis is still often encountered in textbooks). Although wind may, indeed, play some role in their excavation, the correct explanation is linked to duricrust formation (Blackwelder, 1929; Mabbutt, 1977).

Case-hardening of rock surfaces comes at the price of a weakened layer a few centimeters beneath the surface. Thus, when a case-hardened surface is breached for any reason, the underlying weakened rock easily disintegrates. The resulting debris either falls free of the rock face or is removed by wind or rainwash. Once a hollow forms, water remains in its shaded interior longer than on the adjacent rock face and promotes further chemical weathering. Rock deeper in the cavity then disintegrates, creating flakes and small fragments that fall to the floor and are also easily removed from the growing cavity. Where this process is effective, broad vertical rock faces gradually become deeply pitted. As individual cavities continue to grow they may meet behind the original rock face, creating long colonnades sometimes known as "choir stall weathering," some small examples of which can be seen in Figure 7.14.

7.3.5 Desert varnish

Many rock surfaces in arid environments are covered with a thin amorphous coating of black or reddish material known as desert varnish. Its thickness ranges from 0.03 to about 0.1 mm and it is composed mainly of clay and the hydrous oxides of iron and, especially, manganese, which impart its red and black colors. It gradually thickens with time and often possesses an internal stratigraphy. It can, thus, be used to establish relative dates of surfaces (the darker the surface, the older it must be), but its growth rate varies widely from one location to another, so its thickness cannot be used for absolute dating. Occasional wetting seems to be essential for the growth of rock varnish. Much of its material is derived from wind-blown dust, not from the underlying rock, although it often forms on top of weathered duricrust rinds. Black, desert-varnished rocks under the summer Sun can become too hot to touch, so it was a surprise to discover that it has an important biological component (Mabbutt, 1977). Lichens, algae, and soil microflora are active in chemically reducing the oxides in desert varnish. Carbon compounds of biological origin are entombed in the coating. Because of the biological connections of desert varnish, it is considered a prime target for searches for life on Mars (Perry and Kolb, 2003).

7.3.6 Terrestrial soils

Soil on the Earth has, quite naturally, received intense scientific study for as long as science has existed. Human civilization depends crucially upon the quality of the agricultural soil: Civilizations have risen and fallen in response to changes in the soil. It is impossible to do justice here to the immense literature on the subject, but only to indicate some of the important factors in soil formation and maintenance and the major differences between terrestrial soils and those of other planets.

Soils on Earth range in thickness from near zero to as much as 100 m, although a thickness of only a few meters is typical. Rates of soil formation are widely variable: observed rates vary from 3 cm/yr to 3 cm/5000 yr (Leopold *et al.*, 1964). While the nature of the soil depends to some extent on the bedrock from which it forms, climate (moisture and temperature) appears to be the dominant determinant of soil type.

The most compelling aspect of terrestrial soil is its biological activity. Organisms from all the kingdoms of life are at home in the soil, which is an immense reservoir of life ranging from microbes to higher animals and plants. Charles Darwin was so impressed by the importance of biological activity in the soil that he devoted his last book to the humble earthworm and its role in the "economy of nature" (Darwin, 1896). The bulk of Earth's organic carbon resides in the soil; about 1500 Gton total (Amundson, 2001), and turnover from one organism to another is rapid. Because of the presence of life, chemical activity is intense and weathering is faster in the soil than almost anywhere else on the Earth. Indeed, it is difficult to separate purely chemical weathering processes from biologically mediated ones. Experiments on sterile weathering reactions have almost invariably shown that they are slow compared to their naturally observed analogs, and the difference is usually due to biological activity.

Another major factor in soil formation is rainfall. The landscapes of even the driest portions of the Earth show traces of rainfall and no Earthly soil can escape at least an occasional wetting from rain. Rain falling on the surface typically soaks in some distance before it either joins the underground water table, wicks back to the surface or flows laterally to a spring. The result is a two-level structure, where the upper portion is leached by rainwater flowing to deeper levels, while the lower portion receives leachates from above.

These two regimes, along with the organic-rich layer right at or just below the surface, are reflected in the common division of the soil into layers or horizons denoted by letters. Thus, surface organic material is denoted the O-horizon. The typically dark, organic-rich and leached upper layer is termed the A-horizon, while the B-horizon receives solutes leached from above. The C-horizon is composed of weathered parent material and the R-horizon is unweathered material. Other horizon designations and subdivisions are in common use, but for these the reader is referred to specialized texts (Birkland, 1974). Not all of the various horizons may be present in a given soil. Soil naming and classification is a complex affair and depends on composition, color, texture, and other factors that differentiate one soil from another.

7.4 Surface textures

The immediate solid surface of a planet is what we can observe from space and is the major target of any spacecraft imaging system. Although global maps are made of mountain ranges, valleys, and impact craters, the fine-scale structure of the surface is too extensive to receive such attention. Nevertheless, the detailed surface texture, at sizes ranging from meters to microns, gives the surface its distinctive optical properties and has a major effect on the ability of machines and, ultimately, humans to traverse the surface.

7.4.1 "Fairy castle" lunar surface structure

The brightness of a night with a full Moon can be astonishing. Careful observation of the Moon over a lunar cycle shows that the full Moon is not just bright because the illuminated area is large: Each portion of the moon is quite literally brighter during the full Moon than at all other times. This phenomenon has become known as the "opposition surge" and it tells us something about the arrangement of particles on the Moon's surface.

Consider the path of the sunlight that we receive from the Moon's surface. The light starts out at the Sun, travels to a point on the Moon's surface, is reflected (or absorbed or scattered), and then travels in a straight line to our eyes (neglecting refraction in the Earth's atmosphere). The angle between the light ray incident on the Moon's surface and the reflected ray that we receive is called the phase angle. During a full Moon, when the Moon is in opposition to the Sun, this phase angle is very small: When the Moon is half-full this angle is close to $90°$ and it approaches $180°$ at new Moon. The total brightness of the Moon as a function of this angle is plotted in Figure 7.15. It is clear that brightness is neither constant nor a linear function of phase angle. In particular, it increases dramatically (the "surge") at angles less than about $5°$ (Buratti *et al.*, 1996). It is interesting to note that the full range of the opposition surge was not known before spacecraft observations: The moon as a whole cannot be observed from Earth at less than a few degrees phase angle.

Mutual shadowing of surface irregularities causes the opposition surge. Consider an irregular, pitted surface. Looking from nearly any direction, one's field of view includes both bright

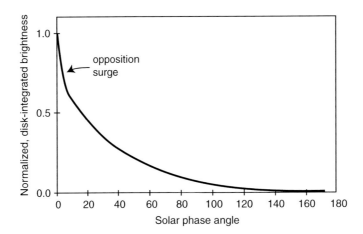

Figure 7.15 Opposition effect of the lunar surface. The Moon is much brighter when incident sunlight is directly backscattered, resulting in a distinct brightening during the full Moon. The brightness peaks at reflection angles less than 5° from exact opposition. Simplified after Figure 5 of Buratti *et al.* (1996).

illuminated high points and shadows in low areas and hollows. However, in one direction, that looking along the direction of illumination (that is, near zero phase angle), the shadows disappear and all surfaces appear bright. One can see this phenomenon by looking at the shadow of one's head on an irregular surface, such as a grassy lawn, or more obviously by looking out of an airplane window at the shadow of the airplane on the ground. The opposition surge is related to the phenomenon known as "heiligenschein," which is the bright halo that appears around the shadow of one's head on dewy grass, although in that case the intensity of the backscattered light is given an extra boost by reflection from dewdrops.

The range of angles over which opposition brightening occurs is related to the average surface slope and the depth to which light penetrates into the surface. Strong opposition effects are typically observed in fine (10 to 20 μm) powders. Detailed models of how surfaces reflect light can be fit to the observed opposition brightening and yield information about the particle size and distribution on the surface (Hapke, 1993). Application of such models to the Moon long ago suggested to astronomers that the Moon's surface is extremely irregular on a fine scale, to which the name "fairy castle" structure was given. Small particles may be sintered together to form tiny towers surrounded by minute deep crevasses. The narrow angle and extent of the opposition surge suggests an extremely pitted, fractal surface on a scale larger than the wavelength of light. This deduction was verified by *in situ* stereoscopic measurements of the undisturbed lunar surface by the Apollo 11, 12, and 14 astronauts (Gold, 1970).

Opposition surges are observed for asteroids, indicating a powdery regolith surface. This effect makes optical detection of asteroids most easy at opposition, as their brightness declines rapidly as they move only a few degrees away from this point. The opposition surge is also apparent in Saturn's rings, and is cited as one of the first direct demonstrations that the rings are composed of a swarm of individual solid particles.

(a)

(b)

(c)

(d)

Figure 7.16 Boulder-strewn surfaces on Venus, Earth, Mars, and Titan. Panel (a) is from the Soviet lander, Venera 13, image VG00261. Panel (b) near Yuma, AZ, looking north toward the Cargo Muchacho Mountains from Indian Pass Road (photo courtesy of Mark A. Dimmitt). (c) is a panorama from the Viking 2 lander. (d) The surface of Titan as viewed by the Huygens lander. The lander evidently set down in a former riverbed, only the "rocks" in the foreground are water ice and the liquid that transported them was liquid methane. The two "rocks" just below the middle of the image are 15 cm and 4 cm in diameter, respectively, and lie about 85 cm from the Huygens probe. The dark fine material on which the "rocks" lie is probably a mixture of water and hydrocarbon ice. Image PIA07232. ESA/NASA/JPL/University of Arizona. See also color plate section.

7.4.2 *Stone pavements: why the Brazil nuts are on top*

Many desert surfaces on Earth, as well as the surfaces of Mars, Venus, and Titan (Figure 7.16), are littered with stones and small boulders. In many places the stones covering these surfaces appear to be fitted together in an almost artificial-looking mosaic (Figure 7.17). Digging into such surfaces, one is surprised to find that the stones form a single layer that is confined to the surface itself: Underneath one finds mostly fine-grained material with few, or no, large stones. Such surfaces are called "desert pavements" or "stone pavements." Their striking appearance has attracted many geologists' attention and has generated a currently raging controversy about their origin.

For at least a century, nearly every writer describing desert pavements has attributed them to deflation. They supposed that the original soil contained a mixture of stones and fine

Figure 7.17 Desert pavement surface showing a surface layer of large and small stones apparently "fitted" together into a layer only one stone thick. The stones are basalt fragments. From central Iceland, photo by the author.

silt. When the wind blows over such a surface, the fine silt is exported from the region and the stones remain behind as a lag deposit that eventually armors the surface against further deflation. The presence of pavements was, thus, taken as evidence that the region is degrading, with preferential removal of the fine-grained material by wind or perhaps water.

Doubts about this explanation began to arise when pavements were found on surfaces that are manifestly aggrading. A revisionist view that is now gaining wide acceptance attributes most pavement formation to upward migration of stones relative to a fine-grained substrate (Dohrenwend, 1987). The explanation of how this counter-intuitive process works is closely related to what is called the "Brazil nut phenomenon" (Rosato *et al.*, 1987). It is easy to show that in a bowl of mixed nuts of different sizes, say, peanuts and Brazil nuts, vigorously shaking the bowl brings the Brazil nuts to the top, where they remain through further shaking (parents may have noted the same phenomenon in children's toy chests: Shaking always brings the larger toys to the top while the small ones disappear into the depths). The tendency for large objects to rise is so strong that it can even invert the density of the shaken objects: In a mixture of large steel ball bearings and small peanuts, shaking still puts the much denser ball bearings on top. The reason for the preferential uplift of large objects is simple: When a large object shifts upward, smaller objects readily fall into the vacated space, whereas it is very unlikely that a number of small objects will shift so as to leave a space large enough for the large object to fall into. As shaking continues, the large object gradually ratchets up relative to the small ones until it emerges onto the surface.

If the Brazil nut phenomenon is the reason that stone pavements form, there must be some process that agitates the large stones and permits them to jostle relative to the surrounding fine-grained sediment. Fortunately, such processes abound near the surface: Wetting and drying of expansive clays, freezing that generates ice pillars beneath the stones followed by thawing, and bioturbation (creatures from lizards to wombats have been invoked) are all plausible and effective types of disturbance. Rapid movement is not necessary, only small, repetitive shifts of the stones relative to the fines. Such agitation is possible only close to the surface. Stones buried deeply enough lose the ability to shift and so become unable to participate in the upward climb. However, the soil surface itself is an active place and stones may continue to "float" on the surface for a long time: 770 000-year-old Australasian tektites are sometimes found on late Pleistocene soils. Although this observation is sometimes used as an objection to the natural occurrence of these objects, this is probably an example of the Brazil nut effect. Likewise, deep-sea manganese nodules with ages of millions of years are found lying on top of contemporary mud surfaces. There is no paradox here if one recognizes the disturbing role of marine bioturbation.

Stone pavements on desert surfaces can remain on the surface even as wind-borne sand and dust accumulate. Fine-grained sediments fall between the stones and eventually disappear beneath the surface as the stones slowly shift and offer them a larger space to fall into. Blocks of basalt that once lay on top of their parent lava flows 12 Myr ago are now separated from them by meters of wind-blown sediment in many places along Arizona's Mogollon Rim. Stone pavement formation is a universal process,which depends upon the restless activity of a planetary surface,for creating organized patterns.

7.4.3 Mudcracks, desiccation features

If the Earth were barren of vegetation, mudcracks would be one of the most common features on the land surface. As it is, they are ubiquitous on bare, exposed clay-rich sediments and are often found as fossils in ancient rocks. Mudcracks appear as networks of open cracks that surround polygonal, sometimes crudely hexagonal, plates. These inter-crack plates range in size from a few centimeters up to hundreds of meters across. The ultimate cause of this cracking is shrinkage upon drying. Clay minerals are especially susceptible to volume changes upon wetting or drying: Water molecules are easily added or lost from the filling of the silicate sandwiches that make up clay minerals.

One of the most surprising characteristics of many mudcrack-covered surfaces is their crudely hexagonal pattern. It has been shown by many researchers that a triple intersection of cracks meeting at mutual angles of 120° minimizes elastic strain energy. However, all three cracks must initiate at the same time if this configuration is to arise at all: If a single tension crack formed first, it would relieve the stress parallel to its length so that a second crack growing to meet it would curve to intersect it at a 90° angle. The result would be a pattern of rectangular plates, not hexagonal ones. One does sometimes see such a rectangular grid of plates, but triple intersections are more common.

The horizontal scale of mudcracks is a complicated function of the strength of the mud, rate of desiccation, and depth of the cracks. Cracks cannot grow too closely to one another, as the formation of the first crack relieves the stress in its vicinity, out to a horizontal distance of several times its depth. Mudcracks are closely spaced where a thin layer dries quickly, while giant mudcracks are occasionally observed in thick playa sediments that dry over long periods of time.

Because the top of drying sheets of sediment typically contracts more than the deeper portions, the plates between mudcracks sometimes break off the underlying sediments and curl up into separate flakes. Such flakes are especially vulnerable to being picked up and transported by the wind. Weakly cemented clumps of fine sediment can then be blown into heaps known as "clay dunes." This process could be the solution to the "kamikaze effect" noted for grains transported by the winds on Mars, discussed in Box 9.1 of Chapter 9, assuming that vapor-phase hydration and freeze-drying can play the same role as liquid water in expanding clays.

Further reading

The best summary of lunar regolith properties is contained in the book by Heiken *et al.* (1991), where the peculiar thermal properties of the lunar regolith are also discussed. French (1977) offers a clear description of the regolith at a semi-popular level, written just after the end of the Apollo missions. Regolith formation and gardening processes are discussed extensively in Melosh (1989). The classic book on terrestrial weathering processes is Ollier (1975), while Bland and Rolls (1998) present a comprehensive overview of weathering from a more fundamental point of view. Few of the other topics in this chapter are treated at book length; however, I have added many references to individual papers in the appropriate places in the body of the chapter.

Exercises

7.1 Gardening asteroids

Like the lunar regolith, asteroid surfaces are constantly bombarded by small meteorites that create craters of all sizes. On average, the depth to which the surface is overturned once in a given period of time is equal to ¼ of the diameter of craters that have, in some sense, just "covered over" the surface. Suppose that "covered over" means that the total area of craters in the diameter range D_c to $2D_c$ just equals the area of the surface itself. Using the cumulative size-frequency distribution, Equation (7.1), show that the following equation expresses this condition:

$$-\frac{\pi}{4}\int_{D_c}^{2D_c} D^2 \frac{\mathrm{d}N_{\mathrm{cum}}}{\mathrm{d}D}\,\mathrm{d}D = \frac{\pi}{4}\frac{bc}{b-2}\left(1-\frac{1}{2^{b-2}}\right)\frac{1}{D_c^{b-2}} = 1$$

so long as $b > 2$.

The cratering rate in the asteroid belt is not well known, so use the better-known lunar cratering rate, $N_{cum}(D) = 1.9 \times 10^{-12} D^{-3.82}$ craters/km² yr for craters < 2 km in diameter (Hartmann, 2005), as an estimate for the cratering rate in the asteroid belt. Note that D is in kilometers. Using your equation for D_c and constants from the lunar cratering rate, compute the overturn time for asteroid regolith to depths of 0.01, 0.1, and 1 m. Watch out for unit conversions!

7.2 Huddling close to the fire

Cool red-dwarf stars (M-type stars) vastly outnumber solar-type stars in our galactic neighborhood. Typical M-class stars have a mass about ½ that of our Sun, a surface temperature of about 3600 K and a luminosity of only 0.08 L_e. The "habitable zone" is the range of distances of a planet from its star throughout which liquid water can occur on its surface. Using the concept of effective temperature, use Equation (7.2) to estimate the semimajor axis of planets in the habitable zone about M-type stars. Assume a planetary albedo similar to that of known planets, such as the Earth or Mars. What would happen to a planet's surface temperature if it became covered with highly reflective snow or ice?

7.3 Cometary heat transfer

During its close approach to Comet Tempel 1 on July 4, 2005, the infra-red spectrometer aboard the Deep Impact spacecraft provided an upper limit to the thermal inertia I of the comet's surface, $I < 50$ J/K m² s$^{1/2}$ (Groussin *et al.*, 2007). Although the comet was just outside the orbit of Mars during the encounter (and also nearly at perihelion), the maximum surface temperatures reached a toasty 340 K because of the comet's exceptionally low albedo. Approximating the heat capacity of the comet's surface material as fresh snow, from Table 7.2, and using the subsequently determined mean density of 400 kg/m³(Richardson *et al.*,2007), estimate the thermal conductivity k of the comet's surface from Equation (7.6). Finally, use this thermal conductivity in Equation (7.4) to compute the thermal skin depth λ during the comet's 6-month-long perihelion passage. What does this imply for the volatiles that drive cometary activity?

7.4 Harold Urey's feedback on planetary habitation

Qualitatively discuss the implications of the Urey reaction, Equation (7.19), for planetary temperatures in the light of the greenhouse effect. Assume that both silicates and carbonates may be present on the surface of the planet. If water is required for this reaction to proceed at an appreciable rate, what are the implications for planets that either possess or lack oceans? What is the role that plate tectonic recycling of carbonates might play? See Kasting and Catling (2003) for an extended discussion of this interesting problem.

7.5 (Methane) frost shattering on Titan?

At present, temperatures on Titan are just a bit too high (by only a few degrees K) to permit methane to freeze anywhere on its surface. Suppose, however, that at some time in the past methane could experience freeze–thaw cycles on Titan's surface. Use Equation (7.20) to discuss the possibility of frost shattering on Titan's surface. Note that the frost-heave coefficient $C_{fh} = 0.32$ MPa/K for methane. Is it plausible that freezing methane could shatter solid ice "rocks"? Could "methane heaves" form in Titan's soil? What properties of the ice/methane interface must be understood to make such arguments quantitative?

7.6 Heaving rocks

The New England region of the United States is mantled with a thin veneer of glacial till deposited by the now-vanished ice sheets of the last continental glaciation. This till is a mixture of fine sand, clay, and decimeter-sized (or more) boulders. Farmers in these states note that every spring a new crop of boulders seems to surface in their fields (some old-timers are convinced that rocks on their farms grow underground like potatoes). As an alternative to the potato theory, consider that the thermal conductivity of solid rock is much larger than that of soil. Coupled with the tendency of soil water to form ice lenses and pillars at freezing fronts, explain how the motion of solid rocks upward with respect to loose fine-grained soil might occur in the annual freeze–thaw layer below the surface. Might any analogous process occur on the surface of Mars? Explain why, or why not, based on your knowledge of Martian surface conditions compared to those in New England.

7.7 Sublime cometary surfaces

In the wake of the Deep Impact missions to comets Tempel 1 and Hartley 2, we know that these comets are composed of fine powder consisting of an intimate mixture of water ice (plus highly volatile CO_2 ice below the surface), silicates, and organic tars. During a comet's perihelion passage, the sun's heat raises low-albedo surface layers to temperatures high enough to rapidly evaporate water ice at pressures well below water's triple point, so that liquid water never forms: Ice simply sublimates into vapor. One of the more bizarre discoveries of the Deep Impact and Stardust missions (Stardust returned images of comet Wild 2) was that of bright, nearly vertical pillars on the comets that range in height from tens up to 100 m. Speculate on how such features might form, given the inherent instability of a surface containing a bright (high-albedo) volatile and dark (low-albedo) non-volatile material. You might also look up references on sun cups, penitentes, and ice pillars on terrestrial snowfields.

8

Slopes and mass movement

Imagine a landmass upon which no rain falls, no rivers flow, no glaciers form, no waves beat, no winds blow. Let chemical and mechanical disintegration disrupt the rocks, and gravity exert its downward pull. On such a landmass earth and rock will move ceaselessly from higher to lower levels, slopes will soften, relief will fade. Given time enough, the whole will be reduced to a featureless plane of disintegrated rock debris.

Douglas Johnson, foreword to the book *Landslides and Related Phenomena* by C. F. S. Sharpe, 1937

In the summer of 1935 young C. F. S. Sharpe undertook an excellent adventure. Having acquired a car, he drove 16 000 miles through 28 American states and 3 Canadian provinces. His quest was unusual: He was out to demonstrate that mass movement of debris over the Earth's surface is an important geological process. After he returned he wrote up his observations in a book (Sharpe, 1938) that forms the basis of our modern understanding of gravity-driven mass motions for the evolution of the Earth's landscape. The current era of space exploration has greatly broadened the reach of the processes he described: Mass motion is important on bodies ranging from tiny asteroids and comets only a few kilometers in diameter up to the largest moons and planets. Its action has been observed on every solid body in the Solar System.

8.1 Soil creep

The mantle of loose debris on slopes everywhere in the Solar System is slowly creeping downhill. This insidious motion is usually too slow to appreciate on human timescales, but the evidence is there for anyone who will look. On Earth, gravestones and old fence posts gradually tilt downhill. Linear trails of fragments lead downslope from distinctive rock outcrops and steep stream banks are gradually overridden by thick sheets of soil that often include intact mats of vegetation and trees. Although slow, the sheer ubiquity of soil creep makes it an important, if not dominant, player in the overall denudation of sloping hillsides. Accurate measurement of creep is difficult, but surface velocities on Earth range from millimeters to centimeters per year, declining gradually to zero, through soil depths ranging

319

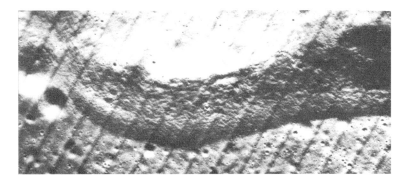

Figure 8.1 Lunar Orbiter image of a steep lunar hillslope illustrating the irregular elephant hide topography with a basal berm that is presumably composed of material that crept downslope and accumulated at the slope base. Note the small number of craters on the slope compared to the adjacent mare surface. Portion of the Flamsteed Ring near 2° 50' S, 42° 40' W. Framelet strip width 220 m. LO III frame H199.

from tens of centimeters to many meters. Although creep is most frequently noted in damp, clay-rich soils on Earth, slope streaks and filled craters on Deimos, the 6 × 8 km smaller moon of Mars, indicate active creep within its regolith, despite its tiny gravity field (2.5 × 10^{-4} g). Slow downslope movement of the regolith occurs everywhere on steep lunar slopes, where it is marked by characteristic "elephant hide" slope textures, the absence of small impact craters, and berms of accumulated debris at the base of steep hillsides (Figure 8.1).

8.1.1 Mechanism of soil creep

The mechanism of soil creep is complex: Many different processes lead to the same overall motion. Sharpe listed seven possible causes in his book and modern research has added several more (Carson and Kirkby, 1972). Many, such as wedging by plant roots and disturbance by burrowing animals (published descriptions range from earthworms to wombats), are distinctly Earth-related. Others, however, are of more general application, such as heating and cooling, wetting and drying of expansive clays, freezing and thawing. Still others, such as meteorite impact, are distinctly extraterrestrial. What they all have in common is mechanical disturbance of the near-surface zone. Repeated disturbances that might only lead to small periodic motions on a level surface are converted by gravity to slow motion in one overall direction – downhill.

One of the earliest theories of downslope creep was put forward by Charles Davison, Esq. (1888). He noted that, in 1855, Canon Moseley had shown that pieces of lead lying loose on the wood roof of Bristol Cathedral moved downslope 18 inches in two years, which Moseley attributed to the alternate expansions and contractions of the lead through diurnal temperature cycles. Davison went further and performed experiments on a pair of bricks and blocks of sandstone tilted at an angle of about 20°, the upper one of which he observed creeping slowly downhill over a period of several months.

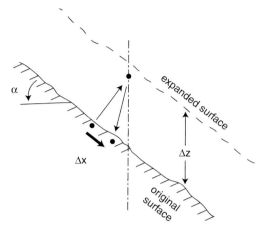

Figure 8.2 The mechanism of creep on a sloping surface. The ground surface first expands perpendicular to the slope and then later contracts along a more vertical direction. The result of many cycles of such expansion and contraction is a net downslope displacement.

The basic mechanism is illustrated in Figure 8.2. Any process that causes the upper soil layer to expand, whether by heating, wetting, or freezing, pushes the land surface up slightly in the direction of least resistance – perpendicular to the sloping surface. Later, when contraction returns the layer to its original thickness, gravity tends to make the soil settle more vertically, which results in a small increment of downslope movement. Early theories of soil creep supposed that the contraction phase was exactly vertical, predicting a downslope movement $\Delta x = \Delta z\ sin\ \alpha$ for each cycle, where α is the angle of the slope from horizontal and Δz is the vertical expansion of the ground surface. These theories predicted creep rates that are much too large, so modern theories of this type assume that the settling phase is not exactly vertical. Mathematically, this introduces a recovery factor r into the equation, $\Delta x = (1-r)\ \Delta z\ sin\ \alpha$, where if $r = 1$ the slope settles back into its original position and if $r = 0$ the slope settles vertically. Naturally, r is an empirical factor that must be determined from observations.

This theory predicts that the creep rate is proportional to the sine of the slope angle, which is borne out by much observational data, and to the number of expansion/contraction cycles, which also seems to be roughly correct. When slopes approach the angle of repose the sine dependence breaks down, and the creep rate becomes a non-linear function of the slope (Roering *et al.*, 1999).

Expansion and contraction is not the only process that causes downslope creep. As early as 1909 G. K. Gilbert noted that the impact of raindrops on steep divides must round off the slopes and give hilltops a convex upward profile (Gilbert, 1909). The extraterrestrial analog of this process is meteorite impact (Figure 8.3). Although each impact initially ejects material symmetrically around the crater, gravity imparts a small downslope increment to the ejecta in flight that results in a net downhill transport. Perhaps more important is the associated effect of seismic shaking induced by each impact. Crater excavation is

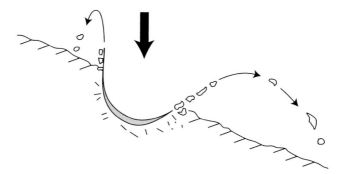

Figure 8.3 Rain splash or impact on a sloping surface ejects more material downslope than upslope, on average. The net result of many such impacts is a net downslope motion of material over a depth approximating the depth of the crater.

accomplished by a shock wave that severely shakes the immediately surrounding terrain. This shaking may induce local movements that cause a net downslope movement of loose surface material (Richardson *et al.*, 2005). On small bodies, such as asteroids, this process is especially effective because the seismic energy is unable to spread out over a large volume. It remains trapped in the asteroid, reverberating until internal friction finally degrades it into heat. This process explains the apparent absence of small impact craters on small asteroids such as 433 Eros (Richardson *et al.*, 2004) or Mars' moons Phobos and Deimos. On steep lunar slopes, which are already vulnerable to downslope movement, seismic shaking by impacts may similarly account for the low density of small craters.

The Moon's surface would also seem to be liable to creep by thermal expansion and contraction. After all, this was the first creep process analyzed by Davison in 1888 and the temperature excursions on the Moon's surface are far larger than those on Earth. Surprisingly, however, this expectation seems to be false. Experiments on strongly heating and cooling basaltic powder at Caltech's Jet Propulsion Lab failed to show any detectable creep (J. Conel, 1988, personal communication). Perhaps the cohesion of particles in fine powders like the lunar regolith prevents this process from being effective on the Moon.

A more exotic mass transport process is electrostatic dust levitation (Lee, 1996). This process is only effective on airless bodies exposed to both direct sunlight and the solar wind. Dust levitation on the Moon was observed by a variety of lunar landers as a "horizon glow" shortly after sunset and was the subject of the LEAM experiment deployed during the Apollo 17 mission. It consists of a light-scattering haze of ~10 μm particles hovering at altitudes up to a few tens of kilometers. This particle cloud is attributed to electrostatic lofting: The Moon's surface acquires a positive charge due to ejection of photoelectrons by solar UV radiation. Dust particles on the surface acquire similar charges and are repelled from the surface, rising until the solar-wind plasma screens the electric field at a height comparable to the plasma Debye length. Although the total mass involved in the dust cloud is very small (fluxes are estimated at about 3×10^{-11} kg/m^2-s), it is possible that, over geologic time, accumulations of dust deposited from this cloud might be significant

Figure 8.4 Terracettes are likely the result of soil creep rather than trampling by cattle. These slopes in Iceland are corrugated into small terracettes ranging from about 30 cm to almost 1 m in height and several meters wide. Photo by Ellen Germann-Melosh.

on the surface. On asteroids, smaller levitated particles may be swept away by the solar wind and lost, while larger (ca. 1–100 μm) particles may settle back onto the surface. Smooth, level "ponds" on the surface of asteroid 433 Eros are attributed to deposition by levitated dust.

8.1.2 Landforms of creeping terrain

Hillslopes dominated by creep are often distinctive. In profile they are typically convex upward with smooth, rounded crests. The slopes themselves often display downhill undulations of slope, forming alternate steep "risers" and less steep "treads" that tend to run parallel to contour lines (Figure 8.4). These features are known as "terracettes" or sometimes "slope garlands" on unconsolidated material such as cinder cones. Streaks of distinctive rocks run straight downhill on creeping slopes, and small landslides may scar the slopes. Where other erosional processes do not remove material at the base of the slope, as on the Moon (Figure 8.1), berms of slope-transported material accumulate.

Terracettes are one of the minor unsolved mysteries of geomorphology. Often attributed to trampling by cattle or sheep, they are sometimes observed in circumstances that make the cattle explanation very unlikely. C. F. S. Sharpe himself was rather skeptical of cattle trampling, although other authors have found substantial support for it: There is no doubt that when cattle or sheep are present they use terracettes and follow trails over them, but it is less clear that animals create them. The observation of terracette-like, elephant-hide features on lunar slopes (Figure 8.1) would seem to resolve the controversy in favor of inanimate causes. Surprisingly little study has been devoted to surface textures on lunar slopes, however, and there are currently no models that describe how this pattern forms.

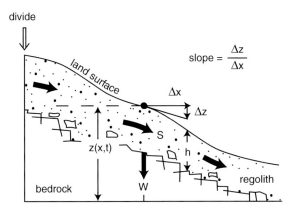

Figure 8.5 The fragmental layer of soil (or regolith) covering the land surface creeps slowly downhill. Mass balance at any location on the hillslope must account for the influx of material from farther uphill, the efflux of material downslope, and the conversion of bedrock into soil. Other mass balance factors, such as mineral dissolution by groundwater, may also have to be taken into account to produce a quantitative model of soil creep and erosion.

Hopefully, with the advent of high-resolution images of the lunar surface from the Lunar Reconnaissance Orbiter mission, this situation will change.

The proportionality between soil creep rate and the sine of the slope angle suggests an appealing analogy between landform degradation by soil creep and thermal diffusion. Noting that the flux of regolith material is proportional to the gradient of the hillslope, W. E. H. Culling (1960, 1963) proposed that soil creep obeys a two-dimensional analog of Fourier's heat equation. Extensions of this work by Carson and Kirkby (1972) and many modern authors have incorporated the diffusion model into comprehensive models of landform evolution (Pelletier, 2008). The basic idea (Figure 8.5) is to combine mass balance with the rate of material transport. In any small area on the slope, regolith is created by weathering, transported into the area from upslope, or exported by flow downslope. Interpolating this balance into an infinitesimal area, the mass balance equation becomes:

$$\frac{\partial S}{\partial x} - (\mu - 1)W = -\frac{\partial z}{\partial t} \tag{8.1}$$

where S is the volume flux of regolith moving along the slope, W is the rate of conversion of bedrock into regolith (weathering rate), μ is the ratio of the volume of regolith to the volume of parent rock, z is the elevation of the ground surface and x is the distance downhill from the uphill divide.

A useful deduction from this equation is the thickness of the soil mantling the surface, $h(x, t)$. Its time rate of change is the difference between the rate of surface elevation change and the weathering rate of the bedrock:

$$\frac{\partial h}{\partial t} = W + \frac{\partial z}{\partial t} = \mu W - \frac{\partial S}{\partial x}. \tag{8.2}$$

If the volume flux S is proportional to the sine of the slope angle, which for small angles is approximately equal to the derivative of z with respect to x, then:

$$S = -K \frac{\partial z}{\partial x}.$$ (8.3)

Using this definition, Equation (8.1) becomes a diffusion equation:

$$K \frac{\partial^2 z}{\partial x^2} + (\mu - 1)W = \frac{\partial z}{\partial t}$$ (8.4)

in which K is a generalized "topographic" diffusion coefficient with the usual units of m^2/s. The sign has changed because a negative slope implies a positive mass transport in the coordinate system defined in Figure 8.5.

The advantage of a diffusion equation is that many solutions to equations of this type already exist. Not only have entire books been written collecting such solutions (Carslaw and Jaeger, 1959), but fast numerical algorithms have been devised to solve this equation under any conceivable set of boundary conditions. This probably explains why the diffusion model is so popular among geomorphologists. Nevertheless, reality intrudes when slope angles approach the angle of repose and the equation acquires new, slope-dependent terms that make it non-linear (Roering *et al.*, 1999). One must also remember that many creep processes depend on the number of freeze/thaw cycles or nearby impact events and so are not direct functions of time, making (8.4) a useful first approximation that must be applied only with due caution.

An interesting consequence of Equation (8.4) is that a hillslope that degrades without change of form (that is, $\partial z / \partial t$ is negative and independent of x and for which the weathering rate W does not depend strongly on distance from the divide) must be convex upward (that is, the second derivative in (8.4) is negative). Hills in a landscape that is being eroded by creep, thus, evolve to be convex upward. Another way of seeing this is to realize that the farther one goes from a divide, the steeper the hill must become to transport the volume of material that has been eroded from all the bedrock between one's position and the divide. Creep-dominated landscapes look "melted," as if a plate of ice cream balls had been left out in the Sun for some time. The diffusive character of Equation (8.4) means that sharp contours are rounded; steep scarps degrade to gentle slopes and deep incisions are filled.

The landscape-softening character of diffusion is in sharp contrast to the effects of fluvial erosion, as will be discussed in more detail in Chapter 10. Fluvial erosion has the opposite effect: It creates incisions and sharpens soft contours. Because the erosive agent in fluvial processes, runoff, increases in intensity as one moves away from a divide, slopes dominated by rainwash are concave upward: They begin steeply near the divide and become more shallow as one proceeds downslope. Landslide-dominated slopes are intermediate in form: Because slope collapse depends on a fixed angle of repose with respect to the horizontal, such slopes are straight. Figure 8.6 illustrates these relationships.

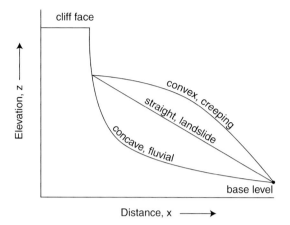

Figure 8.6 Slope profiles can be convex, straight, or concave-upward. Different processes lead to different profiles: Creep usually produces a convex slope profile, landslides produce straight profiles, and fluvial erosion produces a concave profile.

The concatenation of the opposite effects of creep and fluvial erosion leads to much of the topographic variety we enjoy on the Earth's surface. In contrast, the creep-dominated lunar landscape is monotonous on a broad scale.

8.2 Landslides

Slow creep processes are not the only type of mass movement. Much more rapid movements take place when the ability of a slope to resist the force of gravity is exceeded and rock material accelerates downhill until it achieves a new balance with the forces tending to level the landscape.

Some of the most important contributions to understanding landslides were made by Charles-Augustin de Coulomb (1736–1806), who is otherwise famous for his contributions to the electrostatic force law. Coulomb started his career as a military engineer and in 1773 published a memoir treating the conditions of stability of earth retaining walls. He summarized his experimental and theoretical understanding of the resistance of earth materials to collapse in 1776 with the equation (in modernized notation):

$$\sigma_s = c + \sigma_n \tan\phi \tag{8.5}$$

where σ_s is the maximum sustainable shear stress, c an empirical constant called cohesion, σ_n is the stress normal to the plane of shearing, and ϕ is the angle of internal friction. Planetary materials can be classified according to the values of c and ϕ that describe them, and any discussion of landslides is most logically organized around the various terms of this equation.

In the years since Coulomb's revolutionary formula, only one major addition has been necessary: Austrian engineer Karl von Terzaghi (1883–1963) showed that when pore fluids,

Table 8.1 *Angles of internal friction*

Material	Angle of internal friction
Basalt talus	45°
Granitic gneiss talus	31–36°
Alluvium	41–44°
Glacial till	37°
Shale grit	43°
Sand	33–43°
Silt	32–36°
Cold water ice (77–115 K)[a]	29°

Data from Carson and Kirkby (1972) unless otherwise noted.
[a] Beeman *et al.* (1988)

such as water, gas, or oil, are present the pore pressure p must be subtracted from the normal stress (Terzaghi, 1943):

$$\sigma_s = c + (\sigma_n - p)\tan \phi \qquad (8.6)$$

The following sections analyze the effect of each of the terms in this equation.

8.2.1 Loose debris: cohesion c = 0

Loose rock debris has no intrinsic strength: Its cohesion c in Equation (8.6) is zero. Nevertheless, such material is not "strengthless," because of friction. So long as normal stresses, confining pressure, or the weight of overburden act across potential slip planes, the material resists deformation and a finite shear stress must be applied to make it slip. Only when the internal friction ϕ is zero is the material truly without resistance to deforming forces.

Measured values of internal friction are surprisingly uniform (Table 8.1): Even ice near its melting point has an internal friction angle close to that of rock and this persists down to very low temperatures. The internal friction of loose materials depends not only on the mineralogical composition of the material, but also upon such factors as grain size, grain shape, sorting, and degree of interlocking of the grains. Nevertheless, it is a good guess that any geological material has an angle of internal friction between about 30° and 45°.

Angle of repose. The angle of internal friction is an intrinsic property of the given material. For soil materials it is measured in a shear box in the laboratory. The angle of repose of a rock slope is the result of the balance between the intrinsic forces of resistance and the driving force of gravity. Imagine that we have arranged a rock slope so that its tilt can be adjusted. Starting out with a mass of rock debris on a level surface, slowly increase its tilt until rocks on the slope begin to slide. The critical angle at which instability sets in is called the angle of repose, ϕ_r.

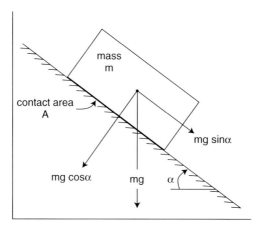

Figure 8.7 A balance between driving forces and resisting forces determines the stability of a block resting on a sloping surface. When the resistance is Coulomb friction, the weight of the block can be decomposed into the force pushing it down the slope and the resisting force, which is proportional to the force pushing it into the slope. The ratio of these two forces when sliding just begins is equal to the coefficient of friction. Because this is a ratio, the acceleration of gravity (and mass of the block) cancel out: The angle of repose is independent of the gravitational acceleration.

The angle of repose is closely related to the angle of internal friction. Referring to Figure 8.7, suppose a block of rock is resting on a surface sloping at angle α from the horizontal. The relevant portion of Coulomb's strength Equation (8.5) is:

$$\sigma_s = \sigma_n \tan\phi. \tag{8.7}$$

The shear stress acting on the base of the rock is equal to its weight mg, where m is its mass, divided by its basal area A and multiplied by the sine of the slope angle: $\sigma_s = (mg/A) \sin \alpha$. The sine of the angle comes from resolving the weight of the rock into its downslope component. The normal stress pressing the rock into the slope is the same expression, but now multiplied by the cosine of the angle: $\sigma_n = (mg/A) \cos \alpha$. Inserting these expressions into (8.7), the common factor mg/A cancels out and we are left with the statement that when instability occurs, $\alpha = \phi_r$, where:

$$\phi_r = \phi, \tag{8.8}$$

that is, the angle of repose is equal to the angle of internal friction!

The fact that the acceleration of gravity cancels out of Equation (8.8) comes as a big surprise to many people. This means that barely stable slopes on a low-gravity body such as the Moon (or an even lower-gravity asteroid) stand at exactly the same angle as they do on Earth. Space artist Chesley Bonestell painted widely publicized landscapes of the Moon for many years, which portrayed lunar slopes much steeper than any slopes possible on Earth. A stickler for accuracy in his paintings, he was reportedly horrified when, in his later years, he was acquainted with the fact that the angle of repose is independent of gravitational acceleration!

The angle of repose need not be exactly equal to the angle of internal friction: Factors such as interlocking of angular clasts and the difference between static friction and sliding friction lead to differences of a few degrees. Nevertheless, the two are very close to one another and laboratory measurements of friction are a good guide to the maximum slopes observed in a planetary landscape. Now that we have good images and shape models of irregular asteroids and comets it is possible to compute the local direction of gravitational acceleration on their surfaces and compare this to the surface slope. In nearly all cases it is found that the slopes stand at less than the angle of repose, indicating that asteroid regoliths can be approximated as loose, cohesionless layers of rock debris.

Pore pressure. Pore-filling fluids in a loose mass of rock debris do not change the coefficient of internal friction, but they can change the angle of repose by supporting some of the weight normal to potential sliding planes. Because fluids partially relieve the normal stress but do not contribute to resisting shear stress, slopes that include fluids fail at shallower angles than dry slopes. The relevant equation is:

$$\sigma_s = (\sigma_n - p) \tan \phi. \tag{8.9}$$

The analysis of this equation for a fluid-saturated slope proceeds in the same manner as for a dry slope, although more care is required in balancing all the forces. If the total density of the saturated granular material is ρ_t and the density of the fluid is ρ_f, it can be shown (e.g. Lambe and Whitman, 1979) that the angle of repose of the slope is:

$$\tan \phi_r = \left(\frac{\rho_t - \rho_f}{\rho_t} \right) \tan \phi. \tag{8.10}$$

The term in brackets is always less than one, so that angle of repose of a fluid-saturated slope is always less than that of a dry slope. For a water-saturated slope in typical terrestrial soil, the angle of repose may be reduced by about a factor of 2: $\phi_r \approx \phi/2$.

The decrease of the angle of repose as a slope becomes saturated with water describes the common observation that small landslides are common after heavy rainfalls. However, the reason that slopes fail is not, as commonly stated, that the water "lubricates" the slope. The coefficient of internal friction of wet rock debris is indistinguishable from that of dry debris. The weight of the water also does not promote failure: it increases the shear resistance σ_n by the same factor that it increases the shear stress σ_s. The entire effect of the water is to increase the pore pressure, effectively "floating" the rock debris off its underlying support and decreasing the shear resistance by decreasing the normal stress.

The importance of pore fluids is not confined to slopes on the Earth. Small dust avalanches on Mars appear to have been triggered by airblasts caused by recent small impacts on its surface (Burleigh *et al.*, 2009). The probable cause of these avalanches is the rapid excursions in atmospheric pressure over the sloping surface. When the external air pressure drops, the Martian atmosphere trapped in the pore spaces of the dust exerts an uncompensated pore pressure and partially lifts the weight of the dust, triggering slope failure. One such event triggered nearly 100 000 small avalanches within a few kilometers of an impact that produced a cluster of 20 m diameter craters.

Slope profiles. As shown in Figure 8.6, slope instability leads to straight slope profiles that stand at the angle of repose. Straight slopes are common on many parts of the Earth.

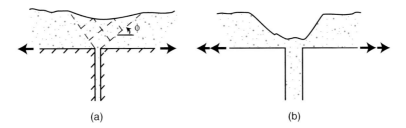

Figure 8.8 Loose surface material drains into gaping fissures along slip surfaces whose slope is determined by the coefficient of friction in the loose material. The result is subsidence pits whose width and spacing are approximately equal to the thickness of the loose layer.

Screes and talus cones at the base of steep cliffs are typically straight, created as blocks from the cliff face fall off and roll some distance downslope. Mountain slopes in seismic areas of southern California, Taiwan, and Tibet, among others, exhibit nearly linear slopes from ridge crest to valley bottom. Measurements after major earthquakes have shown that earthquake-triggered landslides dominate slope degradation in these locations. The slopes of constructional landforms, such as cinder cones and volcanoes, are often limited by slope instability and display correspondingly linear slope profiles.

Subsidence pits and grooves on asteroids. Subsidence pits are observed on Earth when the substrate underlying loose granular material collapses. This occurs naturally over collapsed underground caves in karst terrain and has been observed in Iceland when fissures open up underneath loose volcanic tephra. It has been noted more commonly in association with man-made accidents such as the collapse of old mine excavations, water-main bursts, and the collapse of cavities created by underground nuclear tests.

When loose material drains into a cavity below, it first forms a shallow pit, whose breadth at the surface is comparable to the thickness of the mantle of loose material (Figure 8.8). This is because the shear planes in a granular material form at an angle to the direction of maximum compression that is equal to the angle of internal friction. In the case of draining material the maximum compression axis is vertical, so the shear planes dip steeply, typically at angles near 60° from the horizontal. As drainage continues, the pit deepens while widening slightly until the walls of the pit slope inward at the angle of repose and further widening takes place as material slides down the steep walls of the pit.

Although the formation of drainage pits may seem rather esoteric, it is the favored explanation of a very striking feature that appears to be a common characteristic of asteroid surfaces. Linear arrays of pits and troughs were first observed on Phobos, the larger and inner moon of Mars. With widths of hundreds of meters and lengths stretching from one extremity of Phobos to the other, these pitted grooves are one of Phobos' major topographic features (Figure 8.9). Many theories were concocted to explain them: rolling or bouncing boulders ejected by craters on Phobos (except that they are not all radial to craters and there are no boulders at the ends of the grooves), secondary craters from impacts on the surface of Mars (but why are there no similar chains on the surface of Mars itself?),

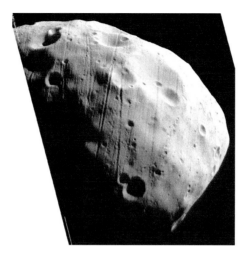

Figure 8.9 Grooves on Phobos are likely to be the result of thick (100 to 200 m) regolith draining into gaping fissures in the body of the moon. Grooves are up to 200 m wide. Mars Express image 20080723, ESA/DLR.

and a number of others. The most plausible and widely accepted interpretation is that they are lines of drainage pits in regolith undermined by gaping fractures in the body of Phobos (Thomas *et al.*, 1978). Large parts of Phobos are currently in an extensional state of stress, caused by tidal forces because it is inside Mars' Roche limit. The width and spacing of the grooves then indicates the depth of the regolith (Horstman and Melosh, 1989).

Although Phobos is in a rather special stress state, grooves have now been reported on nearly every asteroid that has been imaged at high resolution, including 951 Gaspra, 243 Ida, 433 Eros, 2867 Šteins, and the 130 km diameter asteroid 21 Lutetia. None of these is subject to tidal extension, but perhaps large impacts temporarily opened large fissures into which regolith could drain. Once some regolith had fallen into the fissure, it would have jammed the fissure open and more regolith could follow, creating a groove. It may be significant that some of the largest pit chains are observed on Šteins, which was nearly destroyed by a very large impact.

8.2.2 Cohesive materials c > 0

Without cohesion, vertical scarps cannot exist. Even the centimeter-high scarps at the sides of the Apollo astronauts' bootprints imply that the lunar regolith has some degree of cohesive strength. The vertical and even overhanging walls of sand castles at the beach imply a small degree of cohesive strength: Dry sand alone can, at best, stand at the angle of repose. The secret of successful sand castles (known to every sand-castle builder) is that the sand must be damp. The surface tension of the water films between sand grains bonds the grains together and imparts a small cohesive strength to the mass that permits vertical walls. If

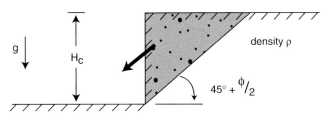

Figure 8.10 Collapse of a vertical cliff in a homogeneous mass of material possessing both internal friction (parameterized by the angle of internal friction ϕ) and cohesion. Failure of such a cliff takes place preferentially along a plane dipping at an angle $45° + \phi/2$ with respect to the horizontal.

the sand becomes completely saturated with water the films disappear and the cohesive strength (and sand-castle ability) disappears.

The most general kind of material strength includes both cohesion and internal friction. There are materials that have cohesion but lack significant internal friction, but they are relatively rare. The internal friction of clays can be neglected for rapid flows (soil engineers call this the "undrained" condition) because clays are so impermeable that water cannot be expelled from between the grains and the pore pressure then equals the overburden pressure, so that the term $(\sigma_n - p)$ in Equation (8.6) vanishes, canceling the dependence on internal friction. The same clays, if deformed slowly (under "drained" conditions) show a strong dependence on internal friction (Lambe and Whitman, 1979).

The failure of many metals is also nearly independent of pressure, hence, these also have negligible internal friction. Although the naked cores of planetesimals could potentially form iron–nickel metal planets, we have no examples yet of landscapes on such bodies. Presumably they would resemble terrain on clay-rich soils, although on a much larger size scale due to iron's much larger strength. Astrophysicists might consider such landscapes on the surfaces of white dwarfs or neutron stars.

Stability of vertical cliffs. Steep or vertical cliffs are the characteristic features of cohesive materials. The most important concept is the relation between cliff height and cohesion. Coulomb, in his 1773 memoir, was the first to analyze this problem in its full generality, including internal friction (Gillmor, 1971). Coulomb, working as a military engineer at that time, had been assigned to construct Fort Bourbon on Martinique and this problem had immediate practical applications. The useful formula he derived is an example of science taking advantage of military technology.

Coulomb's analysis of the stability of a vertical cliff was a classic application of calculus to a practical problem. His method is shown in Figure 8.10. Coulomb supposed that the failure surface was most likely to be a plane extending from the foot of the cliff upward to the surface at the top of the cliff (this supposition was motivated by his observation of actual slope failures). He resolved the weight of the triangular prism above into shear and normal components to the surface and compared these driving forces to the resistance exerted by cohesion and friction on the plane, given by Equation (8.5). The ratio between resistance and driving force is a function of the slope of the failure plane. Coulomb then applied calculus to locate the minimum (that is, the plane on which failure is most likely), which occurs at an angle of $45° + \phi/2$ from the horizontal. He used this angle to solve for the height of a cliff at the limit of stability, H_c:

$$H_c = \frac{2c}{\rho g} \tan\left(45° + \phi/2\right). \tag{8.11}$$

As one might expect, increasing cohesion increases cliff height, as does increasing the coefficient of friction. Higher density ρ and, especially, acceleration of gravity g decreases the possible height of a cliff.

When formula (8.11) is compared to the heights of natural vertical cliff faces it is found that it generally predicts cliffs that are too high. This is mainly because the cohesion of rock in the laboratory is usually measured on unweathered, intact rock samples in an experiment that lasts only seconds. In nature, rocks are filled with joints and other fractures, so their bulk strength is less than that of a small laboratory specimen. Furthermore, the long-term strength of rock is generally less than that measured on a short timescale because of slow processes like stress corrosion cracking. A common procedure, then, is to turn Equation (8.11) around and use the maximum heights of observed cliff faces to deduce the long-term strength of jointed rock. Panama Canal engineer David Gaillard used this scheme in 1907 to deduce the safe angle for hillside cutbacks, a decision that has stood the test of time.

Equation (8.11) has already performed good service in the cause of planetary science. Shortly after Voyager 1 discovered sulfur on Io, Clow and Carr (1980) used a generalized version of this equation to show that the 2 km high scarps of an Ionian caldera were too tall to be composed of pure sulfur and deduced that Io's crust must be made of much stronger material. On the other hand, many journalists were astonished in 2004 when the Stardust spacecraft sent back images of ~100 m high vertical cliffs on comet Wild 2. After all, aren't comets supposed to be very weak? However, application of Equation (8.11) to Wild 2's gravity field indicates that its material need be no stronger than the weakest soufflé: A cliff of the same material under Earth's gravity would collapse if its height exceeded 3 mm!

Figure 8.11 illustrates the varieties of cliff collapse that have been noted on Earth, many of which have now been observed at the base of steep Martian cliffs as well. Cliff collapse varies from spalling of rock slabs that extend from toe-to-crest, to grain-by-grain disintegration. The style of degradation of any individual cliff depends upon its composition, weathering characteristics, and the mechanical condition of the rock (or ice) composing it.

Rotational slumps. How is the stability of a scarp that is not vertical determined? Figure 8.12 illustrates the general case of a scarp of height H_c but with a face sloping at angle α. In this case, observation first indicated that the failure surface is approximately a section of a cylinder, not a plane. The analysis of the stability of such a slope is much more complex than for a vertical scarp and is accomplished by the "method of slices" (Lambe and Whitman, 1979) or the still more sophisticated "slip line analysis" (Scott, 1963). The simplified results of Scott's widely used analysis are shown in Figure 8.13. The right-hand side of this plot, for slope angle 90° (a vertical cliff), agrees with Equation (8.11), but for other slope angles there is no simple analytic formula. This plot makes it

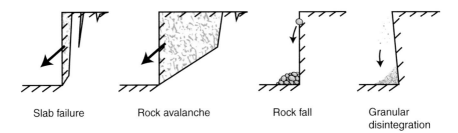

| Slab failure | Rock avalanche | Rock fall | Granular disintegration |

Figure 8.11 Collapse of actual cliffs varies from the ideal because of structures in the rock mass. Slab failure occurs when vertical joints control the strength of the rock face. Rock avalanches are influenced by shallow vertical joints. Rock falls occur where the rock face disintegrates into large blocks and granular disintegration of rock faces occurs where weathering easily releases small rock fragments or grains of sedimentary rocks.

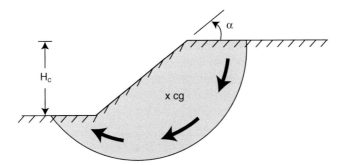

Figure 8.12 Rotational slumps occur in material with cohesion but little internal friction, so that sliding on the deep-seated failure surface is not inhibited by friction. The center of gravity of the mass, the point marked cg, moves downward as the slide progresses.

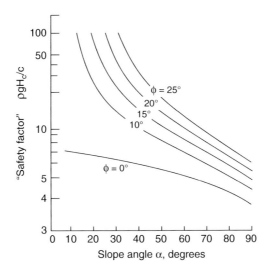

Figure 8.13 Slope stability as a function of slope angle for straight escarpments of the type shown in Figure 8.12 in a material with both internal friction and cohesion. The curves are labeled by the internal friction angle of the material. Greatly simplified after Figure 9–20 of Scott (1963).

clear that rotational slumps are more likely in material with very low angles of internal friction: The stability of a scarp increases rapidly as internal friction increases and slope angle decreases. On Earth, this means that rotational slumps are mostly confined to soils that contain large amounts of wet clay, in which the pore pressure cancels the contribution of internal friction. One would, thus, not predict that rotational slumps would be common on airless bodies such as the Moon, because no pore fluids are available. It therefore comes as a surprise to find that rotational slumps are one of the most common landforms on the Moon – but they are confined to the rims of complex impact craters. Box 8.1 explores this anomaly in more detail.

The most characteristic feature of a rotational slump is its head scarp. As illustrated in Figure 8.14, the steeply dipping failure surface is exposed at the uphill crest of the landslide. One also typically sees several small terraces downhill of the main detachment. The original ground surface rotates backward on the tops of these terraces; it dips toward the detachment and often creates small, closed depressions that can trap water or other liquid, forming ponds. The small ponds of impact melt on top of slump terraces in Copernicus crater on the Moon are good examples of this process.

As the slump rotates along the failure surface, its overall center of mass drops downward, but at its toe the ground is usually upheaved. The material at the foot is pushed upward into an irregular bulbous, hummocky mass that overrides the original surface.

Toreva block landslide. A distinctive variant of landslides in cohesive material is called the Toreva block landslide, after a small town on the Hopi reservation in Arizona where many such landslides occur. First described by Reiche (1937), this type of landslide develops where a strong layer overlies a weak one (Figure 8.15). As the weak layer is eroded it undermines the strong unit above, which eventually collapses as a nearly intact block, often rotating backward like the head of a rotational slump, and plowing up the weak material below it. This kind of detachment may occur several times in succession, leading to stair-step topography at the edges of mesas capped with resistant rock. The continued evolution of escarpments depends crucially upon the activity of other processes to remove the debris that has already collapsed off the cliffs: Otherwise, the mechanical support of the blocks that have already slid off the face will continue to support the scarp.

Although Toreva block landslides are characteristic landforms of the Colorado Plateau in the southwestern Unites States, they may occur wherever strong rock units overlie weaker ones. In volcanic terrains it is not unusual to find strong lava flows interbedded with weak tephra: Large blocks lying below the high scarp at the base of Olympus Mons on Mars may have originated in this manner.

Avalanche chutes. Although not strictly landslides themselves, avalanche chutes and slope flutes are characteristic landforms that develop on steep slopes dominated by rockfalls. Repeated rockfalls eventually erode U-shaped troughs into the slope. Probably starting as small indentations in the slope, these chutes channel further rockslides that then deepen and lengthen the chutes by abrasion and plucking of blocks along their beds. As avalanche chutes grow and encroach on one another, the head of a steep cliff becomes fluted with near-vertical troughs whose spacing is often surprisingly regular. Such small-scale flutings,

Box 8.1 **Crater terraces as slump blocks**

One of the most distinctive features of large, complex impact craters is the wreath of terraces
that step down from the rim into the interior of the crater. The terrace just below the crest of
the rim is usually the widest, and the scarp above it is the highest, of the series. This is the
only terrace visible in many craters because crater floor deposits bury structures closer to the
center. When they can be seen, the terraces become progressively narrower and the scarp offset
smaller as one proceeds from the rim toward the center of the crater.

Crater terraces show the backward-rotated block morphology typical of slump terraces in
landslides. Early Lunar Orbiter images of the Copernicus and Tycho craters on the moon reveal
level ponds of presumed impact melts trapped among its terraces, an observation that now
appears to be common in both lunar and Martian craters (Figure B8.1.1), similar to the water-
filled ponds that form at the head of large terrestrial rotational slump landslides.

Terraces in complex craters, thus, closely resemble the terraces that form at the head of
landslides in wet clay: That is, in materials which possess cohesion but negligible internal
friction. The story of how lunar rock, in the absence of either air or water, can behave like
water-saturated clay is related to the mobility of long-runout landslides and is described in
Sections 6.3.3 and 8.2.4. For the present, this box explores the implications of *assuming* that
the mechanical strength of a planetary surface shortly after crater excavation can be treated
as if the material possesses cohesion but no internal friction. This type of reasoning is called
a "phenomenological model": We do not know *why* the strength behaves this way, but we can
show that, if we assume that it does have this behavior, the model's predictions agree with the
observations.

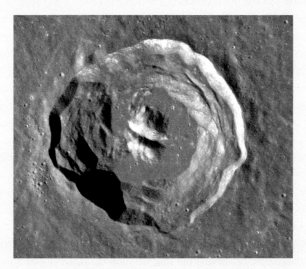

Figure B8.1.1 Terraces beneath the rim of the Bürg crater on the Moon. Bürg is a complex crater
40 km in diameter with clearly defined wall terraces, a smooth floor, and central peak. The lack
of perfect symmetry probably reflects the pre-impact structure in the target. Portion of LROC
WAC monochrome context image M119666881ME, NASA/GSFC/Arizona State University.
Courtesy Mark Robinson.

Box 8.1 **(cont.)**

The events that immediately follow the impact of a meteoroid on a planetary surface are described in Section 6.3. The result of crater excavation is a transient crater, a bowl-shaped cavity with a depth/diameter ratio of about 1/2.7. This ratio implies an average internal slope of about 20°. This is steep, but not steeper than the angle of repose of most rock materials. In other words, such a crater would not normally be expected to undergo much collapse. For lunar craters smaller than about 15 km in diameter, this expectation is borne out: The rim crumbles and slides into the crater, forming a breccia lens on its floor and slightly enlarging its diameter. However, for larger craters the rim slumps down, forming terraces and widening the crater, while the floor of the crater rises, greatly decreasing its depth.

This collapse process can be analyzed by the method of slip line analysis, mentioned in Section 8.2.2. Application of such an analysis to craters with a parabolic profile (Melosh, 1977) shows that parabolic craters are stable until the dimensionless combination $\rho g H/c$ exceeds about 5. This parameter is the same as a soil engineer's "factor of safety," in which H_i is the depth of the transient crater. Slope failure, in which a segment of the rim slides into the crater, produces a single terrace for values of $\rho g H/c$ between 5 and 10. When $\rho g H/c$ exceeds 15, the floor beneath the center of the crater rises almost vertically upward as the rim slumps downward (Figure B8.1.2). In summary:

$$0 \leq H_f \leq 5c/\rho g \qquad\qquad \text{stable} \quad \text{(B8.1.a)}$$

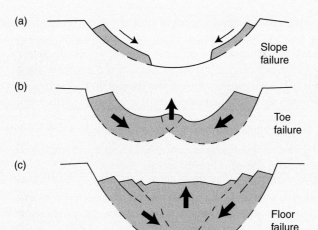

Figure B8.1.2 Three degrees of collapse of crater-like depressions in cohesive material that possesses little or no internal friction. (a) Simple slope or wall collapse in which a layer slides into the crater depressions. This is typical of simple craters, in which the oversteepened rim collapses into the bowl to form a breccia lens. (b) Toe collapse, in which the toes of the sliding masses meet at the base of the crater and uplift a small plug. This is transitional to (c) floor uplift in which the walls slump down and inward, forming terraces, while the floor rises as a semi-rigid plug.

Box 8.1 **(cont.)**

$$5c/\rho g < H_f \leq 15c/\rho g \qquad\qquad \text{slope failure} \quad \text{(B8.1.1b)}$$

$$H_f > 15c/\rho g \qquad\qquad \text{floor failure.} \quad \text{(B8.1.1c)}$$

Because the failure of purely cohesive material is independent of overburden pressure, the collapse threshold is independent of the initial crater's diameter – only its depth is significant. As the crater collapses, material beneath it flows until the stress differences beneath all portions of the crater floor drop below the cohesion, at which point collapse ceases. This model, thus, predicts that the final depth H_f of all complex craters is a constant equal to the depth of a transient crater at the onset of collapse – about 4.5 km on the Moon. The prediction that craters in a cohesive material collapse back to a constant depth also accounts for the development of a broad, flat floor in the interior of large, complex craters. Observationally, the depth of complex craters does depend somewhat upon diameter, but it only varies from about 2.5 to 6 km over a diameter range of 20 to 300 km. This variation could be explained by a very small coefficient of internal friction or by a small dependence of the effective cohesion on crater diameter.

The pure cohesion collapse model also predicts the width of each individual terrace as it forms. A simple analytic formula that gives a good approximation to the numerical slip line results for the terrace width is:

$$w = \frac{c}{\rho g}\left(\frac{1 + 16\lambda^2}{16\lambda^2}\right). \qquad\qquad \text{(B8.1.2)}$$

where λ is the depth/diameter ratio of the crater at the time when the terrace forms. The first terrace to form, when the depth/diameter ratio equals that of the transient crater, is thus the narrowest, whereas the last terrace to form, when the crater's depth/diameter ratio is only slightly larger than its final value, is the largest, in qualitative agreement with observation.

A quantitative test of this equation against the widths of the final terrace in a suite of lunar and Mercurian craters (Leith and McKinnon, 1991; Pearce and Melosh, 1986) indicates an effective cohesive strength of 2–3 MPa for the post-impact strength of both bodies and also agrees with the effective cohesion at the onset of crater collapse.

The phenomenological description of the transient strength of terrestrial planetary surfaces as purely cohesive, with negligible internal friction, thus, agrees well with observations. Application of the same model to the icy satellites indicates that the model also works on these bodies, but with an effective strength about 1/3 that of the rocky planets. This model does not explain the formation of central peaks in complex craters, only the shallow floor. A further mechanical property, viscosity, must be added to the model to explain these features. The rheology needed to more fully model impact-crater collapse is thus more complex – it approximates a Bingham fluid.

spaced tens to hundreds of meters apart, adorn the crests of many of the deep canyons on Mars as well as the Earth (look ahead to Figure 8.17b for good examples).

Granular disintegration produces similar landforms on a much finer scale. The factors that determine the spacing of avalanche chutes are presently unclear: A better understanding

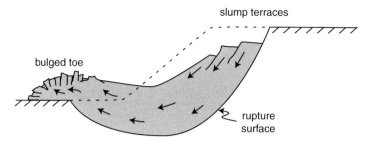

Figure 8.14 The ultimate result of a rotational landslide slump is a terraced head scarp, in which the terraces often rotate the original surface backward into the slope (forming closed depressions), and a bulging toe that pushes the original ground surface upward into an irregular mound.

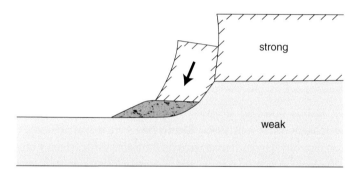

Figure 8.15 Toreva block failure occurs where a mechanically strong layer overlies a weak layer. When failure occurs, the weak layer is compressed and expelled from underneath the strong layer, which subsides as a coherent block. Often multiple failures of this type produce a stair-step structure on the hillside below a mesa capped by the stronger layer.

of this process might give some insight into the properties of the rock that forms the cliff face.

8.2.3 Gravity currents

Once a slope collapses, the mobilized material may move long distances if it becomes fluid. Dry landslide debris usually moves only a short distance before coming to a halt, but when liquids or gases are involved an emulsion of solid and fluid may travel considerable distances at high speed. The driving force behind such flows is gravity, so such flows are known as gravity currents. There are many different types of gravity current, ranging from flows in the atmosphere and the ocean as well as on land (Simpson, 1999). Dry snow avalanches are one type of such flows, as are pyroclastic flows from explosive volcanic eruptions or turbidity currents initiated by slope failures under the ocean. Lava flows can be considered a very slow variety of gravity current. Inertial forces may or may not be important in such flows, depending on their speed. When inertia dominates the flow,

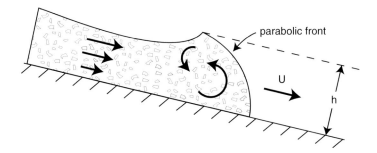

Figure 8.16 Currents of dense fluids flowing down a slope beneath a less dense fluid develop frontal lobes with approximately parabolic profiles, ending in a sharp reversal where the Froude number reaches a critical threshold near $\sqrt{2}$. Fluid behind this front flows in a dense sheet behind the front. Currents of this type are seen both in air (dust storms called haboobs) and under water (turbidity currents).

a characteristic bulge develops at the head of the flow (Figure 8.16). The head, of height h_1, corresponds to a hydraulic jump where the Froude number $U/\sqrt{gh_1}$ is approximately equal to $\sqrt{2}$.

Debris flows. Debris flows are a variety of gravity current that has been analyzed in detail by Arvid Johnson (Johnson, 1970). Debris flows are dense mixtures of mud, rocks, and water that are produced by torrential rains on the Earth. After a heavy rain that saturates and mobilizes the loose debris on a slope, the mixture may roar down a steep canyon with the sound of a speeding freight train. When it emerges from the mouth of a canyon it continues downslope as a dense slug of debris, shedding material from its sides and leaving two parallel ridges behind that are often so regular that they look artificial. It eventually comes to a halt as the slope shallows, leaving a heap of coarse stones armoring its snout, while a rush of muddy water may continue some distance downslope.

A dense mixture of mud, rocks, and water behaves as a Bingham fluid (Section 5.1.3) that flows as a viscous liquid after its yield stress is exceeded. The velocity profiles in debris flows are, thus, very similar to those for lava flows, Figures 5.13 and 5.14, although the Bingham yield stress and viscosity have different values (Rodine and Johnson, 1976).

Debris flows are not entirely confined to the Earth. Many features of the fluidized ejecta blankets around Martian impact craters are suggestive of debris flows and may be due to admixture of subsurface water with the ejected debris (Figure 6.16). The fact that such craters are confined to low latitudes and must be more than a few kilometers in diameter suggest that they have breached a subsurface water table, although this interpretation is not universally accepted (Carr, 2006).

8.2.4 Long-runout landslides or sturzstroms

Mass movements that do not involve a fluid such as water or volcanic gases typically move rather slowly. However, there is a class of landslide that that been observed to move at speeds ranging from 50 to 100 m/s while dry. Although the deposits of such landslides have been known for a long time, their remarkable mobility was not appreciated until a landslide obliterated the Swiss village of Elm in 1881. The disaster was initiated by the villagers' own mining operations, which undermined a steep slope overhanging the town. The occurrence of a landslide was, thus, not a surprise. However, after it began, the dry rock debris traveled several kilometers across the valley floor and even climbed some distance up the opposite slope, overrunning the town as it went. Eyewitnesses measured its velocity to be approximately 45 m/s. The most surprising aspect of this landslide was that the mass of rock debris did not so much slide as flowed "like a torrential flood," as described by geologist A. Heim (1882), who investigated the event shortly after its occurrence. Heim also coined the German name *sturzstrom* (literally, "collapse river") to emphasize the flow of the debris stream. Despite the appearance of flow, geological study indicated that stratigraphic relations between rocks in the debris lobe were preserved throughout emplacement.

Occurrence and morphology. Heim's general observations have been repeated many times since for both historic (e.g. Frank, Alberta 1903) and prehistoric landslides. In the meantime, examples of this sort of landslide have multiplied enormously and the sizes estimated for these floods of rock debris have correspondingly increased (Collins and Melosh, 2003): From the 5 km long prehistoric Blackhawk landslide in Lucerne Valley, CA, to the 50 km Shasta terrain at the foot of Mount Shasta, CA, to gigantic submarine landslides 200 km in length off the coast of Hawaii. As geologists have explored further, the deposits of long-runout landslides seem to be ever more prevalent.

All of these landslides appear to have started as ordinary, although large, rockfalls triggered by a variety of causes: Earthquakes, volcanic eruptions, heavy rainfall, and human activity, including underground nuclear explosions. The 2002 Denali earthquake in Alaska unleashed some of the most recent (Figure 8.17a). All of these landslides are characterized by velocities in the range of 50 to 100 m/s, lack of an obvious fluidizing agent, and preservation of initial stratigraphy (implying laminar flow, in spite of their high velocity). Some are characterized by longitudinal striations, but this is not universal: Transverse ridges cross the surface of the Blackhawk slide.

Long-runout landslide deposits are also among the most ubiquitous features on the surfaces of solid planets. Wherever steep slopes occur, long-runout landslides are likely to be found. They have been documented on Mars (Figure 8.17b), Venus, the Moon, Callisto, Io, and even on Mars' tiny moon Phobos (Collins and Melosh, 2003). The one other requirement is that they always involve large volumes of material: Whatever "magic" allows dry rock debris to flow like a liquid, it does not work for volumes less than about 10^6 m^3 (on the Earth: the minimum is larger on Mars). This large-volume-only constraint makes it difficult to investigate these landslides in the lab, although the Soviet government of Russia did create a few long-runout landslides during construction of debris dams on large rivers, as well as at their Novaya Zemlya nuclear test site.

Low coefficient of friction. Another surprising feature of long-runout landslides is their ability to travel over astonishingly low slopes. This accounts for their tremendous lengths and implies that large volumes of fast-moving rock debris possess very low coefficients of

(a)

(b)

Figure 8.17 Panel (a) is a landslide triggered by the 2002 Denali earthquake in Alaska, looking west toward the divide of the Black Rapids and Susitna glaciers. Image courtesy Dennis Trabant and Rod Marsh, USGS. See also color plate section. Panel (b) shows landslides on the south wall of Valles Marineris of Mars. The image is 60 km across. Viking Orbiter image 14A30.

friction. Because the center of mass of these deposits before and after deposition cannot often be determined, a common metric for describing them is an effective "coefficient of friction" that is equal to the ratio between their height of fall H (measured from the head of the scarp from which they fell to their extreme toe) to the length of runout L (also measured from the head scarp to the toe). The arc-tangent of this ratio can be considered to be a kind of friction angle. For typical small rockslides this ratio is close to 0.6 (implying an angle between scarp and toe of about 30°, close to the normal angle of repose). However, as the data in Figure 8.18 show for both Earth and Mars, this ratio declines as the volume of the slide increases, obeying a crude power law.

$$H/L \propto V^{-0.16}. \tag{8.12}$$

H/L drops to about 0.03 for the largest landslides (implying an effective friction angle of only 1.7°). Note also that this ratio is about a factor of 3 smaller for Earth than it is for Mars, suggestively similar to the ratio between their accelerations of gravity.

The extremely long runout for large landslides also means that they are extraordinarily dangerous on an inhabited Earth. Current building codes do not take this long reach into account, but it is sobering to note that the city of Osaka, Japan, is built upon the debris from an ancient landslide of this type.

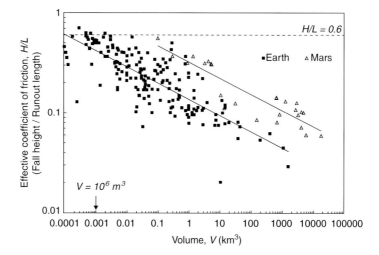

Figure 8.18 The effective coefficient of friction for large landslides depends on the landslide volume. This coefficient of friction is defined as the ratio between the vertical drop H between the head of the landslide and its toe, divided by the horizontal distance L between the head of the landslide and its toe. The square points on this plot are terrestrial landslides and the triangles are Martian. Despite the large scatter, this plot shows that landslides with a volume less than about 10^6 m^3 are limited to H/L = 0.6, but this decreases slowly for larger volumes. Large Martian landslides are less mobile than terrestrial landslides by about a factor of 3. Figure from Collins and Melosh (2003).

Mechanism. The mechanism that permits long-runout landslides to attain such a high fluidity has intrigued geologists and physicists since Heim first described them in 1882. None of these flows shows evidence for excess water. Many writers have sought the action of some fluid other than water, including air, steam generated during sliding, or carbon dioxide evolved from calcined limestone blocks. Geologist Ron Shreve (Shreve, 1966a) proposed the widely accepted idea that the Blackhawk and Sherman Glacier slides floated out across low terminal slopes on a cushion of air trapped underneath the flowing rock debris, but the discovery of long-runout landslides on Mars, the Moon, and airless bodies such as Callisto and Io tends to discount atmospheric gas as a lubricant. Physical mechanisms such as basal melting during sliding are more universal, and glass is found mixed within a few landslides, such as the Köfels slide in Austria. However, examinations of the bases of many other long-runout landslides fail to find any evidence of heating, suggesting that the glass may form during the stopping phase when normal values of rock friction reassert themselves while the mass is still in motion.

I have proposed a process that I call "acoustic fluidization" to explain the mobility of long-runout landslides (Melosh, 1979, 1983). The basic idea is that strong internal vibration in the moving debris builds up during the initial collapse phase to the point that it effectively liquefies the rock debris. Vibration is widely used in industrial operations with granular materials to enhance their flow, so this part of the process is not mysterious. As long as the debris flows down a small slope it gains gravitational energy that can offset the dissipation of vibrational energy (which is lost mainly by radiation through the bottom of the slide). The minimum volume limit arises because energy is lost mainly through the surface of the slide: The moving mass must be big enough that the volume/area ratio is sufficient to retain most of the energy released by sliding. Motion stops when the slide spreads out and becomes too thin: On Earth, this happens when its thickness drops below 10 to 20 m. This mechanism does not require the presence of any fluid and should work well in a vacuum. It leaves no melt or any direct evidence of the former presence of strong vibrations in the rock debris, except perhaps for a peculiar fracture pattern that Shreve identified and called "domino breccia." It is, however, unfortunate that acoustic fluidization does not leave more evidence, for it is difficult to demonstrate that it occurred in any landslide deposit. No active landslide has yet contained the instrumentation necessary to demonstrate the presence of acoustic fluidization, so debate continues about the mechanism that imparts low coefficients of friction to long-runout landslides.

Implications. Long-runout landslides are an effective agent of landform degradation. It appears that the so-called "sector collapse" of large volcanic cones is a major factor in the ultimate leveling of volcanic constructs. These collapse events typically shed long-runout landslides and transport volcanic debris long distances from their source. Landsat images of the Llullaillaco and Socompa volcanoes near the border of Chile and Argentina show broad sheets of landslide debris that stretch up to 100 km from their source volcanoes. The Hawaiian volcanoes also appear to degrade by large catastrophic landslides that, in this case, are under water and so escaped recognition for a long time. The steep scarp that forms the north side of Molokai is one of its most impressive features. We now know that this scarp is merely the head scar of an enormous landslide whose debris blankets the sea floor for 200 km to the north (Moore *et al.*, 1989).

Large long-runout landslide deposits are also seen at the foot of Venusian volcanoes. However, the largest known debris landslides in the Solar System form the "aureole" at the base of Olympus Mons on Mars (McGovern *et al.*, 2004), Figure 8.19. These deposits extend up to 750 km from the base of the volcano. They originated from the 10 km high scarp at the base of Olympus Mons and, once they began, flowed horizontally down a slope averaging only about $0.5°$. McGovern *et al.* (2004) estimated the volume of a single lobe to be about 8.8×10^4 km^3: It was shed from the collapse of a block 10 km thick, 60 km wide, and 150 km long at the northern base of the volcano.

Gravity-driven mass movement, which was an underrated process at the beginning of the last century, is now regarded as a highly effective agent of landform degradation. It is ubiquitous on sloping surfaces and ranges in speed from the imperceptibly slow creep of a thin surficial layer to nearly sonic speeds in immense landslides that can transport debris over significant fractions of a planet's radius.

Further reading

The best quantitative discussion of mass wasting is the now classic book Carson and Kirkby (1972). A more qualitative discussion is part of most books on geomorphology, but

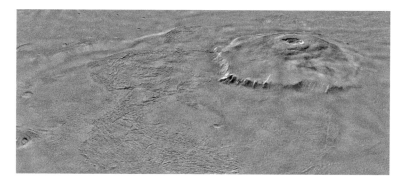

Figure 8.19 Oblique view of Olympus Mons, shown topographically draped over a Viking image mosaic. The topography shows the relationship between the volcano's scarp and massive aureole deposit that was produced by flank collapse. The base of Olympus Mons is 600 km in diameter and the summit caldera is 24 km above the surrounding plains. The vertical exaggeration is 10:1. NASA/ MOLA Science team PIA02805.

a good, focused discussion is in Selby and Hodder (1993). The stability of hillslopes is the traditional topic of rock mechanics (for strong rocks), of which an excellent introduction is Jaeger *et al.* (2007), or soil mechanics (for less cohesive materials), for which I recommend Lambe and Whitman (1979). Turbidity currents of all kinds are discussed in Simpson (1999), although this book is very short on quantitative detail. Sturzstrom and rockfalls are treated in Erismann and Abele (2001). For more information the reader is referred to the references cited in Section 8.2.4.

Exercises

8.1 Creepy lunar regolith

The Apollo 17 heat-flow probe measured a "daily" (that is, 29.5 Earth day) temperature variation of 300 K at the surface of the regolith, which fell to about 30 K at a depth of 10 cm (for the purposes of this exercise it is permissible to approximate the temperature variation as linear – for extra credit, how could you make this more realistic?). If the linear coefficient of thermal expansion of the powdery regolith equals that of basalt (6×10^{-6} K^{-1}), estimate the rate of downhill creep expected for a 30° slope exposed to full sunlight on the moon. Use the theory of creep described in Section 8.1.1 and assume a maximal creep rate with recovery factor $r = 0$. Estimate the rate of slope movement in terms of mass per unit width of the slope. Is this result reasonable in view of the slope in Figure 8.1?

8.2 Diffusing topography

Figure 8.1 shows a bench at the base of a steep-sided lunar ridge viewed at near-vertical incidence. Use the Culling theory of landform degradation described in Section 8.1.2 to *roughly* estimate the topographic diffusion coefficient K for the Moon's surface by assuming that the scarp (part of the Flamsteed Ring) is about 4×10^9 yr old. Make the same type of estimate for the bowl of Meteor Crater, Arizona, using the fact that the crater, which is 50 000 yr old, 1.2 km in diameter, and 180 m deep, has presently been filled to a depth of 30 m by post-impact sediments. Compare the topographic degradation rates of the Moon and the Earth. Does this tell you why we see so few impact craters on the Earth, compared with the Moon?

8.3 Splashback creep

Surface creep can be initiated by either rainsplash impact on the Earth or meteorite impacts on the Moon. When either process excavates a crater of volume V_c, a similar volume of ejecta is launched out of the crater and travels a distance of roughly one crater radius R_c before splashing down onto the surface. The time it remains in flight between ejection and splashdown is given roughly by the ballistic time of flight equation $t_f = \sqrt{2R/g}$ for a particle to travel a distance of R in a gravity field g, assuming a launch elevation angle of 45°. On a flat surface the center of mass of the ejecta lies at the center of the crater. However, on a surface sloping at an angle θ, the center of mass of the ejecta moves downslope under the acceleration of gravity while it is in flight. Derive an equation that expresses the distance the center of mass of the ejecta moves downslope as a function of crater radius, gravity, and the surface slope. Use this equation to derive the downslope volume flux of material (m³ per meter of slope contour distance) moved by a surface-covering barrage of craters whose total volume just equals the volume of the slope's surface layer down to a depth equal to ¼ of the crater diameter $2R_c$. If the duration of this barrage is T, use Equation (8.3) to relate the downslope volume flux to the topographic diffusion coefficient K. Finally, use the fact that the Moon's upper 1 cm of surface material is overturned by micrometeorite impacts about once every 70 000 yr to estimate the numerical value of K in m²/s. At this topographic diffusion rate, how long will a fresh impact crater 1 km in diameter and 200 m deep persist on the surface? What does this tell you about the importance of impact-driven creep for the Moon's surface features?

8.4 Following a French cliff-hanger

Following Coulomb's method of 1773, described in Section 8.2.2, derive Equation (8.11) for the maximum height of a vertical cliff in a homogeneous mass of material that possesses both cohesion and internal friction. Apply this equation to compute the maximum height of a vertical Ionian caldera scarp by using the estimate of Clow and Carr (1980) that the cohesion of Io's porous S-SO$_2$ crust is about $c = 0.3$ MPa and its coefficient of friction $f_f = 1.73$.

8.5 Strong Martian lavas?

The enormous volcanic caldera at the summit of Olympus Mons on Mars is 80 km across and is surrounded by near-vertical cliffs between 2.4 and 2.8 km high. Its floor stands 24 km above the surrounding plains. Use Equation (8.11) to estimate the minimum strength of the lavas in its walls and compare this with the strength of terrestrial lava flows.

8.6 Groovey Lutetia

On 10 July 2010 the European Space Agency's Rosetta spacecraft flew by asteroid 21 Lutetia. Lutetia is a rocky body that is irregular in shape, with dimensions about 132 by 101 by 76 km. Although Lutetia is a main-belt asteroid, its surface is creased by a series of Phobos-like grooves. These grooves have a typically beaded appearance, similar to those on Phobos, but with a width of approximately ½ km and stretching tens of kilometers in length. From these approximate dimensions, estimate the thickness of Lutetia's regolith. Is this a surprise? Discuss what processes may have acted to create this loose blanket of debris.

9

Wind

> Owing to the development of motor transport, it is possible to study in
> the further interiors of the great deserts the free interplay of wind and
> sand, uncomplicated by the effects of moisture, vegetation, or of fauna,
> and to observe the results of that interplay extended over great periods
> of time.
>
> Here, instead of finding chaos and disorder, the observer never fails to
> be amazed at a simplicity of form, an exactitude of repetition and a geo-
> metric order unknown in nature on a scale larger than that of crystalline
> structure.
>
> R. A. Bagnold (1941)

Ralph Bagnold (1896–1990) founded our modern understanding of the interaction between
wind and sand and how that interaction produces dune-covered landscapes in the Earth's
great deserts. He lived to see spacecraft images of the sand seas on Mars and contributed
to our understanding of how universally important wind-driven (eolian) processes are. He
would have been delighted to know about the extensive dune fields of tarry sand on Titan.

Bagnold was a professional soldier and the descendent of a long line of professional sol-
diers (Bagnold, 1990). After an engineering education at Cambridge, he was posted to Egypt
in 1926 and then to other locations in North Africa where he became fascinated by the landscape and decided to devote himself to the study of that region's most abundant commod-
ity – sand. In addition to unprecedented trips deep into the deserts of Sudan and Libya, he
built a wind tunnel out of plywood at Imperial College, London, to further his understanding
of the interaction of wind and sand. His classic book was published in 1941.

One of his major insights about this process is best stated, once again, in his own
words,

> After much desert travel, extending over many years, during which sandstorms of varying intensity
> were frequently encountered, I became convinced that the movement of sand (as opposed to that of
> dust) is a purely surface effect, taking place only within a metre of the ground.
>
> (Bagnold, 1941)

Bagnold used this insight to justify his reliance on wind-tunnel observations. Much more
elaborate and expensive wind tunnels than Bagnold's are now used to simulate sand

transport under both Martian and Venusian conditions. Modern research has also moved outdoors and focuses on the interaction of dunes with both local winds and the planetary boundary layer, but many mysteries still remain and new insights are still needed to fully understand the dynamics of how the wind interacts with granular materials on planetary surfaces.

Following Bagnold's lead, this chapter is more mathematical than any other in this book. Bagnold quite appropriately put the word "physics" into the title of his book and insofar as physics requires the language of mathematics, the study of eolian processes has, since Bagnold, always relied heavily on that mode of expression. As we shall see, however, the mathematics required does not go much beyond algebra: When really intricate analyses are required, such as the structure of turbulent boundary layers, Bagnold himself resorted to empirical models, and we shall do the same.

Eolian processes are concerned with a gas, the atmosphere, and granular solids. A discussion of interactions between the atmosphere and a liquid is reserved for the next chapter.

9.1 Sand vs. dust

The mechanical distinction between sand and dust is simple: Dust is easily suspended in the atmosphere while sand, under a moderately strong wind, hops along the surface. However, quantifying this distinction for a wide variety of solid particles in different planetary atmospheres requires some careful discussion, as does the meaning of a "moderately strong" wind. The first step in this discussion is to understand how solid particles fall through air.

9.1.1 Terminal velocity

Students in freshman physics courses are encouraged to ignore atmospheric drag while they learn Galileo's formulas for falling bodies. However, every skydiver's life depends on the fact that these formulas do not accurately describe his or her descent toward the surface of the Earth. A human body falling out of an airplane rapidly accelerates, initially following Galileo's rules, but within about 15 s achieves a constant velocity known as the terminal velocity. This is about 200 km/hour for a human body in Earth's atmosphere. If this were the end of the story, our skydiver's arrival at the Earth's surface would still be very uncomfortable and skydiving would not be a popular sport. However, the terminal velocity is a balance between the weight of a falling body and the drag force exerted on it by the passing air. This drag force can be greatly enhanced by increasing the area of the falling body – which is what parachutes are intended to do.

Quantitative analysis of the terminal velocity requires some simplifications. In the best tradition of physics, we now consider a spherical sand grain (which is not such a bad approximation as either the traditional spherical cow or a spherical skydiver!). The force accelerating the grain toward the Earth is its weight, w, equal to its mass m times the acceleration of gravity g. Expressing this in terms of the grain diameter d and density σ, the driving force is:

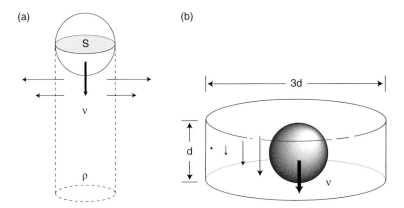

Figure 9.1 Terminal velocity force balance for (a) turbulent flow, where the weight of the grain is balanced against the momentum change of air displaced by the motion of the grain and (b) laminar flow, where the weight of the grain is balanced by viscous drag forces, here approximated as being localized in a cylinder with a diameter three times larger than the grain diameter d. In this case the velocity of the air equals the velocity of the grain at the grain surface but drops to zero (approximately) at the surface of the cylinder.

$$w = \frac{\pi}{6} d^3 (\sigma - \rho) g. \tag{9.1}$$

In this equation ρ is the density of the air, so that the weight is really the immersed weight, which takes account of the buoyancy of the fluid that surrounds the grain. This correction may seem to be negligible for quartz sand (density about 2650 kg/m^3) in the Earth's atmosphere (density 1.2 kg/m^3); however, we shall see that the equations that we derive for sand and air apply almost without alteration for sand and water or any other liquid, where the buoyancy correction may be substantial, so we will retain this distinction in the following analysis.

The drag force is more complex and requires our first empirical injection. Simple consideration of the momentum of the air deflected by the particle (Figure 9.1a) suggests that it should depend on the projected area of the falling grain, $\pi d^2/4$, the density of the air ρ and the square of the relative velocity v between the grain and the air. However, the exact drag force depends on the shape of the grain and its velocity in a more complicated way, so this complexity is absorbed into a mostly empirical constant called the drag coefficient C_D, defined so that the drag force F_D comes out as:

$$F_D = C_D \frac{\pi \rho d^2 v^2}{4}. \tag{9.2}$$

Over a wide range of velocities, $C_D \approx 0.4$ for a sphere.

Equating the weight (9.1) and drag force (9.2) and solving for the velocity yields an expression for the terminal velocity:

$$v = \sqrt{\frac{4}{3} \frac{(\sigma - \rho) d\, g}{C_D\, \rho}}. \qquad (9.3)$$

As one might expect, objects fall faster if they are denser, bigger, or the acceleration of gravity is higher. They fall more slowly if the air density or drag coefficient is higher.

Equation (9.3) does not hold for all grain sizes. In particular, it may give an extremely poor estimate of the terminal velocity for small particles unless the drag coefficient is changed rather substantially. This equation holds best in what is called the turbulent regime, where drag forces are created by the deflection of the air stream. The vigilant reader may also be surprised that there is no dependence on the viscosity of the air in this equation. Viscosity is important only for very small or slow particles. The dividing line between turbulent flow and the low-velocity laminar flow regime is determined by the Reynolds number Re, which is the ratio between inertial and viscous forces,

$$\mathrm{Re} = \frac{\rho\, v d}{\eta}. \qquad (9.4)$$

When the Reynolds number is low, inertial forces are small compared to viscous forces, and viscosity, not the deflection of the air stream, determines the drag force. George G. Stokes (1819–1903) first analyzed the full equations for viscous drag in 1851 and the terminal velocity for a sphere is now called Stokes' law. The full derivation is complex and not very edifying, so I instead present an approximate derivation that captures the essence of the equation.

Suppose that our small falling sphere is surrounded by a cylindrical can of diameter $3d$ and height d (Figure 9.1b). We suppose that the velocity of the air is zero on the surface of the can, but equals the velocity of the sphere at the sphere's surface. This is not really true: The air velocity falls off more gradually with distance away from the sphere, but most of its decline is close to the sphere, so our rigid can is a good first approximation. The air between the sphere and the can is, thus, sheared with a strain rate $\dot{\varepsilon} \approx v/2d$. Remembering that the definition of viscosity, Equation (3.12), is $\sigma_s = 2\eta\dot{\varepsilon}$, where σ_s is the shear stress, we obtain the drag force by multiplying the shear stress times the surface area of the vertical sides of the can, $3\pi d^2$. Equating this drag force to the weight of the grain, (9.1), and solving for the velocity, we obtain the terminal velocity of a small particle for which $\mathrm{Re} \ll 1$,

$$v = \frac{1}{18} \frac{(\sigma - \rho)d^2 g}{\eta} \qquad (9.5)$$

which happens to be exactly Stokes' law, thanks to a clever choice of the dimensions of our cylindrical can. Another way to achieve this result is to note that at low Reynolds number the drag coefficient is given by $C_D = 24/\mathrm{Re}$, which upon substitution into (9.3) yields Stokes' law.

The most notable features of Stokes' law are its inverse dependence on gas viscosity and its dependence on the square of the particle diameter. This means that the terminal velocity of very small particles is very low. Anyone who has wondered why the clouds, composed of tiny water droplets with mean radii of about 10 μm, do not fall out of the sky can answer this question for themselves by evaluating Equation (9.5) for a cloud droplet.

Stokes' law is valid for Reynolds numbers less than about 10, while the turbulent drag equation holds for Reynolds numbers between about 10^3 and 10^5. Between these regimes empirical expressions for the drag coefficient must be used to compute the terminal velocity.

Evaluation of the terminal velocities of small particles in the atmospheres of different planets requires the viscosity of the gas. This can be derived from experiment or looked up in a table. However, it is useful to note a few results on gas viscosity from the kinetic theory of gases. Most importantly, the viscosity of a gas is nearly independent of pressure. Thus, whether we are dealing with air at the Earth's surface or air in the stratosphere, the viscosity η is the same (at the same temperature). J. C. Maxwell first deduced this fact from his kinetic theory of gases and, at first, he could not believe it: Checking this prediction was the motivation for his 1867 experiments that also led him to invent the concept of a Maxwell solid (Section 3.4.3). This surprising behavior comes about because the viscosity of a gas is the consequence of the exchange of momentum between layers of gas moving relative to one another. As pressure decreases there are fewer gas molecules to exchange momentum, but their mean free path increases at the same time, so the exchange takes place between layers of greater relative velocity. The two factors cancel one another and the resulting viscosity is independent of pressure. Maxwell also predicted that the viscosity of a gas depends on the square root of the temperature. This prediction is less accurate: Measurements show that in most gases the viscosity depends more strongly upon temperature. Subsequent research connects the temperature dependence of viscosity to the forces between molecules, so this dependence must usually be determined empirically.

Table 9.1 lists the terminal velocities of small spheres based on Stokes' law for silicate grains near the surface of Earth, Mars, and Venus, along with velocities for tarry organic grains on Titan. It is clear that as particle size decreases the terminal velocity decreases rapidly as well. For a given size particle the terminal velocities are surprisingly similar despite the wide differences between the various bodies. The major determinant of terminal velocity is grain size, not which planet it falls on. Note that for grains larger than 100 μm Stokes' law underestimates the terminal velocity and a more accurate expression for the drag coefficient must be used.

9.1.2 Suspension of small particles

The terminal velocity alone is not enough to estimate whether a particle will be suspended or sink to the surface. In a quiet atmosphere, or if the wind flow was purely laminar, particles of all sizes would eventually settle to the surface. However, the atmosphere of a planet is almost never completely quiet or laminar. Winds are ultimately due to the spherical shapes of planets and the fact that solar radiation is not uniformly distributed over their surfaces. Because of inequalities of heating, currents arise in the atmosphere.

Table 9.1 *Terminal velocities of small particles*

Body	Particle composition, density (kg/m³)	Gas viscosity (10⁻⁶ Pa-s)	100 µm grain diameter (m/s)	30 µm grain diameter (m/s)	10 µm grain diameter (m/s)
Venus	Silicate, 2700	33.0	0.40	0.036	0.0040
Earth	Silicate, 2700	17.1	0.88	0.079	0.0088
Mars	Silicate, 2700	10.6	0.55	0.05	0.0055
Titan	Organic tar, 1500	6.3	0.18	0.016	0.0018

The flow of all known planetary atmospheres is turbulent. The air does not travel smoothly from one point to another. Instead, instabilities in the flow develop that lead to wide fluctuations of the instantaneous velocity about the mean. Turbulent flow develops when the Reynolds number, Equation (9.4), is large, and at the scale of planetary atmospheres it is invariably very large. Turbulence is a broad and complex subject, which is treated in many specialized books devoted to that topic (a good introduction is Tennekes and Lumley, 1972). At the moment what it means to us is that the motion of the atmosphere can be divided into two components: an average wind speed that remains constant for long periods of time, and a fluctuating component that varies widely over short timescales.

The full story of how particles can be suspended in fluids is surprisingly complex, relying on the phenomenon of "bursting" in turbulent fluids to inject momentum from the surface boundary layer into the body of the moving fluid. For the purposes of this book we will bypass this difficult topic and apply Bagnold's rule of thumb, which states that the average velocity of turbulent eddies in a flow of air is equal to 1/5 of the mean velocity. Thus, for a wind velocity of a few meters per second, the turbulence velocity is about 0.5 m/s, so that the dividing line between sand and dust is about 100 µm on the terrestrial planets. It may be somewhat larger on Titan.

9.2 Motion of sand-sized grains

Grains too large to be suspended in the atmosphere may, nevertheless, be quite mobile on the surface under the influence of the wind. The hopping motion of sand grains along a river bed was first described by G. K. Gilbert (1914), who observed the process while studying the ability of water to transport river sediments. Gilbert called this motion "saltation" following McGee (1908), who took the term from the Latin word *saltus*, "leap." Bagnold accepted this term and applied it to the much larger hops that sand grains executed in his wind tunnel.

Modern studies of eolian transport distinguish four modes of wind-driven particle motion. Suspended material is carried aloft by turbulent winds. Saltating grains hop from the surface, travel for some distance (the "saltation length") with the air stream, and then reimpact

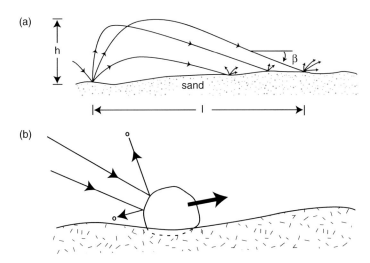

Figure 9.2 (a) Trajectories of saltating sand grains start out steeply as the grain is ejected from the surface. Near the apex of their hops, saltating grains attain a constant velocity and then follow a sloping (at angle β) linear trajectory as they fall back to the surface at terminal velocity, where they may initiate hops of other grains. The hop height is h and the length of each saltation hop is l. (b) Impact creep occurs as objects too large to saltate are struck by saltating grains. The momentum imparted by each impact pushes the larger grain downwind along the surface.

the surface (Figure 9.2a). Sand grains splashed out of the surface when saltating grains touch down either initiate new hops of their own or slither downwind in a snakelike motion called "reptation." Larger sand grains and pebbles too large to hop may still move downwind under the impact of saltating grains, a motion called "impact creep" (Figure 9.2b) that includes rolling along the surface.

9.2.1 Initiation of motion

All the modes of wind transport, except perhaps suspension, require sand grains to be in saltating motion. The problem of how this motion starts from an initial state in which the wind begins to blow over a motionless sand bed has turned out to be both complex and revealing. Before we can begin a full discussion of this process, we need a better understanding of how the wind interacts with the surface.

Wind near the surface. The first real understanding of how a moving fluid interacts with a surface grew out of the studies of German aerodynamicist Ludwig Prandtl (1875–1953). Prandtl spent most of his career at Göttingen University. His major claim to fame is the concept of the boundary layer, a zone of sharply increasing velocity at the interface between a moving fluid and a solid surface. When Bagnold needed more information about how sand grains could begin moving away from a sandy surface, he corresponded with Prandtl and incorporated many of Prandtl's ideas into his work.

The first important concept is the friction velocity v_*. The actual velocity varies with height above the surface in a complicated way that we will explore in a moment, but the friction velocity is meant to express the overall effect of the wind on the surface by a single number that is independent of height. The friction velocity is defined in terms of the shear stress τ that the wind exerts on the surface. You could imagine measuring this wind shear on an ice-covered lake by cutting a small raft of ice free of the cover, then measuring the force with which the raft is pushed downwind while the wind is blowing. The shear stress τ is then the force divided by the area of the raft. The shear stress does not depend on the details of the velocity distribution above the surface – it is a single overall measure of the surface force of the wind. The friction velocity is then *defined* in terms of the shear stress as:

$$v_* \equiv \sqrt{\frac{\tau}{\rho}}. \tag{9.6}$$

The friction velocity is a central concept in theories of wind transport because it is directly related to the force the wind exerts on the surface (and vice versa). However, it is not often practical to measure the shear stress directly. Fortunately, the friction velocity has a simple, almost universal, relationship to the velocity above the surface. The following formula is partly empirical and partly can be derived from Prandtl's mixing length theory of turbulence. It relates the mean wind velocity to height z through the friction velocity and a factor known as "roughness," z_0 :

$$v(z) = 5.75 \, v_* \, \log\left(\frac{z}{z_0}\right). \tag{9.7}$$

The roughness factor is somewhat empirical. It is proportional to the grain size in the surface and approximately equal to 1/30 of the grain size when the grains are tightly packed. For Earth it is typically about 0.2–0.3 mm and it is usually assumed to be about the same on Mars. The roughness of many different surface types has been calculated by measuring wind velocities at different heights above the ground. Given tables of roughness it is possible to convert a wind speed measurement at a single height into a prediction of its value at any other height, as well as to obtain the friction velocity.

Equation (9.7) describes the velocity dependence some distance above the surface. However, it cannot hold down to scales comparable to the roughness because the flow is broken up by the irregularities of the surface. If the roughness is small, another distance scale becomes important, one controlled by the molecular viscosity of the gas. A layer in which turbulence is suppressed then develops next to the surface. Called the viscous sublayer, the thickness of this zone (also empirically determined) is:

$$\delta \approx 5 \, \frac{\eta}{\rho v_*}. \tag{9.8}$$

The numerical factor is empirical, while the dimensional ratio is derived from turbulence theory (Tennekes and Lumley, 1972). Within the viscous sublayer the velocity falls linearly to zero at the surface. The overall dependence of wind velocity on height above the surface is illustrated in Figure 9.3.

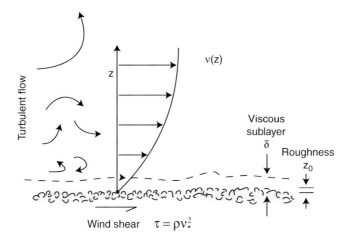

Figure 9.3 Mean velocity profile of wind over a sand surface. Turbulent eddies produce a logarithmic profile of mean velocity above the surface. Very close to the surface, viscous forces dominate and the fluid velocity falls linearly to zero through a viscous boundary layer of depth δ. The surface roughness is characterized by the parameter z_0, while the average wind drag on the surface is related to the friction velocity v_*.

The fluid threshold. When the wind blows over a sand surface where no grains are yet in motion, it has to begin by plucking individual grains out of the surface before the wind stream can accelerate them. Supposing that the grains protrude above the viscous sublayer, each grain experiences a downwind force that is given by the shear stress τ times its projected surface area, $\pi d^2/4$. This drag force is resisted by the weight of the grain (which is decreased by the lift provided by the deflected wind stream), Figure 9.4 and Equation (9.1). Whether the wind actually succeeds in plucking an individual grain out of the surface depends on the geometry of its contacts and the way it deflects the wind, so an exact formula for a real surface is not possible. However, the drag force must equal the weight times some factor of order 1. Calling this factor A^2 and replacing τ by its definition in terms of the friction velocity (9.6), we solve the resulting equation for the threshold friction velocity at which motion just begins:

$$v_{*_t} = A \sqrt{\left(\frac{\sigma - \rho}{\rho}\right) g\, d}. \qquad (9.9)$$

The threshold velocity is proportional to the square root of the particle size, so this expression predicts, not surprisingly, that larger grains are more difficult for the wind to pick up than smaller ones.

Something new happens, however, when we consider very small grains. When the size of the grains becomes smaller than the thickness of the viscous sublayer, Equation (9.8), the wind drag is spread over a large number of particles: it is not localized on a single grain. The grains, thus, become more difficult to pull out of the surface and the wind velocity

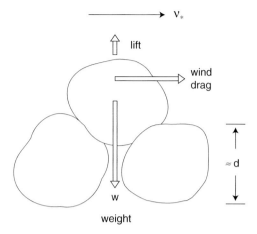

Figure 9.4 Forces acting on a grain resting on a surface of other grains of similar size. Each grain is subject to a drag force from the surface wind, possible lift forces from the deflected wind, friction between grains, and gravity holding it down onto the surface. Drag and lift forces must exceed friction and gravity before a grain begins to move.

must be higher than Equation (9.9) predicts before motion can start. There is a similar extra resistance if the grains stick together by cohesive forces, such as electrostatic or van der Waals forces, which are important for very small grains.

A full analysis of the effects of sublayer resistance is not presently possible. Instead, we must rely on empirical measurements of threshold velocities. Bagnold faced this problem and observed that, although the factor A in Equation (9.9) is roughly equal to 0.1 for large sand grains, it increases rapidly with decreasing particle size below some threshold (Figure 9.5). He found that A is a function of a dimensionless parameter that he called the "friction Reynolds number" Re_*. This Reynolds number is defined like the usual one, Equation (9.4), except that the friction velocity replaces the fluid velocity. For a friction Reynolds number less than about 3.5 the threshold velocity shoots up steeply (Figure 9.6) and appears to depend on Re_* to a large negative power. The appearance of the friction Reynolds number makes a lot of sense in this context because comparison of the definition of Re_* with the thickness of the viscous sublayer indicates that $d/\delta = 5/Re_*$. All this limit says is that the threshold velocity shoots up when the particle size is equal to about 0.7δ.

The approximate dependence of the threshold velocity on grain size can be derived from the dependence of A on Re_*, even though the power n is not well known. If A is a function of $1/Re_*^n$, then it also depends on $1/v_{*_t}^n d^n$ (neglecting other terms in the relationship, which are not important for this argument). Inserting this dependence into Equation (9.9) and solving for v_{*_t}, we find that it must depend on grain size d to the power $(1/2–n)/(n+1)$. But for large n, this ratio approaches $–1$, so that we can say that for small grain sizes, v_{*_t} depends approximately on $1/d$. Clearly, as d decreases the threshold velocity rises. But for $Re_* \gg 3.5$ we know that $A = 0.1$, and v_{*_t} is proportional to $\sqrt{d}$. The threshold velocity, thus, rises rapidly for both large and small particle sizes (Figure 9.6). So we must conclude

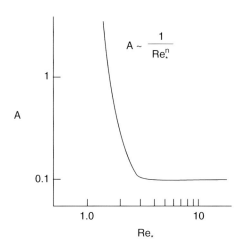

Figure 9.5 Dependence of the friction coefficient A in Equation (9.9) on the friction Reynolds number Re_*. The coefficient is nearly constant with a value 0.1 until Re_* falls below a critical limit, below which it rises rapidly.

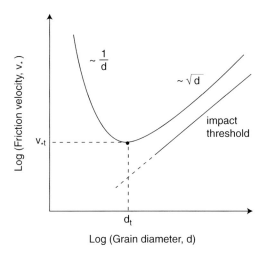

Figure 9.6 The threshold velocity for sand motion over an initially static sand bed has a pronounced minimum as a function of grain size. Small grains are buried in the viscous boundary layer and so are difficult to individually pluck out of the surface, while large grains are too heavy to move easily. This results in a unique size that is most easily entrained by the wind, at a corresponding minimum wind velocity. The threshold is much lower (the impact threshold) when sand grains are already in motion. Inspired by Bagnold (1941, Figure 28).

that there is a minimum for some special grain size that lies between the large particle and small particle limits.

The existence of a minimum in the threshold velocity curve has profound implications for wind- (and water-) transported material on any planet. It means that there is a special grain size, unique to that planet (and process, wind or water), which is most easily moved and which, therefore, characterizes all the deposits emplaced by that process. The diameters of wind-transported sand grains on Earth nearly all lie within a narrow range, 0.1 to 0.3 mm (Ahlbrandt, 1979). This reflects the selection of the most easily moved grains by winds whose strength may vary, but the size of the grains that move first and most often is always close to the minimum of the curve. Water-deposited sands on beaches and in rivers show a wider variety of sizes owing to their frequently complex histories. In principle, we should be able to distinguish wind-blown sands from water-deposited sands because of the difference in threshold diameter. In practice this is not usually possible for a variety of reasons, and other criteria such as shape or surface texture are now used (Siever, 1988).

The actual value of the threshold diameter and velocity can be estimated from the information already presented by setting $A = 0.1$ in Equation (9.9) and using Bagnold's observation that Re$_*$ at the threshold equals 3.5 to determine v_{*t}. The resulting equations can be solved for the grain diameter at the threshold, with the result:

$$d_t = 10.7 \left[\frac{\eta^2}{\rho(\sigma - \rho)g} \right]^{1/3}. \tag{9.10}$$

The corresponding threshold velocity is most easily derived from the Reynolds number at the threshold and the above equation:

$$v_{*t} = 3.5 \frac{\eta}{\rho d_t}. \tag{9.11}$$

These equations only approximately determine the minimum, which is rather broad in practice, so the precise values have to be taken with some skepticism, but using the same formulas to compare the onset of wind transport on different bodies is revealing. Table 9.2 lists the threshold grain diameters and velocities for the wind surface conditions on those Solar System bodies with substantial atmospheres. Where applicable, we also list the initiation conditions for flow in liquids that may be present on the surfaces of these bodies.

In order of ease of transport by wind, we see that the progression is Venus, Titan, Earth, and Mars. The threshold speeds are especially high on Mars, a fact that presents some problems that will be discussed separately. Wind speeds exceeding the threshold have been directly observed on Venus, Earth, and Titan. Dune fields have been observed on all three bodies, consistent with the predictions. The fact that extensive dune fields are also observed on Mars calls for special consideration.

The threshold grain size increases in the same progression, from very fine on Venus to very coarse on Mars. The predicted size for Earth is in fair agreement with the sizes commonly observed for dune sands.

Liquid is much denser than gas and so it can move particles with greater ease than the wind. The threshold equations also predict that it can move coarser particles (Mars, again, is an exception).

Grains at the threshold size are not the only ones that can be moved. The minimum in the curve is, observationally, rather broad (Greeley and Iversen, 1985) and winds higher than the minimum often occur, so that a range of grain sizes is usually transported. Greeley

Table 9.2 *Threshold grain diameters and fluid velocities*

Body	Medium	Viscosity (10^{-6} Pa-s)	Threshold diameter (μm)	Threshold friction velocity (m/s)	Fluid velocity at 1 m[a] (m/s)
Venus	Quartz in CO_2	33.0	94	0.018	0.37
Earth	Quartz in air	17.1	220	0.21	4.50
Earth	Quartz in water	1540	560	0.01	0.21
Mars	Quartz in CO_2	10.6	1100	3.3	69.
Mars	Quartz in water	1540	770	0.007	0.15
Titan	Tar in N_2	6.30	160	0.025	0.53
Titan	Tar in liquid methane	184	410	0.004	0.080
Titan	Ice in liquid methane	184	530	0.003	0.062

[a] Assuming roughness $z_0 = 2 \times 10^{-4}$ m.

and Iversen, in their book, tend to downplay the role of the viscous sublayer in raising the fluid threshold for small particles and instead emphasize cohesive forces between the grains. Electrostatic forces, however, tend to have the opposite effect and strong electric fields in the saltating layer may actually lift small particles off the bed (Kok and Renno, 2009b). It is clear that, even after 70 yr of study, there is still more to be learned about the fluid threshold.

The impact threshold. When sand grains are already in motion, they impact the surface at the end of their saltation hops and often knock other grains into the air. In this case the wind does not have to blow as fast as it does when the surface is quiescent. It is, thus, easier to keep grains in motion, once they have begun moving, than it is to initiate their first motion. This is an example of history-dependence of a process, or hysteresis. Of course, if the wind speed drops too low all of the sand grains fall back onto the surface and the process stops, but there is a range of wind speeds between the threshold speed and this stopping speed where sand motion is possible. The minimum velocity to keep grains in motion is known as the impact threshold and is indicated on Figure 9.6 by the line labeled "impact threshold."

Bagnold's experiments suggested to him that the impact threshold is given by Equation (9.9), with *A* equal to 0.08, rather than 0.1 at the fluid threshold (Figure 9.6). This estimate of Bagnold's has been supported by subsequent research, which places *A* between 0.08 and 0.085 at the fluid threshold (Kok and Renno, 2009a). Naturally there is no upturn reflecting the viscous sublayer, but the decrease in surface velocity due to the grains already in motion may be significant. Saltating grains in water do not approach the surface with as high a velocity as those in air and so impacts may not be a strong factor in this case. However, other phenomena, such as turbulent bursting, may play a role in making the curve different for sediment in motion versus that for clean fluid just initiating motion.

A recent and important contribution to this topic (Kok, 2010) suggests that the impact threshold may be far more important for sand transport on Mars than it is on Earth. This is a consequence of the very high speeds needed to reach the fluid threshold in Mars' thin atmosphere (Table 9.2). Once in motion, grains driven by high winds eject an exceptionally large number of grains from the surface. Taking account of the role of occasional gusts of high speed in initiating grain motion, it seems that winds with speeds ten times slower than the nominal fluid threshold prediction may, nevertheless, be effective in transporting sand and explain the many observations of eolian features on Mars, despite the few instances of winds high enough to reach the fluid threshold.

9.2.2 Transport by the wind

Once a few grains begin to move, the saltating grains knock others out of the surface, which themselves knock still other grains free, until the entire sand surface becomes covered by a low carpet of saltating sandgrains. The dimensions of this carpet and the amount of sand that is in motion are determined by the speed of the wind blowing over the surface and the properties of the atmosphere.

The wind over saltating sand. When Bagnold first tried to analyze this process, he found himself in new territory: He corresponded with Prandtl, the foremost expert of his time, but Prandtl's theories apply to pure fluids and need modification before they can be applied to a carpet of saltating sand. Bagnold, thus, developed a rough-and-ready series of estimates that have mostly stood the test of time and are still used to estimate rates of sand transport, with a few minor modifications. He supposed that the wind velocity over a saltating sand carpet is given by:

$$v = 5.75 \, v'_* \log \left(\frac{z}{z'_0} \right) + v_t \qquad (9.12)$$

where v'_* is the friction velocity when sand is in motion. It reflects the increased drag over the flowing sand carpet and is, therefore, larger than the friction velocity when sand is not in motion. The factor z'_0 is a modified roughness, now known as the "aerodynamic roughness," whose nature puzzled Bagnold. Empirically it is about ten times larger than z_0 and Bagnold suggested that it might correspond to the height of ripples. Similarly, v_t is a threshold velocity at height z'_0. These last two factors must be viewed as empirical fitting factors of obscure physical interpretation. More modern estimates are equally empirical (Greeley and Iversen, 1985). Most recently, however, substantial progress has been made in computing this wind profile using numerical computer codes (Kok and Renno, 2009a).

The flux of wind-driven sand. Bagnold's analysis of the effect of the saltation carpet on the wind near the ground begins with the hop of a single grain. Referring back to Figure 9.2a, a saltating grain first leaps out of the surface and, as it reaches the crest of its trajectory, it is accelerated by the wind to an average horizontal speed u_s. It returns to the bed at its terminal velocity v, so that the angle β at which the saltating grain approaches the bed is given by $\tan \beta = v/u_s$. Each sand grain of mass m thus leaps out of the bed, accelerates from near-zero velocity to final velocity u_s, and then re-impacts the bed a distance l

downwind. This start-stop motion of each saltating grain removes momentum $m\,u_s/l$ per unit distance from the wind. If the net motion of the sand, comprising many sand grains, is expressed as a mass flux, q_s, per unit width perpendicular to the wind (kg/s-m), this momentum loss per second per unit area equals the drag stress τ' required to keep the sand moving,

$$\tau' = \frac{q_s u_s}{l} \equiv \rho v'^{\,2}_* \tag{9.13}$$

where v'_* is the friction velocity in the presence of a carpet of saltating grains. In his wind-tunnel measurements Bagnold found that u_s/l is approximately equal to the vertical ejection velocity, w, divided by g, $u_s/l \approx w/g$. He then supposed that w is proportional to the friction velocity v'_*. These suppositions lead to a useful scaling relation for the saltation hop length l:

$$l \propto \frac{v'_*}{g}. \tag{9.14}$$

The hop height h is proportional to the length through the tangent of the descent angle β.

Inserting the relations in the previous paragraph into Equation (9.13), and solving for q_s, we obtain an expression for the mass flux of the sand in terms of the friction velocity over a moving sand carpet:

$$q_s = C\left(\frac{\rho v'^{\,3}_*}{g}\right) \tag{9.15}$$

where C is a "constant" that Bagnold found depends on the square root of the grain diameter. There are many modern variants of Equation (9.15) for the sand flux: Greeley and Iversen (1985) list 15 of them. An obvious improvement is to subtract a threshold friction velocity from v'_*, so that this equation does not predict non-zero fluxes for arbitrarily low velocities. Nevertheless, the important feature of this equation is its dependence on the friction velocity cubed, and that is widely agreed to be at least approximately correct.

The "constant" C in Equation (9.15) conceals factors only partially considered by Bagnold. The rate of sand transport depends on the surface over which the sand saltates: It is slower over sand surfaces, where the saltating grains lose most of their momentum at the end of each hop, and much faster over stony surfaces, a factor that is important in the accumulation of sand patches. The rate of transport also depends on the slope of the surface, a factor that plays a role in modern theories of sand dune formation.

The major implication of Equation (9.15) is that the flux of sand is not a linear function of the wind speed, but depends on the speed raised to the third power. The most important sand-driving winds are, thus, not the average winds, but the exceptional winds. Meteorological plots of average winds may thus be very deceptive when the direction of sand transport is of interest. Gentle winds may blow from one direction most of the year, but a month of strong winds from another direction may completely dominate the orientation of a dune field. Modern analyses of the wind regime among sand dunes reflect this non-linear dependence of sand transport on the cube of the velocity by plotting a quantity known as the vector drift potential on maps of sand dunes (Fryberger, 1979).

Impact creep and reptation. Although most of the mass moved by the wind travels by saltation, the impulse delivered by the impact of saltating grains contributes up to about 25% of the total. As illustrated in Figure 9.2b, multiple impacts by saltating grains can propel pebbles that are otherwise too big to be moved by the wind. Such pebbles may be seen at the bed of the saltating carpet irregularly jerking or rolling downwind. Recently, an additional type of motion has been distinguished (Lancaster, 1995). Called reptation because the path of the sand grains resembles the slithering of a snake, it describes the motion of coarse sand grains on the bed that are splashed out at low velocity by the impacts of fast-saltating grains.

9.2.3 The entrainment of dust

One of the surprising aspects of wind interaction with the surface is the difficulty of entraining fine particles, dust. Once dust is suspended in the air, it may rise high into the atmosphere, travel over intercontinental distances (Pye, 1987), and may even play a major role in the radiation balance of the atmosphere, as it does on Mars. However, as many observers have noted, dust lying on the surface tends to stay on the surface until something, other than the wind, disturbs it.

Anecdotes about the resistance of fine material to erosion by wind or water abound. Mud mounds in Galveston's harbor survived the catastrophic 1900 hurricane unchanged, while seawalls built of meter-sized rocks were carried away. Gilded Age geologist Raphael Pumpelly was astonished to observe roads in China's loess region entrenched tens of meters into the surface (Figure 9.7), gradually excavated as the dust stirred up by traffic was blown away by the wind. Fine Martian dust that settled over the solar panels of the Spirit and Opportunity rovers was expected to eventually terminate the mission, until dust devils or major sandstorms blasted the dust away.

Bagnold attributed the immobility of dust to its small grain size compared to the thickness of the viscous sublayer in turbulent flow. Greeley and Iversen invoke cohesive forces between small grains. Whatever the cause, neither wind nor water can mobilize fine-grained material without the assistance of some disturbing agency.

Where saltating sand advances over dusty surfaces, the impact of the saltating grains on the surface mixes the dust with the air and thus mobilizes dust. Sand grains and saltating mud chips may play this role in raising dust from terrestrial playas. Sandstorms would be relatively innocuous if they were restricted to the thin carpet of saltating sand, but they are frequently accompanied by thick clouds of dust that do much more damage to the lungs of humans and animals.

On Mars, dust devils seem to play an important role in raising dust from the surface. This has been attributed to the unusually high winds that develop near the core of the dust devil and to the mobilization of low-density aggregates (Sullivan *et al.*, 2008). However, another

Figure 9.7 A road deeply entrenched in the fine-grained loess terrain in China. This illustrates the fact that fine-grained sediments cannot be moved by the wind until disturbed, in this case by traffic. Once entrained in the air, the wind exports the sediment and the road gradually deepens, with the long-term consequences shown here. Reproduced from Pumpelly (1918, p. 468).

A ROADWAY HOLLOWED IN LOESS BY WIND AND
TRAFFIC

From Richthofen's *China*

process known as "backventing" or "reverse percolation" may play a role in both dust-devil mobilization and in association with impact-crater blast waves. This process was first proposed in connection with the "dark halos" that were noted in radar images of craters on Venus (Ivanov *et al.*, 1992). It has also been observed in the vicinity of many explosion experiments, but most discussions of it occur in the "gray" literature (Rosenblatt *et al.*, 1982).

Backventing usually accompanies the passage of a shock wave over a dusty surface. In the shock, the pressure first rises above normal atmospheric pressure, then falls below it during the "negative phase" before returning to normal (Glasstone and Dolan, 1977). As the shock wave passes over a permeable surface, it first drives air into the ground, then, during the negative phase, the air vents out of the surface, often carrying dust that has been entrained by direct lofting in the vertical air stream. Optically dark halos up to 1 km in diameter have been noted about the sites of many recent impacts on Mars (Malin *et al.*, 2006) and probably originate by this backventing process. It seems possible that the sudden drop in pressure in the core of a dust devil might also be responsible for initiating a temporary flow of air out of the Martian soil that could entrain and lift dust off the surface, mixing it into the atmosphere.

9.2.4 Abrasion by moving sand

Pebbles and rocks in the path of saltating sand are frequently polished, faceted, and grooved, sometimes leading to fantastic surface patterns. The name ventifact is given to a rock that has been sculptured by the wind. The flux of moving sand is akin to industrial sandblasting and, over time, considerable abrasion of obstacles may occur. R. P. Sharp (1911–2004), in a series of well-known experiments (Sharp, 1964), found that a brick left in the path of saltating sand was half eroded away after 6 yr of exposure. He also planted Lucite rods vertically in the path of the sand and found that the abrasion varied considerably with height: Whereas very little damage occurred above the normal level of the saltating sand, the greatest losses of material were at about 20 cm above the ground, near the top of the saltation layer, while less was removed close to the ground. Hard igneous rocks, on the other hand, suffered little visible damage during the course of his 10.5-year study. Wind abrasion is evidently slow for resistant rocks. Abrasion rates for different materials were quantified in more detail by abrasion experiments conducted in a wind abrasion simulation apparatus by Greeley and Iversen (1985). These experiments showed the widely variable susceptibility of different rock types to wind abrasion at a given impact velocity.

Wind abrasion is a strong function of the impact velocity of individual sand grains, as well as the total flux encountering an obstacle. Because of the lower threshold velocities on Venus and Titan, wind abrasion may not be of great importance there. On Earth, while wind-abraded pebbles adorn the pages of most geology textbooks, they are rare in the field. Wind abrasion is not a widespread process on our planet except in few especially favored locales. However, because of the much higher threshold velocities on Mars, wind abrasion is expected to be more common and, in fact, it did not take long for the Mars Exploration Rovers to encounter wind-fluted rocks (Greeley *et al.*, 2006).

9.3 Eolian landforms

Landscapes dominated by the wind are rare on Earth: Where they do occur they are so strikingly different from our normal experience that they have received a great deal of attention from geomorphologists. While silicate sand or dust is the material most often moved by the wind, blowing snow may also create similar landforms. Mars seems to be the planet most favored by the wind, in spite of its thin atmosphere, although extensive dune fields have also been imaged on Venus and Titan. In the following discussion we will use the term "sand" in its eolian process sense: Sand is loose granular material that moves by saltation.

9.3.1 The instability of sandy surfaces

Much of the interest we find in eolian landscapes derives from their instability. Dune shapes are constantly shifting and dunes are landforms in slow motion. The formation of dunes themselves is due to the instability of surfaces on which sand and larger rocks are mixed.

The flux of saltating sand depends upon the surface across which it moves. Sand grains saltating over sand on Earth leap only a few tens of centimeters off the surface. However, if the sand moves over a stony surface (or a paved road), the grains rebound from the surface with considerably more energy than if they had struck a loose, sandy surface. The saltation height increases dramatically, the high-flying grains sense a faster wind velocity, and the sand moves faster and farther than before. This rapid motion continues until the sand grains reach another sandy patch, where they lose their energy and slow down, many of them dropping out of the wind stream in the process.

A surface of uniformly mixed sand grains and stones thus quickly separates into sandy patches and stony patches as a result of this feedback. This sorting of the surface into sand and stones persists even as the sand accumulates into dune fields, and in spite of the constant downwind migration of the dunes, as Sharp observed in the Algodones Dunes of California (Sharp, 1979).

This tendency for instabilities to grow and create patterns from an initially uniform landscape is presently recognized as a field of study in itself. "Self-organized" pattern formation occurs when positive feedback regulates systems of this kind. Computer models incorporating idealized versions of the laws of sand transport successfully reproduce many features of the natural system, including sand dunes (Anderson, 1996; Werner, 1995). Although such model building is not considered a part of traditional geomorphology, this approach does expose the most important factors that create the observed forms. Pelletier (2008) describes Werner's model, its philosophy, and provides a computer code for implementing it.

9.3.2 Ripples, ridges, and sand shadows

Ripples. The quintessential eolian feature is the sand ripple (Figure 9.8). Wind ripples develop everywhere that loose sand is exposed, on sand dunes, beaches, and even in children's sand boxes. On Earth, they are typically a few centimeters high, spaced about 10 cm apart, and may extend many meters laterally. They are oriented perpendicular to the predominant wind direction and move slowly downwind. Ripples form within minutes in a strong wind. Nevertheless, even after decades of scientific study, their formation still presents many puzzles, the most serious of which have been raised by the observations of what appear to be wind ripples on Mars.

Bagnold noted a similarity between the saltation hop length and the spacing of ripples and concluded that the regular spacing of ripples reflects the average saltation jump length. He supposed that an initially flat sand surface became wavy under the bombardment of the saltating particles, developing so that most saltation hops start and end on the upwind (stoss) side of the ripple, driving intense upslope surface creep, while suppressing creep on the downwind (lee) side

Although Bagnold's identification of saltation length and ripple spacing has been widely accepted, doubts have arisen. Sharp (1963) noted that during windstorms the ripple spacing gradually increases with time, which seems inconsistent with control by the saltation

Figure 9.8 Asymmetric wind ripples in sand. The pocketknife is ca. 10 cm long. Note poorly developed subsidiary ripples at a steep angle to the main trend. Ripples are topped by somewhat coarser sand grains than average. 2000 photo by Jason Barnes, from longitudinal sand dunes near Tuba City, AZ.

length. He proposed instead that the ripple spacing is controlled by the size of the creeping grains and described extensive observations that tend to support his view.

The origin of ripples has currently reached a crisis precipitated by the observations of surface landers and rovers on Mars. By scaling the saltation length from Earth to Mars using Equation (9.14), both the higher threshold friction velocity and lower gravity suggest that a Martian saltation hop should be more than 100 times longer than on Earth, reaching tens of meters or more. If Bagnold's theory is correct, wind ripples on Mars should be spaced tens of meters apart. However, images of the surface show what appear to be Earth-sized "bedforms" everywhere (Mars scientists are reluctant to use the word "ripple" for these features, even though they strongly resemble the terrestrial feature in size and spacing). It seems that, despite the familiar appearance of Martian eolian features, nothing about them fits at the moment. The grain sizes of the material on top of the ridges are too small (200 to 300 μm near the surface, down to 100 μm a few centimeters below the surface), the dust is observed to move when the sand-sized particles do not, and the millimeter-sized particles predicted by the threshold equation (see Table 9.2) are apparently absent (Sullivan *et al.*, 2008), although see Box 9.1 on this topic.

All of these Martian observations suggest that something is seriously amiss in our understanding of what wind processes on Mars are doing. And if our theories are wrong on Mars, are they wrong on Earth as well? New ideas and approaches are urgently needed, such as the suggestion that a very low impact threshold for sand motion on Mars might account for some of the observations (Kok, 2010), or that ripples are actually due to aerodynamic interactions between the wind and roughness elements on the surface (Pelletier, 2009).

Box 9.1 **Kamikaze grains on Mars**

Wind transport of sand on Mars has long posed a major problem for scientific analysts. Because of Mars' thin atmosphere, the minimum wind velocity required to set particles in motion at the fluid threshold, Equation (9.11) and Table 9.2, is enormous. Note that 69 m/s in Table 9.2 is a wind speed of 250 km/hour one meter above the surface. Do such powerful gales ever blow on Mars? Moreover, the particle size at the threshold of motion is also enormous, of the order of 1 mm, about five times larger than on Earth.

The prospect of a Martian sandstorm is truly fearsome: Anything on the surface would be bombarded by a fusillade of ball-bearing-sized grains flying along at nearly supersonic speeds. Because of this daunting prospect, the cameras of the Viking landers peered out through narrow slits that could be covered in the event of sandstorms. Fortunately, the Viking landers completed their missions before being engulfed in any such destructive sandblasts. But Martian sandstorms must also be highly abrasive to rocks lying on the surface. Indeed, it becomes surprising that there are any intact rocks at all.

Not only would such sandstorms be destructive of hardware on the planet, it should also be destructive of the flying grains themselves. Granting that saltation can get started on Mars, it seems that no sand grain could survive even a single hop before being blasted into dust. Significantly, it seemed, Viking soil-sieve measurements showed a deficit of grain sizes in the millimeter range. This observation has, moreover, been confirmed by the Spirit rover's microscopic imaging system (Sullivan *et al.*, 2008).

This problem became known as the "kamikaze effect": Martian sand grains can hop once, then they die. This begs the question of where sand grains come from in the first place. If this is their fate on Mars, then there cannot be very many of them.

The difficulty of getting sand grains to either move or survive on Mars, coupled with the undoubted observation of sand dunes, led to a variety of proposals for how Mars manages to build sand dunes in spite of its thin atmosphere. So far, no definitive solution has been found. The leading idea is that the sand grains that saltate are not solid particles at all, but instead are aggregates of finer material somehow cemented into millimeter-sized pellets. The lower density of such grains would make entrainment by the wind easier, although it would exacerbate the survival problem. However, even if the grains are destroyed after one hop, their fragments might be re-cemented at a later time to prepare them for another hop. Clay dunes (lunettes) on Earth form in this way, and such dunes, while relatively rare on Earth, have received a lot of attention from the Mars scientific community.

Another possibility is that the Martian sand dunes cannot, in fact, move under current climatic conditions and they are all relicts from a time when Mars' atmospheric pressure was higher. As of this writing, the HiRISE imaging system has not observed any dunes to move, at a resolution of about 30 cm. However, the Spirit rover reported that after strong wind gusts, grains up to 300 μm in diameter appeared on its top deck. While these are smaller than predicted at the fluid threshold, the fact that they were in motion at all is something of a surprise. Because the megaripples investigated by Spirit at El Dorado crater are composed of grains of this size, it seems that at least megaripple-forming materials are mobile during high winds.

Box 9.1 **(cont.)**

A very new idea, mentioned in Section 9.2.1, is that the fluid threshold might not be relevant on Mars. The impact threshold may occur at a much lower velocity on Mars so that the 100 µm to 300 µm particles that make up the megaripples could be the dominant grain size on Mars. Once they are in motion, grains of this size require much lower velocity winds to sustain them and, at this lower velocity, are not imperiled by impact destruction after one saltation hop.

Our understanding of the humble, ubiquitous ripple is certainly in need of improvement. One of the main drivers in this urgent need is the crisis raised by observations made on another planet than the Earth, highlighting the importance of comparative planetology even for those interested only in the Earth.

Ridges. Bagnold also noted that many large ripples (he called them "giant sand ridges") had coarser grains at their crests and suggested that surface creep of the larger grains drove them to the crest and then armored the crest sand against further deflation. Many subsequent observations of sandy ridges that are bigger than ripples but smaller than dunes have agreed with his observation that their grain size distributions are typically bimodal, and that the coarse grains collect at the crest of the ridge. These ridges are long-lasting and, like ripples, form transverse to the prevailing wind.

Some of the puzzling Martian eolian ridges may be of similar origin: The Spirit rover has confirmed the bimodal distribution of grain sizes in several ridges in Gusev crater. These features should, thus, properly be considered ridges, not large ripples, in spite of the unfortunate terminology: "Megaripples" lie everywhere on Mars. Megaripples form transverse to the wind and range from tens of centimeters up to 3 m high and a few to tens of meters apart (Figure 9.9). Many are light-colored, in contrast to the mainly dark dunes of basaltic sand and may be composed of a different material, perhaps low-density dust aggregates.

Sand shadows. Actively moving sand requires wind to keep it in motion. When the wind strength drops, sand accumulates. This simple fact accounts for the accumulations of sand commonly found in the lee of large rocks, brush, and in hollows below the general level of the surface. On Mars, this accounts for the wind-blown material in the lee of crater rims and in their interiors. Small dune fields typically form in the interiors of Martian craters where sand has accumulated and the wind speed is reduced.

Ripples often curve sharply when they cross sand shadows, their curvature indicating the deflection of the surface wind by the obstacle: The wind direction is always perpendicular to the trend of the ripples. One may, thus, note the divergence of the wind upwind of the obstacle and its convergence behind. The antitheses of shadows, sand scours, often develop upwind of obstacles where the wind speed accelerates.

A characteristic pattern of sand shadows and scours develops around small craters, illustrated in Figure 9.10. These organized zones of deposition and scouring are the result of twin eddies shed downwind as the wind parts around the crater rim. Sand accumulations

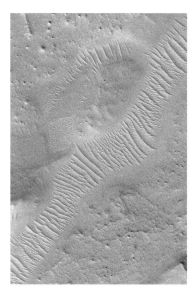

Figure 9.9 Megaripples on Mars. Ripples lie on the floor of a channel at 29.34° N and 299.83° W. Image is 3 km wide. Upper portion of Mars Orbiter Camera (MOC) image 1200991.

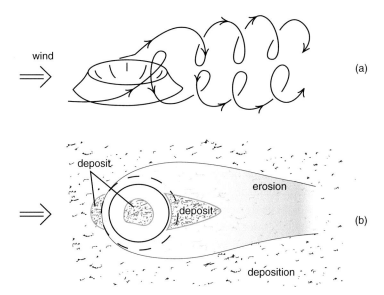

Figure 9.10 Sand shadows behind Martian crater forms. As the wind blows, vortices form and detach from the upwind rim of the crater. Sand is deposited in less windy areas upwind of the crater, in its interior, and immediately downwind of the rim, while the increased wind velocity in the vortices leads to enhanced erosion along the sides and farther downwind of the crater. Interpretation inspired by illustrations in Chapter 6 of Greeley and Iversen (1985).

behind mountain ridges are called "falling dunes," but frequently are merely immobile sand that has collected in the lee of the ridge.

9.3.3 Dunes

Sand dunes are larger accumulations of sand that are too wide for saltating grains to hop across. As dunes grow higher they create zones of quiet air or even reversed flow in their lee and, thus, block the saltation flow across the surface. Such shadow zones lead to the accumulation of more sand behind the crest, building the dune higher until the slope of the lee face reaches the angle of repose. At this point a "slip face" develops where sand that has arrived at the crest avalanches down the face.

Dunes are self-organized forms that grow spontaneously when conditions are right. They may start simply as a patch of sand on a stony plain that tends to capture more sand, which leads to the capture of still more sand in a strong positive feedback. Sand dunes would grow to unlimited height if given an unlimited supply of sand, except that as they grow they deflect the wind over their tops, increasing its velocity and eventually blowing sand off their crests faster than it can continue to build upward.

Dune velocity. All the time that a sand dune is growing in volume it is also moving downwind. The speed of downwind motion is easy to calculate and the result is very instructive. If q_s is the mass rate of sand movement per unit distance perpendicular to the wind, the rate at which it adds volume to the lee side of the sand dune (Figure 9.11) is q_s/σ_s, where σ_s is the density of loose sand in the dune. In each interval of time Δt the sand blown over the brink fills a prism in the lee of the dune of height h, length Δx and volume $h\,\Delta x$. The volume of this prism is equal to the volume of the sand blown in, $q_s\,\Delta t/\sigma_s$, so the dune creeps downwind at a velocity:

$$v_D = \frac{\Delta x}{\Delta t} = \frac{q_s}{\sigma_s\,h}. \tag{9.16}$$

Naturally, the dune velocity increases as the rate of sand transport increases. The interesting thing about this equation is that the velocity depends *inversely* on the dune height. This does make a lot of sense: It takes more sand to build a taller dune one unit of distance downwind than it does for a smaller dune. However, this means that the taller a dune grows, the slower it moves.

In a dune field in which a number of both large and small dunes occur, the small dunes race along while the big dunes lumber downwind. But when a small dune overtakes a big one, it climbs up its upwind face and collapses down the slip face, feeding the big dune while it is itself annihilated. The big dunes, thus, tend to grow at the expense of the small ones, looming larger and slower until the accelerated winds at their tops blow sand off their summits as fast as it accumulates and they reach a fixed height. Even this, however, does not stop their inexorable movement – they continue to crawl downwind until some new fate consumes their sand.

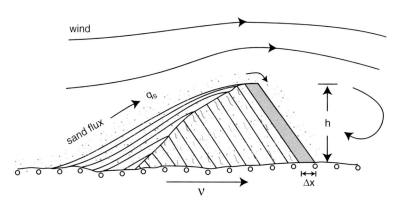

Figure 9.11 Downwind migration of sand dunes. The migration velocity of a sand dune is computed by balancing the sand flux moving up the upwind face of the dune against the volume required to build the dune one unit of length downwind. Sand is immobilized in the lee of the dune after it moves over the crest, where it accumulates at the angle of repose.

An important concept for interpreting the formation of dunes is the timescale over which an individual dune forms (Allen, 1974). In a regime of changing climate, large dunes may reflect conditions from a different climatic era and their morphology may thus be out of harmony with the regime prevailing at the time they are observed. A useful timescale is derived by comparing the volume of a dune with its rate of growth or, equivalently, the length of time required for the dune to move its own length,

$$\tau_D = \frac{\lambda h}{v_D} = \lambda h^2 \frac{\sigma_s}{q_s} \qquad (9.17)$$

where λ is the length/height ratio of the dune, typically equal to about 10. This "dune modification timescale" on Earth ranges from a few years for small dunes to 50 000 yr for large star dunes.

Dunes and dune fields are classified in various ways, but a common division groups them by their orientation with respect to the predominant wind. Dunes are, thus, considered to belong among the transverse, longitudinal, or star classes, depending on whether they are oriented perpendicular to the prevailing wind, parallel to it, or are heaped chaotically with no particular direction evident. There are special classes of dune, such as parabolic dunes, that are dependent upon vegetation to create their form. In this book such forms are ignored, but they are very important in terrestrial geology, particularly along coasts where vegetation is common. In addition to the "pure" forms described below, compounds of different types are almost universal.

Barchan dunes. The classic dune type is the barchan (Figure 9.12a). Shaped like a croissant lying with its convex side upwind, barchan dunes form spontaneously on stony plains where sand is in short supply. Barchans on both the Earth and Mars grow to an approximately constant size in a given area and maintain their characteristic shape as they migrate downwind. On Earth, barchans range from about 3 to 10 m in height and extend about 30

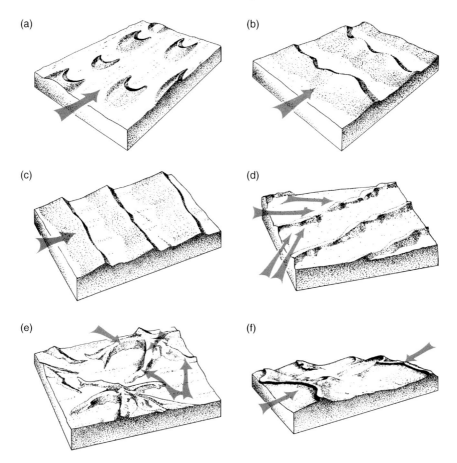

Figure 9.12 Different dune patterns are determined by both the wind regime and sand supply. (a) Barchan dunes form where a unidirectional wind drives a limited supply of sand. (b) Barchanoid ridges with wavy crests form transverse to the wind when more sand is available. (c) Full transverse dunes develop where sand is abundant and the wind is unidirectional. (d) Linear or longitudinal dunes form where the wind blows from more than one direction, but are generally parallel to the dominant wind direction. (e) Star dunes form under conditions of multiple wind directions with none dominant. (f) Reversing dunes are created by unidirectional winds that occasionally reverse themselves. All images from McKee (1979, pp. 11, 13).

to 100 m downwind. Their upwind slopes are gentle, standing at angles of about 5° to 15°, while the slip face stands at the angle of repose, near 32°. Surprisingly, barchan dunes on Mars are about the same size, despite the difference in surface conditions. The upwind face of barchan dunes is symmetrically arched in plan view, while an arcuate slip face develops downwind and lateral horns grade into low mounds lacking slip faces. Because the slip face traps migrating sand and thus prevents it from moving further downwind, barchans shed sand only off their lateral horns. The barchan form seems remarkably stable and these dunes are capable of migrating large distances downwind without change of form.

As the sand supply increases, instead of increasing in size indefinitely, barchan dunes shed sand off their horns, which may then organize itself into small new barchans, a reproductive process that fascinated Bagnold, who described it in the introduction to his book as "vaguely disturbing" in its "grotesque imitation of life."

Barchan dunes are most symmetrical when the wind blows from a single direction. When lateral winds occur, one horn grows larger and longer than the other, with the larger horn developing on the upwind side with respect to the lateral wind direction. With persistent lateral winds barchans may grade into linear dune chains.

Transverse dunes. With increasing sand supply and a constant wind direction, isolated barchan dunes merge into long ridges oriented perpendicular to the wind. Such ridges stretch many kilometers laterally and are repeated downwind by parallel ridges, creating a landscape dominated by parallel ridges with undulating crests. When the crests of these ridges are sinuous in plan they are often called "barchanoid" ridges (Figure 9.12b), but are simply called "transverse dunes" when their crests are more linear (Figure 9.12c). On Earth, there is a crude relation between the height and spacing of these dunes: Dunes a few meters high are spaced about 100 m apart, while dunes 100 m tall are spaced about 1 km apart (Lancaster, 1995).

The regular spacing of transverse dunes is likely due to aerodynamic flow patterns: The presence of a ridge suppresses surface winds for some distance downwind. Such shadow zones extend 12 to 15 times the height of the obstacle, in agreement with the observed spacing. Reverse eddies may also develop between the ridges, and many observers have noted weak winds blowing up the slip face of tall dunes while sand was blowing downwind over the crest.

All of the dune fields imaged on Venus belong to the transverse type, as indicated by their perpendicular orientation to wind streaks (Greeley *et al.*, 1997). Magellan could only resolve the largest dunes, which seem to be spaced about 0.5 km apart with ridges trending 5 to 10 km perpendicular to the wind direction. Magellan could not measure the heights of the dunes.

The transverse dune type also dominates on Mars, although large areas are occupied by sparsely scattered barchan dunes (Greeley *et al.*, 1992). Martian transverse dune ridges are typically spaced 300 to 800 m apart. Most of the large dune fields lie in the northern circumpolar plains. Two varieties of transverse dune are observed in small clusters, mostly lying within craters. The smaller variety is spaced 100–1200 m apart, while the larger variety is spaced at 1600–4000 m. Martian dunes are typically dark and are believed to be composed mostly of basaltic sand.

Linear dunes. Impressively long, linear dunes traverse large areas of the Earth and Titan (Figure 9.13). On Earth, individual linear dunes range from about 20 km to 200 km long, 2–35 m high and are spaced about 200–450 m apart. Compound versions of these dunes may be much larger: 50–170 m high and spaced 1600–2800 m apart. Dunes on Titan cover a vast area of the satellite, occupying about 40% of the low-latitude half of Titan, a larger fractional surface coverage than on any other body in our Solar System (Jaumann *et al.*, 2009). Titan's dunes are estimated to be about 30–70 m high with an average spacing of 2

Figure 9.13 Longitudinal dunes in the equatorial region of Titan, which likely consist of sand-sized particles made of organic material. Notice that the dunes are deflected by the brighter (and presumably higher) terrain, following the wind direction. The imaged area is 225 x 636 km from the T-25 pass of the Cassini orbiter Synthetic Aperture Radar. North is to the right. PIA12037: NASA/JPL.

km. Chains of dunes can be traced hundreds of kilometers. The dominance of linear dunes on Titan may be connected with its unique wind regime, in which the surface winds are mainly driven by tides rather than thermal gradients.

Linear dunes, called seif dunes by Bagnold, form when the sand-moving winds blow from more than one direction (Figure 9.12d). One direction must predominate, which determines the trend of the dunes, but they require a crosswind from another direction for at least part of the year. Because of this inconstant wind regime, the crests of linear dunes are often irregular and small slip faces alternate from one side to another.

The impressively parallel trend and spacing of linear dunes may be maintained by the formation of alternate helical eddies aligned with the prevailing wind. This idea was first proposed by Bagnold and has been popular with many subsequent authors. However, direct measurements of the sizes of helical eddies often do not agree with the observed dune spacing. At the moment, there is no universally agreed process that controls the lateral spacing of linear dunes.

Star dunes. Where no predominant wind direction exists, the direction of sand transport shifts constantly and sand piles up into huge, complex heaps known as star dunes. Strictly speaking, a star dune must have at least three arms containing slip faces (Figure 9.12e). The tallest accumulations of sand on Earth are star dunes, with heights that reach up to 4 km, but are more typically a few hundred meters high. Star dunes are spaced at distances equal to about ten times their height, a common dimensional ratio among dunes.

Reversing dunes. In areas where the wind regularly reverses direction, slip face directions may alternate from one side of a dune to another. Strictly speaking, reversing features occur at the crest of larger accumulations of sand: Because the reversal timescale is short, the volume of the reversing portion of a dune cannot be large. Nevertheless, the crests of reversing dunes are distinctively symmetrical, with triangular profiles (Figure 9.12f). Barchan dunes will also tolerate an exact reversal of the wind direction without major effects on their morphology, so long as one direction predominates.

Many other dune forms and features have been identified and named. Lunettes, for example, develop around the downwind margins of playa lakes and are often composed

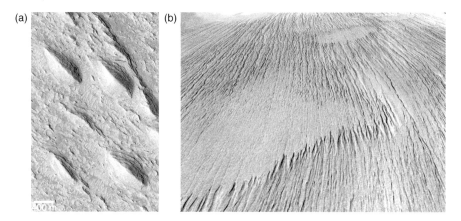

Figure 9.14 Yardangs on Mars. (a) Closeup view of yardangs showing the classic "inverted boat hull" morphology, located near 1° N, 214.4° W. Scale bar at the lower left is 400 m. MOC image PIA04677 NASA/JPL/Malin Space Science Systems. (b) Panoramic view of a soft rock unit dissected into yardangs south of Olympus Mons. The three flat regions in the fore-, middle-, and background measure about 17 x 9 km in this oblique view. Mars Express HRSC image, orbit 143, ESA/DLR/FU Berlin. (G. Neukum). See also color plate section.

of fine-grained material, even clay, whose particles are normally too small to saltate as individuals. However, when cemented by salts or after wetting, these fine particles become weakly cemented into aggregates that can be moved short distances by the wind.

Climbing dunes, as opposed to falling dunes, are special forms that develop on steep slopes upwind of resistant ridges over which the sand eventually passes, to accumulate into falling dunes on the lee side. For more detail on the range of types recognized, the reader is referred to the specialized literature, such as the book by Lancaster (1995).

9.3.4 Yardangs and deflation

Dunes are depositional landforms in which material accumulates. Yardangs are the characteristic erosional form created by the wind (Blackwelder, 1934). On Earth, yardangs are a minor geomorphic curiosity, seldom seen by anyone but travelers in arid regions. Where they occur, they are usually small, only fractions of a meter to at most tens of meters high and up to 20 m apart. They usually develop in weakly cohesive rocks such as clay, silt, or weakly cemented sandstones. Yardangs appear as long, linear ridges with streamlined upwind edges whose shape many observers have compared to overturned boat hulls (Figure 9.14a). Saltating sand erodes the troughs between yardangs only along their floors, so the ridges are often undercut along their sides and upwind edges. The troughs are typically flat-floored or U-shaped.

Whereas yardangs are only minor features on Earth, they dominate some landscapes Mars (Figure 9.14b). Weak, perhaps pyroclastic, rock layers near the Martian equator are deeply entrenched by linear grooves aligned with the prevailing wind. The edges of mesas are often deeply etched by yardangs and crater ejecta deposits are dissected into linear ridges. Yardangs on Mars are enormous by terrestrial standards, tens of kilometers long and separated by troughs averaging 200 m wide.

Yardangs may also be extensive on Venus. Although difficult to identify unambiguously from Magellan radar images, Greeley *et al.* (1997) described an extensive field of yardangs about 500 km southeast of Mead crater. These yardangs, if that is what they are, average 25 km long by 0.5 km wide and are spaced from 0.5 to 2 km apart.

Dust-raising winds may also erode shallow, closed depressions, in strong contrast to fluvial processes, which tend to fill in closed basins. Called deflation hollows or pans on Earth, such depressions range from "buffalo wallows" only a few tens of meters across to the Qattara Depression in Egypt, a shallow, crescent-shaped basin more than 100 km wide and 200 km long that has been excavated 134 m below sea level. The Qattara Depression may have formed as recently as the past 2 Myr, suggesting a very high rate of eolian erosion, even by terrestrial standards. Deflation hollows have not yet been definitively identified on other planets, perhaps due to the difficulty of ruling out other processes that create shallow depressions. Any persistent source of eolian dust must eventually evolve into a shallow basin of this kind.

9.3.5 Wind streaks

Images of wind streaks returned by the Mariner 9 orbiter provided the first evidence for wind action on Mars (aside from the global dust storms themselves). Seen as variable albedo patterns in the lee of obstacles such as crater rims, Martian wind streaks puzzled early observers: some were bright and some were dark. Sometimes both bright and dark streaks formed around the same feature (Figure 9.15). Martian wind streaks change with time: They are especially likely to have changed after dust storms. We now believe that most Martian sand materials are dark, of mainly basaltic composition, whereas the ubiquitous red or orange dust is bright, a simple fact that resolves many puzzles involving albedo markings on Mars.

Wind streaks have no topographic expression. They result from the deposition or removal of thin deposits of wind-blown material whose color or brightness contrasts with the underlying surface. They are common on Earth as well as Mars. They appear often in snow-covered terrains, but more permanent streaks have formed behind obstacles such as cinder cones. The well-studied wind streak downwind of Amboy Crater in southern California is a good example that can be readily seen in Google Earth images. This particular streak is more permanent and appears to result from differential trapping of light-colored sand on dark desert pavements near the cinder cone.

Both bright and dark wind streaks were observed downwind of Venusian craters by Magellan's radar. In radar images, "bright" means "rough" (at the radar wavelength, 12.6 cm) and may, in this case, imply removal of fines by enhanced turbulence downwind of the crater. Dark streaks imply deposition of smooth material. Analysis of Magellan images revealed almost 6000 wind streaks (Greeley *et al.*, 1997) that were used to map global atmospheric circulation patterns.

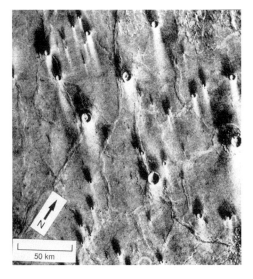

Figure 9.15 Bright and dark wind streaks are simultaneously present around craters on Mars. These presumably formed during different wind regimes and are the result of the differential vulnerability of dark basaltic sand and bright dust to wind erosion or deposition. Image is located at 28° S, 245° W in Hesperia Planum. Viking Orbiter frame 553A54. NASA/JPL.

9.3.6 Transient phenomena

Patterns resulting from eolian deposition can often be observed on the surfaces of planets. Global dust storms on Mars redistribute dust on an almost annual basis, obscuring areas previously cleared of dust by impact blast waves or dust devils. Dust-devil tracks a few tens of meters wide and many kilometers long criss-cross dusty plains, appearing dark after the removal of the bright dust.

Volcanic eruptions create plumes of airborne volcanic dust that settles out downwind for distances that may reach thousands of kilometers on Earth. Close to the volcano where deposits of volcanic tephra are thick, winds may heap the fine material into dunes.

The ejecta from impact craters also interact with the atmosphere to create deposits similar to those of volcanoes, although organized somewhat differently. Notable impact-related features are the dark crater parabolas on Venus (Figure 9.16). Radar-dark, parabola-shaped features surround about 60 fresh Venusian craters. With blunt ends facing almost due east, parabolas stretch a few hundred to more than 2000 km from east to west and extend up to 1000 km from north to south at their widest extremities. These features are successfully explained as the fine-grained ejecta deposits of impact craters (Schaller and Melosh, 1998; Vervack and Melosh, 1992). Although the impact throws out ejecta basically symmetrically, most of the distal ejecta fall back into the atmosphere after a short ballistic flight. After re-entry it drifts with the wind as it settles toward the surface. The upper atmosphere of Venus flows steadily from east to west at speeds up to 60 m/s at 50 km altitude. The ejecta drifts downwind, traveling westward as it settles. Because the ejecta falling closer to the

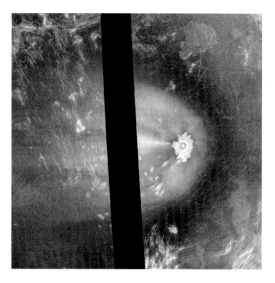

Figure 9.16 Adivar Crater on Venus and its parabolic ejecta deposits. Adivar is 30 km in diameter, located at 9° N, 76° E, and is surrounded by an inner radar-bright (rough) parabola, which is in turn surrounded by a radar-dark (smooth) parabola. This image is 674 × 674 km, showing the enormous extent of these deposits, which probably represent coarse ejecta close to the crater and finer material that fell back into the atmosphere farther away and was blown westward by strong prevailing winds in the upper atmosphere. Portion of Magellan Radar Image C2-MIDR.00N080;1. NASA/JPL.

crater is coarser than that falling farther away, the closest ejecta is deposited first, while the more distal ejecta blows farther downwind, creating the parabolic shape. The details of the parabolas' shapes even permits the particle size to be inferred and yields a general relation between ejecta particle size, ejection velocity, and crater size for large impact craters. This information could probably not be gained in any other way, short of arranging for a series of large impacts.

Further reading

The fundamentals of the transport of sand and dust by wind are well treated by, obviously, Bagnold (1941) as well as in a later USGS report (Bagnold, 1966). A more modern treatment from the planetary perspective can be found in Greeley and Iversen (1985). Dust deposition and transport is the subject of Pye (1987), while the standard work (which includes many spectacular images) is McKee (1979). A more modern treatment of dunes is given by Lancaster (1995). Wind erosion and deposition by wind on the Earth is well described in an older classic (Mabbutt, 1977). The "planetary connection" is made by a series of long papers: For Venus, see Greeley *et al.* (1997); for Mars see Carr (2006). Our knowledge of Titan is relatively recent, but a good summary of what we do know is in Jaumann *et al.* (2009).

Exercises

9.1 The looming clouds?

Clouds on Earth are composed of water droplets whose density, about 1000 kg/m^3, is much larger than the density of air (about 1.3 kg/m^3 at the surface of the Earth). So why don't the clouds simply fall out of the sky? Discuss this problem in the light of Stokes' law for the rate of fall of small particles. You may find it useful to know that the average size of cloud droplets is about 10 μm.

9.2 Blowing in the wind or falling like a brick?

Small meteorites (weighing a few milligrams) that enter the Earth's atmosphere typically evaporate at altitudes near 70 km as they flare up briefly as "shooting stars." The refractory silicates that compose most of these small meteorites then condense into "smoke" particles with sizes that range from 1 to 3 nm. An estimated total of about 100 tons/day of such material currently falls into the Earth's atmosphere. Estimate how long particles of this size might stay aloft in the atmosphere. Compare this to the time that a large, solid, meteorite fragment, 20 cm in diameter, takes to fall to the surface.

9.3 Do the math!

Derive Equation (9.9) in the text for the threshold velocity of the wind to just move a sand grain of diameter d lying on the surface.

9.4 Sand dunes on Triton?

Triton, Neptune's largest moon, possesses a very thin atmosphere that is composed mainly of N$_2$ gas at a chilly 38 K. Nevertheless, geysers spout plumes 8 km high into the atmosphere. Suppose that loose "sand" grains of ice (perhaps from impact ejecta) lie on the surface. How fast do the winds of Triton have to blow to just entrain such ice grains, and how big are these grains? Compute both the minimum friction velocity needed to loft these grains and the minimum wind speed 1 m above the surface. Compare this velocity to the speed of sound in Triton's atmosphere. What can you conclude about the probability of finding "sand" dunes on Triton if it is visited by a spacecraft with an imaging system capable of resolving such features? Facts that you may find useful: The viscosity of nitrogen gas at 38 K is about 2.2×10^{-6} Pa-s and its density at Triton's atmospheric pressure of 1.5 Pa is 1.3×10^{-4} kg/m^3. The acceleration of gravity at the surface of Triton is 0.78 m/s^2.

9.5 The marching dunes

The Algodones Dunes along the eastern margin of southern California's Imperial Valley exhibit slip faces up to 40 m high. Large complex transverse dunes alternate with interdune flats that are typically about 500 m to 1 km wide. The dunes themselves are about the same width as the interdune flats. R. P. Sharp (1979) measured the rate of the downwind migration of these dunes to be about 35 to 40 cm/yr. From this data estimate the average sand flux along the dune chain and the lifetime of an individual dune in the field. The California Department of Transportation is contemplating the construction of a new highway about 50 m downwind of a small outlier dune with a slip face 5 m high. Discuss the wisdom of this plan.

10

Water

All indurated rocks and most earths are bound together by a force of cohesion which must be overcome before they can be divided and removed. The natural processes by which the division and removal are accomplished make up erosion. They are called disintegration and transportation. Transportation is chiefly performed by running water.

...A portion of the water of rains flows over the surface and is quickly gathered into streams. A second portion is absorbed by the earth or rock on which it falls, and after a slow underground circulation reissues in springs. Both transport the products of weathering, the latter carrying dissolved minerals and the former chiefly undissolved.

G. K. Gilbert, *Geology of the Henry Mountains* (1880)

The Earth's surface is dominated by landforms that have been carved by running water. Fluvial landforms are usually apparent in even the driest deserts. Running water is such an effective agent of erosion because of its density: Almost 1000 times denser than air, it exerts greater shear stress, buoys the weight of entrained particles, and is driven more forcefully by gravity than an equivalent volume of air.

Where rainfall is possible, even small amounts of water trump any other agent of erosion. Although rain is not possible on Mars under current conditions, its landscape plainly bears the scars of rainfall in the distant past. Some things are different: Mars has seen enormous floods that are comparable to the largest floods known on Earth, and groundwater sapping plays (or played) a far larger role than it does on our planet. We still do not understand how Mars' floods originated or how such large volumes of water came to be suddenly released. Nevertheless, once released, the water followed the same laws as water on Earth and produced landforms for which terrestrial analogs exist.

Recent exploration of Titan reveals that active fluvial landscapes are not unique to Earth: Although the fluid on Titan is liquid methane at 95 K and the rocks are hard-frozen water ice, Titan's landscapes of stream valleys, riverbeds, and enormous lakes are eerily familiar. The materials are different but the processes are similar and can be analyzed by the same methods as the Earth's water-carved surface. In the following chapter, as in its title, the word "water" is used to keep the discussion focused, but most of the concepts apply to any fluid interacting with a solid substrate.

382

As we explore fluvial processes in this chapter we will emphasize general laws and relationships that hold where any fluid interacts with a solid surface. Whether on Earth, Mars, or Titan, similar physical laws lead to similar landforms. Process trumps contingency here.

10.1 "Hydrologic" cycles

It is a familiar concept that rain falling on the land suffers a number of possible fates, returning eventually to the sea or evaporating into the air to continue the cycle that made rivers seem eternal to our ancestors. Rain that falls on the surface can infiltrate into the ground, where it joins the volume of groundwater beneath the surface, evaporate immediately back into the air, or run over the surface to collect into streams and possibly large bodies of standing water. On Earth, transpiration by plants is also an important factor, but plants, so far as we know, play no role on either Mars or Titan.

The relative quantities of these factors, along with the movement of water underground, are the subject matter of the field of *hydrology*. We will be mainly interested in that part of the cycle which results in runoff and can, thus, perform work on the landscape.

10.1.1 Time, flow, and chance

The climates of the Earth are usually characterized by their temperature and annual rainfall. The annual rainfall statistic specifies how much water arrives as rain in an "average" year. However, as everyone knows, the amount of rainfall in a given year can fluctuate greatly. Moreover, the pattern in which the rainfall arrives is also important – clearly, 30 cm of rain per year that arrives in several large, brief thundershowers is more effective in eroding the landscape than 30 cm of rain arriving in gentle, continuous showers.

In general, most geologic work is done by large, relatively infrequent events. Most mountain streams run clear over large boulders that they obviously cannot move. Such boulders are transported only during extreme rains when the clear rivulet may become a muddy, churning torrent for a few hours. The level of water in streams may reach as high as their banks only once a year, but that annual flood determines the form of their channel. Much less frequent floods may sculpt their higher floodplains.

It is important to realize the significance of fairly rare events in surface processes. Consider the example shown in Figure 10.1. The upper panel depicts a typical relationship between streamflow (which is proportional to the friction velocity) and the amount of sediment moved (e.g. Equation (9.15) for wind-blown sand: Sediment in streams obeys a similar equation). The middle panel is a probability distribution showing the likelihood of a given streamflow. It peaks at the mean discharge. Discharges both lower and higher than average are less likely. The curve becomes increasingly less certain as the discharge increases: Very rare events that are, nevertheless, possible may never have been observed within the time frame over which records are available. The product of the two curves is shown in the lower panel. This yields the probability of a given sediment discharge.

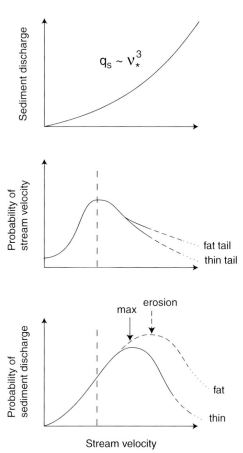

Figure 10.1 The amount of geological work done by flowing water depends on the cube of the velocity, as shown in the top panel. The probability of finding a given velocity peaks at some stream velocity between zero and infinity, as shown in the middle panel, although the high-velocity behavior is somewhat uncertain, depending on whether the probability falls exponentially (thin tail scenario) or as a power-law function of the velocity (fat tail scenario). The lower panel shows that the product, the probability of a given sediment discharge, peaks at a higher stream velocity than the maximum velocity.

Because the amount of sediment moved by a given flood is a strongly increasing function of the streamflow, the peak of the lower curve is displaced away from the mean flow, toward a higher than average streamflow. Thus, the event that is most likely to move a large amount of sediment, and so, cause a large amount of geologic change, is not the annual average but a rare event that corresponds to an unusual flood.

The extreme end of the probability curve is the subject of major debates at the moment. Because it corresponds to very rare events, there is not a lot of data to tie down the precise form of the right-hand end of the curve, and these uncertainties are greatly magnified by the large amount of sediment moved in extreme events. This debate is connected with the general "fat tail" problem that haunts large, rare events. If one assumes that the shape of the probability curve is Gaussian, the probability of events far from the mean drops exponentially fast with distance from the mean (a "thin tail"). However, if the curve falls as a power law (a "fat tail"), then the probability drops much more slowly. In an extreme case, if the curve falls as a power law less steep than $1/v^3$, where v is the stream velocity, then there is no maximum at all and erosion is dominated by the most rare, but largest events! Water floods do not seem to work this way, although meteorite impact events do, because there is no meaningful upper limit to the size of a potential impactor.

As an example, the mean annual discharge of the Columbia River is about 7800 m^3/s. It fluctuates during an "average" year from a low of about 2800 m^3/s to a high of about 14 000 m^3/s in the late spring. However, there are wide fluctuations about these means: The lowest flow ever recorded is about 1000 m^3/s, while the peak is about 34 000 m^3/s. Nothing in these records, however, prepares one for the fact that between 15 000 and 13 000 yr ago the Columbia River drainage suffered a series of catastrophic floods that increased the discharge to about 10^7 m^3/s for a few days. These floods were associated with the collapse of ice dams holding back the waters of Glacial Lake Missoula. The outpouring of water from these floods scarred the surface of eastern Washington, digging huge channels, transporting blocks of basalt tens of meters across, and leaving enormous deposits of gravel in a series of catastrophic floods that did more geologic work in a few days than millions of years of normal erosion could accomplish (Baker and Nummedal, 1978).

Geologist Gene Shoemaker reached a similar conclusion in 1968 after he repeated Powell's historic trip down the Colorado River through the Grand Canyon. Shoemaker reoccupied 150 sites that Powell's photographers had recorded in 1871–1872 and compared the images of the canyon taken almost 100 yr apart. In looking at these images (Stephens and Shoemaker, 1987), one is struck by the fact that, apart from predictable changes in vegetation, very few differences are seen in most images. Individual rocks can be recognized in the same places as 97 yr previously. However, in a small number of comparisons, the scene has changed utterly: almost nothing has remained the same. These drastic changes are due to unusual floods originating in side canyons of the river that cleared away massive heaps of rock and deposited new piles of rock debris from further upstream.

The lesson Shoemaker learned about stream erosion is that it is episodic: little may change for a long time until an unusual event comes along that makes sudden, large alterations. The fluvial landscape does not change gradually, but evolves in a series of jumps whose effects accumulate over time.

This kind of evolution is called catastrophic: The strict definition is that a catastrophe is a large event that causes more change than all smaller events combined. Floods fulfill this definition within limits: For floods, the curve in the bottom panel of Figure 10.1 eventually turns over (although the Lake Missoula floods do cause some concern about the validity of this claim). This is not the case for meteorite impacts. Similar probability curves have yet to be established for many other processes.

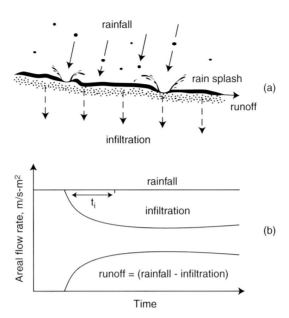

Figure 10.2 When rain begins to fall on a porous surface it initially all soaks into the ground. Once the capacity of the surface to absorb water is saturated, however, the water collects in a thin film on the surface and runs off. Panel (a) shows this process schematically, as well as the disturbance of the granular surface by the small impacts of raindrops. Panel (b) indicates the volumetric division of rainfall into infiltrated water and runoff, which reaches a steady state after an infiltration relaxation time t_i.

10.1.2 Rainfall: infiltration and runoff

Rainfall is ubiquitous on the Earth wherever temperatures permit liquid water to exist, although some regions may receive more, others less. Rain also seems to have fallen on the ancient cratered highlands of Mars to create valley networks and, although it has not been observed directly, must be occurring in near-current times on Titan (except that it is methane drops that fall there, not water).

The first drops of water that fall on the ground surface are trapped in small surface depressions and irregularities – no runoff is produced. However, as more rain falls, the capacity for depression storage is usually exceeded after some period of time. This moment is easily observed while watching rain fall on a bare soil surface, for as soon as a connected sheet of water forms the entire surface seems to suddenly glisten and water begins to flow down whatever slope is present. Some water soaks into the ground: The amount depends upon the permeability of the soil, among other factors. The remainder begins to flow along the surface. This runoff is responsible for most geologic work (Figure 10.2a).

Runoff is equal to the difference between the rate at which rainwater arrives on the surface and the rate at which it infiltrates into the surface (Figure 10.2b). If the infiltration capacity is high enough or the rain gentle enough, there may be no runoff at all. However, as the rainfall intensity increases, the amount of infiltration remains approximately constant while the runoff necessarily increases.

The process of infiltration has received much study, and there are entire books devoted to it (Smith *et al.*, 2002). A great deal of this study has practical ends: It is often considered desirable to increase infiltration to both reduce erosion and increase groundwater supplies. Because infiltration involves only the partly saturated flow of water into loose soil that already contains air, the process is very complex and depends on many factors. Chief among them is the intensity of the rainfall (volume of water per unit time per unit area) and the duration of the rain, along with its integrated volume. Raindrops falling on a bare surface usually beat down and compact the soil, reducing the infiltration capacity as time goes on. Equilibrium between rainfall and infiltration is reached after a relaxation time t_i, after which the rate of infiltration becomes approximately constant, whatever the rainfall intensity or duration might be.

One factor that most geologists take for granted is the ability of water to "wet" silicate minerals. Capillary forces play an important role in infiltration and the surface contact energy between water and minerals is crucial to the ability of water to flow into the pores between mineral grains in the soil. But what about methane on Titan? Does liquid methane flow into the pores between grains of cold ice? If methane beads up on the ice surface like water on a well-waxed car, Titan might not have a subsurface hydrologic cycle at all. Fortunately, recent experiments (Sotin *et al.*, 2009) show that not only does liquid methane wet ice, it soaks into the tiniest cracks and pores on contact. Quantitative data on surface energy is still lacking, but the viscosity of liquid methane is only about 10% that of water (Table 9.2), so that liquid methane may readily infiltrate into the Titanian surface.

On Earth, the infiltration capacity depends upon the condition of the soil surface. Vegetation plays a big role here, as does the type of soil. Clays, for example, are very impermeable and, thus, have a low infiltration capacity, which promotes runoff and hence surface erosion. Gullying on clay-rich badlands is intense. On the other hand, coarse sands and loose volcanic cinders on the sides of fresh cinder cones are highly permeable and may entirely suppress runoff. A frequent observation is that fresh cinder cones stand for long periods of time without any sign of gullies, in spite of being composed of loose, often cohesionless, volcanic lapilli. However, with time, weathering eventually converts the glassy lapilli to impermeable clay and wind-blown dust settles between the lapilli, filling the pores between them. When this has gone far enough, runoff finally begins and the cone is removed in a geologic instant: In a field of cinder cones it is common to see ungullied fresh cinder cones, but rare to see gullied cones. Instead, one finds lava flows whose original vents lack cinder cones – they have been removed by fluvial erosion. Cinder cones in volcanic fields on Venus are far more abundant than on Earth, presumably because of the greater efficacy of fluvial erosion on our planet.

On Earth, other highly permeable deposits may display a strong resistance to erosion, not because of intrinsic strength but because of their ability to soak up rainfall and prevent runoff. Highly fractured lava plains often have very little runoff and may, thus, be very long-lasting. Even gravel may be highly resistant to erosion for this reason (Rich, 1911).

Water that infiltrates into the soil may percolate down to join the water table, flow laterally to seep out as springs or into streams (meanwhile initiating sapping erosion that may undermine the slopes out of which it flows), or evaporate from the surface, depending upon the climate, geologic structure beneath the surface, and rock permeability. The fate of rainfall is, thus, complex and depends upon many factors, so that simplified models of the type presented here may not be realistic: Much current work in hydrology and geomorphology is focused on overcoming the drawbacks of overly simplistic models.

10.2 Water below the surface

On the Earth, every vacant space in rocks underground is occupied by fluid: water, brine, oil, natural gas, or, very close to the surface, air. Most of these fluids reside within the upper few kilometers of the surface, but significant porosity and permeability may extend to depths of tens of kilometers. When piezometric pressures differ in these fluids they tend to flow from areas of high pressure to low pressure. Underground flows can be as important as flows above ground in moving dissolved substances, contaminants, and potential resources. On Mars, where surface water is not stable under the present climate, a global underground hydrologic cycle has been proposed to explain sapping and catastrophic outbursts of water on the surface. Seeping brines may account for some of the modern-day gullies on crater walls.

An understanding of how underground water moves and interacts with the surface was first achieved in 1856 by French engineer Henry Darcy (1803–1858), who investigated the water supply of Dijon, France. Percolation of fluids through fractured rocks has been vigorously studied ever since, driven largely by economic concerns with water supply, oil exploration, and, at present, pollution remediation. In more modern times, M. King Hubbert (1940) made vital contributions to this field.

10.2.1 The water table: the piezometric surface

When a well is drilled into the surface of the Earth, water is usually not immediately encountered. Although some water is often found adhering to mineral grains, the zone in which all the pores are filled with water is only reached at some depth. The top of this saturated zone is known as the water table. It is defined by the level at which water stands in an open well. Above this fully saturated zone is a thin, partially saturated layer (the "capillary fringe") in which capillary forces raise water a small distance against gravity.

If ground water is stagnant and suffers neither gain nor loss, the water table beneath the surface is level, conforming to a surface of constant gravitational potential (Figure 10.3a). However, when ground water is recharged by infiltration or drained by discharge, its upper surface becomes complex, crudely reflecting the overlying topography in regions where the permeability is uniform (Figure 10.3b).

To describe how water moves underground it is necessary to define the force that makes it move. Simply citing a pressure is not enough, for the pressure varies with depth in a

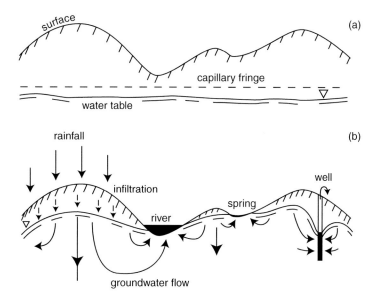

Figure 10.3 If there is no recharge of the groundwater, as in panel (a), the water table with its capillary fringe relaxes to parallel an equipotential surface, whatever the topographic complexity above. When rainfall recharges the groundwater, as in (b), the water table tends to mimic the topographic variations. Springs and streams of water emerge from the ground where the water table intersects the surface. Extracting water from a well, shown on the right of this figure, locally draws down the water table and creates a "drawdown cone."

stagnant fluid because of the weight of the fluid. The most meaningful measure of the tendency of a fluid to move is its total energy per unit mass, which is the sum of its gravitational potential energy, kinetic energy, and the work done by the pressure on the fluid volume. Energy is defined only up to an arbitrary constant, Φ_0. Assuming that we can treat water as incompressible, the total energy of a unit mass of water underground is:

$$\Phi - \Phi_0 = gz + \frac{P}{\rho} + \frac{v^2}{2} \qquad (10.1)$$

where z is elevation above some datum (measured positive upwards), P is the pressure, ρ the (constant) density, and v the fluid velocity. Because water underground usually moves slowly, the last term can be dropped and the energy equation is written:

$$\Phi - \Phi_0 = gz + \frac{P}{\rho}. \qquad (10.2)$$

In a stagnant fluid Φ is constant and the pressure $P = P_0 - \rho gz$, increasing as z becomes more negative (deeper).

Hydrologists usually do not deal directly with fluid energy, but instead work with the *hydraulic head*, h, which is the level above a datum to which a fluid would rise in a tube inserted at the point of interest. By definition, $h = (\Phi - \Phi_0)/g$. Variations in head express variations in the potential energy capable of driving fluid flow. The head is constant if the fluid is stagnant, but if the head varies laterally, the fluid tends to flow from regions with higher head to lower head.

A piezometric surface is the hypothetical surface to which water would rise in a well drilled into an aquifer. It coincides with the water table if the aquifer is unconfined, that is, not closed in by impermeable layers. However, if impermeable layers do exist the piezometric surface can be very different from the surface topography.

When two fluids are present in the same aquifer, for example, oil and water or fresh water and brine, their potential energies are different even at the same elevation and pressure because their densities in Equation (10.2) are different. Thus, oil will displace water downward because it is less dense and brine will displace fresh water upward because it is more dense. The density of a fluid may also vary for other reasons, such as the content of a solute or temperature. Potentials thus exist for driving convective motions of the fluid due to either compositional or thermal differences. There are numerous specialized treatments for these cases and we do not consider them further in this book.

10.2.2 Percolation flow

It is important to distinguish the porosity of a rock from its permeability. Porosity measures the fraction of void space in a rock. It is usually designated by the symbol ϕ. Porosity ranges from 0 (a perfectly dense rock) to 1 (a hypothetical "rock" that is all void). If the density of a non-porous rock is ρ_0, then the density of a rock with porosity ϕ is $(1-\phi)\,\rho_0$. Permeability, on the other hand, measures how readily a fluid can move through a rock. A rock must have some porosity to be permeable, but it is possible for a rock to be porous without having any permeability at all – it depends on how well connected the pores are.

Henry Darcy first determined the relation that is now named after him. He measured the rate at which water flows through a cylinder full of sand as a function of the difference in head between the water at its entrance and exit and wrote an equation very similar to Fourier's law for heat conduction:

$$\mathbf{Q} = -\frac{k\,\rho}{\eta}\nabla\Phi = -\frac{k\,\rho\,g}{\eta}\nabla h = -K\,\nabla h \qquad (10.3)$$

where $\mathbf{Q}$ is the vector volume flux of water (which has the same units as velocity), η is its viscosity, and k is the permeability, which has dimensions of (length)^2. Permeability k is *defined* by the rate at which a fluid (water) moves through a rock. The permeability is, thus, a kind of fluid conductivity. $K = k\rho g/\eta$ is known as the hydraulic conductivity. The equations describing the percolation of a fluid underground can be put into the same form as the heat conduction equation, which immediately makes a host of solutions to the heat conduction equation applicable to the problem of fluid flow (Bear, 1988; Hubbert, 1940).

The magnitude Q of the volume flux of water in Equation (10.3) is not equal to the velocity of the water in the rock, despite having the same units. Sometimes called the "Darcy velocity," it is the average rate at which a volume of water flows through the rock. However, because water fills only a fraction of the total rock's volume, the water must move faster than Q to deliver the observed flux. If all of the porosity ϕ contributed to the permeability, the actual velocity of water in the rock would equal Q/ϕ. In practice, not all of the porosity contributes to the permeability and the local velocity may be much higher. The local velocity may become sufficiently high in some circumstances that the flow becomes turbulent, in which case the permeability depends on the flow rate in a complicated manner.

Equation (10.3) can be elaborated to describe the unsteady flow of fluids as well as co-transport of heat and dissolved substances. Combined with other relations describing the storage of fluid in the rock and the rate of chemical reactions, many geologic problems can be related to the transport of water, or other fluids, through fractured rock (Lichtner *et al.*, 1996). Problems of this type are important for understanding the origin of ore bodies, underground motion of contaminants, and the transformation of sediments into rock (diagenesis and metamorphism).

Computation of the permeability of a given rock is a very complex affair and many different equations for estimating the permeability of a rock exist. Indeed, much of the uncertainty in the field of hydrology centers about permeability and its distribution. Permeability depends strongly on the size and distribution of the connected passages through rock. A simple model in which the rock is filled with a cubic array of tubes of diameter δ spaced at distances b apart can be solved to give a permeability (Turcotte and Schubert, 2002):

$$k = \frac{\pi}{128} \frac{\delta^4}{b^2}. \tag{10.4}$$

Note the strong dependence of permeability on the size of the narrowest passages through the rock. This is a typical behavior: Permeability is a very strong function of the grain size of the rock. Thus, it is large for coarse sands, but becomes very small for fine-grained silts and clays. Table 10.1 lists some "typical" values for the permeability of various rock types and indicates the enormous range of permeability found in nature. Note that permeability is commonly listed in the more convenient units of "darcys": 1 darcy = 9.8697×10^{-13} m^2.

It is very important to note that, although most of the values for permeability in Table 10.1 apply to "intact" rocks, the actual permeability of a rock mass may be very different (higher) because of fractures or even macroscopic cavities. Permeability is usually measured for small specimens that are often selected for their integrity. A rock mass in nature, however, is inevitably cut by large numbers of joints and faults that can have a major effect on the permeability of the mass as a whole: open joints allow fluids to flow along them easily, while faults may be lined with fine-grained gouges that inhibit fluid motion. Extrapolations of permeability from a small sample to an entire rock mass may, thus, give highly misleading results.

Where possible, permeability is measured directly in the field. The first such method was developed by Charles Theis (1935) and involves measuring the level of water in observation wells adjacent to a test well from which water is actively pumped. Other methods involving suddenly displacing the water in a single well by a cylindrical "slug" and

Table 10.1 *Permeability of rocks*

Rock type	Permeability k (m^2)
Gravel	10^{-9}–10^{-7}
Loose sand	10^{-11}–10^{-9}
Permeable basalt[a]	10^{-13}–10^{-8}
Fractured crystalline rock[a]	10^{-14}–10^{-11}
Sandstone	10^{-16}–10^{-12}
Limestone	10^{-18}–10^{-16}
Intact granite	10^{-20}–10^{-18}

After Turcotte and Schubert (2002), Table 9–1.
[a] Lichtner *et al.* (1996)

measuring the recovery of the water level, or by injecting water between pressurized packers in a borehole.

The porosity and permeability of rocks generally decrease with increasing depth. Near the surface, wide variations of permeability are common, which lends hydrology much of its variety and complexity. However, with increasing depth, overburden pressure rises and pores are crushed closed. Temperature also increases with depth and this promotes pore closure by viscous creep, so that porosity and permeability in terrestrial rocks are essentially absent at depths greater than a few kilometers in most places.

On the Moon and planets such as Mercury and Mars, large ancient impacts initially fractured the surface rocks to great depth, but overburden pressure has closed most pores at a depth that depends on the resistance of the rock to crushing. Seismic measurements on the Moon suggest that most lunar porosity disappears at a depth of about 20 km, or at a pressure of about 0.1 GPa. Under Martian surface gravity this implies that most porosity is crushed out at a depth of about 6.5 km. The fractured Martian regolith is estimated to be capable of holding the surface equivalent of between 0.5 and 1.5 km of water.

10.2.3 Springs and sapping

When water percolating through rock reaches the surface it leaves the fractures and pores of the rock and flows out onto the surface, either joining a saturated body of water already there (a streambed, lake, or the ocean), or flowing out onto the land surface as a spring. Because of the low permeability typical of rocks, the flow of water underground is usually slow and springs may continue to flow long after rains have ceased (see Box 10.1 to estimate how long a streamflow may continue without recharge). Water underground constitutes a reservoir that buffers changes in surface runoff. The total volume of groundwater on the Earth is only about 1.7% of the total surface water (this includes the oceans), but 30% of the Earth's fresh water resides underground. Other planets may have substantial inventories of groundwater: The principal uncertainty in estimates of how much water is

Box 10.1 **How long can streams flow after the rain stops?**

The question of how streams can continue to flow even when they are not being fed directly by rainfall is as old as the science of hydrology. Pierre Perrault (1608–1680) was the first to show that the rain- and snowfall in the basin of the Seine River is about six times larger than the river's discharge and could, thus, more than account for the flow from all of the springs and streams in the region (he neglected evaporation and transpiration by vegetation, which balances the net water supply).

Some streams, particularly those in arid climates, do not flow unless they have been recently filled by rain. The water in such streams rapidly infiltrates and percolates downward to a deep water table. In more humid climates, however, streams may continue to flow for weeks or months between rains. Such streams are fed by groundwater seeping into their beds. Because groundwater moves slowly, it may take a considerable period of time for groundwater to move from the area where it infiltrates to its emergence in the bed of a stream, during which time the stream continues to flow.

Using Darcy's law for a uniform permeability, we can estimate the relaxation time over which an elevated water table continues to feed a stream or spring from which the water drains. Let h be the height of the water table above its outflow point, located a horizontal distance L from the groundwater divide (Figure B10.1.1). The total volume of water contained above the outflow point is of order $hL\phi$ per unit length along the stream, where ϕ is the permeability of the aquifer. If the width of the zone along which water seeps out is w, and the groundwater discharge is Q per unit area, then water is discharged from the groundwater reservoir at a rate Qw per unit length of the stream (perpendicular to the plane of Figure 10.B1.1). The rate at which the height of the water table declines is thus $\dot{h}L\phi = Qw$, from conservation of volume. The discharge Q is, thus, related to the height of the water table (which is equal to the head in this case) by $Q = \dot{h}L\phi/w$. This expression can now be inserted into Darcy's Equation (10.3) to give:

$$Q = \frac{\dot{h}L\phi}{w} = -\frac{k\rho g}{\eta}\frac{h}{L} \qquad (B10.1.1)$$

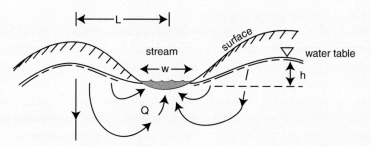

Figure B10.1.1 The capacity of subsurface water to support a surface streamflow depends on the quantity of water available, measured by the height of the water table h above the elevation of the stream and the horizontal extent of the water supply, L. This water flows to the stream, discharging a volume flux Q over a stream of width w. The rate of flow is regulated by the average permeability of the rock beneath the surface.

Box 10.1 (**cont.**)

where the gradient ∇h has been approximated by h/L. This equation can be rearranged into a first-order differential equation for the height of the water table h:

$$\dot{h} = -\left(\frac{k\rho g w}{\eta \phi L^2}\right) h. \tag{B10.1.2}$$

The solution to an equation of this type is well known:

$$h = h_0 e^{-t/\tau_R} \tag{B10.1.3}$$

where the *relaxation time*, τ_R, is given by:

$$\tau_R = \frac{\eta \phi L^2}{k \rho g w}. \tag{B10.1.4}$$

Inserting some typical values, $L = 1$ km, $w = 10$ m, $\phi = 0.1$, $k = 10^{-10}$ m² (sand), and the density and viscosity of water at 20°C, we find a relaxation time of about 4 months. It, thus, appears that even for relatively permeable aquifers like sand, the time for discharge of groundwater may be a large fraction of a year and streams may continue to flow even after months of drought. Conversely, recharge of a depleted aquifer may take a similarly long period. The timescales for groundwater flow may be very long because of the small permeability of many rocks that form aquifers.

present on Mars is its subsurface inventory of ice and liquid water. Ice appears to lie within tens of centimeters of the surface at high latitudes, suggesting that much of Mars' crustal porosity may be saturated with water. Similarly, the volume of methane observed in Titan's lakes might be only a fraction of the total residing below the surface. The Huygens lander detected methane that evaporated from the regolith beneath the warm lander, suggesting a methane table just below the surface of the landing site.

Water flowing from springs may undermine the mechanical stability of the surface. Just beneath the surface, the exit flow creates a pressure gradient that is connected to the fluid discharge by Darcy's law (10.3). This extra pressure adds to the pore pressure in the rock and may greatly weaken it, for reasons described in Section 8.2. In addition, the flow of water through the surface may carry fine-grained silt and clay out of the rock matrix, further weakening it. This selective removal of fine-grained constituents or dissolution of grain-binding minerals may disintegrate the rock or soil and excavate tunnels through which the water flows out readily. Known as piping, such tunnels develop in regions of heavy rainfall and the cavities it creates can undermine the soil surface (Douglas, 1977). Piping also plays a major role in undermining dams through which water is seeping. Water seepage may, thus, cause wholesale collapse of the rock face from which it flows. This erosion process is known as sapping.

Sapping may produce large-scale landforms. Such landforms are rare on Earth, where overland flow and stream transport usually cause erosion. However, in restricted locales where a permeable layer is underlain by an impermeable zone, surface runoff is limited and most of the outflow occurs at the interface between the permeable and impermeable layers. This situation is common, for example, on the USA's Colorado Plateau where permeable eolian sandstones overlie clay-rich fluvial formations (Howard *et al.*, 1988). Former inter-dune areas in the sandstone units also create extensive impermeable layers. Springs form everywhere at the contact between permeable and impermeable layers.

Long-continued spring flow at the base of cliffs undermines the overlying rock by dissolving the minerals that cement the sand grains into sandstone. The constant seepage at the base of cliffs also encourages chemical weathering that further weakens the rock. Alcoves form where the cliffs collapse and spring-fed streams carry the fallen debris away. As time passes, the alcoves grow deeper because they serve as drains for the subsurface water flow and, thus, intensify the flow at their heads as they lengthen. Given sufficient time, a canyon system develops that is characterized by stubby, sparsely branched tributaries with steep amphitheater-like headwalls. Canyon de Chelly is a terrestrial example of this kind of development. Many canyons on Mars also appear to have formed by groundwater sapping, of which the Nirgal Vallis is a prime example.

Groundwater sapping is of particular interest for Mars because it does not require overland flow initiated by rainfall. However, excavation of a canyon does require some process that removes collapsed material from the headwall. Streamflow cannot presently occur on the surface of Mars and so even these valleys may be relics from a time when Mars possessed a higher atmospheric pressure.

10.3 Water on the surface

The terrestrial hydrologic cycle is dominated by the flow of water over the Earth's surface. Water falls as rain or snow, of which a portion flows over the surface from high elevations to low. In the process, liquid water entrains solid material, transports it, and eventually deposits it at lower elevations. Because of the prevalence of this process, undrained depressions on the Earth's surface are rare and, where they occur, draw immediate attention as indicators of an unusual geologic situation.

The processes and products of fluvial erosion, transport, and deposition are so familiar to Earth-bound geologists that it comes as a shock to realize that no other body in our Solar System, with the possible exception of Titan, is so completely sculptured by fluvial erosion as the Earth. The "rock cycle" as envisioned by James Hutton is essentially a fluvial cycle, in which rock debris is removed from the land by rainfall, washed to the oceans by streams, and converted back into rock after deep burial by other sediments. No other planet in the Solar System experiences a cycle of this type. Because of the importance of fluvial processes to terrestrial geology, each part of this cycle has been examined in great detail by many researchers. Most texts that deal with surface processes on the Earth devote one or more chapters to each portion of the fluvial cycle. This book attempts to put surface processes into a broader Solar System context, so our coverage of fluvial processes is necessarily more cursory, being compressed into a single chapter. References to more detailed treatments are given at the end of this chapter.

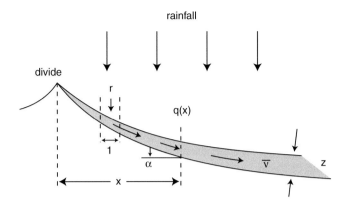

Figure 10.4 Runoff from rainfall on a sloping surface accumulates as an increasing function of distance x from the divide. As the volume $q(x)$ of runoff increases the flow both moves faster and becomes deeper. The flow velocity is regulated by an equation that incorporates both these factors.

10.3.1 Overland flow

When rain falls on a slope, that portion that does not infiltrate or evaporate flows off the surface as runoff. An important characteristic of runoff is that it increases steadily with distance downhill from the crest of the slope. The slope crest is an important location for fluvial processes and in this context it is called a divide because runoff moves in different directions on either side of it. If the rate at which runoff is generated per unit area (projected onto a horizontal surface) is r, measured in units of m³/s per m², or m/s, then the volume flux of water flowing off the slope per meter of contour distance q (m²/s) is given by:

$$q(x) = r\,x = zV \qquad (10.5)$$

where x is distance from the divide measured along the slope gradient, z is the average depth of the flow (measured perpendicular to the slope), and V is the mean velocity of the flow (Figure 10.4). The runoff r is roughly equal to the rate of rainfall minus the infiltration rate. The volume flux of water, or slope discharge, increases with distance from the divide because the amount of water passing each contour of the slope includes all of the runoff generated between that contour and the divide. Because of the increasing volume of water, the potential for erosion also increases downslope. This is one of the distinguishing characteristics of fluvial erosion: The farther one travels from the divide, the greater the ability of water to erode. The consequences of this relationship for slope profiles will be explored in more detail in Section 10.4, but it should already be clear that it leads to the concave-up profile typical of landscapes dominated by fluvial erosion (see Figure 8.6).

Fluvial erosion. The process of sediment entrainment by overland flow was first analyzed by Robert Horton (1875–1945) in a landmark paper published just months before his death and is now known as "Horton overland flow" (Horton, 1945). Horton assumed that runoff begins as a thin sheet of nearly uniform thickness before concentrating into rilles farther downslope. Although thin sheets of water may flow in this way, during heavy rainfall on smooth slopes the flow organizes into waves, now known as roll waves, that develop when the Froude number, $\bar{v}^2/gd$, exceeds approximately 4. Under the wave crests, erosion proceeds faster than expected for a uniform flow. Horton described such waves, but he neglected the process of rain splash, which can mobilize soil particles in the zone that Horton considered to be erosion-free as well as generating turbulence in the flowing sheet of water. Nevertheless, Horton's model agrees qualitatively with observations and is still widely used to explain the onset of water erosion.

Horton supposed that runoff close to a divide would form a sheet too thin to entrain surface material and so a "belt of no erosion" would develop. Such belts are observed in badlands that lack vegetation, although rain splash may actually entrain some material in the flow as well as redistributing it through creep. Carson and Kirkby (1972) point out that natural surfaces are seldom smooth enough to permit an actual sheet of uniform depth to flow over the surface and that the flow very rapidly becomes concentrated into small rilles and channels. When the flow is thin, however, rain splash and surface creep constantly rearrange these channels so that persistent rilles do not form. Farther downslope, with increasing discharge, rilles do form that evolve into drainage networks.

Nevertheless, as Horton described, there is a finite distance between a divide and the location of the first permanent rille. This distance depends upon the infiltration capacity of the surface: It is larger for materials with high infiltration capacity, such as sand and gravel, and small for materials with low infiltration capacity, such as clay or silt.

The capacity of the flowing water to entrain surface material and erode the slope is proportional to the shear stress exerted on the surface, as discussed in Section 9.2.1. The shear stress τ exerted by a sheet of flowing water of depth z on a slope α is given by:

$$\tau = \rho g z \sin \alpha \qquad (10.6)$$

where ρ is the density of the fluid (water). Flowing water begins to erode its substrate when the shear stress exceeds a threshold that depends on the grain size and cohesion of the surface material as well as on the properties of the fluid. Modern research on this topic has centered on the semi-empirical Shields criterion, originally proposed by A. Shields in 1936 (Burr *et al.*, 2006). This criterion can be expressed in a non-dimensional form that is applicable to any planet. The threshold shear stress is given in terms of a non-dimensional parameter θ_t:

$$\tau_t = (\sigma - \rho) g d \theta_t \qquad (10.7)$$

where σ is the density of the entrained material and d is the particle diameter. Defining a friction velocity v_* in terms of the shear stress as in Equation (9.6), and a friction Reynolds number Re_* as in Section 9.2.1, the Shields threshold is given by (Paphitis, 2001):

$$\theta_t = \frac{0.188}{1 + \mathrm{Re}_*} + 0.0475\left(1 - 0.699 e^{-0.015 \mathrm{Re}_*}\right). \qquad (10.8)$$

This expression is good up to $\mathrm{Re}_* \approx 10^5$. This curve lacks the steep upturn at small Re_* described by Bagnold (discussed in Section 9.2.1) and by Shields in his original work. This

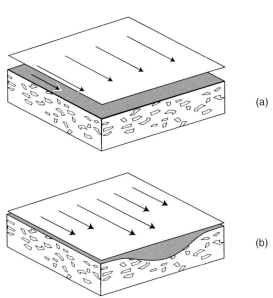

(a)

(b)

Figure 10.5 Overland flow of a thin sheet of water over an erodible bed is not stable. Panel (a) shows a uniform sheet flowing over the surface of the soil. Panel (b) indicates that any small, accidental, increase of depth concentrates the flow, increasing the shear stress under the deeper flow (and at the same time thinning the flow over the adjacent regions), which increases the erosion rate beneath the deeper portion, producing a positive feedback that quickly concentrates the flow into channels.

is reportedly because this equation takes into account rare but intense turbulent fluctuations in the boundary layer. It seems unclear at the moment whether this equation applies to eolian transport as well as water transport: The data on which it is based becomes sparse and somewhat contradictory below Re_* less than about 0.1, although it is stated to be valid down to $Re_* = 0.01$. The experimental material may also have contained a mixture of grain sizes. It seems that this equation should be applied with caution to very small grains until this situation is clarified.

When the threshold stress is exceeded, water begins to entrain material and carry it away, eroding the underlying slope. Erosion is enhanced by any factor that increases the shear stress: Deeper flows and increased slopes both contribute to the erosion rate, in addition to the erodibility of the material.

Rille networks. As runoff moves downslope it collects into channels, and the channels merge into larger channels. This process has been observed both on natural slopes exposed to gullying by vegetation removal and in experimental rainfall plots. Horton described the evolution of rilles into a drainage network in detail. He began by asking the question "Why, then, do rille channels develop?" He supposed that accidental variations in slope topography first concentrate sheet flow into slightly deeper than average proto-channels. Because these concentrations are deeper than average, they also exert greater than average shear stress on their beds and, thus, grow still deeper. That is, a flowing sheet of water of

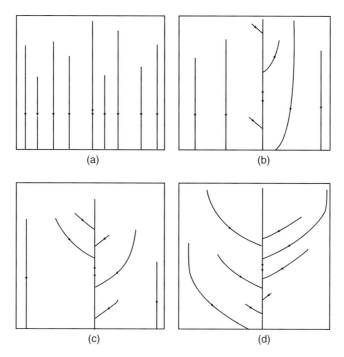

Figure 10.6 Straight, parallel rilles flowing off of an initially uniform slope are themselves unstable. Side branches develop that collect the water from adjacent rilles, deepening the main channel while starving the adjacent rilles. This process, called "cross-grading," operates by the principle that "the rich get richer while the poor get poorer." A dendritic stream network is the result of this kind of natural selection, which can be seen as a variety of stream piracy. Redrawn after Carson and Kirkby (1972, Figure 8.1).

uniform thickness on an erodible bed is unstable: Any slightly deeper pocket erodes faster than the adjacent regions and becomes still deeper, soon concentrating all of the flow in one channel (Figure 10.5). Horton observed, however, that the actual evolution of a system of rilles is more complex, because deepening channels initiate lateral flow that diverts water from adjacent rilles, "pirating" their headwaters in a process Horton termed "cross-grading." This lateral evolution continues until a network of initially parallel rilles becomes a branched network, illustrated in Figure 10.6.

The process of rille development and cross-grading proceeds at progressively larger scales as the landscape evolves. Undrained hollows are filled in, steep scarps are eroded down, and the stream network adjusts itself until, as described in 1802 by John Playfair (1964),

Every river appears to consist of a main trunk, fed from a variety of branches, each running in a valley proportioned to its size, and all of them together forming a system of [valleys], communicating with one another, and having such a nice adjustment of their declivities that none of them join the principal valley either on too high or too low a level.

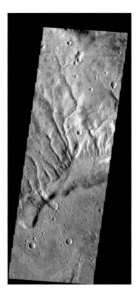

Figure 10.7 Dendritic stream network on Mars. Warrego Valles at 19.1° S and 244.0° E. This is part of the densest drainage network on Mars and strongly suggests that it was created by surface runoff from precipitation. Image is 24.8 km wide. THEMIS image, PIA05662. NASA/JPL/ASU.

Because of the tendency of water to collect in channels, the landscape becomes dissected into a fractal pattern of valleys and hillslopes on many different scales (Figure 10.7). In strong contrast to soil creep, which tends to simplify contours and smooth slopes, fluvial erosion sharpens irregularities.

Drainage basins. As stream networks develop, the landscape becomes divided into a hierarchical set of units known as drainage basins. In a fluvial landscape, drainage basins are the most natural geographic unit from the viewpoint of water supply or transport of sediment and solutes. The drainage basin associated with some point on a stream or river is that area of the land that contributes runoff to the stream. Divides – hillcrests down which water flows in different directions – separate drainage basins. The size of a drainage basin varies with the size of the stream: Small rilles may have drainage basins of only a few hundred square meters, while major rivers have drainage basins that encompass most of a continent.

One of the most regular quantitative relationships in fluvial geology relates drainage basin area A_b to the length L_b of the basin. Over a range of 11 orders of magnitude, from tiny rilles to continental drainages, the relation states (Montgomery and Dietrich, 1992):

$$L_b \simeq \sqrt{3A_b}. \tag{10.9}$$

Drainage networks are self-similar over this enormous range. That is, a map of a drainage network does not allow one to infer the actual scale of the image. However, this regular relationship breaks down at the smallest scale where the "belt of no erosion" asserts itself at a drainage length of a few tens of meters. Channels disappear below this scale. Another way of looking at this is to note that there is a threshold for channelized stream erosion that must be exceeded before network processes become important. Once this threshold is crossed, however, the processes that shape drainage networks become self-similar and any hint of a scale disappears from the system.

10.3.2 Streamflow

After his success in understanding eolian processes, Ralph Bagnold, our hero of Chapter 9, spent the rest of his career working on sediment transport in streams, applying ideas of fundamental physics to streamflow. He wrote a series of classic papers that still form the basis of our understanding of how fluids affect their beds (Bagnold, 1966). Before Bagnold, G. K. Gilbert approached the problem of sediment transport in streams from an experimental perspective, having constructed an enormous flume on the campus of the University of California at Berkeley to understand how streams transport sediment (Gilbert, 1914). Gilbert's interest had been piqued by the very practical problem of how to deal with the fluvial debris produced by hydraulic gold mining in California's Sierra Nevada. By 1905, that debris was advancing down California's rivers and had begun blocking the mouth of San Francisco Bay, creating a classic confrontation between the interests of gold miners, farmers along the river margins, and ocean shipping (Gilbert, 1917). Although Gilbert's meticulous flume experiments produced data that are still cited today, he failed to come up with any simple laws describing the interaction of sediment transport with the flowing water. His results were summarized by a large number of empirical correlations that, to a large extent, still characterize this field.

Sediment transport. Once established, steams continue to erode their beds and transport sediment delivered to them from upstream tributaries. The load of material transported by a stream is divided into several components. The dissolved load is composed of chemically dissolved species or colloids that are uniformly mixed with the water. The bedload consists of coarse material that slides, rolls, or saltates along the bed. Finer sediment moves as a "suspended load" that is concentrated near the bed but may be found higher in the water column, while the "washload," composed of still finer sediment, is uniformly mixed with the water. The criterion that distinguishes these categories is based on the dimensionless ratio ζ between the terminal velocity v of the sediment (Section 9.1.1) and the friction velocity of the flow, v_*:

$$\zeta = \frac{\text{terminal velocity}}{\text{friction velocity}} = \frac{v}{v_*}. \tag{10.10}$$

The transition between bedload and suspended load can be taken at $\zeta = 1.0$ to 1.8, while that between suspended load and washload occurs about $\zeta = 0.05$ to 0.13 (Burr *et al.*, 2006). Figure 10.8 illustrates these transitions and the threshold of motion using the modern version of the Shield's curve, Equation (10.8).

Stream velocity and discharge. A frequently asked question is, "What is the discharge (or velocity) of a stream or river given its depth, width and slope?" This question has engaged

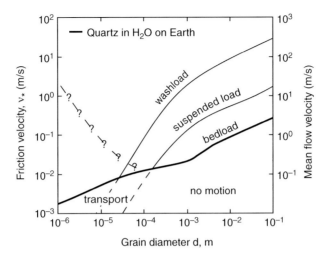

Figure 10.8 The relation between friction velocity and grain diameter for quartz grains in water on Earth. The heavy curve for grains moving along the bed has no minimum, although the queried extension at small grain sizes indicates much uncertainty in this conclusion, because the data from which the heavy curve was compiled may include sediment with a range of sizes and, thus, represents the impact threshold. Other curves show the thresholds for suspending grains in the lower water column or mixing it entirely through the water mass. The limits for either transport or no bed motion at low velocities are also shown. Greatly simplified after Figure 3 in Burr *et al.* (2006).

hydraulic engineers for centuries and several widely used equations can be found in the literature. French engineer Antoine de Chézy (1718–1798) gave the first useful answer to this question in 1775. His formula, as written today, is:

$$V = C\sqrt{RS} \qquad (10.11)$$

where V is the mean velocity of the flow (equal to the discharge Q divided by the cross-sectional area A of the stream), S is the slope of the channel, equal to $\tan \alpha$, and R is the hydraulic radius (equal to the area A divided by the perimeter of the wetted surface P). C is a dimensional constant that Chézy determined by comparing the velocity of one stream with that of another. Although this equation was adequate for Chézy's canal-design efforts, it contains many empirical constants and it is unclear how to scale this to a planet with a different gravitational field and to fluids other than water.

The next major improvement in an equation for streamflow came from the Irish engineer Robert Manning (1816–1897) in 1889. The "Manning" equation that we now write was neither recommended nor even devised by Manning himself, who actually did include the acceleration of gravity in his original formula. As usually written, the equation is:

$$V = \frac{k_{\mathrm{M}}}{n} R^{2/3} S^{1/2}. \qquad (10.12)$$

Where k_{M} is a dimensional factor equal to 1.49 ft$^{1/2}$/sec or 1.0 m$^{1/2}$/sec.

Table 10.2 *Manning roughness for terrestrial rivers*

Bed material	Grain size (mm)	Manning roughness n ($m^{1/6}$)
Sand	0.2	0.012
Sand	0.4	0.020
Sand	0.6	0.023
Sand	0.8	0.025
Sand	1.0	0.026
Gravel	2–64	0.028–0.035
Cobbles	64–256	0.03–0.05
Boulders	>256	0.04–0.07

Data from Arcement and Schneider (1989).

The factor n, called the "Manning roughness," has dimensions of $(length)^{1/6}$. This factor is widely tabulated for different channel conditions (smooth concrete, gravel, rock, etc.; usually in units of $ft^{1/6}$). It, like the Chézy coefficient, does not indicate how to scale to other fluids or planetary surface gravities. Table 10.2 gives some representative values of n for terrestrial rivers in metric units.

The most fundamental approach to this problem makes use of the balance between driving forces and resisting forces through the Darcy–Weisbach coefficient f. Unfortunately, this generality comes with the price of an equation that cannot be expressed as a simple analytic formula. Julius Weisbach (1806–1871) was a German engineer who focused on fundamental equations in hydraulics. In 1845 he published his major work on fluid resistance.

Consider a straight section of a stream channel that we will, for the moment, take to be rectangular in section with depth h and width w. The weight of the water per unit length of channel is ρgwh. The component of the force acting downstream is $(\rho gwh) \sin \alpha$, where α is the channel slope. The shear stress on the bed and sides of the stream is just this force divided by the area of the streambed and wetted banks, equal to the perimeter $P = (w + 2h)$. The stress τ_b on the wetted bed of the stream is, thus:

$$\tau_b = \frac{\rho gwh \sin \alpha}{w + 2h} = \rho g \left(\frac{wh}{w + 2h} \right) \sin \alpha = \rho g \left(\frac{A}{P} \right) \sin \alpha = \rho gR \sin \alpha \qquad (10.13)$$

where R is, again, the hydraulic radius. Note that for a stream much wider than its depth the hydraulic radius is nearly equal to its depth. Weisbach related this shear stress, the driving force, to the flow resistance, which he expressed in terms of the mean velocity V of the stream:

$$\tau_b = \frac{f}{4} \frac{\rho V^2}{2}. \qquad (10.14)$$

Equating (10.13) and (10.14), then solving for V, we obtain the Darcy–Weisbach equation for the mean flow velocity:

$$V = \sqrt{\frac{8}{f} g R \sin \alpha}. \tag{10.15}$$

If we ignore the difference between $\sin \alpha$ and $\tan \alpha$ for small angles, comparison of (10.15) and (10.11) shows that the Chézy coefficient is:

$$C = \sqrt{\frac{8g}{f}}. \tag{10.16}$$

This indicates how the average stream velocity depends on the acceleration of gravity, but we still lack an expression for the Darcy–Weisbach friction coefficient f. Determination of this coefficient requires the solution of a transcendental equation, for which the reader is referred to a clear and detailed discussion in the book by Rouse (1978). The friction coefficient is a function of the Reynolds number of the streamflow, $Re = \rho V h / \eta$. For $Re \gg 1000$ the usual practice is to relate f to the widely tabulated Manning roughness n and to use these empirical tables to calculate f:

$$\frac{1}{\sqrt{f}} = \frac{k_M \, R^{1/6}}{\sqrt{8g} \, n}. \tag{10.17}$$

It may come as a surprise that, although this expression depends explicitly on the acceleration of gravity, it does not contain the viscosity of the liquid. The flow velocity does, in fact, depend somewhat on the fluid viscosity, but only through the Reynolds number Re. At very low Reynolds number this equation becomes equivalent to the expression for the velocity of a flowing viscous liquid, Equation (5.15), which depends on the inverse viscosity, $1/\eta$, but at high Reynolds number the viscosity of the fluid does not matter much because the resistance to fluid flow depends mainly upon the exchange of momentum by turbulence in inertial flow.

Floods. The discharge of a stream or river varies enormously over a seasonal cycle and from season to season. Furthermore, catastrophic floods have occasionally scoured the surface of both Earth and Mars. Most sediment transport takes place during floods. There is no simple rule that relates sediment concentration to discharge, but rivers do typically carry more sediment in flood than during average flows. However, as illustrated in Figure 10.1, the peak sediment transport takes place during greater than average floods because the transport capacity increases non-linearly with increasing flow velocity. The morphology of a river valley is, thus, controlled by large, rare events, a fact that makes many fluvial features difficult to understand unless this is taken into account.

The level of water carried in a river channel varies with the discharge. It is difficult to estimate the depth of a natural flow of water from a given discharge without detailed

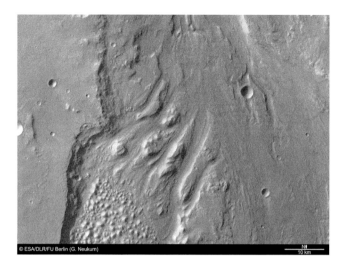

Figure 10.9 Ares Vallis is one of the large outflow channels on Mars. This image shows the transition between the Iani Chaos region to the lower left and the plains of Xanthe Terra to the top (north). The spurs between the individual channels have been shaped into crude streamlined forms by massive floods of water. 10 km scale bar is at lower right. Mars Express images by ESA/DLR/FU Berlin. (G. Neukum). See also color plate section.

knowledge of the topography. The Darcy–Weisbach equation, (10.15), can be written in terms of the total discharge $Q = VA$ and a geometric factor $A\sqrt{R}$:

$$Q = A\sqrt{R}\,\sqrt{\frac{8}{f}\,g\sin\alpha}. \qquad (10.18)$$

As the discharge increases in a flood, the product of channel area and the square root of hydraulic radius increases, but to make definite statements about the depth of the flow we have to know how the depth and width vary with each other. In a wide, shallow channel, $A\sqrt{R} \approx wh^{3/2}$, but unless the width is known we cannot solve for the depth as a function of the discharge. In general, the width of the channel increases as its depth does, so all we can say is that increased discharge leads to increased depth.

Catastrophic floods. It is somewhat easier to estimate the discharge in the aftermath of a flood when we know both the depth achieved by the flow and its width. This method has been applied to the outflow channels on Mars to estimate the discharge during the height of the floods. Some of these channels are hundreds of kilometers wide (Figure 10.9) and, if they were once completely filled, must have carried enormous volumes of water. Martian channels present the problem that the depths of the flows are not known, but estimates based on the elevation of water-modified surfaces adjacent to the channels suggest that discharges ranged from 10^7 to 10^9 m^3/s, compared with about 10^7 m^3/s for the largest terrestrial floods (Carr, 1996). The total volume of water is estimated from the volume of sediment removed. Assuming a maximum sediment/water ratio (typically about 40%), one can estimate the water volume. Combined with the peak discharge rate, this gives the duration of the flood. For example, the Ares Vallis flood was estimated to have moved 2×10^5 km^3 of material. If the flow was 100 m deep, then the flood must have lasted 50 days. If it was 200 m deep, then it lasted 9 days.

Floodplains. Apart from catastrophic floods of the type that created the Martian outflow channels or the Channeled Scabland on Earth, much smaller floods occur regularly on terrestrial rivers and streams. During times of larger than average discharge the water spills over the banks and floods the adjacent terrain, putting the excess water into temporary storage. Unless steep canyon walls confine the channel, the water spreads out over a broad area, where its velocity decreases. Sediment previously in suspension settles out into a thin deposit of fine-grained material adjacent to the channel. The grain size of the deposited sediment falls off rapidly with distance from the channel, grading laterally from coarse near the former banks to fine silt farther away.

Repeated overbank flooding eventually builds up a gently sloping plain adjacent to the channel, known as the floodplain. Gentle ridges and swales, traces of former channels, usually curve across floodplains as their parent rivers shift. "Oxbow" lakes (abandoned former channels) and wetlands occur locally. The material that underlies floodplains is generically referred to as "alluvium." Alluvium is generally fine-grained material: gravel, sand, and silt, that is only temporarily at rest. Deposited by floods, its fate is to eventually become re-mobilized as the channel shifts and move on downstream, traveling inexorably toward the sea by slow leaps and bounds.

Levees. After repeated cycles of floods, the relatively coarse deposits close to the channel build up broad natural levees that stand above the level of the surrounding floodplain and tend to confine the water to the main channel. In subsequent floods the water rises higher in the channel and, when it eventually breaks through a levee, causes more violent floods. The location where the water breaks through a levee is known as a crevasse and the fan-shaped deposit of coarse material laid down after a breakthrough is known as a splay.

Floods are a normal part of the hydrologic cycle, although humans who have built structures on the floodplain often treat them as a major calamity. The floodplain is an active and necessary part of a river system, but its operation is unfortunately sporadic, making it difficult for many people to appreciate its essential role in the river system.

Alluvial fans. When a sediment-laden stream debouches onto a land surface, as opposed to a body of water, the flow spreads out, slows down, and the sediment is deposited in a conical heap known as an alluvial fan. In plan view, the contours of alluvial fans form circular arcs that center on the mouth of the stream. At any given time the stream flows down a radial channel over the surface of the fan, but as sediment is deposited and the channel builds up, the active stream eventually shifts to another direction, in time covering the entire conical pile. On alluvial fans in California's Death Valley one can easily recognize multiple deposits of different ages by the different degrees of desert varnish on the fan debris. The slope of an alluvial fan is typically steep at its head and gentler with increasing distance downslope, in the concave-upward pattern characteristic of fluvial landforms.

The area of an alluvial fan A_f is related to the area A_b of the drainage basin of the stream that feeds it by a simple equation (Bull, 1968):

$$A_f \approx c A_b^{0.9} \qquad (10.19)$$

where the dimensional coefficient c varies with the climate and geology of the source area. This equation has been established both in the field and in small-scale laboratory experiments. Alluvial fans are often fed by debris flows and, high up on the fan surface, one can often recognize boulder-strewn debris flow levees. Lower down, the surface is covered with finer silt where muddy splays of water separated from the boulders as the fan slope decreased. When many alluvial fans form close together against a mountain front they may merge into a sloping surface known by its Spanish name, a bajada.

10.3.3 Channels

Channels develop where the flow of water over the surface persists over long periods of time. The morphology of the channel itself is often distinctive and, thus, indicates the action of a fluid flowing over the surface and excavating the channel, although the nature of the fluid itself is often unclear. When the Mariner 9 images of Mars first revealed giant outflow channels in 1971, many planetary scientists did not believe that they could have been cut by water, because Mars' atmospheric pressure is too low to sustain liquid water on the surface. Many different fluids were proposed, ranging from low-molecular-weight hydrocarbons to ice or mixtures of ice and water. Although it is still not possible to entirely rule out ice or brines, in the wake of the discovery of channel networks that seem to require overland flow, most planetary scientists now accept the likelihood of liquid water on the surface of Mars under climatic conditions that differ greatly from those prevailing today. Channels on Titan (Figure 10.10) were probably cut by liquid methane and channels on Venus were formed by lava flowing over its hot surface.

Channel features. Streamflow over a granular bed produces a variety of distinctive features from the interaction of the fluid and the deformable bed. Streamflow in a fixed channel is difficult enough to analyze at high Reynolds number because of the complex nature of fluid flow, but when coupled with the additional complexity of a deformable bed it poses problems that have yet to be completely solved. Nevertheless, extensive experimental study by many researchers using flumes and field observations of streamflow and its consequences has produced some understanding of how flow affects the bed of a stream and what features are produced by the interaction of the streamflow and its bed.

The major factor in the formation of bed features such as ripples, dunes, and larger accumulations of sediment is grain size. In addition, some measure of the velocity of the flow is needed. Much research has shown that to understand channel features the best measure of the flow is stream power (Allen, 1970), given by the shear stress on the bed τ_* multiplied by the mean velocity V, $\tau_* V$, which is measured in W/m². Low-power streams do not transport sediment at all, but as the stream power increases characteristic fluvial features such as

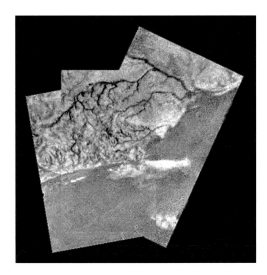

Figure 10.10 Mosaic of three images from the Huygens Descent Imager/Spectral Radiometer showing a dendritic drainage system on the surface of Titan. Each panel of the mosaic is about 7.5 km wide. Image PIA07236 NASA/JPL/ESA/University of Arizona.

ripples and dunes develop as a function of both grain size and stream power (Figure 10.11). A peculiar finding is that, at certain combinations of stream power and grain size, the bed is flat, even though intense sediment transport may be occurring. There is a small such plane bed field for intermediate stream power and large grains, and a much larger field at high stream powers for all grain sizes. These are designated, respectively, the lower and upper plane bed regimes.

Antidunes are unique to shallow, rapid flows. Unlike dunes, antidunes move upstream as flow proceeds. Although the form itself moves against the flow, sediment continues to move downstream: Only the wavelike shape of this feature moves counter to the current direction. Antidunes form when the Froude number of the flow, $V/\sqrt{gh}$, approaches 1. Their wavelength is approximately $2\pi h$, where h is the depth of the flow.

Ripples, dunes and antidunes can be distinguished in sedimentary deposits by means of cross-bedding. The ability to "read the rocks" and infer the nature of a flow, whether fast or slow, unidirectional or alternating, shallow or deep, is an important tool in the kit of a sedimentary geologist (Allen, 1982; Leeder, 1999). Observations of apparent cross-bedding in Martian sedimentary deposits by the Opportunity rover have led to the important inference of the former presence of shallow lakes on the surface of Mars (Grotzinger *et al.*, 2006b).

Bedforms develop even in streams that flow over solid rock. Channels often excavate into the underlying bedrock by quarrying away small joint blocks through differential pressures and cavitation behind obstacles. Gravel and cobbles carried along the bed may, over time, gradually erode the bed by abrasion. The result is flutes and, where circular motion is maintained over long periods of time, potholes. Potholes reach impressive depths of tens of meters as they are ground into the bed by swirling cobbles and debris below the streambed.

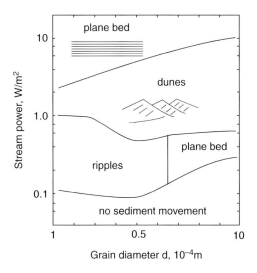

Figure 10.11 The type of bedform that develops beneath flowing water depends upon both stream power and the grain size of the sediment. This figure applies to water flowing over quartz sand on Earth, but it shows the variety of forms that can develop. No movement occurs below the lowest line. Note that the bedform pattern is not unique: Plane beds occur for both high stream power (upper regime flow) and low stream power over coarse sediments (lower regime flow). Simplified after Allen (1970, Figure 2.6).

In addition to the abrasion of the bed, the material carried by a river or stream is itself abraded as it travels downstream. Angular grains of sand become rounder, gravel becomes finer and cobbles are smoothed, rounded and reduced in size as they move downstream. Such downstream variations in the size of transported sediment make it difficult to associate a particular grain size with fluvial processes, because grains that might begin at the threshold of transportation move into suspension as they are broken and abraded while large, initially immovable rocks are broken down and begin to slide and roll along the bed.

Streamlined forms. Large floods create characteristic bedforms. One that is considered particularly diagnostic of floods are teardrop-shaped, streamlined islands that develop behind obstacles that divert the flow. These features are bluntly rounded in the upstream direction and taper to a point downstream. Once diverted by the obstacle, the flow closes back in around it downstream, but because this takes some time, the obstacle shields a tapered triangle from the flow, creating this shape. The faster the flow, the more gradual is this closure and the longer the island becomes. Streamlined islands in the Channeled Scablandsof Washington State are typically about three times longer than their maximum width, whereas streamlined forms on Mars are a little longer, about four times their width (Baker, 1982).

Hydraulic geometry. A widely used metric relates quantitative descriptors of the channel to its discharge. Discharge is chosen as the independent variable because it is believed that, while a stream may adjust its width, depth, or velocity by moving sediment from one

Table 10.3 *Hydraulic geometry of selected rivers (Leopold et al., 1964, p. 244).*

Location	Type	b	f	m
Midwest, USA	Fixed station	0.26	0.40	0.34
Semiarid, USA	Fixed station	0.29	0.36	0.34
Rhine River	Fixed station	0.13	0.41	0.43
Midwest, USA	Downstream	0.5	0.4	0.1
Semiarid, USA	Downstream	0.5	0.3	0.2

place to another, the discharge is determined by the climate and area of the drainage basin and so cannot be adjusted by the interaction between the stream and its bed. The method is entirely empirical: Data are collected from many rivers and streams, plotted against log-log axes, and lines are fit to the data that typically form fuzzy linear arrays. These fits must thus be regarded as approximate, but they do indicate general trends. Because straight lines on log-log plots indicate power laws, the relations for stream width w, depth h, and velocity V are written:

$$\left.\begin{array}{l} w = aQ^b \\ h = cQ^f \\ V = kQ^m \end{array}\right\} \tag{10.20}$$

where a, c, k are dimensional fitting parameters and b, f, m are dimensionless exponents. There are two constraints on these parameter sets because $Q = whV$: $a\,c\,k = 1$ and $b+f+m = 1$. Because river discharge even at one location is not constant over time, there are two ways in which these parameters are compiled: either at a fixed location as a function of time, or at different locations downstream on the same river. The exponents, b, f, m are considered the most significant parameters and they are tabulated in Table 10.3 for a small number of river systems.

Ratios illustrate the utility of this parameterization. For example, the aspect ratio of a river is the ratio between its depth and width. The aspect ratio h/w is proportional to Q^{f-b}, so at a fixed location the aspect ratio equals $Q^{0.14}$ for rivers in both the Midwest and semiarid USA. Thus, as discharge increases the relative depth increases slowly at a given station. On the other hand, going downriver the aspect ratio decreases slowly with discharge as $Q^{-0.1}$ for Midwestern rivers: Near its mouth the Mississippi is much shallower for its width than it is upstream. But, despite folklore, the Mississippi is not actually a "lazy" river – its velocity continues to increase with discharge either downriver or as a function of time at one location.

Meandering rivers. The tendency of rivers and streams to deviate from a straight course has long fascinated observers. Rivers flowing without constraint over an erodible bed quickly develop a meandering course. In an initially straight channel, the meanders begin as gentle bends as the water swings from side to side. The meanders grow in amplitude until they develop such extreme hairpin curves that they loop back on themselves until the water finally cuts through the narrow neck, temporarily shortening the course of the river. The abandoned meander loop forms an oxbow lake on the floodplain adjacent to the new main channel. The wavelength of a meander is a function of the channel size. A careful regression of meander wavelength λ_m and channel width w shows that they are nearly (but perhaps not exactly) proportional to one another (Leopold *et al.*, 1964):

$$\lambda_m = 10.8 \, w^{1.01} \qquad (10.21)$$

where the wavelength and width are both in meters. Although meander loops on a river floodplain are continually growing and being cut off, there is correlation between meander amplitude A_m and width as well:

$$A_m = 2.4 \, w^{1.1}. \qquad (10.22)$$

In addition to meandering laterally, rivers also meander vertically: Rhythmic variations of depth develop as deep pools alternate with shallow riffles with the same periodicity as the lateral meanders. The pools develop on the outside of meander bends, while the riffles form between them. These rhythmic depth variations develop even when the channel is confined between rocky walls that suppress lateral meanders, such as in the Colorado River confined within its canyon in Arizona.

One occasionally finds a river channel meandering through a valley that itself mean-ders on a much larger scale. In such cases one can infer that the smaller stream (called an "underfit" stream) carries a much smaller discharge than its former counterpart. This relationship is often observed in channels that once drained the meltwater from retreating Pleistocene ice sheets on Earth. A similar relationship, but in a sinuous lava channel, is observed in Schröter's Rille on the Moon.

The outside of meander bends is usually a steep bank that is often undercut and is obvi-ously undergoing erosion. A gently sloping bar on which sand and fine gravel is deposited, called a point bar, occupies the inside of the bend. As the channel shifts laterally, the flood-plain is consumed at the outer part of the bend and rebuilt on the inner bend. Cross-sections of the migrating channel show coarse-grained material (often gravel) at the former channel floor, fining upward into sands where the point bar is deposited, then silts where former point bars are buried by floodplain silts. Such fining-upward sequences, when they can be recognized in ancient fluvial deposits, provide a direct indication of the depth of the former river channel (Figure 10.12). From the depth, the correlations of hydraulic geometry yield an estimate of the discharge of the ancient river that created the floodplain.

Meanders do not form simply because water flowing in the straight sections between meanders impinges on the outside of the bend. Many authors, including James Thomson (William Thomson's brother) in 1876 and Joseph Boussinesq in 1883 independently dis-covered the helical flow of water in meander bends. However, the most famous re-dis-coverer of this effect was Albert Einstein, who perceived the effect while stirring a cup of tea (Einstein, 1954). Originally publishing in 1926, Einstein noted, as many others have

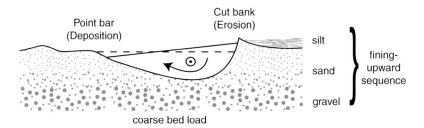

Figure 10.12 Water on the outside of meander bends rises higher than the water on the inside of the bend because of centripetal acceleration. This drives a secondary circulation that moves water from the outside of the bend toward the inner bend to create an overall helical circulation as the water moves downstream. Sediment is eroded from the outside of the bend and deposited on point bars on the inside. At the same time, coarse material remains near the bed while sand is deposited higher up on the point bar. Silt is deposited on top of the sand during overbank floods to produce a fining-upward sequence of sediment sizes.

done, that the tealeaves in the bottom of a stirred cup of tea gather together at the center of the cup. He inferred that the rotating liquid must create a helical flow in addition to its rotation. This flow descends along the outside of the cup and ascends in its center, sweeping the tealeaves into the center as it flows. Einstein realized that this flow could account for river meanders and point bars by transporting sediment down along the outside of the meander bends and depositing it on the inner point bar. The reality of such helical flows in river bends has now been demonstrated many times and this spiral flow is an accepted part of river hydraulics.

The helical flow is due to two factors. The first is centrifugal force. As the water flows around the bend, it rises higher along the outside of the bend. By itself, this would not cause a secondary flow if the water in the river were in rigid-body rotation. However, friction reduces the water velocity along the contact between the water and the outside wall of the bend and the excess pressure caused by the elevated water surface drives a flow downward along the outside wall.

Meanders are not restricted to just rivers flowing over granular material. They are commonly observed in glacial meltwater streams flowing over solid ice and in lava channels (they are called "sinuous rilles" in lava feeder channels on the Moon). In these cases one must presume that the same helical flow enhances channel migration on the outside of bends, perhaps by thermal erosion as the flow brings hotter material to bear against the outside of bends, but the precise mechanism is currently unclear.

Braided rivers. Steeply sloping streams heavily loaded with coarse sediment do not flow in well-defined channels. Instead, the flow divides into a complex network of shallow short-lived channels that diverge and rejoin many times as the water and sediment move downstream. Such channels are known as braided rivers.

Unlike rilles on a slope, the constantly shifting channels in the bed of a braided river do not unite to create channels progressively more capable of carrying the available load of sediment. Such streams are sometimes described as "overloaded," in the sense that the union of two sediment-laden channels is less capable of transporting the load than the individual channels, so some of the load is deposited when channels join, creating a temporary bar that eventually diverts the flow to a new location.

The factors that decide whether a stream channel is meandering or braided are still poorly understood (Schumm, 1985). Several rivers have been observed to alternate in style between meandering and braided or vice versa in historical time, a process termed "river metamorphosis" by fluvial geologist Stanley Schumm. For example, the channel of the South Platte River in Nebraska changed from braided in 1897 to meandering in 1959 in response to a large decrease in mean annual discharge due to irrigation projects that extracted water from the river (Schumm, 1977, p. 161). Likewise, the lower reaches of the Pleistocene Mississippi River were braided because of its greater slope down to the lowered sea level during the ice ages, as well as to the greater discharge it carried as the ice sheets melted. As the sea level rose and the ice sheets disappeared, its channel changed from braided to meandering.

An often-cited criterion that divides meandering from braided rivers on Earth is expressed in terms of channel slope S and bankfull discharge Q_{bf}, where the discharge is in m³/s (Schumm, 1985). The river is usually braided when the average channel slope exceeds:

$$S \geq 0.0125 \, Q_{bf}^{-0.44}. \tag{10.23}$$

Thus, any factor that increases either slope or discharge favors braiding over meandering. The frequent observation that braided rivers typically carry coarse debris is implicit in Equation (10.23) through the connection between average channel slope S and the grain size of bedload: Rivers that carry coarse debris are steeper than those that carry fine sediment.

The paleohydrologic hypothesis. Schumm also noted an apparent connection between channel stability and the presence of vegetation. Vegetation growing on islands in shifting channels tends to stabilize them. Root mats add cohesion to channel banks and hold fine sediment in place until undercut by powerful currents. Both of these factors tend to favor meandering channels rather than braided. Schumm noted the lack of evidence for meandering river channels before the Late Paleozoic era when vegetation first covered Earth's land surface. He elaborated a "paleohydrologic hypothesis" that suggests that all river channels were braided before the evolution of land plants. If this connection between vegetation and channel form is correct, we should not expect to find meandering river channels on Titan or Mars. On Titan, present resolution is too poor to be sure, but meandering channels do appear to cross the surface of Xanadu. On the other hand, the boulder-strewn surface at the Huygens landing site is consistent with a braided river channel (Figure 7.16d). On Mars, there are now HiRISE images of indisputable meandering channels in Aeolis Planum (Burr *et al.*, 2009), Figure 10.13. It must, thus, be possible for meanders to develop in the absence of vegetation (Howard, 2009), perhaps because of cohesion from clay or permafrost that binds the sediment together. Channel meanders, while evidently not requiring the presence of vegetation, may, nevertheless, indicate special mechanical conditions in the sediment adjacent to the channel.

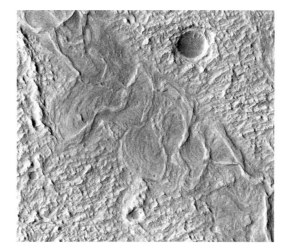

Figure 10.13 A meandering channel in Aeolis Planum, Mars, that belies the proposal that meanders develop only when banks are stabilized by vegetation. These highly sinuous meanders actually stand as ridges at the present time, an example of inverted topography. Gravel in the channel presumably made it more resistant to erosion than fine-grained surrounding material. Portion of HiRISE image PSP_006683_1740. Image is approximately 2.3 km wide. NASA/JPL/University of Arizona.

River terraces. Gently scalloped scarps are often found parallel to the active flood-plains of large river systems. The downstream slope of the relatively flat surface behind such scarps is similar to that of the active floodplain. These surfaces, which often look like treads on a giant staircase stepping up away from the river, are known as river terraces. Because of their distinctive appearance and their importance for land use, terraces have received a great deal of attention in the terrestrial geologic literature (e.g. Ritter, 1986).

River terraces are abandoned floodplains of a river system that has eroded deeper into its valley. Geomorphologists distinguish paired terraces, which appear at the same elevation on opposite sides of the river, from unpaired terraces. Terraces are the result of the lateral migration of the river channel back and forth across the floodplain as the channel slowly erodes downward into the floor of the river valley. Terraces are important because they indicate changing conditions, although they are not usually diagnostic of exactly what conditions are changing. For example, the erosion they record could be due either to tectonic uplift of the rock underneath the stream, or increased downcutting of the stream. Downcutting can be due to increasing water supply, decreasing sediment load, or a lower base level at which the river discharges.

Although it is sometimes stated that the existence of discrete terraces indicates that downcutting must be episodic rather than steady, this is not necessarily true. Because a long interval of time may separate the impingement of the main channel on one valley wall during its slow lateral swings, the change in the level of the stream between two terrace-cutting events reflects the accumulated erosion between cutting events. Discrete terraces form even if the rate of downcutting is uniform because of this interval between terrace-cutting events. Although the ages of river terraces on the Earth can now be determined through the measurement of cosmogenic isotopes, it is still extremely difficult to discriminate episodic versus steady downcutting from such data.

Tributary networks. The most familiar pattern of drainage networks is one in which smaller channels join into larger ones that, in turn, join still larger ones, forming a network formally called a tree or dendritic pattern (Box 10.2). This tributary pattern persists on the Earth's land surface over most of its area because of the increasing capacity of downstream water to carry sediment. Rivers that branch downstream and then rejoin do occasionally occur on Earth, but they are relatively rare. Such non-tree-like patterns are more common in Martian channels, for reasons not currently understood.

The junction angle in tributary networks is such that the acute angle between links usually occurs upstream of the junction. This is presumably because the momentum of the joining currents tends to carry both in the same direction – it is unusual for a tributary to discharge its water upstream into the channel it joins.

Distributary networks. When a stream or river can no longer carry its sediment load, due either to loss of water by infiltration into a substrate (as on an alluvial fan), or because of a decreasing gradient (as when it encounters a lake or the ocean), its sediment is deposited and a system of dividing channels develops. The branching pattern of such a distributary network may resemble the tree-like form of a tributary network, but the slope in this case is reversed: The largest channels are upslope of the smallest channels. The acute angle of the junctions is downstream in this case, again tending to preserve the momentum of the dividing channels. Similar networks develop among the channels actively feeding lava flows spreading over flat terrain.

Unusual networks develop where the flow direction alternates, such as in tidal marshes where the surface is alternately flooded and drained. In this case the same channels serve alternately as a distributary and a tributary network. In such networks the channels tend to divide and rejoin frequently and junction angles are typically close to 90°, perhaps because of the frequent collisions between incoming and outgoing streams of water.

Venusian channels. No one seriously expected to find fluvial channels on Venus. The surface temperature is far too high to permit liquid water to flow over its surface. However, images returned by the Magellan radar (Figure 10.14) reveal channels that are remarkably similar to those of terrestrial river systems. Meandering channels with natural levees, streamlined islands within the channel, even crevasse splays and abandoned meanders can all be identified in the images. On Venus, however, we must certainly be looking at channels that once carried lava, not water. Because of Venus' high surface temperature and large eruption volumes, lava cools relatively slowly compared to terrestrial lava flows; furthermore, the flows may have continued over such long periods of time that "fluvial" features developed. Lava, like water, is capable of eroding its bed by plucking and of transporting "sediment" in the form of more refractory minerals, so this may be a case in which similarity of physical process promotes similarity of form, even though the materials involved are very different.

Box 10.2 **Analysis of stream networks**

Rilles and gullies join to form larger streams, which join again to form still larger streams, and so on up to large rivers. The result is a branching or *dendritic* (tree-like) network that extends from the smallest rilles to the largest trunks: rivers like the Mississippi or the Amazon. Water and sediment eroded from the land are flushed down these channels, eventually to be deposited in the oceans.

Most sediment is fed into the network from overland flow at the level of rilles. Larger and larger streams mainly serve to transport it. Most erosion, thus, occurs on the scale of small drainage basins, grading into transportation at larger scales, although deeply incised rivers such as the Colorado may be fed large amounts of material directly along the main channel by mass movement.

Robert Horton (1945) brought order to stream network analysis by proposing a simple numbering scheme, which has been slightly improved by other authors. Horton assigned the smallest recognizable rilles to order number 1. When two order 1 channels join they become a channel of order 2. The union of two order 2 channels is of order 3, and so on. When a channel of lower order joins a channel of higher order, the order of the higher channel does not change. Thus, when an order 3 channel is joined by an order 2 channel, it remains of order 3. See, for example, the fourth-order network shown in Figure B10.2.1.

To appreciate the success of Horton's idea of numbering from the smallest rilles to the largest river, consider the opposite scheme in which the main trunk of a large river is assigned order 1. Proceeding upstream to smaller and smaller tributaries, we would fined that the last recognizable rilles in most basins had a different order, even though their function in the

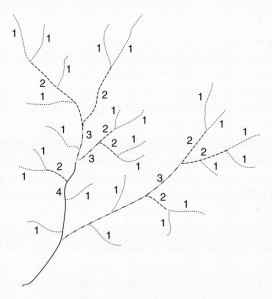

Figure B10.2.1 A typical stream network, which illustrates the ordering of stream segments. This network is of order 4. After Figure 10.1 of Morisawa (1968).

Box 10.2 **(cont.)**

hydrologic system is identical. Apparently similar rilles on opposite sides of a divide would, in general, be assigned different orders.

Some North American examples of high-order river systems are the Mississippi of order 10, the Columbia of order 8, the Gila of order 8, and the Allegheny of order 7. Unfortunately, tables of the orders of all the world's rivers seem to be hard to find, although ordering is so suitable for computer computation that the ArcGIS program includes a tool that assigns orders to streams.

The major problem with this scheme is that the definition of the first-order rille is uncertain: Depending on the map scale, this could be the smallest recognizable rille (as Horton supposed), or the smallest perennial stream (which means its assignment depends on climate). Inadequate resolution caused a problem with the initial ordering of Martian channel networks: At the lower resolution of Mariner 9 and Viking images, Martian valley networks appeared to have much lower drainage densities than terrestrial networks, for which a variety of causes were cited (Carr, 1996). However, once higher-resolution Mars Orbiter Camera (MOC) images became available, true first-order rilles could be recognized and it was realized that Martian and terrestrial networks have similar densities (Carr, 2006). Given this history, no one has seriously tried to assign orders to Titan networks (except at the Huygens landing site) because of the low resolution at which they are currently seen.

Stream ordering would be an amusing but mechanical pastime, except that ordering clarifies the statistical properties of networks for practical applications in flood wave prediction and sediment yield estimates (as only two examples). There are useful quantitative relations between landscape properties, such as average slope or stream discharge, as a function of order. It was also originally hoped that the statistical properties of stream orders might be diagnostic of network origin.

Some properties of a fourth-order drainage basin are listed in Table B10.2.1. Several characteristics are clear upon inspection: The number of streams decreases sharply with increasing order. In addition, the average channel slope decreases regularly with increasing order, the average channel length and basin area increases with order, and the drainage density is nearly independent of order.

Drainage density is defined as the sum of the lengths of all channels in a basin divided by its area. Its inverse is approximately the distance between streams in the basin. For first-order streams this is also the width of the belt of no erosion, as defined by Horton. The drainage density

Table B10.2.1 *The Mill Creek, Ohio, drainage network (Morisawa, 1959, Table 12).*

Stream order	Number of streams	Average length (m)	Average basin area (km²)	Average channel slope	Stream density (km/km²)
1	104	111	0.065	21.6°	3.39
2	22	303	0.313	7.01°	4.36
3	5	1046	1.505	2.23°	3.77
4	1	1915	6.941	0.57°	3.52

Box 10.2 **(cont.)**

is, thus, closely related to infiltration capacity: Large drainage densities imply a low infiltration capacity, as often seen in badlands, whereas low densities imply that water sinks in readily.

Plots of the logarithms of different characteristics of drainage networks versus stream order generally form straight lines. This suggests power-law relationships among the different quantities. For example, the number of streams of a given order p, expressed as $N(p)$, can be written:

$$N(p) = r_b^{(s-p)} \qquad (B10.2.1)$$

where s is the order of the main stream in the network. The constant r_b is known as a bifurcation ratio. In the fourth-order network of Table B10.2.1, $s = 4$. There are five third-order streams, and $s - p = 4 - 3 = 1$, so $r_b = 5$. So far, this is nothing new. But now note that for $p = 2$ this formula predicts that there should be 5^2 or 25 second-order streams (there are really 22). For $p = 1$ the formula predicts 5^3 or 125 first-order streams (there are really 104). The fit is not perfect, but it is fairly close.

Similar power-law formulas can be constructed for other quantities, such as stream length and slope. Over large, high-order drainage basins fits can be adjusted by least squares to obtain best estimates for each of these quantities, which then describe the branching properties of a drainage network.

This type of fitting was popular between 1945 and about 1970 when it was believed that such fits and bifurcation ratios reveal important information about how a drainage network develops and could, in some way (no one knew quite how) be related to the fluvial processes that created the network. Unfortunately, geomorphologist Ron Shreve dashed most of these hopes in 1966 when he showed that relations of this type develop in *any* dendritic network, including networks generated by random walks in a computer (Shreve, 1966b). Shreve's arguments have been confirmed and extended by later work (e.g. Kirchner, 1993). Nevertheless, many workers are convinced that stream networks statistics indicate *something* about the organization of fluvial processes and attempts have been made to assign a kind of entropy to stream networks and show that actual networks maximize that entropy (Rinaldo *et al.*, 1998) or expend the least work. Other, more complex, numbering schemes have been developed that claim to have genetic significance, but their success is presently unclear.

Statistical descriptions of drainage networks do have practical value in estimating the numbers and lengths of links at different levels without actually having to measure the entire network, but it is not easy to relate the parameters to genetic processes. Networks developed by sapping seem to be less branched and possess shorter links than those developed by overland flow, but this type of distinction can probably be made without the aid of detailed statistical analyses.

10.3.4 Standing water: oceans, lakes, playas

Running water tends inexorably downhill. When it reaches the lowest possible level it accumulates into a body of standing water that may range in size from tiny temporary ponds to global oceans. The most important geologic fact about running water is its ability to transport sediment from higher levels to lower. When it enters a large accumulation

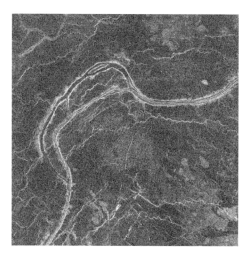

Figure 10.14 Sinuous lava channel on the plains of Venus. The overall channel flows from Fortuna Tessera in the north, south to Sedna Planitia. Channel is about 2 km wide and is interrupted by streamlined islands. The channel pattern illustrates the formation of an alternative channel during flow. Frame width is 50 km. Magellan F-MIDR 45N019;1, Framelet 18. NASA/JPL.

of water its velocity drops (although currents are never completely absent in any body of water: Underwater gravity currents are discussed in Section 8.2.3) and its sediment is dropped somewhere near the shore. In contrast to the land surface, bodies of standing water are the locales of sediment accumulation rather than erosion.

Standing water, however, possesses its own distinctive ability to move material. This ability depends on the action of waves, so that sediment transport occurs mainly at the level of the water surface. Wave action produces distinctive landscape features that remain even long after a body of water disappears.

Aside from the prevalence of coastlines on the Earth, ancient Mars may have possessed extensive bodies of water whose former shorelines, if found, would demonstrate their presence and dimensions. Titan is now known to possess ephemeral lakes of liquid methane and ethane, making beach and lake processes of prime interest to planetary geologists.

An appreciation of the landforms created by waves began with eighteenth-century British geologists who initially attributed most former geologic activity to the waves that they observed crashing around the edges of their sea-girdled isle. In the nineteenth century, American geologist G. K. Gilbert took a large step forward with his study of the now nearly extinct Lake Bonneville in Utah (Gilbert, 1890). The present Great Salt Lake in Utah is a small remnant of a much greater lake that existed during the Pleistocene. When it drained about 14 500 yr ago, it exposed the beaches, deltas, spits, and bars that formed during its brief existence of about 17 000 yr.

Gilbert was impressed that most of these features are the consequence of waves breaking against the shore. His research, as does much modern research, therefore focused on the generation, propagation, and interaction of waves with the shore.

Waves on water. The generation of waves has received a great deal of attention for its own sake and we can only touch on the basics in this book. The reader who wishes to go further should consult the treatise by Kinsman (1965). Waves on the ocean, lakes, or even ponds are created by wind blowing over the surface. William Thomson (who became Lord Kelvin) and German physicist Hermann von Helmholtz were the first to understand how wind can generate water waves by an aerodynamic instability, now called the Kelvin–Helmholtz instability. The interface between two fluids, such as air and water, cannot remain flat if the fluids move with different velocities. Waves develop on the interface, beginning with small, short-wavelength waves for which the restoring force is surface tension, then growing into much larger waves that are dominated by the weight of the water, called gravity waves. The overall wave-generation process transfers energy from the wind to waves on the water surface.

Wind must blow over the surface of the sea for some time, and continue over some distance, before a fully developed set of waves develops. The size and wavelength of water waves, thus, depend on the speed of the wind, its duration, and the distance, or fetch, over which the wind acts. Higher wind speeds develop higher waves of longer wavelength and period. These simple facts permitted Gilbert to understand why the beaches of Lake Bonneville are best developed along the Wasatch Mountain front on the eastern side of the former lake: The prevailing wind blows from west to east, to the extent that shoreline features are hardly recognizable on Lake Bonneville's western side where few waves ever beat.

Although the shape of a water wave may move at high speed over the water surface, anyone observing the motion of an object floating in the water knows that the water itself moves very little. There are two velocities relevant to waves. They both depend on the period T, wavelength L, and water depth H, in addition to the acceleration of gravity g for a gravity-dominated wave. In general, wave speed $c = L/T$. The first important speed is the phase speed, c_p, which is the speed at which some part of the waveform, its crest or trough, moves across the water. For waves of small amplitude (compared to their wavelength) the general expression for this phase speed is:

$$c_p = \frac{gT}{2\pi}\tanh\frac{2\pi H}{L}. \tag{10.24}$$

In deep water, $H \gg L$, this simplifies to $c_p = gT/2\pi$. Similarly, in shallow water this equation simplifies to $c_p = \sqrt{gH}$. Note that the speed of deep-water waves depends on their period, so after traveling some distance long-period waves arrive earlier than short-period waves. This explains why long, slow waves are the first to arrive at the shore after a distant storm, followed by shorter, choppier waves. The dependence of wave speed on period is called dispersion: Wave packets tend to disperse as they propagate, spreading out and changing shape.

The other velocity associated with waves is called the group velocity c_g. This velocity determines how fast the energy associated with a packet of waves propagates. It can be derived from the phase velocity by a simple derivative:

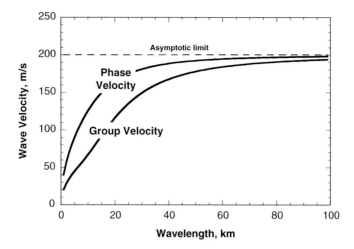

Figure 10.15 Group velocity and phase velocity for water waves on the Earth in water 4 km deep, equal to the average depth of the Earth's oceans. Long-wavelength waves approach a "shallow-water wave" limit when their length is much longer then the depth of the water. Short-wavelength wave velocities are increasing functions of their length in the "deep-water wave" limit.

$$c_g = \frac{c_p}{2}\left[1 + \frac{\frac{4\pi H}{L}}{\sinh\left(\frac{4\pi H}{L}\right)}\right]. \qquad (10.25)$$

Figure 10.15 illustrates the relations between the phase velocity, group velocity, and wavelength for waves in 4 km deep oceans on the Earth. The important feature to note is that both wave velocities are highest in deep water, and that the group velocity is always less than the phase velocity at a given wavelength. Thus, the energy from a disturbance on the ocean propagates more slowly than the leading waves. Waves from deep water thus slow down on approaching the shore. Because energy is conserved, energy piles up in shallow water. More energy means higher waves, so we can deduce that the wave height must increase as the water shoals.

The energy E (per unit area of ocean surface) in a wave of amplitude A_0 (one-half of the vertical distance from crest to trough) is made up of equal contributions of the gravitational potential energy and the kinetic energy of motion. Its magnitude is given by:

$$E = \frac{1}{2}\rho g A_0^2. \qquad (10.26)$$

Because energy propagates at speed c_g, the energy flux P in a wave is given simply by $P = c_g E$.

The path of a particle of water as a wave passes by is approximately a circle of radius A_0 near the ocean surface. At greater depths below the ocean surface the amplitude of the

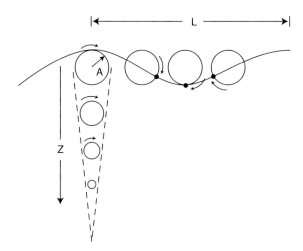

Figure 10.16 The orbits of particles of water in deep-water waves are nearly circular. The amplitude of this motion decreases exponentially with greater depth. Although the wave appears to progress from left to right, the water itself only moves in small circles, whose maximum diameter at the surface is equal to the distance between the crest and trough of the wave.

circle decreases exponentially. If z is the depth below the surface, the wave amplitude $A(z)$ in deep water decreases as:

$$A(z) = A_0 e^{-\frac{2\pi z}{L}}. \tag{10.27}$$

Figure 10.16 illustrates the rapid decline in wave amplitude with increasing depth. In water shallower than the wavelength, $H < L$, the velocity of the water on the bottom is still appreciable. The orbits of the water particles in shallow water are ellipses, not circles, which degenerate to a straight line parallel to the bottom on the seabed itself.

The rapid decrease of the amplitude of water oscillation with increasing depth below the surface defines the geologic concept of "wave base." It is well known that a submarine can ride out even the most violent storm by submerging to a depth comparable to the wavelength of the largest seas in the storm. In a similar manner, the seabed below some critical depth is unaffected by waves that may break up rocky shorelines. The ability of waves to erode the shoreline is, thus, limited in depth. The short-lived island of Surtsey off the southern coast of Iceland provides a fine example of the limited power of the waves. Surtsey was built by a series of volcanic eruptions in 1963. Well observed and widely reported in the news media, Surtsey was immediately attacked by the waves and within a few years most of its original area had disappeared below the waves. The eroded base of the island is still there, but it was planed off by waves to a depth of about 30 m below the surface, a depth that represents the effective wave base at this location. In a similar manner, volcanic Graham Island appeared in the Mediterranean in 1831. Its ownership was hotly disputed by Britain, Spain, and Italy, but wave erosion cut it down below the sea surface by 1832. Normal waves can move sand down to a depth of about 10 m, so the concept of wave base is somewhat fuzzy: The exact limit of erosion depends on the frequency of storms, the wavelength (and, hence, the exposure to wave-generating winds), and the wave amplitude. The important concept is that waves act only close to the surface of a body of standing water.

As waves approach the shoreline the water shoals and, as mentioned before, the waves increase in height, eventually breaking. Wave breaking is a complex phenomenon for which many theories have been proposed. A good summary can be found in the book by Komar (1997). A simplified way of looking at wave breaking is that it occurs when the velocity of the water particles at the crest of the wave exceeds the phase velocity. When this happens, the steepness of the wave front exceeds the vertical and the wave crest cascades over its front, dissipating much of its energy as turbulence. There are several ways in which waves break, each type distinguished by the steepness of the beach face. In order of increasing beach slope, these styles are called spilling, plunging (the iconic breaking wave is a plunging breaker), collapsing, surging, and, in the case of vertical seawalls, a reflected wave that does not break at all.

The principal consequence of wave breaking is that the wave energy, ultimately originating from the wind, is focused on the beach, where it is dissipated in turbulence. Where the waves impinge directly on rocky cliffs, the hydraulic pressures generated by the breaking waves may drive air or water into joints, loosening joint blocks or abrading the rock by dashing smaller boulders and sand against it. Wave action moves sand up and down the beach face and alternately offshore into bars, then onshore onto the beach again. Beach sand is suspended by each breaking wave and becomes vulnerable to transport by longshore or rip currents. Overall, wave energy makes the beach a highly dynamic environment in which erosion, deposition, and sediment transport are all active processes.

Coastal processes have received a great deal of study and limited space prevents a detailed treatment in this book: The interested reader is referred to a number of excellent texts on this subject in the further reading section at the end of this chapter. For our brief survey here the only other processes of major importance are wave refraction and longshore drift, as these are chiefly responsible for building beaches that might be seen from orbiting spacecraft.

Wave refraction. Wave refraction refers to the bending of wave fronts in water of varying depth. Once waves begin to "feel bottom" at a depth H equal to about $L/2$, the phase speed is proportional to the square root of the depth, $c_p = \sqrt{gH}$. Thus, the shallower the water becomes, the more slowly the wave fronts move. Consider a linear wave approaching a uniformly sloping shoreline at an oblique angle (Figure 10.17a). Because the wave moves more slowly in shallow water, the oblique wave gradually rotates to become more parallel to the shoreline as it approaches. It cannot turn exactly parallel to the beach, but the angle it makes with the beachfront is greatly decreased before it reaches the beach and breaks.

An oblique wave arrival means that the momentum transported in the wave is not completely cancelled when the wave breaks on the beach. A component of this momentum remains and generates a current, the longshore drift, which moves sediment in the direction of the acute angle between the wave front and the beach.

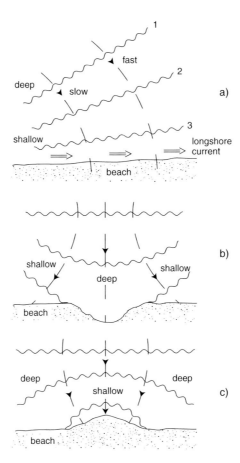

Figure 10.17 Behavior of wave crests approaching a shoreline. Panel (a) illustrates an oblique approach of the wave crests (wavy lines) to the shore. As the water becomes shallower, the wave velocity decreases so that wave crests near the shore travel more slowly than those farther out in deeper water. The net result is that the wave crests rotate and tend to become parallel to the beach as they approach. The oblique convergence also transfers a component of momentum along the beachfront to produce a longshore current. The dashed lines indicate the direction of energy flow perpendicular to the wave crests. Panel (b) illustrates the refraction of wave energy away from an offshore trough. The wave crest over the deep water in the trough moves ahead of the wave crests over its shallow flanks, turning the wave crests away from the axis of the trough and directing energy away from the trough and onto the adjacent portion of the beach. Panel (c) illustrates the opposite effect, when the waves approach over a ridge perpendicular to the shoreline. In this case the waves move more slowly over the shallow ridge and the wave energy is concentrated on the ridge crest. The combination of the focusing actions shown in (b) and (c) tends to even out submarine irregularities near the shore by wave erosion of ridges and filling of troughs.

If the bottom is not uniform, wave refraction acts to fill in hollows and erode protuberances. Figure 10.17b shows how waves approaching a shore are refracted over a submarine canyon running perpendicular to the beach. Because the canyon is deeper than the surrounding seafloor, waves moving down its axis travel more rapidly to the shore. Waves

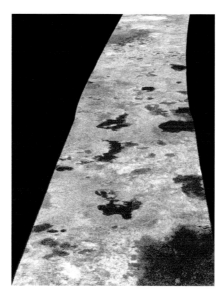

Figure 10.18 Liquid methane lakes near the North Pole of Titan imaged by the Cassini synthetic aperture radar. Dark regions are smooth lake surfaces and brighter regions are the surface. Intermediate brightness levels near the lake shores indicate some radar return from the lake bottoms. Image is centered near 80° N and 35° W and the strip is 140 km wide. Smallest details are about 500 m across. The radar strip was foreshortened to simulate an oblique view from the west. Image PIA09102. NASA/JPL/USGS. See also color plate section.

to either side move more slowly and refract the wave fronts away from the canyon axis. Because wave energy flows perpendicular to the wave crests, the wave energy is refracted away from the canyon axis toward its edges. The margins of the canyon are, thus, more heavily eroded by wave action than its axis and so sediment tends to accumulate in the canyon, evening out the bottom contours parallel to the shoreline.

On the other hand, a submarine ridge, which might reflect the presence of a headland on shore, tends to focus wave energy over its crest, as shown in Figure 10.17c. The approaching waves collapse onto the shore right over the ridge, leading to intense wave action and erosion of the ridge, again evening out the bottom contours parallel to the shoreline.

Wave refraction, thus, tends to straighten out complex shorelines, focusing wave energy on promontories and diverting it from inlets. A newly flooded landscape, such as might be created by erecting a dam at the mouth of a river (Lake Powell on the Colorado River is a good example), presents a fractal shoreline of great complexity with protruding spurs and deep inlets everywhere. However, if the water level remains constant, in time these spurs and inlets are battered back and filled up, leading to a much more even shoreline.

The highly convoluted, fractal shorelines of the methane lakes on Titan constitute a puzzle (Figure 10.18). These shorelines show no sign of wave action; no bars, no spits, or eroded headlands. On the other hand, an observation of sunlight reflected from the surface of one lake appears to be perfectly mirror-like, showing no sign of the glitter typical of reflections from wave-ruffled liquid surfaces. Do lakes on Titan lack waves? If so, why? Titan's equatorial region is notable for its broad expanses of sand dunes, so winds must exist. Or are the levels of methane in the lakes constantly changing so that there is no time for erosion to straighten out the shorelines? At the moment the answers to these questions are unknown.

Longshore drift. Longshore drift is another powerful force that tends to straighten out shorelines. Beachgoers often confound longshore drift with the along-beach motion of sand particles in the back-and-forth swash of waves breaking on the beach face. This motion does drive some sand in the general direction of the longshore current illustrated in Figure 10.17a, but the current that flows parallel to the beach just offshore transports a far larger flux of sand. This current is driven by the uncompensated component of the momentum of the waves parallel to the beach. It is localized near the beach within the breaker zone. Ocean bathers are often unaware of its existence until they suddenly notice that they are far down the beach from where they thought they should be.

Sediment suspended by waves breaking in the surf zone is caught up in the longshore drift and transported parallel to the beach. This sediment-laden current is a true "river of sand" that moves large volumes of material along the shore. The direction of the longshore drift varies with the shoreline topography and the local direction of approaching waves. Coarse sediment deposited by rivers flowing into a lake or the sea is often caught up by the longshore drift and moved "down drift" to nourish beaches and build bars across inlets or spits out from headlands. G. K. Gilbert noted many gravely bars and spits created during the high stands of Lake Bonneville. These bars and spits are now dry, level ridges standing in the Utah desert to bear witness to the former existence of a large lake.

Many other currents and interactions occur close to the beachfront. Rip currents develop outside of the surf zone to return water pushed up onto the beach by shoaling waves that, unlike deepwater waves, transport water in addition to energy and momentum. Rip currents are often spaced periodically along the beach, their spacing determined by the excitation of edge waves, a variety of trapped wave that can exist only along the beach face. Beach cusps are rhythmically spaced beach features whose origin is still debated, but appear to be related to standing edge waves. All of these currents have complex interactions with tides and the material that makes up the beach. For more information, however, the reader should refer to the references at the end of the chapter.

Playas. Playas are shallow, ephemeral lakes that develop in regions dominated by interior drainage. Playas are flooded after rain falls in the drainage basins that discharge into them. The flowing water carries fine silt and dissolved minerals into the lake basin and then evaporates, leaving this non-volatile material behind. Playas are, thus, accumulation points for evaporite minerals. These minerals often form concentric rings around the center of the basin, ranging from the most soluble minerals in the center (usually halite and other chlorides on Earth) to less soluble minerals at the edges (typically carbonates on the outside and sulfates in an intermediate position between carbonates and chlorides). The edges of playas grade upslope into alluvial fans, which trap most coarse sediment moving from adjacent mountain fronts.

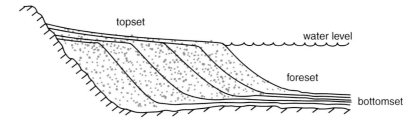

Figure 10.19 Formation of a delta near the mouth of a river or stream discharging into a larger body of water. As the stream loses its momentum in the lake or a sea it deposits much of the sediment it carries. This initially produces steep foreset beds of coarse material close to the shore and thin beds of finer sediment farther away in deep water. As the delta continues to build outward, topset beds are laid down on a shallow slope, while the foreset and bottomset beds build farther from the mouth of the stream. After Figure 14 in Gilbert (1890) and based on Gilbert's observation of deltas left behind when Lake Bonneville drained.

Because the water that floods playa lakes is often only a few tens of centimeters deep, the surfaces of playas deviate only very slightly from an equipotential surface – large playas form some of the most level (but not flat!) surfaces on Earth. Playa surfaces are usually devoid of rocks or coarse sediment, except in circumstances where other processes move rocks across them. A famous example is Racetrack Playa in Death Valley, whose surface is criss-crossed by the trails of boulders weighing many kilograms that somehow move across the level playa. No one has yet seen these boulders in motion, but they do shift between repeated surveys, perhaps during winter storms with high winds when the playa surface is wet and slick.

Playas may also serve as sources of fine dust, as they do in the American southwest. High winds drive sand grains and mud chips across their surfaces, raising dust that may be exported in suspension from their immediate vicinity.

Deltas. Deltas form where a river transporting sediment enters a larger body of water, decreases its velocity, and drops its sediment load near the shoreline. If this sediment is not carried away immediately by longshore drift, it builds up into an accumulation called a delta. The eponymous delta is that of the Nile River, which is triangular in plan like the Greek letter Δ. Because of interaction with waves and currents, deltas can be of many different shapes and sizes ranging from a few meters across to hundreds of kilometers, but all are sediment accumulations built out into a body of standing water.

The sedimentary layers that compose a delta are divided into three general types: Bottomset beds that underlie the delta, foreset beds that compose most of its interior and topset beds that cap it near water level (Figure 10.19). Bottomset beds, as their name implies, are laid down at the foot of the delta. They are typically composed of fine-grained sediment that formerly traveled in suspension and may be deposited by density currents that carry their sediment load far out into the body of water. Bottomset beds are usually thin and their sediments are graded from coarse at the bottom of each bed to fine near the top. Foreset beds are laid down in more steeply dipping sets. They are composed of coarse material originally carried as bedload that avalanches down the front of the delta. The dip of foreset beds may approach the angle of repose in small lakes, but in the deltas of major river systems they may dip as little as 1°. Topset beds are extensions of the floodplain. They come to overlie the foreset beds as the delta advances into the lake or ocean. Topset beds are composed of sand- and silt-sized material typical of the floodplain and typically show cut-and-fill channel features.

Sediment deposition in Earth's oceans is complicated by the mixing of fresh water from rivers with salt water in the oceans. Fine-grained sediments such as clay particles flocculate upon mixing with salt water and settle out more rapidly then they would if they had entered fresh water.

Turbidity currents. Upon arriving in a large body of quiet water, the suspended load from rivers usually forms a mixture of water and sediment that is denser than the surrounding water. If the time required for the sediment to settle out of the mixture is long compared to the time for the mixture to flow down the face of the delta, it moves downslope as a density current, called a turbidity current. Turbidity currents act somewhat as underwater rivers: They gouge underwater channels that may possess levees and create distributary networks on the lower parts of deltas or deep-sea fans.

Deep submarine canyons that head on the continental shelves off the mouths of major rivers, such as the Hudson River of New York, were initially thought to require enormous fluctuations in sea level when they were first discovered. Only after a great deal of research was it realized that turbidity currents, not subaerial rivers, cut these canyons.

Turbidity currents usually flow episodically. After a period of accumulation near a sediment source, a threshold is passed and the sediment pile collapses, mixing sediment with water and generating a muddy, underwater density current. Storms and earthquakes may also trigger the release of large and powerful turbidity currents.

The deposits of turbidity currents are called turbidites. Turbidites are layered accumulations of sediment that grade from coarse at the bottom of each layer to fine near the top. They often show evidence of high-velocity deposition, such as incorporation of rip-up clasts and upper-regime planar bedding. The thickness of individual turbidite beds is usually highly variable, reflecting statistically random triggering processes. Although turbidites are usually deposited in deep water, many of them incorporate shallow-water fossils and other debris acquired in the near-surface source area, before being carried to much greater depths.

10.3.5 *Fluvial landscapes*

Long-continued fluvial transportation and erosion create distinctive landscapes. We have discussed branching channel networks, but fluvial processes have additional characteristics. Undrained depressions are rare on fluvial surfaces: Lakes and other depressions are quickly filled by transported sediment. Terrestrial geomorphologists regard any undrained depression as an anomaly needing explanation. Even our ocean basins are anomalous: If it were not for plate tectonic recycling, all land surfaces would eventually be cut down below the level of the sea (to wave base) and the ocean basins would be partially filled.

Timescales are important when considering fluvial erosion. Annual floods and large multi-annual inundations adjust the forms of the channel and floodplain, but the landscape as a whole responds on a much longer timescale. This timescale can be estimated by comparing the volume of material that can be eroded from a drainage basin to the rate at which sediment is carried out of the basin. The sediment discharge has been measured for many watersheds on Earth and the result is about 10 cm of land surface (averaged across the entire basin) per thousand years for the pre-industrial Earth – present erosion rates are much higher. These rates depend upon relief in the basin as well as climate, so this is a very rough average. If we take the average elevation of the continents to be a few hundred meters, the terrestrial erosion timescale is thus a few million years. This is roughly the time required for fluvial erosion to strongly affect the Earth's topography.

Base-level control. The base level for large fluvial systems on Earth is mean sea level. This is the level to which long-term fluvial erosion tends to reduce the land, because erosion below the level of the sea is very slow. Of course, base-level control on the Earth has been complicated by hundreds of meters of sea-level change during the glacial cycles of the past 3 Myr, a fact that must be taken into account when interpreting modern fluvial landscapes. Local base levels may develop above long-lived lakes or other obstructions to fluvial downcutting. The base level may occasionally change drastically in extraordinary events, such as the nearly complete evaporation of the Mediterranean Sea about 6 Myr ago, which caused rivers such as the Rhone, Po, and Nile to excavate kilometer-deep gorges that are now buried by modern sediment.

The base level changes whenever a dam is built; we now have considerable experience with the changes that such disturbances engender. Aside from artificial dams, landslides and lava flows create natural dams that may block an existing drainage, creating a temporary lake and inducing changes in river flow that gradually work their way upstream.

Older discussions of fluvial erosion supposed that after some change takes place, conditions remain constant for a long period and the landscape has time to adjust to the change. It has become clear that the Earth's landscape is too dynamic for such long-term equilibrium. Changes usually occur on timescales that are short compared to the equilibration time, and so the landscape is constantly adjusting to perturbations.

Graded rivers. The concept of the graded river was famously introduced by J. Hoover Mackin (1948). Mackin proposed that, over the long term, the slope of a river is adjusted to the volumes of both the water and the sediment it carries. The volume of water increases downstream (for perennial rivers) as more and more tributaries feed their water into the main trunk river. At the same time the size of the sediment carried usually decreases downstream, permitting more of the load to travel in suspension. Mackin's idea was that the river strives toward a balance between the load to be carried and the water that carries it, such that the "long profile" of the river (its elevation as a function of distance from its mouth) tends toward a final state that is steep near its sources and gentle near its mouth.

Mackin proposed that the long profile of a river is self-adjusting: If some reach of a river should suffer a decreased slope for any reason, the sediment that was formerly in transit is deposited, building up the bed of the river upstream while the water downstream, relieved of its sediment load, erodes into its bed (this is currently happening around many artificial dam sites). Both processes tend to increase the slope and oppose the original perturbation of the profile. Similarly, if any reach becomes steeper for some reason, the capacity of the river to erode its bed increases and the river cuts into its bed, forming a step in the long profile that gradually propagates upstream until an equilibrium is again established.

A change in the nature of the sediment carried by a river has analogous effects. As described above, hydraulic mining in California's Sierra Nevada in the mid-1800s added a large volume of coarse debris to the rivers flowing into the Pacific (Gilbert, 1917). The rivers responded by steepening their gradients until the coarse gravel could move downstream. The extra load of gravel built up the riverbeds downstream, causing the rivers to overflow their previous banks and deposit gravel on the adjacent farmland. If the injection of coarse gravel into the headwaters had continued, the net result would have been a river system with a much steeper gradient from the ocean to the mountains, although this would have required burial of most of the interior valley of California – an outcome considered highly undesirable by the residents of the Golden State, which is why hydraulic mining is now strictly banned.

Landscape evolution. Theories of how the Earth's landscape evolves under the influence of fluvial processes are as old as geology itself. James Hutton in 1795 attributed river valleys to erosion by the streams that flow in them, an idea that did not sit well with his contemporaries or even with his intellectual heir Charles Lyell nearly a century later. Even after the fluvial origin of landscapes was accepted, ideas on how they evolved were qualitative and made few testable predictions. American geographer William Morris Davis (1850–1934) is widely remembered for his classification of landscapes as young, mature, and old, based on a presumed rapidity of tectonic movements, which create initial landscapes that are later dissected by fluvial erosion during an era of tectonic quiescence. Davis supposed that, following a long period of erosion, landscapes are reduced to a surface of low relief near the base level that he christened a "peneplain." Davis and German geologist Walther Penck (1888–1923) engaged in a bitter but somewhat fuzzy controversy over whether landforms "wear down" (Davis), with slopes everywhere declining as time passes, or "wear back" (Penck) who suggested that steep slopes retreat from their initial positions while retaining their steepness.

Classical ideas on fluvial processes, to which the ubiquitous G. K. Gilbert made many contributions, especially in his report on the geology of the Henry Mountains (Gilbert, 1880), divided the fluvial system into erosion, transportation, and deposition. Erosion takes place mainly at the level of valley sides and first-order rilles, which yield sediment that is carried through the stream system and is finally deposited in an alluvial fan or body of standing water.

Research in the modern era has focused on quantitative descriptions of each of the parts of this cycle. Until recently, most effort has gone into understanding individual processes, such as Bagnold's work on the physics of sediment entrainment and transportation, or have focused on hillslope processes, of which the book by Carson and Kirkby (1972) is a fine example. "Process geomorphology" has now become so large a subject, and the synthetic landscape evolution models of Davis and Penck have such a reputation for imprecision, that many recent textbooks avoid the topic of landscape evolution altogether.

Most recently, perhaps driven by the immense increase in computer power, quantitative syntheses of fluvial processes into landscape models have become possible and are now achieving impressively realistic results. These results are being subjected to quantitative tests through our recent ability to date landscapes through cosmogenic isotope methods. The first quantitative model of fluvial evolution of this kind was constructed by geophysicist Clem Chase (1991), who built a model of landscape evolution on a two-dimensional grid that evolved by simple rules suggested by fluvial processes. Simple as this pioneer model was, it produced very realistic landscapes that evolved in ways similar to those inferred for actual landscapes. Models of this kind are now reaching a high level of sophistication: The time has clearly come for our detailed understanding of individual processes to be synthesized into descriptions of how entire landscapes evolve. A recent review of progress in this area is by Willgoose (2005).

Although the evolution of fluvial landscapes has seemed a quintessentially terrestrial process (Martian fluvial landscapes are clearly not highly evolved), Cassini images of integrated drainage networks on Titan have created a new field for application of these models. Many of the basic parameters that control fluvial processes on Titan are unknown: How much precipitation falls and how it varies with time, what materials are being eroded, how Titanian "bedrock" (very cold water ice) weathers to sand-sized particles, and many other important facts are presently unknown. However, landscape evolution models themselves may shed some light on these questions. For example, how much sediment must be moved before an integrated drainage network forms? How much erosion does it take to turn a densely cratered landscape into a fractal landscape of connected drainage basins?

Further reading

Hydrology is an enormous subject on its own. One of the founding papers that can still be read with profit is Hubbert (1940). This paper takes a quantitative analytical approach that is difficult to find even in the modern literature. A comprehensive look at the older literature that still has much of value is Meinzer (1942). A full treatment of the percolation of fluids through porous media is found in Bear (1988). Students looking for a quick but mathematical overview of flow through porous media and its application to geodynamic problems can do no better than to consult Turcotte and Schubert (2002).

Fluvial processes are also the subject of an enormous literature, although many of the modern treatments focus more on societal problems such as pollution and water supply than on fundamental science. Two of the classic, science-oriented treatments are Schumm (1977) and Leopold *et al.* (1964). A thorough examination of the role of water in all surface processes, not just rivers and streams, is by Douglas (1977). The fluid mechanics of flow in open channels and an in-depth discussion of the various resistance formulas can be found in Rouse (1978). The interaction between flowing water (and other fluids) and its channel is covered by Allen (1970), while the best treatment of the physics of sediment entrainment is still Bagnold (1966). The sedimentological aspects of transport by water are well reviewed in the massive book by Leeder (1999). Coastal and wave processes in general are lucidly discussed in Komar (1997), while the more geomorphological aspects of shorelines and coasts are treated by Bird (2008). The modern synthetic approach to landscape evolution is too new to have texts describing it: The interested reader is referred to the short review of Willgoose (2005), but readers wishing insights into the history of landform analysis will be delighted by Davis (1969), or for a shorter and more comprehensive introduction by Kennedy (2006).

Exercises

10.1 Underground plumbing on Titan

During a particularly hot, dry spell on Titan, when temperatures rose to a balmy 97 K, over a period of one Earth year about a meter of methane evaporated from a (hypothetical) 100 km wide lake. In spite of the methane loss, no change in the level of the lake was observed (to a precision of ± 10 cm). If no surface methane flowed into the lake at this time, the loss must have been compensated by subsurface flow. Estimate the minimum permeability required in the subsurface to supply this loss. Comparing this to permeability of rocks on Earth (Table 10.1), what can you infer about Titan's subsurface?

10.2 Take Manning to new worlds

Show how the Manning equation (10.12) scales with the acceleration of gravity g by assuming that the Darcy–Wiesbach coefficient f is independent of gravity. Using this equation, compute the flow velocity of a flood on Mars that moved down a channel 30 km wide and scoured hills up to 100 m high above the channel floor. MOLA measurements indicate that the slope in this region was about 1 m/km. Assume that the material of the valley floor was large cobbles more than 20 cm in diameter. Use the same method to calculate the velocity of a methane stream on Titan that flowed in a channel 100 m wide and 3 m deep down a slope of 0.3°. Estimate the grain size of the material using available data/reasonable guesses.

10.3 Swing wide, lazy river

A meandering river on Earth flows through a channel 100 m wide and moves along at an average velocity of 3 m/s. It swings through a meander bend with an inner radius of 0.5 km. The centripetal acceleration in moving around the bend raises the level of the water on the outside of the bend relative to that inside. What is this elevation difference and is this large enough, in your opinion, to measure?

10.4 Entrenched meanders on Mars?

One of the major problems of analyzing ancient rivers is figuring out how much water once flowed through them. Nirgal Vallis on Mars is a sinuous trough nearly 500 km long and about 8 km wide in its lower reaches. In this area the sinuous undulations and alcoves along the walls have a wavelength of about 15 km. Although Nirgal has been attributed to sapping, suppose that its undulations were formed by a meandering river that gradually eroded down into the surface. Use the information given here to estimate the original width of its channel. Given this width, how might you estimate the discharge of the river that cut this gorge?

10.5 Low-gravity surfing

Evidence is accumulating for the former presence of lakes and perhaps oceans on Mars. Discuss, so far as you are able, how the acceleration of gravity affects waves on the surface of standing bodies of water (considerations of wave velocity and energy transport are particularly relevant here). In particular, how do you think low gravity might affect shoreline processes such as erosion and transport on Mars or Titan?

11

Ice

> My theory of glacier motion then is this – A glacier is an imperfect fluid, or a viscous body, which is urged down slopes of a certain inclination by the mutual pressure of its parts.
>
> J. D. Forbes, *Travels Through the Alps of Savoy* (1846), p. 365

Scottish physicist J. D. Forbes (1809–1868) was an inveterate mountaineer, observer, and scientific quarreler. His direct observations of glacier flow led him to describe glaciers as highly viscous fluids, a conclusion that could not be reconciled with the mechanical understanding of his day. He spent many years ferociously defending his view against the opposing opinions of another mountaineering physicist, John Tyndall (1820–1893), an Irishman who argued that glaciers deform by regelation, a process in which ice melts and refreezes under pressure.

Both Forbes and Tyndall had become fascinated by the mechanics of glacial phenomena at a time when Swiss paleontologist Louis Agassiz (1807–1873) was announcing his deduction that Europe had once been smothered under immense glaciers during a former age of ice. Although Forbes' views are now accepted as being closer to the truth, a full understanding of how crystalline ice can flow like a viscous fluid was not achieved until the 1950s. Tyndall's regelation idea still finds a place in glacier mechanics, although it is now relegated to sliding at the base of warm-based glaciers.

Ice on the surface is mirrored by ice in the ground. The surfaces of cold regions are often underlain by frozen ground that, because of seasonal thermal cycling, is unusually active and highly productive of distinctive organized patterns.

The slow flow of ice, whether frozen water or other substance, as an "imperfect fluid or a viscous body" creates landforms that are different from those produced by wind or water. Ice occurs either disseminated through the regolith, in small bodies properly called glaciers, and in large ice sheets that can cover large areas.

11.1 Ice on planetary surfaces

Glaciers are the counterparts of rivers in a cold hydrologic cycle. Ice accumulates as snow at high elevations, slowly flows down to lower elevations, and there melts, eroding and transporting the adjacent rock as it moves. Glaciers on Earth currently underlie about 10% of its surface and incorporate about 3% of its water, a number that is rapidly declining in the modern anthropogenic world. During the Earth's recent Ice Age glaciers and ice sheets occupied 30% of the surface and locked up 8% of the water. Water-ice glaciers are apparently active on Mars at the present time and, on Mars as on the Earth, seem to have been much more extensive in the past, although the former Martian Ice Age occurred much farther back in time than the Earth's. Mars may also host glaciers composed of solid CO_2, which require temperatures lower than any achieved on Earth. So far, no glaciers have been found elsewhere in the Solar System: Titan's surface is just a bit too warm for methane to freeze into ice. Ammonia glaciers are theoretically possible but no examples have yet been discovered. Although some researchers have speculated about solid nitrogen glaciers on Triton, no glacial features have been identified there. Earth, however, hosts a very unusual type of "glacier" composed of salt (halite), NaCl, that flows by virtue of interaction with small amounts of water.

11.1.1 Ice within the hydrologic cycle

The fluvial hydrologic cycle begins with water falling on the surface as rain, running downhill and picking up sediment, then flowing to an accumulation point where the sediment is dumped into a delta or moved further by longshore and marine currents. In an analogous manner, snow that falls on mountaintops or cold regions, where snowfall exceeds melting, accumulates into a permanent snowfield. If this accumulation were to continue without limit, the Earth's water would soon end up frozen into huge ice mountains. However, snow metamorphoses into ice, which can flow off the land surface as a thick, viscous liquid and, thus, returns the water to lower elevations, where it melts back into water to close the cycle.

Different portions of a glacier or ice sheet can, thus, be distinguished by their function in the overall mass balance as either accumulation areas or ablation areas, which are connected by flowing ice (Figure 11.1). These regions are readily recognized in a terrestrial glacier by color and texture (or thermal inertia) at the end of the summer season: Accumulation areas are bright white, underlain by coarse snow (firn), which is undergoing the transformation into ice, while ablation areas appear as dense blue ice that is in the process of melting. Typical glacier flow velocities range from 0.1 to 2 m/day, exceptionally up to 6 m/day. Large outlet glaciers from ice sheets may flow up to 30 m/day. Some glaciers are observed to suddenly speed up to rates of 70 m/day in brief episodes known as surges. The mechanical basis of surges was once mysterious, but we now know that surges are due to changes in the subglacial plumbing of warm-based glaciers, discussed below.

The heads of glaciers and ice sheets are in areas where more snow accumulates each year than melts, and they flow down to elevations where melting exceeds snowfall, often to altitudes well below the snowline at which accumulation and melting are in balance. A glacier is, thus, in a state of dynamic balance between accumulation and melting.

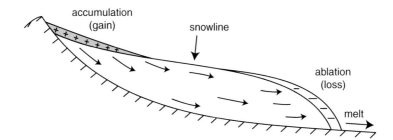

Figure 11.1 Schematic plan of a valley glacier, illustrating the accumulation zone at high altitude, flow of glacier ice downslope, and the melting of the ice in the low-altitude ablation zone. Depending on its temperature, the ice at the base of the glacier may or may not be able to slide over the bed. Inspired by Figure 1 of Sharp (1960).

Accumulation may occur by means other than snowfall directly onto a glacier or ice sheet. Avalanches from adjacent highlands or wind-blown snow (especially important on icecaps where precipitation is low) may deposit snow or even ice directly onto the accumulation area. Valley glaciers, like rivers, may be fed by tributaries that themselves originate in merging ice streams, although the number of links in such networks is usually small.

Once snow falls on a glacier, it undergoes a regular series of changes as it metamorphoses from new snow, to old snow, to firn (density about 550 kg/m^3), then finally into glacier ice (density 820 to 840 kg/m^3). These processes, discussed in Section 7.2.2, involve the interaction with seasonal liquid meltwater and vapor-phase transport within the snowpack. Atmospheric gas bubbles, presently of great importance for measuring the composition of Earth's pre-industrial atmosphere, may be trapped in the process and preserved for many thousands of years.

Ablation, or mass wasting, of the ice is usually by melting. Evaporation is generally unimportant except for tropical glaciers and in the dry valleys of Antarctica. Tidewater glaciers and continental ice sheets, however, may lose most of their mass by calving of icebergs into the sea.

11.1.2 *Glacier classification*

A common classification of glaciers is based on their morphology. There are three general types: *Icecaps* or *ice sheets* are continuous sheets of ice. Their flow is centripetal, from a high-standing center toward their edges. Terrestrial examples are the Greenland and Antarctic ice sheets or the former Pleistocene ice sheets. A former Martian ice sheet may have covered large areas in its Southern Highlands. *Valley glaciers* are ice streams that have heads in mountainous terrain. They are common in the Earth's high mountains, such as the Alps of Europe, Himalayas of Asia, or the northwest coast of North America. The North American glaciers are noteworthy as most of them head at low elevations – often

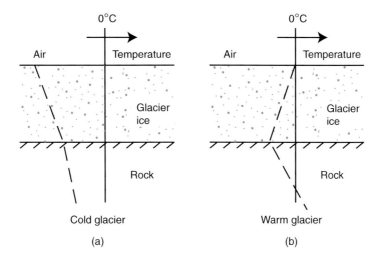

Figure 11.2 Thermal regimes of glacier ice. (a) Illustrates a cold-based (or polar) glacier, in which the temperature is everywhere below the freezing point of water and geothermal heat is conducted to the surface. (b) Illustrates a warm-based (or temperate) glacier in which the temperature is at the pressure-controlled melting point of ice. Because of the inverse dependence of the ice's melt temperature on pressure, the temperature is slightly colder at the base than at the surface, which implies that heat is conducted to the glacier bed. After Figure 5 of Sharp (1960).

only 2000 m – where they are nourished by the heavy winter snowfall from the adjacent Pacific Ocean. *Piedmont glaciers* are ice sheets formed by the coalescence of several valley glaciers on flat terrain at the base of mountains. An example is the Malaspina Glacier in Alaska.

A second classification is based on temperature. Cold glaciers (alternatively called polar glaciers) are easiest to understand. The temperature throughout the ice body of the glacier is below the pressure melting point of ice (Figure 11.2a). The normal gradient of the internal temperature of the Earth is continued through the ice, from warmer below to colder near the surface, although the slope is different because the thermal conductivity of ice is somewhat different from that of rock (Table 4.2). Cold glaciers are frozen to their beds. Their motion occurs by internal deformation of the ice itself through solid-state creep (Section 3.4.3). They are among the most slow-moving of terrestrial glaciers and they are generally ineffective in eroding their beds. Often, where a polar glacier has melted away, there is little evidence of its former existence.

Warm glaciers (alternatively called temperate glaciers) are at the pressure melting point of ice throughout their mass. Because of the inverse slope of the melting curve of water ice, $dT_m/dz = -0.65$ K/km of ice, the temperature actually *decreases* with increasing depth in the glacier (Figure 11.2b). This inverse gradient means that thermal conduction moves heat from the surface *downward*, toward the glacier bed. At the same time, the normal geothermal gradient in the rock below the glacier moves heat upward toward the glacier bed with a slope of about 30 K/km. This creates a thermal crisis for the glacier, which responds by

(a) 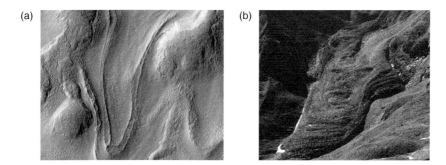 (b)

Figure 11.3 (a) A tongue-shaped flow on Mars located on the eastern wall of Hellas Planitia. This flow is about 5 km long and 1 km wide. It is likely to be a Martian analogue of terrestrial rock glaciers. Image PIA09594_fig 1, portion of HiRISE image PSP_002320_1415. NASA/JPL/ University of Arizona. (b) Jungtal rock glacier in the Swiss Alps (image courtesy of Dr. Jan-Christophotto, 2011).

melting at its base, converting about 1 cm of ice into water each year. Warm-based glaciers are saturated with liquid water, which is in equilibrium with ice throughout the body of the glacier. Temperatures remain everywhere at the pressure melting point, but a great deal of heat is, nevertheless, transferred in such glaciers by the latent heat from the conversion of solid ice to water, which can flow readily from place to place transferring its latent heat as it moves and then freezes.

Warm-based glaciers can slide over their beds and, with the aid of rocks and debris frozen into the ice, are very effective at abrading and quarrying out the underlying bedrock. These glaciers also deform internally: Typically about half their surface velocity is due to internal deformation and half is due to basal sliding.

Temperature is not, however, always a good classification for the entire glacier, because the thermal regime can change with position in the glacier. Thus, a glacier's upper reaches could be "cold" while its lower parts are "warm." Moreover, parts of the Antarctic ice sheet have meltwater near their beds, indicating a warm-based regime, while their surfaces are cold, well below the pressure melting temperature.

11.1.3 Rock glaciers

Although sometimes not considered "proper" glaciers, rock glaciers are dense mixtures of rock and ice that, despite being mostly composed of rock, nevertheless show clear evidence of flow, albeit moving much more slowly than the mostly pure ice glaciers familiar to glacial geologists. Rock glaciers creep along at rates of centimeters to meters per year, but exhibit the lobate margins, drapery-like ridges, and lateral moraines typical of valley glaciers. Their margins are typically steep, at or close to the angle of repose. They are included here because recently discovered Martian glaciers may be rock glaciers, not solid ice (Figure 11.3).

Mixtures of ice and rock in glaciers form a continuum, running from nearly pure ice, through ice carrying small quantities of rock and debris, to rock glaciers, which are mainly composed of rock debris. Some rock glaciers have ice-rich cores mantled with ice-free rock debris, while in others the ice merely fills the interstices between boulders. Rock glaciers have not received much study, partly because of their rarity and partly because of the difficultly of probing into their interiors: Unlike glaciers, one cannot simply melt boreholes through them with electric heaters or hot steam.

The detailed mechanism by which rock glaciers deform internally is not well understood, in spite of finite element modeling of their flow (see the review by Whalley and Azizi, 2003). The slow creep of the rock/ice mixtures must be due to the included ice, but models of how the heavy burdens of rock debris affect the flow rate and its dependence on factors such as shear stress are not well developed.

11.2 Flow of glaciers

Many theories of how glaciers flow have been proposed since Agassiz brought the importance of glaciers to the attention of geologists. These historical theories include crevasse filling and refreezing, regelation within the mass of ice, and many others. The reason for so many theories was the apparent paradox of a crystalline solid (you can easily see the crystals in partially melted specimens of glacier ice) that, nevertheless, flows like a fluid.

We now understand that "solid" is a poor description of a crystalline material near its melting point, because any material can flow, although the motion is perceptible only very near the melting point. This was first demonstrated for ice by glaciologist J. W. Glen in 1955, although this sort of "creep" had been observed in metals and some rock materials long before. Individual crystals flow by the generation and movement of a peculiar sort of crystal defect known as a dislocation. Dislocations and their dynamics were some of the most important discoveries of twentieth-century materials science.

Figure 11.4 shows how a dislocation (an "edge" dislocation in this figure) can move across a crystal with minimal distortion of the crystal lattice and yet accommodate a net shear displacement. This mode of deformation is common to all crystalline substances, so all can deform plastically and, thus, creep. There is only a minimal threshold stress, so no material has finite strength at high temperatures or over long time periods. Figure 11.4 shows how a dislocation can "glide" through a crystal lattice. In any real material many dislocations are present at the same time and, by gliding across one another, they create kinks in one another that effectively pin the dislocations at the crossing locations. When this occurs, glide ceases after a few percent of strain. A new step is required to free the dislocations from their pinning points. That step requires the bulk motion of atoms through the lattice – diffusion. The process of dislocation glide coupled with diffusive untangling of dislocations is known as dislocation climb. Because diffusion is a thermally activated process, so is the rate of creep. The flow of crystalline solids is, thus, strongly temperature-dependent, with an activation energy similar to that of the bulk diffusion of the solid. This explains why creep is rapid only at high temperatures.

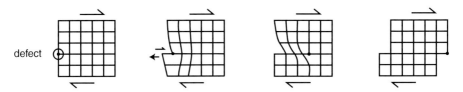

defect

Figure 11.4 Motion of a dislocation through a crystal lattice leads to shear deformation of the crystal. Starting at the left, a line defect is created when the atomic bond of one row of atoms in the upper half of the crystal is shifted one lattice spacing to the right, creating a line of local disorganization known as a dislocation. The atoms in the crystal shift partners as the dislocation moves to the right. After the dislocation finally emerges from the crystal on the right, the upper half of the crystal is shifted one lattice spacing to the right, accommodating an increment of shear strain. This is the mechanism by which ice crystals deform in a creeping glacier.

11.2.1 Glen's law

J. W. Glen (1955) was the first to make careful measurements of the relation between stress and strain in polycrystalline ice and to apply it to the flow of glaciers. Unlike the flow of viscous fluids, he found that the strain rate is not a linear function of stress, but depends upon stress raised to a power n greater than 1. As described in Section 3.4.3, this is the general behavior of creeping hot solids, materials whose flow is dominated by dislocation motion. Glen expressed his rheological law in terms of the strain rate:

$$\dot{\varepsilon} = A\sigma^n = B\,e^{-Q/RT}\,\sigma^n \tag{11.1}$$

where Q is the creep activation energy, R is the gas constant, T the temperature in K and σ is the applied shear stress. Glen found that the power n ranged between 2 and 4, with a preferred value of 3.2 (the modern value of n is 4). His estimate of the activation energy was $Q = 134$ kJ/mol, less than the modern value of 181 kJ/mol listed in Table 3.3. His experiments gave the constant $B = 3.5 \times 10^{20}$ MPa$^{-3.2}$s^{-1}. Nevertheless, Glen clarified the important differences between the flow of Newtonian viscous liquids and glacier ice (Forbes was not *completely* right) and showed that very cold ice should flow less readily than ice near its melting point. He also realized that the creep rate changes as the ice recrystallizes during flow, a factor that is still not fully incorporated into modern creep laws.

The implication of a power-law dependence of strain rate on stress is that as stress increases the strain rate increases faster than a direct proportion. Thus, doubling the stress for $n = 4$ means that the strain rate increases by a factor of 16. Most of the strain, thus, becomes concentrated in high-stress areas; at the bottom of an ice sheet on an inclined surface, for example. Non-linear rheological laws such as Equation (11.1) generally resist easy analytic solutions and require numerical methods to get quantitative results. Thus, the creep law for an infinite sheet of power-law fluid flowing down a surface of constant slope can be readily obtained, but it is quite remarkable that there is also an analytic solution for the much less straightforward case of a power-law fluid flowing down a constant slope in a semicircular channel of constant width (Nye, 1952). We give this more interesting result here, leaving the much simpler case of an infinite plane sheet as an exercise for the interested reader.

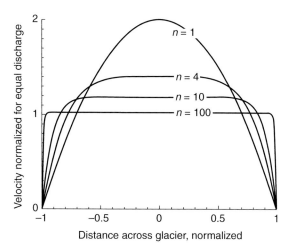

Figure 11.5 Flow profiles across different power-law fluids flowing down a trough-shaped channel with a semicircular cross section. Newtonian fluids with $n = 1$ attain a parabolic profile, while fluids with increasingly high n attain more plug-like profiles, with uniform velocity in the center and steep gradients near their walls.

Let r be the radial distance from the centerline of a power-law fluid flowing through a trough-like semicircular channel of radius R that slopes downhill at angle α. For this case, the downstream velocity $u_z(r)$ is given by:

$$u_z(r) = \frac{2A}{n+1}(\rho g \sin \alpha)^n \left[\left(\frac{R}{2} \right)^{n+1} - \left(\frac{r}{2} \right)^{n+1} \right]. \tag{11.2}$$

For flow in a parabolic or elliptical channel the coefficient of the right-hand side changes slightly, and the scales for vertical and horizontal velocity gradients are different, but the overall behavior is similar. Naturally, this equation also applies to Newtonian flow when $n = 1$.

Plots of this equation in Figure 11.5 show that as n increases the flow becomes more plug-like. Comparison of this equation to the profiles of actual glacier velocity profiles in a straight reach of Saskatchewan Glacier shows good agreement with $n = 3$ and are inconsistent with a Newtonian, $n = 1$ curve (Meier, 1960). Deformation of vertical boreholes through a glacier can also be compared against theoretical velocity profiles such as Equation (11.2). Measurements of this kind show good agreement between the non-Newtonian flow theory and observation (Paterson, 1999).

For more complicated channel geometries with varying cross sections, bends and obstructions, numerical solutions to the flow equations must be constructed. A great deal of progress has been made in the application of finite element methods to prediction of glacier velocity patterns.

The actual rheology of ice is more complicated than Glen's law alone would suggest. Recent investigations indicate that a variety of mechanisms in addition to dislocation climb are important for the deformation of ice, especially at low stress. Such processes as intra-grain diffusion, diffusion through grain boundaries, and grain boundary sliding all contribute to the deformation of ice under various conditions of stress and temperature. The interested reader will find an excellent summary of the complex rheology of ice with planetary applications in the review by Durham and Stern (2001).

11.2.2 The plastic-flow approximation

Figure 11.5 shows that as the exponent n increases, the corresponding flows become more plug-like. For very large n the result would approximate the flow of a perfectly plastic material, one that remains rigid until a yield stress Y is exceeded, after which it flows to an extent determined only by external constraints on the displacement. This observation led to the idea that a power-law fluid with large n can be approximately represented as a perfectly plastic material. Application of this idea to ice suggests adoption of an effective yield stress of about 0.1 MPa for the "strength" of ice.

The plastic model of ice rheology gives a fairly good prediction for the height profile of an ice sheet as well as the cross section of some glacier tongues, as discussed in Section 5.3.2 for lava flows and below for ice sheets. However, there is a serious internal problem with this model and it should be used only with some caution and understanding of this issue. This problem can be seen from dimensional considerations. The fundamental rheological law (11.1) for ice relates a strain rate to a stress, so the coefficient relating the two has the dimensions of inverse stress (to the nth power) and inverse time. However, the plastic law contains no dimension of time, only stress. These two relations, thus, do not transition smoothly over into one another as n becomes large unless some quantity with dimension of time is present. That is, one must quote a timescale τ in addition to a yield stress for this relation to be meaningful. To see this explicitly, rewrite (11.1) as:

$$\dot{\varepsilon} = \frac{1}{\tau}\left(\frac{\sigma}{Y}\right)^{n}. \tag{11.3}$$

This rewrite is always possible once Y is given. In that case τ is defined implicitly by (11.1). In the limit $n \to \infty$ this equation approaches the plastic yield condition:

$$\begin{aligned} \dot{\varepsilon} &\to 0 \quad \sigma < Y \\ \dot{\varepsilon} &\to \infty \quad \sigma > Y. \end{aligned} \tag{11.4}$$

Equation (11.4) is commonly cited in connection with the plastic approximation (e.g. Paterson, 1999), but without a timescale, τ, Equation (11.3) is not dimensionally correct. The creep rate according to (11.1) is never really zero at low stress – it is only "almost as good as zero" over some timescale that the user considers important.

The widespread neglect of this timescale in the glaciological literature is probably why there is quite a bit of variation in estimates quoted for the yield stress, although it is usually given as "about" 0.1 MPa in round numbers. One can make this approximation precise for Glen's flow law if $\tau = 1.4$ yr is the relevant timescale over which deformation is considered important. This is also the time required for a stress equal to the yield stress to produce a strain of 100%.

11.2.3 Other ices, other rheologies

Deformation by dislocation climb and other mechanisms is not confined to water ice. Ices of other substances, such as solid CO_2, ammonia, and nitrogen, are all of interest for the formation of glaciers on other Solar System bodies. Unfortunately, the rheology of such ices is not as well explored as that of water ice. Table 3.5 lists recent data on the flow law of solid CO_2, but the flow rates of other ices can only be conjectured at the present time. It is, thus, important that theoretical models of steady-state creep exist and, in many cases, are successful in predicting the rheological behavior of complex substances. The detailed examination of creep mechanisms is a large area of research that is beyond the scope of this book. For more information, the reader is referred to the recent monograph by Karato (2008). However, a few simple scaling arguments can be stated that allow crude estimates of the relative creep rates of different substances.

Nearly all rheological creep laws depend upon thermally activated diffusion to permit slow deformation of crystalline material. This explains the observed exponential temperature dependence of relations like Glen's law (11.1). The actual temperature dependence is a function of the binding energy of atoms in the material, the species diffusing, and whether the diffusion occurs along grain boundaries or through the body of the crystal. The dependence of strain rate on stress is a function of whether the deformation is by dislocation climb or by pure diffusion. Dependence of the creep rate on grain size is also a function of the mechanism: Dislocation climb rates do not depend on grain size, whereas diffusion creep processes are functions of grain size because it determines the length of the diffusion path.

The individual equations for different creep mechanisms are, thus, functions of many different variables, not all of which may be known for a new substance. Nevertheless, the overwhelming influence of diffusion makes it possible to create order of magnitude estimates of creep rates. We can crudely write the strain rate at a given stress as:

$$\dot{\varepsilon} = (\text{a bunch of complicated stuff}) \, D_0 e^{-Q/RT} \, \sigma^n. \tag{11.5}$$

The complicated term in the parenthesis is a function of the many variables we have mentioned (and others that we have not). The second and third terms, combined, constitute the temperature-dependent diffusion coefficient, while the last term is the shear stress, raised to some exponent ranging between 1 (for pure diffusion creep) and about 5 (for generic dislocation climb).

We can now form the ratio between the creep rate of an unknown substance and that of a known material (water ice, for example) and compare the two. It should be clear that the relative creep rate depends mainly on the relative diffusion coefficients:

$$\frac{\dot{\varepsilon}^{\text{unknown}}}{\dot{\varepsilon}^{\text{ice}}} = (\text{number of order 1}) \frac{D_0^{\text{unknown}}}{D_0^{\text{ice}}} e^{(Q_{\text{ice}} - Q_{\text{unknown}})/RT}. \tag{11.6}$$

This allows us to evaluate the relative effectiveness of different substances as candidate glacier materials if we can make some guess about their diffusion coefficients relative to ice. We may, if we are desperate and in complete ignorance, take one more step into the world of wild and wooly approximations (a world in which planetary scientists must dwell all too frequently) and invoke *Shewmon's rule of thumb*, which states that the diffusion coefficient of *any* substance at its melting point is 10^{-12} m^2/s (Shewmon, 1963). This leads one to expect that, at their melting points, all substances should flow roughly like glacier ice. At lower temperatures, flow rates are slower and depend upon the activation energy for whatever form of diffusion is most effective in permitting the crystal to deform.

Interactions of ices with other substances, especially interstitial fluid, may greatly enhance the rate of creep. Box 11.1 describes the strange case of salt glaciers on Earth, in which the presence of small amounts of intergranular water greatly enhances the creep rate and permits salt to form kilometer-long flows that look superficially like glaciers. Similar enhancements might occur in water ice at low temperatures when small quantities of ammonia are present, as occurs in the outer Solar System.

11.2.4 Basal sliding

The sliding of a glacier over its bed is a quintessentially water-ice process. Warm-based glaciers with melt at their beds are a consequence of water's nearly unique negative-slope melting curve. The process of regelation that makes glacial erosion so effective in warm-based glaciers also depends on this peculiarity of water's melting curve.

When it can occur, basal sliding is an important contributor to the overall motion of a glacier or ice sheet. Many estimates suggest that approximately half of the surface velocity of a warm-based glacier is due to basal sliding. The warm-based Antarctic ice streams similarly depend on basal slip for their high velocities. In these streams the basal ice interacts with water-saturated deformable sediments, not rock. The mechanics of this soft-sediment interaction is complex and not presently well understood. The mechanical behavior of water-saturated sediments themselves is complicated and their confinement beneath a moving ice sheet introduces complex feedbacks that are the subject of current research.

Regelation. Basal sliding over a rigid bed is better understood. When moving ice at the base of a glacier encounters an obstacle on the bed, such as a rock protuberance or wedged boulder, the pressure on the upstream side of the obstacle increases, while that on the downstream side decreases. Because of the negative slope of water ice's melting curve, this lowers the melting point on the upstream side. A small amount of ice melts and the local temperature declines slightly to the pressure melting point. Upstream ice thus melts, but not instantaneously: The rate of melting is regulated by the rate at which the latent heat of melting, 334 kJ/kg, is supplied to the compressed ice by conduction from the adjacent ice and rock. However, as heat is conducted to the upstream ice, the melt-water flows around the obstacle and freezes behind it at a slightly higher than ambient

---

Box 11.1 **Salt glaciers and solution creep**

Salt on the surface of the Earth would seem to be one of the least likely materials to flow as a glacier. Measurements of the creep of pure halite (NaCl) show that, although it does creep more readily than most rocks, it still requires temperatures in the vicinity of 550 °C for it to creep at a rate comparable to that of glacier ice.

Unlikely as it might seem, glaciers of salt several kilometers long were described from the dry Zagros Mountains of Iran in 1929. The discoverers did not believe that salt at normal surface temperatures could flow at rates comparable to glacier ice and supposed instead that the salt had erupted hot, at temperatures near 300 °C, and that the glaciers are not moving at present.

However, salt is highly soluble in water and a small amount of rain does fall in this region. The theoretical possibility that small amounts of water could enhance the flow rate of salt by the mechanism of pressure solution creep was investigated by Wenkert (1979). Pressure solution creep occurs when a crystal subject to differential stress preferentially dissolves on faces under compression and is deposited on faces under extensional stresses. The dissolved crystal material diffuses much more readily through the solvent than through the body of the crystal. The shear strain rate is given by:

$$\dot{\varepsilon} = 21 \frac{V_0 C_L D_L f}{kTd^2} \sigma \qquad \text{(B11.1.1)}$$

where V_0 is the volume of the diffusing species, C_L its molar solubility in the solvent, D_L is its diffusivity in the solvent, k the Boltzmann constant, T the absolute temperature, d the grain size in the solid, and f is the fraction of liquid wetting the solid.

Applied to salt glaciers, this equation predicts a strain rate 10^8 faster than that expected for pure halite. This prediction was verified by both direct observations of the creep of salt glaciers following rare rainfall events and by laboratory measurements of damp halite (Urai *et al.*, 1986).

Study of the creep of salt has attracted a large amount of attention because of its importance for proposed nuclear waste storage in salt deposits. Aside from this practical application, solution creep is expected to greatly enhance the flow of limestone in the Earth. It has also been proposed as an agent in enhancing the creep of cold water ice in the outer Solar System through the solution of ice in interstitial ammonia.

---

temperature, there releasing its latent heat, which is now available to be conducted to the upstream face.

This process of melting under compression, followed by meltwater flow and freezing in the adjacent low-pressure zone, all regulated by the conduction of heat from the freezing water to the melting zone, is called regelation. Regelation is easily demonstrated in the kitchen (or classroom) by hanging a wire loop weighted at both ends over an ice cube supported at its ends like a beam. The wire very quickly slices through the ice cube and emerges on the other side, leaving the ice cube apparently intact (actually, it is not quite intact – examination under polarized light shows that the ice along the path of the wire has recrystallized).

Regelation at the base of a glacier is very efficient for small obstacles, through which heat is rapidly conducted, but inefficient for large obstacles. On the other hand, the glacier can easily deform around long-wavelength obstructions, but deformation is difficult for small wavelengths, requiring high strain rates. There is, thus, some intermediate wavelength that is maximally obstructive – the expectation is that this wavelength accounts for most of the resistance to basal sliding. Estimates of the size of this most obstructive obstacle indicate that it is about 10 cm.

Glacier Surges. Most glaciers move down their valleys at a sedate speed of a few meters per day. However, a few glaciers are observed to suddenly accelerate to many tens of meters per day in rapid advances known as glacier surges. A surging glacier rapidly lengthens and thins, overrunning forests and roads in its path. Its surface breaks up into a wilderness of crevasses separating large blocks that topple as the glacier moves, making it nearly impossible to cross or even remain safely in one spot on the ice for more than short periods of time. Because they are so difficult and dangerous to study, little was known about the mechanics of surging glaciers until a heroic effort with massive helicopter support was mounted during the 1982–1983 surge of the Variegated Glacier in Alaska (Kamb *et al.*, 1985).

It was discovered that the immediate cause of the Variegated Glacier's acceleration was a large increase in the water pressure at its base, which occurred in conjunction with a rearrangement of the system of subglacial cavities and tunnels that drain the glacier. This increase of water pressure lifted much of the glacier's weight off its bed, decreasing basal friction and greatly enhancing the rate of basal sliding.

Surges are evidently restricted to warm-based glaciers and it may be that, given enough time, all warm-based glaciers will exhibit surge activity. An interesting question is whether the warm-based Antarctic ice sheet is also subject to surges and, if it is, what conditions must be met to cause a surge.

11.3 Glacier morphology

Glaciers are tongue-like masses of ice that flow down valleys, whereas ice sheets are broad plains of ice that spread centripetally from their high centers. Valley glaciers typically carry masses of rock debris along their margins, material that has avalanched from the valley sides onto their surfaces. Where ice streams meet, these lateral moraines merge into long trains of debris within the body of the compound glacier and are known as medial moraines.

The terminus of a glacier or ice sheet may remain at the same location for a long period of time, but this does not mean that the ice is not moving. Instead, the ice is continually pushing forward while it melts back at the same rate, making the terminus a dynamic location that is constantly subject to small oscillations as the balance between flow and melting shifts slightly. Because new ice is constantly arriving at the terminus, debris frozen into the ice melts out and gradually builds up into what may become a large heap – the terminal moraine. Even when the ice front retreats because melting predominates over flow, the internal movement of the ice is still downward: The flow velocity never reverses.

11.3.1 Flow velocities in glaciers and ice sheets

A widespread misconception supposes that the ice in a glacier is squeezed out by the weight of the overlying ice, somewhat like toothpaste from a tube that has been accidentally stepped upon. Called "extrusion flow," this idea is imbedded in many older texts on glacier flow. Unfortunately, it is not supported by observation: Intensive studies of glacier deformation in boreholes, starting during the International Geophysical Year in 1957–1958, have uniformly shown that the maximum velocity in a glacier occurs at its surface. Because of friction on the walls and bed of a glacier, the velocities near the contact between ice and rock are lower than elsewhere. Velocity contours on a transverse section across a glacier are concentric arcs around the maximum, which is on the surface. In bends of the ice stream the position of the maximum shifts from the centerline toward the outside of the bend.

The former beds of vanished glaciers sometimes slope uphill in what was obviously the downglacier direction and observers wonder how a glacier could have been flowing uphill. However, careful consideration of the equilibrium of a block of ice with surface slope α_s and basal slope α_b show that the shear stress on the base of a glacier of thickness H is given by $\rho g H \sin \alpha_s$. That is, the basal shear stress depends only on the surface slope of the glacier ice and is independent of the basal slope – even of its sign. This concept is not as paradoxical as it might seem: The beds of many rivers also slope uphill in reaches between deep pools and shallow riffles, but as long as the surface slope is downhill, the water flows in the expected direction. Thus, so long as the surface of a glacier slopes downhill, the shear forces driving it along continue to urge it downhill, even though its bed might have the opposite slope.

The vertical component of the ice velocity varies systematically along the course of a glacier. In the accumulation area the vertical velocity is downward. A marker placed on the surface of the ice is gradually buried by snow, sinking into the glacier as the snow metamorphoses into new ice and more snow accumulates on top of it. Once our marker (whether it be a meteorite that has fallen onto the snow or the body of some unfortunate early mountaineer) moves into the ablation region, melting ice gives the vertical velocity an upward component and markers once frozen into the ice emerge onto the surface.

The longitudinal velocity of an ice stream does not follow any simple rule, responding instead to variations in the underlying topography and the local thickness of the ice. Where topographic steps occur the ice may flow particularly fast and the surface becomes very steep in reaches known as icefalls. Extensive crevasse systems as well as locally high velocities characterize icefalls.

The downstream velocity of a wide ice sheet of thickness H and surface slope α_s is given by the expression (derived from Glen's law):

$$u_z(y) = u_{bs} + \frac{2A(\rho g \sin \alpha_s)^n}{n+1} \left[H^{n+1} - (H - y)^{n+1} \right] \tag{11.7}$$

where y is the height above the glacier bed, ρ the density of the ice and u_{bs} is the velocity of basal sliding.

As shown in Figure 11.5, Equation (11.7) predicts that as the creep power law n becomes larger, the flow is more concentrated toward the base of the glacier where the shear stress is higher. In the limit of very large n the flow approximates that of a perfectly plastic material and all of the deformation occurs at the bed.

11.3.2 *Longitudinal flow regime and crevasses*

As the longitudinal velocity of an ice stream varies along the glacier due to variations in ice thickness and bed slope, the overall strain rate at the surface of the glacier alternates from compressional to extensional. These strain-rate changes are accompanied by corresponding changes in the longitudinal stress. Stress variations might remain unknown to a visual observer, except for the fact that extensional stresses open crevasses that are readily seen in images. The crevasse pattern on the surface of a glacier thus contains clues about the stress state and flow regime of the ice.

Crevasses are gaping open fissures that cut the brittle upper surface of an ice sheet. Their depths are limited (unless they become filled with water) because increasing overburden pressure eventually overcomes the tensile stress and squeezes the crack closed. A crude means of estimating the maximum depth of crevasses derives from the plastic approximation to power-law flow. If the maximum stress in a glacier is limited by a plastic yield limit of 0.1 MPa, then the maximum depth of a crack is reached when the plastic yield stress Y equals the overburden pressure, divided by 2 (this factor of 2 comes from the relation between shear stress and the unidirectional stress exerted by the overburden). Thus, the maximum crevasse depth is roughly:

$$h_{crevasse} \approx \frac{2Y}{\rho g}. \tag{11.8}$$

Substituting numerical values for ice on Earth, this comes out to about 20 m. Measured crevasse depths in glaciers seldom exceed 25 to 30 m, so this result is the right order of magnitude. A more sophisticated approach incorporating the full rheologic equation is given in Paterson (1999).

Glacier reaches where the flow rate accelerates are in extension, while those in which the flow decelerates are under compression. Crevasses occur on the surface where the stress is tensional, that is, where the flow is accelerating. They are rare where the flow is decelerating. Near the snout of a glacier the flow is typically decelerating as the ice is thinned by

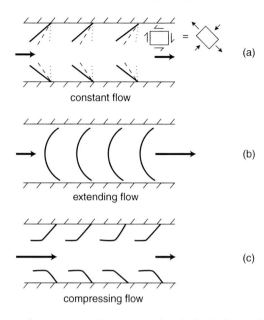

Figure 11.6 The pattern of crevasses on the margin of a glacier indicates the state of longitudinal strain. (a) A glacier flowing in a uniform channel with uniform longitudinal velocity. Friction against the side walls creates shear stress that tends to open crevasses at 45° to the flow in the direction of greatest extension (the inset shows how shear stress can be resolved into extensional and compressional principal stresses). With time, these crevasses rotate down the glacier (dashed lines). (b) Where the flow velocity increases downglacier (extending flow) the crevasses may extend all the way across the glacier nearly perpendicular to its flow direction. (c) Where the longitudinal velocity decreases downglacier (compressing flow) crevasses are suppressed in the center of the glacier stream and curve upglacier. After Figure 9.8 of Paterson (1999).

ablation, the glacier surface is under compression and the enterprising mountaineer may ascend the *Gesundheitstrasse* (German for "healthy route") onto the glacier surface.

Diagonal crevasses often form along the margins of glaciers, where friction against the wall creates shear stresses. Resolving the shear into its diagonal components of compression and extension, one expects the crevasses to form at a 45° angle to the wall of the glacier, with the acute angle facing upstream. However, with time these 45° crevasses rotate to become more transverse to the trend of the glacier because of the differential flow of the ice. Figure 11.6 illustrates the typical crevasse patterns expected for glacier reaches where the flow is (a) neither compressing nor extending, but wall friction is important, (b) extending, and (c) compressing. Moreover, if the glacier spreads out over a broad region, as in a piedmont glacier, crevasses often form perpendicular to the direction of lateral expansion.

11.3.3 Ice-sheet elevation profile

The elevation profile of ice sheets is often well approximated by a parabola, or at least a simple curve resembling a parabola. The parabolic form is a direct consequence of the

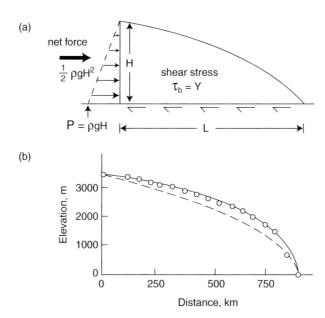

Figure 11.7 The topographic profile of ice sheets can be approximately computed from a plastic-flow model. (a) Illustrates the balance of forces on a mass of ice in which the pressure of the ice mass to the left is balanced against the shear stress at its base, resulting in a parabolic relation between the ice thickness H and the distance to the ice margin L. (b) Comparison between the plastic-flow model (dashed line) and the profile of the East Antarctic Ice Sheet between Vostok and Mirny (circles). A better fit, indicated by the solid line, incorporates a uniform accumulation of ice. After Figure 11.4 of Paterson (1999).

plastic approximation to power-law flow. This same model was used in Section 5.3.2 to argue that the profile of a lava flow consisting of a Bingham fluid should be close to a parabola. As shown in Figure 11.7a, the argument proceeds by balancing the total force from the base of the sheet, YL, against the pressure driving it outwards, $\frac{1}{2} \rho g H^2$. The resulting equation for the thickness of the sheet H as a function of distance from the edge L along a line running radially outward from the center of the sheet is:

$$H = \sqrt{\frac{2YL}{\rho g}}. \tag{11.9}$$

For example, for a point 750 km from the edge of the ice sheet and a yield stress of 0.1 MPa on the Earth, Equation (11.9) predicts an ice elevation of about 3900 m, not far from the observed elevations of the East Antarctic Ice Sheet.

The plastic-flow model is only an approximation and better fits can be attained using the full flow law, coupled with the recognition that temperatures in the upper part of the Antarctic ice sheet are below 0°C and so the ice there is less fluid. Figure 11.7b shows the elevations along a profile from Vostok to Mirny Stations along with a parabola (dashed) and a more exact model (solid line). The parabolic model gives a good first approximation, but the more accurate treatment improves the fit considerably. The simple model also does not take changes in mass balance into account, which is important for most ice sheets.

Because of concerns over the effects of global climate change, modeling of the Antarctic and Greenland icecaps is reaching a high degree of sophistication and incorporates details well beyond the scope of this book. The interested reader is referred to the recent monograph by Greve and Blatter (2009).

11.4 Glacial landforms

Glaciated landscapes betray themselves to the knowledgeable viewer by a variety of characteristic features. Glaciers and ice sheets are effective agents of both erosion and deposition. The present landscapes in high-latitude regions of the Earth bear many scars of the recent series of ice ages. Earth has experienced other episodes of glaciation even farther back in time: during the Permian Era and in the Neoproterozoic. We are aware of these ancient episodes by the changes they produced in rocks exposed at that time, although these are certainly not as apparent as the changes dating from 12 000 yr ago.

Features that seem to indicate glacial erosion also occur on Mars. These appear to be much more ancient than glacial features on the Earth, but thanks to the very slow rate of surface modification on Mars they remain to betray their origin. Acceptance of widespread ice sheets on Mars has been slow, but very recent discoveries of relict ice masses near the equator of Mars point clearly to a former era of extensive ice. In addition, eskers and many other landforms are consistent with a previous age of ice on Mars.

11.4.1 Glacial erosion

Valley glaciers modify the stream valleys they initially followed, grinding the original V-shaped cross profiles into a U-shaped trough. Glacial erosion truncates spurs, creates bowl-like cirques at the head of canyons, and leaves tributary valleys hanging high above their normal level of junction. The longitudinal form of glacial valleys is converted into a giant staircase of treads and risers. The treads frequently slope against the general trend of the valley and, after the glaciers have melted away, trap small lakes called tarn lakes.

The process of glacial erosion proceeds largely by the removal of large blocks from the bed of the glacier, a process called quarrying or plucking. As the glacier slides over irregularities in its bed, it may move fast enough that the ice cannot close in behind the obstacles, leaving open cavities in the lee that fill with meltwater. However, the pressure in the meltwater is far below the overburden weight of the glacier, except perhaps during glacier surges. The pressure gradient between the upstream side of the block on which the ice is impinging and the downstream water pocket is often enough to slide the block out of the bed and incorporate it into the ice. The blocks may be directly fractured from intact bedrock by this pressure differential or may be pre-existing joint blocks. In either case, once the ice mobilizes these blocks they are incorporated into the basal ice and dragged downglacier along the bed, contributing to further erosion by abrading the bedrock still in place.

Abrasion occurs between rock debris already incorporated into the ice and the rock bed of the glacier. Its importance can be judged from recently deglaciated surfaces, which are typically striated, smoothed and, in places, even polished by the action of debris moving along with the ice. Abrasion of solids is a reasonably well-understood process, at least in its dependence upon the force and velocity of the grinding surfaces, so it was a surprise when Geoffrey S. Boulton in 1974 discovered an unanticipated aspect of glacial abrasion that goes far toward explaining some of the details of glacial bedrock erosion.

Boulton (1979) studied glacial erosion by inserting plates of rock directly on the beds of several glaciers beneath which tunnels had been excavated to collect glacial meltwater for water supply. He found, as one might expect, that the thicker the glacier, the harder any rocks frozen into the glacier ice bore down on the bed and the more material was abraded from the bed. However, when the overburden pressure of the ice exceeded about 2 MPa, the increase in the normal pressure of rock on rock ceased, because the embedded rocks simply punched back into the ice instead of transmitting more pressure. When the ice overburden exceeded 3 MPa, the rocks in the glacier simply stalled against the bed and the ice flowed around them, bringing erosion to a halt. At greater pressures the basal debris was deposited beneath the glacier as till.

There is, thus, a limit to how much pressure ice-entrained debris can exert: Abrasion is possible beneath ice about 300 m thick or less. This limit depends somewhat on glacier speed, with the pressure at the peak of abrasion ranging from about 1 MPa at speeds of 5 m/yr up to about 3 MPa at speeds of 100 m/yr. Nevertheless, the qualitative limit to erosion goes far toward explaining the U-shape of glacier valleys.

Consider a glacier initially flowing in a V-shaped fluvial valley. It is deepest at its center, but if its overall thickness approaches the limit of abrasion, it is relatively ineffective at eroding its deepest portion, while removing more material from the walls higher up. The shape then gradually changes from a V to a U as the rate of erosion is equalized across the valley and the glacier continues to grind deeper into the bedrock (Harbor *et al.*, 1988).

The centers of continental ice sheets easily exceed Boulton's abrasion limit, so that most of their work in plucking and grinding their beds is done within a few hundred kilometers of their margins, where the thickness of the ice is relatively low. This prediction accords well with the observation that the continental ice sheets eroded most deeply near their edges, for example excavating the Great Lakes and Finger Lakes beneath the edges of the Laurentide Ice Sheet of North America. Streamlined, ice-shaped hills, such as *roche moutonnée* are best developed near the former margins of the great ice sheets.

11.4.2 *Glacial deposition*

Moraines. Glaciers are unselective agents of transport. They can and do carry everything from multi-meter blocks of rock to the finest silt. Valley glaciers transport any debris that happens to fall on their surfaces, an occurrence that is common because of rock avalanches from their over-steepened walls. Moving ice of all types picks up material from its bed and carries it along as it moves. When this material in transit reaches the terminus where melting exceeds the rate of ice motion, this debris is dumped in an unsorted heap. This material is called glacial till and the landform it creates is called a moraine.

Besides containing a miscellaneous collection of boulders, pebbles, and silt, glacial till is also compounded of rock flour, an unusual type of sediment unlike that produced by other processes. Rock flour is finely pulverized but otherwise fresh bedrock. Chemical weathering of rock flour is minimal because it is produced by grinding of rock upon rock at low temperatures beneath a glacier. It is composed of grains mostly less than 100 μm in diameter that are easily suspended in meltwater streams. The bluish, milky color of glacial streams and lakes is due to heavy loads of this material. When deposited in front of a glacier it is easily picked up by the wind and blown in dense clouds that make the terminus of a glacier a dirty, gritty place to work. During the ice ages the entire atmosphere of the Earth was laden with dust from rock flour. It was laid down in thick deposits known as loess in extensive plains in China and the midwestern United States that are today valued for their agricultural potential.

Moraines also contain large amounts of sand-sized material that may be mobilized by the high winds that often accompany glacial climates. During the Earth's recent ice age a great sand sheet formed the Sand Hills of Nebraska, created by sand washed out of glacial meltwater streams. The high winds, lack of vegetation, and abundant sand-sized sediment in the Polar Regions led to the surprising development of dune fields in this environment. Given the evidence for former ice sheets on Mars it seems possible that some of the sand-sized material there has a glacial origin.

Glaciologists distinguish several types of moraine, depending on where they form. Terminal moraines pile up at the ends of glaciers, becoming large during times when the ice margin remains at a nearly fixed location. Lateral moraines form at the edge of ice streams. Ground moraines form when the ice retreats rapidly, leaving a thin, loose deposit on the surface. Lodgement moraines form beneath the glacier and their till is often strongly compacted by the weight of the glacier.

While *roche moutonnée* are streamlined hills eroded from underlying bedrock that range in size from kilometers to a few meters, drumlins are streamlined depositional forms molded out of till, whose size range is similar. From morphology alone it is difficult to separate these two features: Indeed, some *roche moutonnée* have downglacier tails of streamlined till and so are hybrid forms. Flowing ice may also produce very elongated hills that grade into fluted surfaces with alternating hills and troughs aligned in the direction of ice flow.

Most glacially deposited material is reworked to some extent by water, for melting and runoff are ubiquitous in the vicinity of glacier ice. A large variety of names have been applied to these deposits depending on the special circumstances of their formation: The interested reader is referred to the more specialized discussions cited at the end of this chapter. In this chapter we refer to only some of the most important features for planetary observations.

Kettles or kettle holes are small, sometimes circular, pits that form in the wake of retreating ice. They have occasionally been mistaken for impact craters, although they almost always lack rims. They form around blocks of ice left stranded by the retreating ice front. These ice blocks are then partially or completely buried by outwash. After the ice melts, a pit remains.

Water at the base of a glacier or ice sheet flows as films along the rock–ice interface, fills pockets and cavities bridged by the moving ice, and eventually collects into streams that form a subglacial drainage network. Subglacial streams are far more dynamic than those flowing over a landscape because the ice flows to fill cavities where the pressure is low. Any tunnel drilled into the ice closes in rapidly until it meets resistance, so the water within a glacier travels in tunnels that tailor themselves to fit the volume of the flow. Increased water pressure opens the passage until it is in balance with the ice pressure, while a decreased head causes the conduit to shrink, always keeping the water pressure in balance with the pressure of the encasing ice.

Eskers. As water moves beneath the glacier it picks up silt and debris and carries it along, depositing it when the current slackens. The deposits of such englacial or subglacial streams are known as eskers. After the ice melts away, eskers stand as branching, sinuous ridges on the land surface. The material that composes eskers is clearly water-laid, with the graded bedding, bedforms, and sorting typical of fluvial deposits. The bedding planes in eskers tend to be anticlinal in cross section, rather than horizontal, as a result of collapse along their margins as the confining ice melted away. An apparently enigmatic feature of esker deposits is that these river-like ridges can travel up and over hills, apparently paying little attention to the slope of the land surface. This is partly because some eskers were draped over the topography after the ice melted away, but a more important difference between esker networks and those of open streams is that they flowed in pressurized conduits for which local slopes are less important than regional pressure (head) variations.

Possible eskers were first recognized on Viking images of Mars and with increasing image resolution in subsequent missions the esker interpretation has become increasingly secure (Figure 11.8). They have been found at both low and high latitudes in the southern hemisphere, but those in the high-elevation Southern Uplands are the most prominent (Banks *et al.*, 2009).

Eskers lead out from underneath the ice sheet and sometimes can be seen to connect with water-laid deposits that form an outwash plain in front of the ice margin. Outwash plains are effectively low deltas or alluvial cones that represent the material deposited by streams or rivers draining from the ice. They are complex deposits in their own right, with many characteristics different from those typical of long-term fluvial deposition

11.5 Ice in the ground

Water in the ground must be frozen wherever the mean surface temperature is below freezing. On the present Earth, about 35% of the land surface satisfies this condition. The mean temperature over the entire surface of Mars is below the freezing point of water. If the

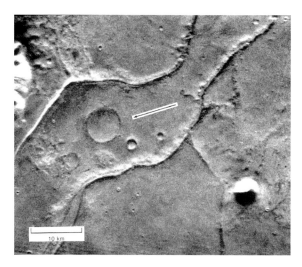

Figure 11.8 Eskers on Mars. It is suggested that these 1 km wide sinuous ridges on the floor of the Argyre Basin at 55.5° S and 40.2° W are eskers formed by subglacial streams that were deposited as the ice sheet melted away. The degraded crater in the center is 7 km in diameter. Arrow points north. Frame is 50 km wide. Viking Orbiter frame 567B33.

temperature remained permanently below freezing, water in the soil would simply stay frozen and not much of interest would occur. Temperature fluctuations, particularly those that cycle about the freezing point, are what produce most landforms and lend geomorphic interest to frozen landscapes.

11.5.1 Permafrost

The strict, but rather pedantic, definition of permafrost is that permafrost is ground that is below 0°C, *whether or not water is present*, independent of rock type. This is pedantic because if no water is present then absolutely nothing of interest happens and there are no special landforms to talk about. All of the following discussion focuses on what happens when water is present and on the effects of freezing and thawing water in the soil. Most of this discussion holds equally well for water and any other substance that undergoes a liquid–solid transformation during temperature excursions that occur on a planetary surface. Methane on Titan freezes at 91 K, just a few degrees below its average surface temperature of 94 K. However, Titan's massive atmosphere prevents temperature variations of even a few degrees from this average, so, at the moment, we do not believe that periglacial processes are relevant to Titan (unless some other common substance on its surface undergoes freeze/thaw cycling).

 The nature and behavior of permafrost were not well known in the United States until the later years of World War II. At that time, a Russian-speaking geologist named Siemon W. Muller was employed by the US Army to read and translate the extensive Russian literature

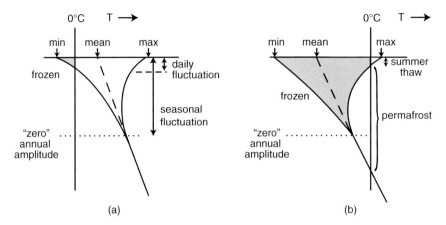

Figure 11.9 Annual temperature variations below the ground surface. (a) Indicates a temperate climate in which freezing takes place only in winter and mean temperatures are above freezing. (b) Illustrates a cold climate in which the mean temperature is below freezing. Thawing takes place only in summer. Ice is stable below the depth of the summer thaw, down to a maximum depth determined by warming due to geothermal heat flow. Permafrost encompasses the entire range over which temperatures remain below 0°C.

on the subject. After the end of the war he published a summary of his gleanings as a book (Muller, 1947) that formed the basis for our modern understanding of permafrost. Although very dated, this book can still be read with profit. Muller coined the word "permafrost" during his research.

Thermal Regime. Figure 11.9a illustrates the subsurface temperature at a location where seasonal cycles allow some freezing temperatures, but the mean temperature is above freezing, and Figure 11.9b illustrates a location where the mean annual temperature is below freezing and a permafrost zone is present. Although the top of the frozen zone is subject to seasonal temperature variations, these become negligible below the level of "zero" annual amplitude (temperature variations are never actually zero, but at this depth they are so small that they can be neglected). Below this, the bottom of the permafrost zone is determined by the planetary geothermal gradient. On Earth the geothermal gradient is about 30 K/km and the thickness of the permafrost in Siberia ranges from 200 to 400 m at 70° N down to a few tens of meters (where it is discontinuous) at 50° N. Unfrozen patches within the permafrost are known as *talik*. Talik occurs for many reasons even deep within permafrost zones; under deep lakes and rivers, for example.

Active Zone. The zone from the surface down to the level where the soil thaws annually is known as the "active zone" for reasons that will shortly be apparent. Seasonal temperatures still vary noticeably below this zone and, because not all the soil water is frozen at 0°C (or even at −30°C; see Section 7.3.2), there is still some movement of liquid water even below the permafrost table. The depth of freezing and amplitude of thermal fluctuations depend sensitively upon the nature of the surface. Surface covers of insulating material such as grass, peat, or snow have a large effect on the thermal regime of the ground underneath.

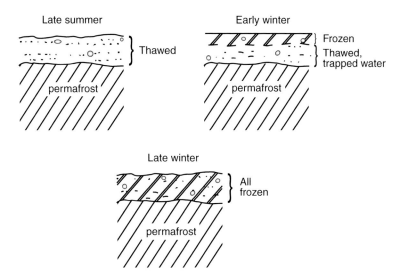

Figure 11.10 Seasonal evolution of the *active layer* overlying a permafrost terrain. In late summer this layer is completely thawed, although the permafrost below creates an impermeable layer that prevents water from draining into the subsurface and so this layer is usually saturated with water. In early winter the top of the active layer freezes, trapping water in the lower part of the active layer between the frozen water above and the permafrost below, a circumstance that promotes many kinds of instability as water pressures rise. By late winter the active layer is entirely frozen. When spring arrives the layer thaws from the top down.

Because such covers depend upon the details of surface topography and exposure, large lateral variations in thermal regime are common in permafrost areas.

Where summer temperatures rise above the freezing point of water, a seasonal cycle of freezing and thawing develops that ranks permafrost terrains among the most unstable on Earth. Permanently frozen ground is highly impermeable to liquid water: Any water that reaches the permafrost table quickly freezes and seals any cracks through which it may have originally entered. The soil overlying permafrost is, thus, commonly saturated with water when temperatures are above freezing, leading to the concept of an "active layer" (Figure 11.10). In late summer a warm thermal wave has propagated to its maximum annual depth. The soil overlying the permafrost table is thawed and often saturated with water. It forms a sea of mud that ranges from tens of centimeters deep in the far north to meters deep farther south. As winter arrives, the upper part of the active layer freezes over, trapping water in the lower portion of the layer between the impermeable permafrost table and the similarly impermeable frozen upper soil. On level ground the trapped water is stable, unless the heavy tread of a caribou or human breaks through to the mud below. However, on sloping ground the trapped water migrates laterally and high pressures can build up only a few tens of centimeters below the surface. Should the surface layer rupture for any reason, near-freezing water flows out in large volumes and quickly freezes in a low mound or sheet on the surface. Such surface layers of ice are known as "icings." Muller, in his book, delighted in showing pictures of cabins built directly on the ground in perma-frost regions. Heating the cabin destroys the upper layer of ice, so in the early winter such cabins filled suddenly with icy water that spilled out the windows before freezing solid. Eventually, by late winter, all the water in the active layer freezes and this layer becomes quiescent until melting begins again in the spring.

Solifluction. Soil creep is rapid in the active layer and freeze/thaw processes lower slope angles quickly. High pore pressure in the active layer during early winter greatly enhances the probability that thin landslides develop. Creep is caused by the alternate growth and melting of ice crystals under the surface. Lenses of ice forming above the permafrost table cause intense frost heaving. When large (tens of meters broad and meters high), these ice lenses are called frost blisters: They may tilt overlying trees in the boreal forests and prod-uce what the Russians fondly call "drunken forests." All of these processes mobilize the soil, which flows downhill in a process called solifluction or sometimes gelifluction. This soil motion often organizes into lobe-shaped steps in the surface that range from tens of centimeters to meters in height.

Pingos. Pingos are small, dome-like hills cored with ice. Internally they possess lenses of more or less pure ice beneath a layer of soil. Their mechanics of formation resem-bles that of igneous laccoliths, and they are sometimes called "hydrolaccoliths." They may reach heights of a few tens of meters (rarely a hundred meters) and diameters of nearly a kilometer. They often exhibit gaping radial dilation cracks at their crests from the uplift and stretching of the overlying sediment as they grew. Pingos are classified as either open-system types, in which the growing ice lens is fed by water flowing from beneath the permafrost layer, or closed-system types that develop where a former lake has frozen and fed water into the near-surface ice lens. Pingos in which the ice lens has melted resem-ble small volcanoes, with a central collapsed "caldera" surrounded by uplifted sediments. Pingos have been reported on Mars, but it is extremely difficult to differentiate pingos from small volcanic cones ("rootless cones" or hornitos) on morphologic characteristics alone and so these identifications are presently somewhat dubious.

Permafrost on Mars. After many years of conjecture, the presence of permafrost (in the extended sense of including frozen water) has been confirmed on Mars. Permafrost should be stable down to depths of several kilometers in Mars' polar regions, although it is not expected to be stable over the long term near the equator. Mars Global Surveyor studies of thermalized neutrons from primary cosmic rays striking the surface revealed water (more strictly, hydrogen atoms, irrespective of chemical bonding) within a few tens of centimeters of the surface in 2001. This near-surface ice extends poleward from about 50° latitude. The thrusters of the Phoenix spacecraft, which landed at 68° N, dir-ectly excavated an ice table about 5 cm below the surface. Fortuitously, five small clus-ters of impacts imaged by the HiRISE camera aboard the Mars Reconnaissance Orbiter also revealed rather pure ice close to the surface at five locations north of about 45° N, which includes the Viking 2 landing site. Evidently, if the Viking lander had dug just 10 cm deeper it might have uncovered water ice during its operational period from 1976–1979.

11.5.2 Patterned ground

The repeated thermal cycling of the active layer in permafrost terrain affords surface features many opportunities for self-organization. Freezing and thawing in the active layer leads to a poorly understood kind of slow convective motion that sorts fine-grained silts from rocks and organizes them into repeating patterns. Early explorers of permafrost terrains on Earth were astonished at the regular patterns of polygonal troughs, sorted stone circles, stripes, and other forms that develop with such regularity they often appear to be artificial. The size scale of these features is of the order of a few times the depth of the active layer, a few to perhaps ten meters. Larger-scale features are exceptional and require special explanations.

Ice-wedge polygons. The best understood of these features are ice-wedge polygons, thanks to the efforts of Arthur Lachenbruch, whose study of these ice wedges has become a classic of geomechanics (Lachenbruch, 1962). The implications of his study extend far beyond that of ice-wedge polygons themselves. His report should be read by anyone interested in the application of mechanical thinking to geology.

Ice-wedge polygons form networks that cover vast areas of permafrost terrain. The individual polygons may have either high centers bordered by troughs, or low centers surrounded by ridges that are split by troughs. In either case the distance across polygons is typically a few to tens of meters. The intersections of the troughs may be either close to 90°, in which cases the polygons are rectangular, or close to 120°, in which case they approach a hexagonal shape, very much like mudcracks caused by desiccation. A wedge of ice lies beneath the polygonal troughs in active terrain. These ice wedges may extend 30 m below the surface and are a few tens of centimeters to a meter wide at their tops.

Lachenbruch's explanation for how ice wedges form is illustrated in Figure 11.11. During the coldest part of the winter the active layer is frozen. Cold snaps of many days' duration often occur, at which time the frozen ground contracts and strong tensional stresses develop in a layer between the surface and the depth to which the thermal wave from the cold snap extends. If the tensile stress becomes great enough a crack opens. This crack may propagate several times deeper than the depth of the cooling from the cold snap itself, penetrating into the permafrost below the depth where seasonal temperature variations are negligible. The opening of a crack relieves tensional stresses in its vicinity out to a horizontal distance comparable to its depth. A second crack is, thus, unlikely to form close to the first: Cracks tend to be evenly spaced at distances comparable to their depths.

When the cold snap ends, the surface ice expands again, but the crack never quite closes: Dirt and small stones prop it slightly open for the rest of the winter. Upon arrival of the spring thaw, water from the active layer trickles down into the crack and freezes there, forming a thin sheet of ice along the crack surface. During the next winter the cycle repeats, but the crack is now a weak zone: Pure ice is weaker than frozen soil, so cracking during subsequent cold snaps occurs preferentially along the first-formed crack.

The next summer a second layer of water flows into the crack and freezes. After hundreds or thousands of seasonal cycles the thin crack grows into a massive wedge of ice. The

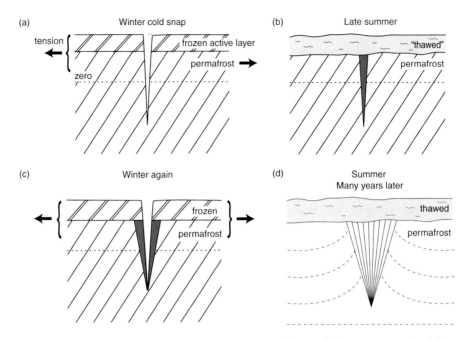

Figure 11.11 Formation of ice wedges, according to the theory of Art Lachenbruch (1962). (a) During late winter a cold snap causes the ground to contract, creating enough tensional stress to open a vertical crack that propagates some distance into the permafrost. The dashed "zero" line indicates the depth below which annual temperature variations are negligible. (b) During the late summer thaw, water percolates into the open crack and fills it, freezing at the subzero temperatures in the permafrost. (c) The next winter another cold snap re-opens the same crack because water ice is weaker than the surrounding permafrost. (d) After many such cycles of crack-opening and water-filling a broad wedge of ice has grown in the original crack, slowly enlarging by forcing adjacent sedimentary layers to deform as it grows.

soil stretches and thins over the opening wedge, which also presents a mechanical and thermal contrast to the rest of the permafrost, being composed of nearly pure ice. Soil adjacent to the growing wedge is slowly pushed aside by the wedge and heaped up into a flanking ridge or thrust farther into the center of the polygon. Such soil deformation is frequently noted in exposed sections of ice wedges.

If the climate changes and the permafrost warms and melts, melting the ice wedges with it, relicts of the ice wedges still remain. As the ice wedges melt away, soil from the active layer flows into the vacated wedge and may be recognized long after by the interruption of the original stratigraphy of the permafrost, distorted layers, and textural differences in the ice-wedge filling. Such fossil ice wedges are frequently discovered in former permafrost terrains and serve to indicate the extent of cold conditions during the Earth's recent ice age era.

The size scale of ice-wedge polygons reflects the depth of penetration of the thermal wave from cold snaps (times a factor of a few to account for the deeper penetration

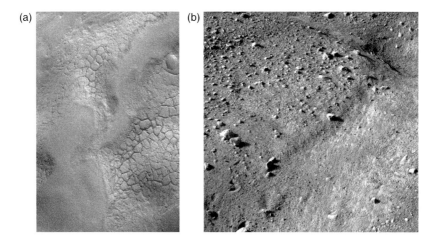

Figure 11.12 Ice-wedge polygons on Mars. (a) Patterned ground seen on the ejecta from Lyot Crater at 54.6° N and 326.6° W. In this case the polygon margins are ridges on which lie large boulders. This image is 3 km wide and is illuminated from the lower left. Image MOC2–564. NASA/JPL. (b) Troughs are spaced 1.5 to 2.5 m apart near the Phoenix landing site at 68°N and 26° W. They are believed to represent ice-wedge activity. On Earth, ice wedges may also be manifested by either ridges or troughs. NASA/JPL/University of Arizona. See also color plate section.

of the cracks beyond the depth of actual tension). It is, thus, a combined function of the duration and intensity of a cold snap and the thermal conductivity of the soil. One might then wonder if more long-continued eras of cold create larger polygons. This is precisely what many geologists thought when 5–10 km scale polygonal troughs were discovered on Viking images of the northern plains of Mars. However, the theory of ice-wedge formation, coupled with the rheology of ice (Glen's law) show that this cannot be the case.

Tensile stresses in ice due to contraction endure only so long as creep does not relax them. Slow freezing that extends to great depths requires a long period of time, during which the ice has time to flow under the applied stress and zero the stress. If the cold snap is not quick, it cannot generate tensile stresses and deep cracks cannot develop. This argument was applied to the polygonal terrain on Mars to show conclusively that such large polygons cannot have been created by thermal contraction (Pechmann, 1980). They may instead be due to draping of compacting sediments over a pre-existing cratered terrain (McGill, 1992). Only recently, with the advent of the very high-resolution imaging possible with the HiRISE camera system and the Phoenix lander have true ice-wedge polygons been observed on Mars (Figure 11.12). These polygons have the expected dimensions of 2–3 m across and closely resemble terrestrial ice-wedge polygons.

The observation of apparently fresh ice-wedge polygons on Mars suggests that liquid water on the surface is not, in fact, necessary for their formation. Perhaps the crack fillings on Mars are dust that has drifted into gaping cracks, not frozen water, and we are really looking at dust-wedge polygons in the Martian permafrost.

Block fields. Block fields are enigmatic periglacial features that have been found in many areas formerly occupied by permafrost. They appear as gently sloping, nearly planar surfaces up to many kilometers in extent, which are covered with boulders of roughly the same size. Block fields lack any obvious matrix material, although when active they may have contained interstitial ice. They may be related to frost heaving in some way as they show little sign of lateral movement: The blocks in these fields seem to have formed in place from jointed bedrock.

11.5.3 Thermokarst

Permafrost betrays its presence most clearly when it is about to disappear. The most dramatic landforms created by permafrost are formed as it is melting during a period of climatic warming. Permafrost does not melt uniformly: Small variations in surface thermal conductivity and exposure become amplified by positive feedback and are expressed as topographic features. The most obvious result of melting permafrost is a volume change. Thawed permafrost expels water and contracts, sagging downward into small ponds that collect more water and enhance melting. Such thaw lakes are common, creating landscapes packed with kilometer-diameter circular to elliptical ponds that are often aligned with the prevailing wind. Such lakes constitute the infamous Carolina Bays, which impact-crater enthusiasts persistently claim to be of impact origin in spite of the complete lack of evidence for impacts. Extensive fields of active thaw lakes occur in lowland areas of northern Alaska, Yukon Territory, and northern Russia. Depressions believed to be thaw lakes have been recognized in the catastrophic outflow channels on Mars, suggesting at least one era of warming on that currently chilly planet.

Once depressions are created by melting permafrost, the scarps that form at the interface between the thawed and still-frozen ground are subject to rapid denudation that removes the insulating surface layer and accelerates the disintegration of the permafrost. These small scarps retreat rapidly, forming shallow cirques that cut into the frozen ground.

Asymmetric valleys are common in permafrost terrains and may be accentuated by its decay. North–south asymmetry develops predominantly because of differences of exposure to solar radiation. East–west asymmetry may also develop because of differing exposure to the prevailing wind and the resulting differences in the depth of wind-drifted snow and chilling by the wind.

The Southern Polar Cap of Mars is subject to another kind of weathering akin to thermokarst disintegration, but not involving liquid water. Thin layers (~10 m) of residual CO_2 ice overlying water ice sublime away to create coalescing circular pits informally dubbed "Swiss-cheese terrain." These pits range from a few hundred meters up to a kilometer across and are approximately 10 m deep.

Further reading

The classic American study of the geological effects of the Pleistocene continental ice sheets is the fat book of Flint (1971). The basic physics of glacier flow, mass balance, and ice sheet formation is Paterson (1999), a book that has gone through many editions (a fourth has just appeared), but remains the most lucid of several such books. Readers seeking to visually feast on glacial features and phenomena should peruse the picture book by Post and LaChapelle (2000), which also contains much wisdom in addition to its magnificent photographs. The ability of glaciers and ice sheets to create landscapes is treated in Sugden and John (1976), a book that has unfortunately not been updated in recent years. The mechanics of cold soil and its implications for landform evolution is the topic of Williams and Smith (1991), while the more observational aspects of periglacial environments is well covered by Washburn (1980). The geomorphology of both glaciated regions and periglacial environments is the subject of a pair of books by Embelton and King (1975a, b).

Exercises

11.1 Rheology in space

The Maxwell relaxation time τ_M for ice at 273K (0°C) is 100 minutes. Use the approximately universal relation:

$$\dot{\varepsilon} = A(\sigma)e^{-gT_m/T}$$

where $g \sim 26$ for non-metals. Compute τ_M for the following substances on the indicated planetary body (suppose σ is the same as for ice at the τ_M given, and that all materials have nearly the same shear modulus):

  Methane, ammonia on Triton, surface temperature 45K
  Ammonia, CO_2 on Ganymede, surface temperature 145K
  CO_2, water on Mars polar caps, mean temperature ca. 170K
  Salt (NaCl), olivine (forsterite) on Venus, surface temperature 750K.

Melting points are:

  Methane $T_m = 91$K
  Ammonia $T_m = 196$K
  CO_2 $T_m = 216$K
  Salt (NaCl) $T_m = 1100$K
  Olivine (Fo) $T_m = 2200$K.

Which of the two materials is more likely to show evidence of flow in surface deposits ("glaciers")? Compare your computed flow rates to those of a terrestrial glacier.

11.2 The inner heat

Compute the rate of basal melting of a warm-based glacier on the Earth and on Mars (use some plausible means of estimating the heat flow of Mars, perhaps by assuming that the rate of heat generation per unit volume on Mars is the same as on the Earth). The average heat flow on the Earth is 80 mW/m². Does this suggest a way of estimating the minimum rate of snow precipitation necessary to support warm-based glaciers? If so, what is this minimum for Mars?

11.3 Infinite flowing ice

Derive the velocity profile for an infinitely wide sheet of power-law material (Glen's law, Equation (11.1)) of uniform thickness H creeping down a surface with a constant slope α. That is, show how Equation (11.7) comes about in the case of a zero basal sliding velocity. Using the data on Glen's law given in Section 11.2.1, estimate the surface velocity of an ice sheet 3 km thick with a surface slope of 0.01 (about half a degree) at a temperature of 0°C and at −40°C.

11.4 Only the surface matters!

Demonstrate the assertion in Section 11.3.1 that, in a straight reach of a glacier, the basal shear stress depends only upon the surface slope α_s, not the basal slope α_b. Hint: Consider the forces acting on the left and right sides of a block of downglacier length Δx, then get the basal shear stress from the force per unit area (if you get really stuck on this, see Patterson, 1999, p. 241).

11.5 Permafrosty Mars

Estimate the thickness of the permafrost layer on Mars by computing the depth of the 273 K isotherm below the surface at the equator and at 50° N. Assume a global average heat flow of 30 mW/m^2 and a thermal conductivity of the Martian surface layers of about 1.5 W/m-K. The average surface temperature of Mars at the equator is 240 K, but falls to a chilly 200 K at 50° N. Compute the depth to the bottom of the permafrost at these two latitudes. How does this depth compare to the depth of the (Martian) seasonal temperature fluctuation?

References

Aharonson, O. and Schorghofer, N. (2006). Subsurface ice on Mars with rough topography. *J. Geophys. Res.*, **111**, 2005JE002636.

Ahlbrandt, T. S. (1979). Textural parameters of eolian deposits. In *Desert Sand Seas*, ed. E. D. McKee. Washington, DC: US Geological Survey, pp. 21–51.

Allen, J. R. L. (1970). *Physical Processes of Sedimentation*. London: Allen & Unwin.

(1974). Reaction, relaxation and lag in natural sedimentary systems: General principles, examples and lessons. *Earth Science Reviews*, **10**, 263–342.

(1982). *Sedimentary Structures: Their Character and Physical Basis*. Amsterdam: Elsevier.

Amante, C. and Eakins, B. W. (2009). *ETOPO1 1 Arc-Minute Global Relief Model: Procedures, Data Sources and Analysis*. Report.

Amundson, R. (2001). The carbon budget in soils. *Annu. Rev. Earth Planet. Sci.*, **29**, 535–62.

Anderson, D. L. (1995). Lithosphere, asthenosphere, and perisphere. *Rev. Geophys.*, **33**, 125–49.

Anderson, E. M. (1951). *The Dynamics of Faulting*. Edinburgh: Oliver and Boyd.

Anderson, R. S. (1996). The attraction of sand dunes. *Nature*, **379**, 24–5.

Araki, H., Tazawa, S., Noda, H., Ishihara, Y., Goossens, S., Sasaki, S., Kawano, N., Kamlya, I., Otake, H., Oberst, J. and Shum, C. (2009). Lunar global shape and polar topography derived from Kaguya-LALT laser altimetry. *Science*, **323**, 897–900.

Arcement, G. J. and Schneider, V. R. (1989). Guide for selecting Manning's roughness coefficients for natural channels and flood plains (metric version). Report *USGS Water Supply Paper 2339*.

Artyushkov, E. V. (1973). Stresses in the lithosphere caused by crustal thickness inhomogeneities. *J. Geophys. Res.*, **78**, 7675–708.

Ashby, M. F. and Sammis, C. G. (1990). The damage mechanics of brittle solids in compression. *Pure and Appl. Geophysics*, **133**, 489–521.

Atkinson, B. K. (1987). Fracture mechanics of rock. In *Fracture Mechanics of Rock*. San Diego, CA: Academic, p. 534.

Bagenal, F., Dowling, T. E. and McKinnon, W. B. (2004). *Jupiter*. Cambridge, UK: Cambridge University Press, p. 719.

Bagnold, R. A. (1941). *The Physics of Blown Sand and Desert Dunes*. London: Methuen & Co.

(1966). An approach to the sediment transport problem from general physics. *USGS Prof. Pap.*, **422I**, 1–37.

(1990). *Sand, Wind and War*. Tucson, AZ: University of Arizona Press.

Baker, V. R. (1982). *The Channels of Mars*. Austin, TX: Texas University Press.

Baker, V. R. and Nummedal, D. (1978). *The Channeled Scabland*. Washington, DC: NASA.

Baldwin, R. B. (1963). *The Measure of the Moon*. Chicago, IL: University of Chicago Press.

Banerdt, W. B., Golombek, M. P. and Tanaka, K. L. (1992). Stress and tectonics on Mars. In *Mars*, ed. H. H. Kieffer, B. M. Jakowsky, C. W. Snyder and M. S. Matthews. Tucson, AZ: University of Arizona Press, pp. 249–97.

Banks, M. E., Lang, N. P., Kargel, J. S., McEwen, A. S., Baker, V. R., Grant, J. A., Pelletier, J. D. and Strom, R. G. (2009). An analysis of sinuous ridges in the southern Argyre Planitia, Mars using HiRISE and CTX images and MOLA data. *J. Geophys. Res.*, **114**, E09003, doi:10.1029/2008JE003244.

Bass, J. D. (1995). Elasticity of minerals, glasses, and melts. In *Mineral Physics and Crystallography: A Handbook of Physical Constants*, ed. T. J. Ahrens. Washington, DC: American Geophysical Union, pp. 45–63.

Bate, R. D., Mueller, D. D. and White, J. E. (1971). *Fundamentals of Astrodynamics*. New York, NY: Dover.

Bear, J. (1988). *Dynamics of Fluids in Porous Media.* New York, NY: Dover.

Beatty, J. K., Petersen, C. C. and Chaikin, A. (1999). *The New Solar System*, 4th edn. Cambridge, UK: Cambridge University Press.

Beeman, M., Durham, W. B. and Kirby, S. H. (1988). Friction of ice. *J. Geophys. Res.*, **93**, 7625–33.

Ben-Avraham, Z. and Nur, A. (1980). The elevation of volcanoes and their edifice heights at subduction zones. *J. Geophys. Res.*, **85**, 4325–35.

Bergstralh, J. T., Milner, E. D. and Shapley, M. (1991). *Neptune*. Tucson, AZ: University of Arizona Press, p. 1076.

Bingham, E. C. and Green, H. (1919). Paint, a plastic material and not a viscous liquid: The measurement of its mobility and yield value. *Am. Soc. Testing Materials Proc. Pt. II*, 641–75.

Bird, E. (2008). *Coastal Geomorphology*, 2nd edn. New York, NY: Wiley.

Birkland, P. W. (1974). *Pedology, Weathering and Geomorphological Research.* New York, NY: Oxford.

Blackwelder, E. (1929). Cavernous rock surfaces of the desert. *Am. J. Sci.*, **17**, 393–9.
 (1934). Yardangs. *Bull. Geol. Soc. Amer.*, **45**, 159–66.

Bland, M. T. and Showman, A. P. (2007). The formation of Ganymede's grooved terrain: Numerical modeling of extensional necking instabilities. *Icarus*, **189**, 439–56.

Bland, W. and Rolls, D. (1998). *Weathering*. London: Arnold.

Bolt, B. A. (1970). Jottings from Australia: Australian seismographic stations, earthquake engineering and the 1968 Meckering earthquake. *Bull. Seis. Soc. Amer.*, **60**, 675–83.

Bottke, W. F., Cellino, A., Paolicchi, P. and Binzel, R. P. (2002). *Asteroids III*. Tucson, AZ: University of Arizona Press, p. 785.

Bougher, S. W., Hunten, D. M. and Phillips, C. B. (1997). *Venus II*. Tucson, AZ: University of Arizona Press, p. 1362.

Boulton, G. S. (1979). Processes of glacier erosion on different substrata. *J. Glaciology*, **23**, 15–38.

Brace, W. F. (1960). An extension of the Griffith theory of fracture to rocks. *J. Geophys. Res.*, **65**, 3477–80.

Branney, M. J. and Kokelaar, P. (2002). *Pyroclastic Density Currents and the Sedimentation of Ignimbrites*. London: Geol. Soc. London.

Brown, C. D. and Grimm, R. E. (1997). Tessera deformation and the contemporaneous thermal state of the plateau highlands, Venus. *Earth Planet. Sci. Lett.*, **147**, 1–10.

Brown, R. H. and Matson, D. L. (1987). Thermal effects of insolation propagation into the regoliths of airless bodies. *Icarus*, **72**, 84–94.

Brown, R. H., Lebreton, J.-P. and Waite, J. H. (2009). *Titan from Cassini–Huygens.* Dordrecht: Springer, p. 535.

Bucher, W. H. (1933). *Deformation of the Earth's Crust.* Princeton, NJ: Princeton University Press.

Bull, W. B. (1968). Alluvial fans. *J. Geol. Educ.*, **16**, 101–6.

Buratti, B. J., Hillier, J. K. and Wang, M. (1996). The lunar opposition surge: Observations by Clementine. *Icarus*, **124**, 490–9.

Burchfield, J. D. (1990). *Lord Kelvin and the Age of the Earth.* Chicago, IL: University of Chicago Press.

Burleigh, K. J., Melosh, H. J., Tornabene, L. L. and McEwen, A. S. (2009). Small impacts trigger dust landslides on Mars. In *40th Lunar and Planetary Science Conference.* Abs. #1431. Houston, TX: Lunar and Planetary Institute.

Burr, D. M., Emery, J. P., Lorenz, R. D., Collins, G. C. and Carling, P. A. (2006). Sediment transport by liquid surficial flow: Application to Titan. *Icarus*, **181**, 235–42.

Burr, D. M., Enga, M.-T., Williams, R. M. E., Zimbelman, J. R., Howard, A. D. and Brennand, T. A. (2009). Pervasive aqueous paleoflow features in the Aeolis/Zephyria Plana region, Mars. *Icarus*, **200**, 52–76.

Byerlee, J. (1978). Friction of rocks. *Pageoph.*, **116**, 615–26.

Cailleau, B., Walter, T. R., Janle, P. and Hauber, E. (2005). Unveiling the origin of radial grabens on Alba Patera volcano by finite element modeling. *Icarus*, **176**, 44–56.

Carey, S. W. (1953), The rheid concept in geotectonics. *Aust. J. Earth Sci.*, **1**, 67–117.

Caristan, Y. (1982). The transition from high-temperature creep to fracture in Maryland diabase. *J. Geophys. Res.*, **87**, 6781–90.

Carr, M. H. (1996). *Water on Mars.* New York, NY: Oxford.
  (2006). *The Surface of Mars.* Cambridge, UK: Cambridge University Press.

Carslaw, H. S. and Jaeger, J. C. (1959). *Conduction of Heat in Solids*, 2nd edn. Oxford: Oxford.

Carson, M. A. and Kirkby, M. J. (1972). *Hillslope Form and Process.* Cambridge, UK: Cambridge University Press.

Cathles, L. M. (1975). *The Viscosity of the Earth's Mantle.* Princeton, NJ: Princeton University Press.

Chandrasekhar, S. (1961). *Hydrodynamic and Hydromagnetic Stability.* New York, NY: Oxford.
  (1969). *Ellipsoidal Figures of Equilibrium.* New Haven, CT: Yale University Press.

Chapman, C. R. and McKinnon, W. B. (1986). Cratering of planetary satellites. In *Satellites*, ed. J. A. Burns and M. S. Matthews. Tucson, AZ: University of Arizona Press.

Chase, C. (1991). Fluvial landsculpting and the fractal dimension of topography. *Geomorphology*, **5**, 39–57.

Chase, M. W. (1998). *NIST-JANAF Thermochemical Tables*, 4th edn. Woodbury, NY: American Chemical Society.

Chopra, P. and Paterson, M. S. (1984). The role of water in the deformation of dunite. *J. Geophys. Res.*, **89**, 7861–76.

Clark, B. E., Hapke, B., Pieters, C. and Britt, D. (2002). Asteroid space weathering and regolith evolution. In *Asteroids III*, ed. W. F. Bottke, A. Cellino, P. Paolicchi and R. P. Binzel. Tucson, AZ: University of Arizona Press, pp. 585–99.

Clark, S. P. (1966). Thermal conductivity. In *Handbook of Physical Constants*, ed. S. P. Clark. Denver, CO: Geological Society of America.

Clauser, C. and Huenges, E. (1995). Thermal conductivity of rocks and minerals. In *Rock Physics and Phase Relations*, ed. T. J. Ahrens. Washington, DC: American Geophysical Union, pp. 105–26.

Clow, G. D. and Carr, M. H. (1980). Stability of sulfur slopes on Io. *Icarus*, **44**, 268–279.

Collins, G. S. and Melosh, H. J. (2003). Acoustic fluidization and the extraordinary mobility of sturzstroms. *J. Geophy. Res.*, **108**, doi: 10.1029/2003JB002465.

Collins, G. C., Head, J. W. and Pappalardo, R. T. (1998). The role of extensional instability in creating Ganymede grooved terrain: Insights from Galileo high-resolution stereo imaging. *Geophys. Res. Lett.*, **25**, 233–236.

Collins, G. S., Melosh, H. J. and Marcus, R. A. (2005). Earth Impact Effects Program: A Web-based computer program for calculating the regional environmental consequences of a meteoroid impact on Earth. *Meteoritics and Planet. Sci.*, **40**, 817–840.

Comer, R. P., Solomon, S. C. and Head, J. W. (1979). Elastic lithosphere thickness on the Moon from mare tectonic features: A formal inversion. *Proc. Lunar Planet. Sci. Conf. 10th*, 2441–63.

Croft, S. K. (1980). Cratering flow fields: Implications for the excavation and transient expansion stages of crater formation. *Proc. Lunar Planet. Sci. Conf. 11th*, 2347–78.

Cruikshank, D. P. (1995). *Neptune and Triton*. Tucson, AZ: University of Arizona Press, p. 1249.

Cruikshank, D. P. and Morbidelli, A. (2008). *The Solar System Beyond Neptune*. Tucson, AZ: University of Arizona Press, p. 632.

Culling, W. E. H. (1960). Analytical theory of erosion. *J. Geology*, **68**, 336–44.
  (1963). Soil creep and the development of hillside slopes. *J. Geology*, **71**, 127–61.

Currie, J. B., Patnode, H. W. and Trump, R. P. (1962). Development of folds in sedimentary strata. *Geol. Soc. Amer. Bull.*, **73**, 655–74.

Cyr, K. E. and Melosh, H. J. (1993). Tectonic patterns and regional stresses near Venusian coronae. *Icarus*, **102**, 175–84.

Darwin, C. (1896). *The Formation of the Vegetable Mould through the Action of Worms*. New York, NY: Appleton.

Dash, J. G., Rempel, A. W. and Wettlaufer, J. S. (2006). The physics of premelted ice and its geophysical consequences. *Rev. Mod. Phys.*, **78**, 695–741.

Davies, A. G. (2007). *Volcanism on Io: A Comparison with Earth*. Cambridge, UK: Cambridge University Press.

Davies, G. F. (1980). Thermal histories of convective earth models and constraints on radiogenic heat production in the Earth. *J. Geophys. Res.*, **85**, 2517–30.
  (1982). Ultimate strength of solids and formation of planetary cores. *Geophys. Res. Lett.*, **9**, 1267–70.
  (1999). *Dynamic Earth: Plates, Plumes and Mantle Convection*. Cambridge, UK: Cambridge.

Davies, M. E., Colvin, T. R., Oberst, J., Zeitler, W., Schuster, P., Neukum, G., McEwen, A. S., Phillips, C. B., Thomas, P., Veverka, J., Belton, M. J. S. and Schubert, G. (1998). The control networks of the Galilean satellites and implications for global shape. *Icarus*, **135**, 372–6.

Davis, G. L. (1969). *The Earth in Decay*. New York, NY: American Elsevier.

Davison, C. (1888). Note on the movement of scree-material. *Quater. J. Geol. Soc. London*, **44**, 232–8.

Dodds, P. S. and Rothman, D. H. (2000). Scaling, universality and geomorphology. *Ann. Rev. Earth Planet. Sci.*, **28**, 571–610.

Dohrenwend, J. C. (1987). Basin and range. In *Geomorphic Systems of North America*, ed. W. L. Graf. Boulder, CO: Geological Society of America, pp. 303–42.

Dones, L., Chapman, C. R., McKinnon, W. B., Melosh, H. J., Kirchoff, M. R., Neukum, G. and Zahnle, K. (2009). Icy satellites of Saturn: Impact cratering and age determination. In *Saturn from Cassini-Huygens*, ed. M. K. Dougherty, L. W. Espositto and S. M. Krimigis. Berlin: Springer, pp. 613–35.

Douglas, I. (1977). *Humid Landforms*. Cambridge, MA: MIT Press.

Durham, W. B. and Stern, L. A. (2001). Rheological properties of water ice – Applications to satellites of the outer planets. *Annu. Rev. Earth Planet. Sci.*, **29**, 295–330.

Durham, W. B., Kirby, S. H. and Stern, L. A. (1993). Flow of ices in the ammonia–water system. *J. Geophys. Res.*, **98**, 17 667–82.

(1999). The rheology of solid carbon dioxide: New measurements. *LPSC XXX*, Abstract, # 2017.

Ehlers, E. G. (1987). *The Interpretation of Geological Phase Diagrams*. New York, NY: Dover.

Einstein, A. (1954). The cause of the formation of meanders in the courses of rivers and of the so-called Baer's law. In *Ideas and Opinions*, ed. A. Einstein. New York, NY: Random House, pp. 249–53.

Eisenberg, D. and Kauzmann, W. (1969). *The Structure and Properties of Water*. Oxford: Oxford University Press.

Embelton, C. and King, C. A. M. (1975a). *Glacial Geomorphology*, 2nd edn. New York, NY: Wiley.

(1975b). *Periglacial Geomorphology*, 2nd edn. New York, NY: Wiley.

Engelder, T. (1993). *Stress Regimes in the Lithosphere*. Princeton University Press.

England, P. C., Molnar, P. and Richter, F. M. (2007). Kelvin, Perry and the age of the Earth. *Am. Scientist*, **95**, 342–9.

Erismann, T. H. and Abele, G. (2001). Dynamics of Rockslides and Rockfalls. Berlin: Springer.

Ernst, R. E. and Buchan, K. L. (2003). Recognizing mantle plumes in the geological record. *Annu. Rev. Earth Planet. Sci.*, **31**, 469–523.

Ernst, R. E., Grosfils, E. B. and Mége, D. (2001). Giant dike swarms: Earth, Venus, and Mars. *Annu. Rev. Earth Planet. Sci.*, **29**, 489–534.

Evans, B. and Kohlstedt, D. L. (1995). Rheology of rocks. In *Rock Physics and Phase Relations: A Handbook of Physical Constants*, ed. T. J. Ahrens. Washington, DC: AGU, pp. 148–65.

Fegley, B., Klingelhöfer, G., Lodders, K. and Widemann, T. (1997). Geochemistry of surface-atmosphere interactions on Venus. In *Venus II*, ed. S. W. Bougher, D. M. Hunten and R. J. Phillips. Tucson, AZ: University of Arizona Press, pp. 591–636.

Ferrill, D. A., Wyrick, D. Y., Morris, A. P., Sims, D. W. and Franklin, N. M. (2004). Dilational fault slip and pit chain formation on Mars. *GSA Today*, **14**, 4–12.

Festou, M. C., Keller, H. U. and Weaver, H. A. (2004). *Comets II*. Tucson, AZ: University of Arizona Press, p. 780.

Fleitout, L. and Froidevaux, C. (1982). Tectonics and topography for a lithosphere containing density heterogeneities. *Tectonics*, **1**, 21–56.

Fletcher, R. C. and Hallet, B. (1983). Unstable extension of the lithosphere: A mechanical model for basin-and-range structure. *J. Geophys. Res.*, **88**, 7 457–66.

Flint, R. F. (1971). *Glacial and Quaternary Geology*. New York, NY: Wiley.

Fourier, J. (1955). *The Analytical Theory of Heat.* New York, NY: Dover.

Francis, P. and Oppenheimer, C. (2003). *Volcanoes.* New York, NY: Oxford.

Freed, A. M., Melosh, H. J. and Solomon, S. C. (2001). Tectonics of mascon loading: Resolution of the strike-slip fault paradox. *J. Geophys. Res.*, **106**(20), 603–20.

French, B. M. (1977). *The Moon Book.* New York, NY: Penguin.

Frenkel, J. (1926). Zur Theorie der Elastizitätsgrenze und der Festigkeit kristallinischer Körper. *Z. Physik*, **37**, 572–609.

Fryberger, S. G. (1979). Dune forms and wind regime. In *Desert Sand Seas*, ed. E. D. McKee. Washington, DC: US Geological Survey, pp. 137–69.

Garland, G. D. (1965). *The Earth's Shape and Gravity.* Pergamon.

Gault, D. E. (1970). Saturation and equilibrium conditions for impact cratering on the lunar surface: Criteria and implications. *Radio Science*, **5**, 273–91.

Gault, D. E., Hörz, F., Brownlee, D. E. and Hartung, J. B. (1974). Mixing of the lunar regolith. In *Proceedings of the Fifth Lunar Science Conference*. Houston, TX, pp. 2365–86.

Gilbert, G. K. (1880). *Report on the Geology of the Henry Mountains*. Washington, DC: Government Printing Office.

  (1890). *Lake Bonneville*. Washington, DC: US Government Printing Office.

  (1909). The convexity of hilltops. *J. of Geology*, **17**, 344–50.

  (1914). *The transportation of debris by running water*. Prof. Paper 86, USGS.

  (1917). *Hydraulic-Mining débris in the Sierra Nevada*. Prof. Paper 105, USGS.

Gillmor, C. S. (1971). *Coulomb and the Evolution of Physics and Engineering in Eighteenth-Century France*. Princeton, NJ: Princeton University Press.

Gilman, J. J. (1969). *Micromechanics of Flow in Solids.* New York, NY: McGraw-Hill.

Glasstone, S. and Dolan, P. J. (1977). *Effects of Nuclear Weapons*. United States Departments of Defense and Energy.

Glen, J. W. (1955). The creep of polycrystalline ice. *Proc. Roy Soc. London A*, **228**, 519–738.

Goehring, L., Mahadevan, L. and Morris, S. W. (2009). Nonequilibrium scale selection mechanism for columnar jointing. *Proc. Nat. Acad. Sci.*, **106**, 387–92.

Gold, T. (1970). Apollo 11 and 12 close-up photography. *Icarus*, **12**, 360–75.

Goldreich, P. and Soter, S. (1966). Q in the Solar System. *Icarus*, **5**, 375–89.

Golombek, M. P. (1979). Structural analysis of lunar grabens and the shallow crustal structure of the Moon. *J. Geophys. Res.*, **84**, 4657–66.

Gooding, J. L., Arvidson, R. E. and Zolotov, M. Y. (1992). Physical and chemical weathering. In *Mars*, ed. H. H. Kieffer, B. M. Jakosky, C. W. Snyder and M. S. Matthews. Tucson, AZ: University of Arizona Press, pp. 626–51.

Gordon, J. E. (2006). *The New Science of Strong Materials*, 2nd edn. Princeton, NJ: Princeton University Press.

Greeley, R. and Batson, M. (2000). *Planetary Mapping*. Cambridge, UK: Cambridge University Press.

Greeley, R. and Iversen, J. D. (1985). *Wind as a Geological Process*. Cambridge, UK: Cambridge University Press.

Greeley, R., Lancaster, N., Lee, S. and Thomas, P. (1992). Martian aeolian processes, sediments and features. In *Mars*, ed. H. H. Kieffer, B. M. Jakowsky, C. W. Snyder and M. S. Matthews. Tucson, AZ: University of Arizona Press, pp. 730–66.

Greeley, R., Bender, K. C., Saunders, R. S., Schubert, G. and Weitz, C. M. (1997). Aeolian processes and features on Venus. In *Venus II*, ed. S. W. Bougher, D. M. Hunten and R. J. Phillips. Tucson, AZ: University of Arizona Press, pp. 547–89.

Greeley, R., Arvidson, R. E., Barlett, P. W., Blaney, D., Cabrol, N. A., Christensen, P. R., Fergason, R. L., Golombeck, M. P., Landis, G. A., Lemmon, M. T., McLennan, S. M., Maki, J. N., Michaels, T., Moersch, J. E., Neakrase, D. V., Rafkin, S. C. R., Richter, L., Squyres, S. W., de Sousz, P. A., Sullivan, R. J., Thompson, S. D. and Whelley, P. L. (2006). Gusev crater: Wind-related features and processes observed by the Mars Exploration Rover Spirit. *J. Geophys. Res.*, **111**, E02S09.

Greenberg, R., Geissler, P., Hoppa, G., Tufts, B. R., Durda, D. D., Papplardo, R., Head, J. W., Greeley, R., Sullivan, R. and Carr, M. H. (1998). Tectonic processes on Europa: Tidal stresses, mechanical response, and visible features. *Icarus*, **135**, 64–78.

Greene, M. T. (1982). *Geology in the Nineteenth Century*. Ithaca, NY: Cornell University Press.

Gregg, T. K. P. and Fink, J. H. (1999). Quantification of extraterrestrial lava flow effusion rates through laboratory simulations. *J. Geophys. Res.*, **101**, 16 891–900.

Greve, R. and Blatter, H. (2009). *Dynamics of Ice Sheets and Glaciers*. Berlin: Springer.

Grieve, R. and Therriault, A. (2000). Vredefort, Sudbury, Chicxulub: Three of a kind? *Ann. Rev. Earth Planet. Sci.*, **28**, 305–38.

Griggs, D. T. (1935). The strain ellipsoid as a theory of rupture. *Am. J. Science*, **30**, 121–37.
  (1936). The factor of fatigue in rock exfoliation. *J. Geology*, **44**, 783–96.
  (1939). Creep of rocks. *Jour. Geol.*, **47**, 225–51.

Griggs, D. T. and Handin, J. (1960). Observations on fracture and a hypothesis of earthquakes. In *Rock Deformation, Geol. Soc. Amer. Memoir 79*, ed. D. T. Griggs and J. Handin. Boulder, CO: Geological Society of America, pp. 347–64.

Griggs, R. F. (1922). *The Valley of Ten Thousand Smokes*. Washington, DC: National Geographic Society.

Grotzinger, J., Jordan, T. H., Press, F. and Siever, R. (2006a). *Understanding Earth*, 5th edn. San Francisco, CA: W. H. Freeman.

Grotzinger, J., Bell, J. F., Herkenhoff, K., Johnson, J., Knoll, A., McCartney, E., McLennan, S. M., Metz, J., Moore, J. M., Squyres, S. W., Sullivan, R., Ahronson, O., Arvidson, R. E., Joliff, B. L., Golombeck, M. P., Lewis, K., Parker, T. and Soderblom, L. A. (2006b). Sedimentary textures formed by aqueous processes, Erebus crater, Meridiani Planum, Mars. *Geology*, **34**, 1085–8.

Groussin, O., A'Hearn, M. A., Li, J.-Y., Thomas, P. C., Sunshine, J., Lisse, C. M., Meech, K. J., Farnham, T. L., Feaga, L. M. and Delamere, W. A. (2007). Surface temperature of the nucleus of Comet 9P/Tempel 1. *Icarus*, **191**, 63–72.

Haberle, R. M., Forget, F., Colaprete, A., Schaffer, J., Boynton, W. V., Kelly, N. J. and Chamberlain, M. A. (2008). The effect of ground ice on the Martian seasonal CO_2 cycle. *Planetary & Space Sci.*, **56**, 251–5.

Handin, J. (1966). Strength and ductility. In *Handbook of Physical Constants*, ed. S. P. Clark, Jr. New York, NY: Geological Society of America, pp. 223–89.

Hapke, B. (1993). *Theory of Reflectance and Emittance Spectroscopy*. Cambridge, UK: Cambridge University Press.

Harbor, J. M., Hallet, B. and Raymond, C. F. (1988). A numerical model of landform development by glacial erosion. *Nature*, **333**, 347–9.

Hartmann, W. K. (1972). *Moons and Planets*. Belmont, CA: Wadsworth Pub. Co.

Hartmann, W. K. (2005). Martian cratering 8: Isochron refinement and the chronology of Mars. *Icarus*, **174**, 294–320.

Hauber, E., Broz, P. and Jagert, F. (2010). Plains volcanism on Mars: Ages and rheology of lavas. In *41st Lunar and Planetary Science Conference*. Abstract #1298. Houston, TX: Lunar and Planetary Institute.

Haxby, W. F. and Turcotte, D. L. (1976). Stresses induced by the addition or removal of overburden and associated thermal effects. *Geology*, **4**, 181–4.

Heiken, G. H., Vaniman, D. T. and French, B. M. (1991). *Lunar Sourcebook*. Cambridge, UK: Cambridge University Press.

Heim, A. (1882). Der Bergsturz von Elm. *Zeit. der Deutschen Gesellschaft*, **34**, 74–115.

Heiskanen, W. A. and Vening Meinesz, F. A. (1958). *The Earth and its Gravity Field*. New York, NY: McGraw-Hill.

Helfenstein, P. and Parmentier, E. M. (1983). Patterns of fracture and tidal stresses on Europa. *Icarus*, **53**, 415–30.

Hertz, H. (1884) Ueber das Gleichgewicht schwimmender elasticher platten. *Wiedemann's Annalen der Physik und Chemie*, **22**, 449–55.

Hiesinger, H., Head, J. W. and Neukum, G. (2007). Young lava flows on the eastern flank of Ascraeus Mons: Rheological properties derived from High Resolution Stereo Camera (HRSC) images and Mars Orbiter Laser Altimeter (MOLA) data. *J. Geophys. Res.*, **112**, E05011 10.1029/2006JE002717,2007.

Hill, R. (1950). *The Mathematical Theory of Plasticity*. Oxford.

Hobbs, B. E., Mühlhaus, H. B. and Ord, A. (1990). Instability, softening and localization of deformation. In *Deformation Mechanisms, Rheology and Tectonics, Special Publication 54*, ed. R. J. Knipe and E. H. Rutter. London: Geological Society, pp. 143–65.

Hon, K., Kauahikaua, J., Denlinger, R. and Mackay, K. (1994). Emplacement and inflation of pahoehoe sheet flows: Observations and measurements of active lava flows on Kilauea Volcano, Hawaii. *Geol. Soc. Amer. Bull.*, **106**, 351–70.

Hoppa, G., Tufts, B. R., Greenberg, R. and Geissler, P. (1999). Strike-slip faults on Europa: Global shear patterns driven by tidal stress. *Icarus*, **141**, 287–98.

Horstman, K. C. and Melosh, H. J. (1989). Drainage pits in cohesionless materials: Implications for the surface of Phobos. *J. Geophys. Res.*, **94**, 12 433–41.

Horton, R. E. (1945). Erosional development of streams and their drainage basins. *Geol. Soc. Amer. Bull.*, **56**, 275–370.

Hörz, F., Cintala, M. J., Rochelle, W. C. and Kirk, B. (1999). Collisionally processed rocks on Mars. *Science*, **285**, 2105–7.

Houghton, J. T. (2001). *The Physics of Atmospheres*, 3rd edn. Cambridge, UK: Cambridge.

Howard, A. D. (2009). How to make a meandering river. *Proc. Nat. Acad. Sci.*, **106**, 17245–6.

Howard, A. D., Kochel, R. C. and Holt, H. E. (1988). *Sapping Features of the Colorado Plateau*. Washington, DC: NASA.

Howard, K. A. and Larsen, B. R. (1972). Lineaments that are artifacts of lighting. In *Apollo 15 Preliminary Science Report*: NASA SP-289, pp. 25–58 to 25–62.

Hubbert, M. K. (1940). The theory of ground water motion. *J. Geology*, **48**, 785–944.

Hubbert, M. K. and Rubey, W. W. (1959). Role of fluid pressure in overthrust faulting I. Mechanics of fluid-filled porous solids and its application to overthrust faulting. *Geol. Soc. Amer. Bull.*, **70**, 115–66.

Hull, D. and Bacon, D. J. (2001). *Introduction to Dislocations*, 4th edn. Oxford, UK: Butterworth-Heinemann.

Hurford, T. A., Beyer, R. A., Schmidt, B., Preblich, B., Sarid, A. R. and Greenbert, R. (2005). Flexure of Europa's lithosphere due to ridge-loading. *Icarus*, **177**, 380–96.

Iess, L., Rappaport, N. J., Jacobson, R. A., Racioppa, P., Stevenson, D. J., Tortora, P., Armstrong, J. W. and Asmar, S. W. (2010). Gravity field, shape and moment of inertia of Titan. *Nature*, **327**, 1367–9.

Ito, E. and Takahashi, E. (1987). Melting of peridotite at uppermost lower-mantle conditions. *Nature*, **328**, 514–17.

Ivanov, B. A. and Melosh, H. J. (2003). Impacts do not create volcanic eruptions. *Geology*, **31**, 869–72.

Ivanov, B. A., Nemchinov, I. V., Svetsov, V. A., Provalov, A. A., Khazins, V. M. and Phillips, R. J. (1992). Impact cratering on Venus: Physical and mechanical models. *J. Geophys. Res.*, **97**, 16 167–81.

Jaeger, J. C., Cook, N. G. W. and Zimmerman, R. W. (2007). *Fundamentals of Rock Mechanics*. London: Blackwell.

Janes, D. M. and Melosh, H. J. (1990). Tectonics of planetary loading: A general model and results. *J. Geophys. Res.*, **95**, 21 345–55.

Jardetzky, W. S. (1958). *Theories of Figures of Celestial Bodies*. New York, NY: Interscience.

Jaumann, R., Kirk, R. L., Lorenz, R. D., Lopes, R., Stofan, E. R., Turtle, E. P., Keller, H. U., Wood, C. A., Sotin, C., Soderblom, L. A. and Tomasko, M. G. (2009). Geology and surface processes on Titan. In *Titan from Cassini-Huygens*, ed. R. H. Brown, J.-P. Lebreton and J. H. Waite. Dordrecht: Springer, pp. 75–140.

Jeffreys, H. (1950). *Earthquakes and Mountains*. London: Methuen.

  (1952). *The Earth*, 3rd edn. Cambridge, UK: Cambridge University Press.

  (1962). *The Earth*, 4th edn. Cambridge, UK: Cambridge University Press.

  (1976). *The Earth: Its Origin, History and Physical Constitution*, 6th edn. Cambridge University Press.

Johnson, A. M. (1970). *Physical Processes in Geology*. San Francisco, CA: Freeman, Cooper.

Johnson, A. M. and Fletcher, R. C. (1994). *Folding of Viscous Layers*. New York, NY: Columbia University Press.

Johnson, R. E., Carlson, R. W., Cooper, J. F., Paranicas, C., Moore, M. H. and Wong, M. C. (2004). Radiation effects on the surfaces of the Galilean satellites. In *Jupiter: The Planet, Satellites and Magnetosphere*, ed. F. Bagenal, T. E. Dowling and W. B. McKinnon. Cambridge, UK: Cambridge University Press, pp. 485–512.

Kamb, B., Raymond, C. F., Harrison, W. D., Engelhardt, H., Echelmeyer, K. A., Humphrey, N., Brugman, M. M. and Pfeffer, T. (1985). Glacier surge mechanism: 1982–1983 surge of Variegated Glacier, Alaska. *Science*, **227**, 469–79.

Karato, S.-I. (1998). A dislocation model of seismic wave attenuation and micro-creep in the Earth: Harold Jeffreys and the rheology of the solid Earth. *Pure Appl. Geophys.*, **153**, 239–56.

  (2008). *Deformation of Earth Materials: An Introduction to the Rheology of Solid Earth*. Cambridge, UK: Cambridge University Press.

Kargon, R. and Achinstein, P. (1987). *Kelvin's Baltimore Lectures and Modern Theoretical Physics*. Cambridge, MA: MIT, p. 547.

Kasting, J. F. and Catling, D. (2003). Evolution of a habitable planet. *Annu. Rev. Earth Planet. Sci.*, **41**, 429–63.

Kasting, J. F. and Toon, O. B. (1989). Climate evolution on the terrestrial planets. In *Origin and Evolution of Planetary and Satellite Atmospheres*, ed. S. K. Atreya, J. B. Pollack and M. W. Matthews. Tucson, AZ: University of Arizona Press, pp. 423–49.

Kaula, W. (1968). *An Introduction to Planetary Physics*. New York, NY: Wiley.

Keller, W. D. (1957). *The Principles of Chemical Weathering*. Columbia, MO: Lucas Brothers.

Kennedy, B. A. (2006). *Inventing the Earth*. London: Blackwell.

Kinsman, B. (1965). *Wind Waves.* New York, NY: Dover.

Kirby, S. H. and Kronenberg, A. K. (1987a). Rheology of the lithosphere: selected topics. *Rev. Geophys.*, **25**, 1219–44.

(1987b). Correction to "Rheology of the lithosphere: selected topics." *Rev. Geophys.*, **25**, 1680–1.

Kirchner, J. W. (1993). Statistical inevitability of Horton's laws and the apparent randomness of stream channel networks. *Geology*, **21**, 591–4.

Knopoff, L. (1964). Q. *Rev. Geophys.*, **2**, 625–60.

Kok, J. F. (2010). Difference in the wind speeds required for initiation versus continuation of sand transport on Mars: Implications for dunes and dust storms. *Phys. Rev. Lett.*, **104**, 074502.

Kok, J. F. and Renno, N. O. (2009a). A comprehensive numerical model of steady state saltation (COMSALT). *J. Geophys. Res.*, **114**, D17204.

(2009b). Electrification of wind-blown sand on Mars and its implications for atmospheric chemistry. *Geophys. Res. Lett.*, **36**, L05202.

Komar, P. D. (1997). *Beach Processes and Sedimentation*, 2nd edn. Englewood Cliffs, NJ: Prentice-Hall.

Kreslavsky, M. A. and Head, J. W. (2000). Kilometer-scale roughness of Mars. Results from MOLA data analysis. *J. Geophys. Res.*, **105**, 26 695–711.

Lachenbruch, A. H. (1962). *Mechanics of Thermal Contraction Cracks and Ice-Wedge Polygons in Permafrost.* New York, NY: Geological Society of America

Lambe, T. W. and Whitman, R. V. (1979). *Soil Mechanics, SI Version.* New York, NY: John Wiley & Sons.

Lambeck, K. (1988). *Geophysical Geodesy.* Oxford: Oxford University Press.

Lancaster, N. (1995). *Geomorphology of Desert Dunes.* London: Routledge.

Landau, L. D. and Lifshitz, E. M. (1970). *Theory of Elasticity*, 2nd edn. Oxford: Pergamon.

Langseth, M. G., Keihm, S. J. and Chute, J. L. (1973). Heat flow experiment. In *Apollo 17 Preliminary Science Report*, ed. L. B. J. Space Center. Washington DC: NASA, pp. 9–1 to 9–24.

Lawn, B. R. and Wilshaw, T. R. (1975). *Fracture of Brittle Solids*, 2nd edn. New York, NY: Cambridge University Press.

(1993). *Fracture of Brittle Solids.* Cambridge, UK: Cambridge University Press.

Lee, P. (1996). Dust levitation on asteroids. *Icarus*, **124**, 181–94.

Leeder, M. (1999). *Sedimentology and Sedimentary Basins.* Oxford: Blackwell.

Leith, A. C. and McKinnon, W. B. (1991). Terrace width variations in complex Mercurian craters, and the transient strength of the cratered Mercurian and lunar crust. *J. Geophys. Res.*, **96**, 20923–31.

Leith, C. K. (1923). *Structural Geology.* New York, NY: Henry Holt & Co.

Leopold, L. B., Wolman, M. G. and Miller, J. P. (1964). *Fluvial Processes in Geomorphology.* San Francisco, CA: Freeman.

Lewis, J. S. (1995). *Physics and Chemistry of the Solar System*, revised edn. San Diego, CA: Academic Press.

Li, S.-I., Miller, N., Lin, D. N. C. and Fortney, J. J. (2010). WASP-12b as a prolate, inflated and disrupting planet from tidal dissipation. *Nature*, **463**, 1054–6.

Lichtner, P. C., Steefel, C. I. and Oelkers, E. H. (1996). Reactive transport in porous media. In *Reviews in Mineralogy*, ed. P. H. Ribbe. Washington, DC: Mineralogical Society of America.

Lilljequist, R. and Henkel, H. (1996). The Uppland structure: A suspected 300-km-diameter impact crater in Sweden. *Meteoritics and Planet. Sci.*, **31**, A59.

Lockner, D. A. (1995). Rock failure. In *Rock Physics and Phase Relations*, ed. T. J. Ahrens. Washington, DC: Am. Geophys. Union, pp. 127–47.

Lowell, J. D. (1972). Spitsbergen tertiary orogenic belt and the Spitsbergen fracture zone. *Geol. Soc. Amer. Bull.*, **83**, 3091–102.

Lundborg, N. (1968). Strength of rocks and rock-like materials. *Int. J. Rock Mech. and Mining Sci.*, **5**, 427–54.

Mabbutt, J. A. (1977). *Desert Landforms*. Cambridge, MA: MIT Press.

Mackin, J. H. (1948). Concept of the graded river. *Bull. Geol. Soc. Amer.*, **59**, 463–512.

Mackwell, S. J., Zimmerman, M. E. and Kohlstedt, D. S. (1998). High-temperature deformation of dry diabase with application to tectonics on Venus. *J. Geophys. Res.*, **103**, 975–84.

Malin, M. C., Edgett, K. S., Posiolova, L. V., McColley, S. M. and Noe Dobrea, E. Z. (2006). Present-day impact cratering rate and contemporary gully activity on Mars. *Science*, **314**, 1573–7.

Mandl, G. (1988). *Mechanics of Tectonic Faulting*. Amsterdam: Elsevier.

Mark, K. (1987). *Meteorite Craters*. Tucson, AZ: University of Arizona Press.

Matson, D. L. and Brown, R. H. (1989). Solid state greenhouses and their implications for icy satellites. *Icarus*, **77**, 67–81.

Matsuyama, I. and Nimmo, F. (2008). Tectonic patterns on reoriented and despun planetary bodies. *Icarus*, **195**, 459–73.

Maxwell, D. E. (1977). A simple model of cratering, ejection, and the overturned flap. In *Impact and Explosion Cratering*, ed. D. J. Roddy, R. O. Pepin and R. B. Merrill. New York, NY: Pergamon, pp. 1003–8.

Maxwell, J. C. (1867). On the dynamical theory of gases. *Philosophical Transactions of the Royal Society of London*, **157**, 49–88.

McBirney, A. R. and Murase, T. (1984). Rheological properties of magmas. *Ann. Rev. Earth Planet. Sci.*, **12**, 337–57.

McGarr, A. and Gay, N. C. (1978). State of stress in the Earth's crust. *Ann. Rev. Earth Planet. Sci.*, **6**, 405–36.

McGee, W. J. (1908). Outlines of hydrology. *Geol. Soc. Amer. Bull.*, **19**, 193–200.

McGetchin, T. R. and Head, J. W. (1973). Lunar cinder cones. *Science*, **180**, 68–71.

McGill, G. E. (1971). Attitude of fractures bounding straight and arcuate lunar rilles. *Icarus*, **14**, 53–8.

  (1992). Origin of Martian polygons. *J. Geophys. Res.*, **97**, 2633–47.

McGovern, P. J., Smith, J. R., Morgan, J. K. and Bulmer, M. H. (2004). Olympus Mons aureole deposits: New evidence for a flank failure origin. *J. Geophys. Res.*, **109**, E08008.

McGreevy, J. P. (1981). Some perspectives on frost shattering. *Prog. Phys. Geog.*, **5**, 56–75.

McKee, E. D. (1979). *Desert Sand Seas*. Washington, DC: Prof. Paper 1052, US Geological Survey.

McKenzie, D. and Bickle, M. J. (1988). The volume and composition of melt generated by extension of the lithosphere. *J. Petrology*, **29**, 625–79.

McKinnon, W. B., Schenk, P. M. and Dombard, A. J. (2001). Chaos on Io: A model for formation of mountain blocks by crustal heating, melting and tilting. *Geology*, **29**, 103–6.

McPherson, G. J., Kita, N. T., Ushikubo, T., Bullock, E. S. and Davis, A. M. (2010). High-precision ^{26}Al/^{27}Al isochron microchronology of the earliest Solar System. In *41st LPSC*, Abstract #2356, Houston, TX.

McSween, H. Y., Richardson, S. M. and Uhle, M. E. (2003). *Geochemistry: Pathways and Processes*, 2nd edn. New York, NY: Columbia University Press.

Meier, M. F. (1960). *Mode of flow of Saskatchewan Glacier*. Alberta, Canada. Report.

Meinzer, O. E. (1942). *Hydrology.* New York, NY: Dover.

Mellon, M. T., Fergason, R. L. and Putzig, N. E. (2008). The thermal inertia of the surface of Mars. In *The Martian Surface: Composition, Mineralogy and Physical Properties*, ed. J. F. Bell. Cambridge: Cambridge, UK, pp. 399–427.

Melosh, H. J. (1977). Crater modification by gravity: A mechanical analysis of slumping. In *Impact and Explosion Cratering*, ed. D. J. Roddy, R. O. Pepin and R. B. Merrill. New York, NY: Pergamon Press, pp. 1245–60.

  (1977). Global tectonics of a despun planet. *Icarus*, **31**, 221–43.

  (1978). The tectonics of mascon loading. In *Proc. 9th Lunar Science Conf.*, pp. 3513–25.

  (1979). Acoustic fluidization: A new geologic process? *J. Geophys. Res.*, **84**, 7513–20.

  (1980a). Tectonic patterns on a reoriented planet: Mars. *Icarus*, **44**, 745–51.

  (1980b). Tectonic patterns on a tidally distorted planet. *Icarus*, **43**, 334–7.

  (1983). Acoustic Fluidization. *Am. Sci.*, **71**, 158–65.

  (1989). *Impact Cratering: A Geologic Process.* New York, NY: Oxford University Press.

  (1990a). Giant impacts and the thermal state of the early Earth. In *Origin of the Earth*, ed. J. H. Jones and H. E. Newsom. New York, NY: Oxford University Press, pp. 69–83.

  (1990b). Mechanical basis for low angle normal faulting in the Basin and Range province. *Nature*, **343**, 331–5.

Melosh, H. J. and Dzurisin, D. (1978). Mercurian tectonics: A consequence of tidal despinning? *Icarus*, **35**, 227–36.

Melosh, H. J. and Ivanov, B. A. (1999). Impact crater collapse. *Ann. Rev. Earth Planet. Sci.*, **27**, 385–415.

Melosh, H. J. and Janes, D. M. (1989). Ice volcanism on Ariel. *Science*, **245**, 195–6.

Melosh, H. J. and Williams, C. A. (1989). The mechanics of graben formation in crustal rocks: A finite element analysis. *J. Geophys. Res.*, **94**, 13 961–73.

Minton, D. A. (2008). The topographic limits of gravitationally bound, rotating sand piles. *Icarus*, **195**, 698–704.

Molnar, P. and Lyon-Caen, H. (1988). Some simple physical aspects of the support, structure, and evolution of mountain belts. In *Processes in Continental Lithosphere Deformation, GSA Special Paper 218*, ed. S. P. Clark, B. C. Burchfiel and J. Suppe. Geological Society of America, pp. 179–207.

Montgomery, D. R. and Dietrich, W. E. (1992). Channel initiation and the problem of landscape scale. *Science*, **255**, 826–30.

Moore, H. J., Arthur, D. W. G. and Schaber, G. G. (1978). Yield strengths of flows on the Earth, Mars, and Moon. In *Proc. Lunar Planet Sci. Conf. 9th*. Houston, TX: Lunar and Planetary Institute, pp. 3351–78.

Moore, J. G., Clague, D. A., Holcomb, R. T., Lipman, P. W., Normark, W. R. and Torresan, M. E. (1989). Prodigious submarine landslides on the Hawaiian Ridge. *J. Geophys. Res.*, **94**, 17 465–84.

Moore, J. M., Chapman, C. R., Bierhaus, E. B., Greeley, R., Chuang, F. C., Klemaszewski, J., Clark, R. N., Dalton, J. B., Hibbitts, C. A., Schenk, P. M., Spencer, J. R. and Wagner, R. (2004). Callisto. In *Jupiter: The Planet, Satellites and Magnetosphere*, ed. F. Bagenal, T. E. Dowling and W. B. McKinnon. Cambridge, UK: Cambridge University Press, pp. 397–426.

Moores, J. E., Pelletier, J. D. and Smith, P. H. (2008). Crack propagation by differential insolation on desert surface clasts. *Geomorphology*, **102**, 472–81.

Morisawa, M. E. (1959). Relation of the quantitative geomorphology to streamflow in representative watersheds of the Appalachian Plateau province. Office of Naval Research, Project NR 389–042, Tech. Rept. 20, p. 94.

(1968). *Streams.* New York, NY: McGraw-Hill.

Morrison, D. (1970). Thermophysics of the planet Mercury. *Space Science Reviews*, **11**, 271–307.

Mueller, K. and Golombek, M. (2004). Compressional structures on Mars. *Annu. Rev. Earth Planet. Sci.*, **32**, 435–464.

Muller, S. W. (1947). *Permafrost, or Permanently Frozen Ground and Related Engineering Problems.* Ann Arbor, MI: J. W. Edwards.

Murray, C. D. and Dermott, S. F. (1999). *Solar System Dynamics.* New York, NY: Cambridge.

NASA (1978). *Standard Techniques for Presentation and Analysis of Crater Size-Frequency Data.* Technical Memorandum 79730. Washington, DC: NASA.

Neumann, G. A., Zuber, M. T., Smith, D. E. and Lemonine, F. G. (1996). The lunar crust: Global structure and signature of major basins. *J. Geophys. Res.*, **101**, 16 841–3.

Neumann, G. A., Zuber, M. T., Wieczorek, M. A., McGovern, P. J., Lemonine, F. G. and Smith, D. E. (2004). Crustal structure of Mars from gravity and topography. *J. Geophys. Res.*, **109**, E08002, doi:10.1029/2004JE002262.

Nevin, C. M. (1942). *Principles of Structural Geology.* New York, NY: J. Wiley & Sons.

Newsom, H. E. (1995). Composition of the Solar System, planets, meteorites and major terrestrial reservoirs. In *Global Earth Physics*, ed. T. J. Ahrens. Washington, DC: American Geophysical Union, pp. 159–89.

Newton, I. (1966). *Principia* (2 vols. Cajori version). Berkeley: University Calif. Press.

Nichols, B. L. (1939). Viscosity of lava. *J. Geology*, **47**, 290–302.

Nimmo, F. and Pappalardo, R. T. (2006). Diapir-induced reorientation of Saturn's moon Enceladus. *Nature*, **441**, 614–16.

Nye, J. F. (1952). The mechanics of glacier flow. *J. Glaciology*, **2**, 82–93.

Oberbeck, V. R. and Quaide, W. L. (1968). Genetic implications of lunar regolith thickness variations. *Icarus*, **9**, 446–65.

Ollier, C. D. (1975). *Weathering.* London: Longman.

Paige, D. A., Wood, S. E. and Vasavada, A. R. (1992). The thermal stability of water ice at the poles of Mercury. *Science*, **258**, 643–6.

Paphitis, D. (2001). Sediment movement under unidirectional flows: An assessment of empirical threshold curves. *Coast. Eng.*, **43**, 227–45.

Paterson, M. S. and Wong, T.-F. (2005). *Experimental Rock Deformation – The Brittle Field*, 2nd edn. New York, NY: Springer.

Paterson, W. S. B. (1999). *The Physics of Glaciers*, 3rd edn. Oxford: Butterworth-Heinemann.

Pauling, L. (1988). *General Chemistry.* New York, NY: Dover.

Pearce, S. J. and Melosh, H. J. (1986). Terrace width variations in complex lunar craters. *Geophys. Res. Lett.*, **13**, 1419–22.

Pechmann, J. C. (1980). The origin of polygonal troughs on the Northern Plains of Mars. *Icarus*, **42**, 185–210.

Pelletier, J. D. (2008). *Quantitative Modeling of Earth Surface Processes.* Cambridge, UK: Cambridge University Press.

(2009). Controls on the height and spacing of eolian ripples and transverse dunes: A numerical modeling investigation. *Geomorphology*, **105**, 322–33.

Perry, R. S. and Kolb, V. M. (2003). Biological and organic constituents of desert varnish: Review and new hypotheses. In *Instruments, Methods and Missions for Astrobiology VII*, ed. R. B. Hoover, R. R. Paepe and A. Y. Rozanov. Bellingham, WA: SPIE, pp. 202–17.

Petford, N. (1996). Dykes or diapirs? *Trans. Roy. Soc. Edinburgh: Earth Sci.*, **87**, 105–14.

Phillips, R. J., Zuber, M. T., Solomon, S. C., Golombek, M. P., Jakowsky, B. M., Banerdt, W. B., Smith, D. E., Williams, R. M. E., Hynek, B. M., Ahronson, O. and Hauck, S. A. (2001). Ancient geodynamics and global-scale hydrology on Mars. *Science*, **291**, 2587–91.

Pierazzo, E. and Melosh, H. J. (2000). Understanding oblique impacts from experiments, observations and modeling. *Annu. Rev. Earth Planet. Sci.*, **28**, 141–67.

Pike, R. J. and Rozema, W. J. (1975). Spectral analysis of landforms. *Annals Assoc. Am. Geograph.*, **65**, 499–516.

Playfair, J. (1964). *Illustrations of the Huttonian Theory of the Earth.* New York, NY: Dover.

Poirier, J.-P. (1985). *Creep of Crystals.* Cambridge: Cambridge University Press.
  (1991). *Introduction to the Physics of the Earth's Interior.* Cambridge, UK: Cambridge University Press.

Pollard, D. D. and Fletcher, R. C. (2005). *Fundamentals of Structural Geology.* Cambridge, UK: Cambridge University Press.

Post, A. and LaChapelle, E. (2000). *Glacier Ice.* Seattle, WA: University of Washington Press.

Presley, M. A. and Christensen, P. R. (1997). Thermal conductivity measurements of particulate materials 1. A review. *J. Geophys. Res.*, **102**, 6535–49.

Prockter, L. M. and Pappalardo, R. T. (2000). Folds on Europa: Implications for crustal cycling and accommodation of extension. *Science*, **289**, 941–3.

Project, B. V. S. (1981). *Basaltic Volcanism on the Terrestrial Planets.* New York, NY: Pergamon.

Pumpelly, R. (1918). *My Reminiscences.* New York, NY: Henry Holt and Co.

Pye, K. (1987). *Aeolian Dust and Dust Deposits.* London: Academic Press.

Quitté, G., Latkoczy, C., Halliday, A. N., Schönbächler, M. and Günther, D. (2005). Iron-60 in the Eucrite parent body and the initial $^{60}Fe/^{56}Fe$ of the Solar System. *LPSC XXXVI*, Abstract 1827.

Ramberg, H. (1967). *Gravity, Deformation and the Earth's Crust.* London: Academic Press.

Ramsay, J. G. (1967). *Folding and Fracturing of Rocks.* New York, NY: McGraw-Hill.

Ranalli, G. (1995). *Rheology of the Earth*, 2nd edn. London: Chapman & Hall.

Rayleigh, Lord (1916). On convection currents in a horizontal layer of fluid, when the higher temperature is on the under side. *Philosophical Magazine*, **32**, 529–46.

Reiche, P. (1937). The Toreva block, a distinctive landslide type. *J. Geology*, **45**, 538–48.

Rice, J. R. (1976). The localization of plastic deformation. In *14th International Conference of Theoretical and Applied Mechanics*, pp. 207–20.

Rich, J. L. (1911). Gravel as a resistant rock. *J. Geol.*, **19**, 492–506.

Richardson, J. E., Melosh, H. J. and Greenberg, R. (2004). Impact-induced seismic activity on asteroid 433 Eros: A surface modification process. *Science*, **306**, 1526–9.

Richardson, J. E., Melosh, H. J., Greenberg, R. and O'Brien, D. P. (2005). The global effects of impact-induced seismic activity on fractured asteroid surface morphology. *Icarus*, **179**, 325–49.

Richardson, J. E., Melosh, H. J., Lisse, C. M. and Carcich, B. (2007). A ballistics analysis of the Deep Impact ejecta plume: Determining comet Tempel 1's gravity, mass and density. *Icarus*, **190**, 357–90.

Rinaldo, A., Rodriguez-Iturbe, I. and Rigon, R. (1998). Channel networks. *Ann. Rev. Earth Planet. Sci.*, **26**, 289–327.

Ritter, D. F. (1986). *Process Geomorphology.* Dubuque, IA: Wm. C. Brown.

Rodine, J. D. and Johnson, A. M. (1976). The ability of debris, heavily freighted with coarse clastic materials, to flow on gentle slopes. *Sedimentology*, **23**, 213–34.

Roering, J. J., Kirchner, J. W. and Dietrich, W. E. (1999). Evidence for non-linear, diffusive sediment transport on hillslopes and implications for landscape morphology. *Water Resources Res.*, **35**, 853–70.

Rosato, A., Strandburg, K. J., Prinz, F. and Swendsen, R. H. (1987). Why the Brazil nuts are on top: Size segregation of particulate matter by shaking. *Phys. Rev. Lett.*, **58**, 1038–40.

Rosenblatt, M., Hassig, P. J. and Orphal, D. L. (1982). Soil–air interactions during airblast-induced ground motions. Report. DNA–TR–81–103.

Rouse, H. (1978). *Elementary Mechanics of Fluids.* New York, NY: Dover.

Rubin, A. M. (1993). Dikes vs. diapirs in viscoelastic rock. *Earth Planet. Sci. Lett.*, **119**, 641–59.

  (1995). Propagation of magma-filled cracks. *Annu. Rev. Earth Planet. Sci.*, **23**, 287–336.

Ryan, M. P. (1987). Neutral Buoyancy and the mechanical evolution of magmatic systems. In *Magmatic Processes: Physicochemical Principles*, ed. B. O. Mysen. The Geochemical Society, pp. 259–87.

Sagan, C. (1970). *Planetary Exploration.* Eugene, OR: Oregon State System of Higher Education.

Sanchez-Levega, A. (2010). *An Introduction to Planetary Atmospheres.* Francis and Taylor.

Schaller, C. J. and Melosh, H. J. (1998). Venusian ejecta parabolas: Comparing theory with observation. *Icarus*, **131**, 123–37.

Schenk, P. M. and Bulmer, M. H. (1998). Origin of mountains on Io by thrust faulting and large-scale mass movements. *Science*, **279**, 1514–17.

Schenk, P. M., Wilson, R. R. and Davies, A. G. (2004). Shield volcano topography and the rheology of lava flows on Io. *Icarus*, **169**, 98–110.

Schmincke, H.-U. (2003). *Volcanism.* Berlin: Springer.

Scholz, C. H. (1990). *The Mechanics of Earthquakes and Faulting.* Cambridge, UK: Cambridge University Press

Schramm, D. N., Tera, F. and Wasserburg, G. J. (1970). The isotopic abundance of ^{26}Mg and limits on ^{26}Al in the early Solar System. *Earth Planet. Sci. Lett.*, **10**, 44–59.

Schubert, G. (1975). Subsolidus convection in the mantles of the terrestrial planets. *Annu. Rev. Earth Planet. Sci.*, **7**, 289–342.

Schubert, G., Turcotte, D. L. and Olson, P. (2001). *Mantle Convection in the Earth and Planets.* Cambridge University Press.

Schulte, P., Alegret, L., Arenillas, I., Arz, J. A., Barton, P. J., Bown, P. R., Bralower, T. J., Christeson, G. L., Claeys, P., Cockell, C. S., Collins, G. S., Deutsch, A., Goldin, T. J., Goto, K., Grajales-Nishimura, J. M., Grieve, R. A. F., Gulick, S. P. S., Johnson, K. R., Kiessling, W., Koeberl, C., Kring, D. A., MacLeod, K. G., Matsui, T., Melosh, J., Montanari, A., Morgan, J. V., Neal, C. R., Nichols, D. J., Norris, R. D., Pierazzo, E., Ravizza, G., Rebolledo-Vieyra, M., Reimold, W. U., Robin, E., Salge, T., Speijer, R. P., Sweet, A. R., Urrutia-Fucugauchi, J., Vajda, V., Whalen, M. T. and Willumsen, P. S. (2010). The Chicxulub asteroid impact and mass extinction at the Cretaceous-Paleogene Boundary. *Science*, **327**, 1214–18 10.1126/science.1177265.

Schumm, S. A. (1977). *The Fluvial System.* New York, NY: Wiley.

  (1985). Patterns of alluvial rivers. *Ann. Rev. Earth Planet. Sci.*, **13**, 5–27.

Sclater, J. G., Jaupart, C. and Galson, D. (1980). The heat flow through oceanic and continental crust and the heat loss of the earth. *Rev. Geophys.*, **18**, 269–312.

Scott, R. F. (1963). *Principles of Soil Mechanics*. Reading, MA: Addison-Wesley.
   (1967). Viscous flow of craters. *Icarus*, **7**, 139–148.
Segatz, M., Spohn, T., Ross, M. N. and Schubert, G. (1988). Tidal dissipation, surface heat flow, and figure of viscoelastic models of Io. *Icarus*, **75**, 187–206.
Seidelmann, P. K., Abalakin, V. K., Bursa, M., Davies, M. E., DeBergh, C., Lieske, J. H., Oberst, J., Simon, J. L., Standish, E. M., Stooke, P. and Thomas, P. (2002). Report of the IAU/IAG working group on cartographic coordinates and rotational elements of the planets and satellites: 2000. *Celest. Mech. Dyn. Astron.*, **82**, 83–110.
Selby, M. J. and Hodder, A. P. W. (1993). *Hillslope Materials and Processes*, 2nd edn. New York, NY: Oxford University Press.
Self, S., Keszthelyi, L. P. and Thordarson, T. (1998). The importance of pahoehoe. *Annu. Rev. Earth Planet. Sci.*, **26**, 81–110.
Sharp, R. P. (1960). *Glaciers*. Eugene, OR: Oregon State System of Higher Education.
   (1963). Wind ripples. *J. Geology*, **71**, 617–36.
   (1964). Wind-driven sand in the Coachella Valley, California. *Geol. Soc. Amer. Bull.*, **75**, 785–804.
   (1979). Interdune flats of the Algodones chain, Imperial Valley, California. *Geol. Soc. Amer. Bull.*, **90**, 908–16.
Sharpe, C. F. S. (1938). *Landslides and Related Phenomena*. Paterson, NJ: Pageant Books.
Shelton, J. S. (1966). *Geology Illustrated*. San Francisco, CA: Freeman.
Shewmon, P. G. (1963). *Diffusion in Solids*. New York, NY: McGraw-Hill.
Shoemaker, E. M. (1962). Interpretation of Lunar craters. In *Physics and Astronomy of the Moon*, ed. Z. Kopal. New York, NY: Academic Press, pp. 283–259.
Shoemaker, E. M., Hackman, R. J. and Eggleton, R. E. (1963). Interplanetary correlation of geologic time. *Adv. in Astronaut. Sci.*, **8**, 70–89.
Shoemaker, E. M., Batson, R. M., Hold, H. E., Morris, E. C., Rennilson, J. J. and Whitaker, E. A. (1967). Surveyor V: Television pictures. *Science*, **158**, 642–52.
Shoemaker, E. M., Batson, M., Holt, H. E., Morris, E. C., Rennilson, J. J. and Whitaker, E. A. (1969). Observations of the lunar regolith and the Earth from the television camera on Surveyor 7. *J. Geophys. Res.*, **74**, 6081–119.
Shreve, R. L. (1966a). Sherman landslide, Alaska. *Science*, **154**, 1639–43.
   (1966b). Statistical law of stream numbers. *J. Geology*, **74**, 17–37.
Shumskii, P. A. (1964). *Principles of Structural Glaciology*. New York, NY: Dover.
Siever, R. (1988). *Sand*. New York, NY: Freeman.
Sigurdsson, H. (1999). *Melting the Earth*. Oxford: Oxford.
Simpson, J. E. (1999). *Gravity Currents: In the Environment and the Laboratory*, 2nd edn. Cambridge, UK: Cambridge University Press.
Smith, R. E., Smetten, K. R. J., Broadbridge, P. and Woolhiser, D. A. (2002). *Infiltration Theory for Hydrologic Applications*. Washington, DC: American Geophysical Union.
Smrekar, S. E., Kiefer, W. S. and Stofan, E. R. (1997). Large volcanic rises on Venus. In *Venus II*, ed. S. W. Bougher, D. M. Hunten and R. J. Phillips. Tucson, AZ: University of Arizona Press, pp. 845–78.
Soderblom, L. A. (1970). A model for small-impact erosion applied to the lunar surface. *J. Geophy. Res.*, **75**, 2655–61.
Solomon, S. C. and Head, J. W. (1982). Mechanisms for lithospheric heat transport on Venus: Implications for tectonic style and volcanism. *J. Geophys. Res.*, **87**, 9236–46.

(1984). Venus banded terrain: Tectonic models for band formation and their relationship to lithospheric thermal structure. *J. Geophys. Res.*, **89**, 6885–97.

Sonder, L. J. and Jones, C. H. (1999). Western United States extension: How the West was widened. *Ann. Rev. Earth Planet. Sci.*, **27**, 417–62.

Sonnett, C. P., Kvale, E. P., Zakharian, A., Chan, M. A. and Demko, T. M. (1996). Late Proterozoic and Paleozoic tides, retreat of the Moon, and rotation of the Earth. *Science*, **273**, 100–4.

Sotin, C., Mielke, R., Choukroun, M., Neish, C., Barmatz, M., Castillo, J., Lunine, J. I. and Mitchell, K. (2009). Ice–hydrocarbon interactions under Titan-like conditions: Implications for the carbon cycle on Titan. *LPSC Conference*, **40**, Abs. # 2088.

Squyres, S. W., Janes, D. M., Baer, G., Binshadler, D. L., Schubert, G., Sharpton, V. L. and Stofan, E. R. (1992). The morphology and evolution of coronae on Venus. *J. Geophys. Res.*, **97**, 13 611–34.

Stacey, F. D. (1992). *Physics of the Earth*, 3rd edn. New York, NY: John Wiley & Sons.

Stephens, H. G. and Shoemaker, E. M. (1987). *In the Footsteps of John Wesley Powell*. Boulder, CO: Johnson Books.

Stern, S. A. and Tholen, D. J. (1997). *Pluto and Charon*. Tucson, AZ: University of Arizona Press, p. 728.

Stesky, R. M., Brace, W. F., Riley, D. K. and Bobin, P. Y. (1974). Friction in faulted rock at high temperature and pressure. *Tectonophysics*, **23**, 177–203.

Stevenson, D. J. (1981). Models of the Earth's core. *Science*, **214**, 611–19.

Stewart, S. A. and Allen, P. J. (2002). A 20-km-diameter multi-ringed impact structure in the North Sea. *Nature*, **418**, 520–52.

Strom, R. G. (1964). Analysis of lunar lineaments I: Tectonic maps of the Moon. *Comm. Lunar Planet. Lab*, **2**, 205–16.

(1984). Mercury. In *The Geology of the Terrestrial Planets*, ed. M. H. Carr. Washington, DC: US Govt. Printing Office, pp. 13–55.

Strom, R. G. and Sprague, A. L. (2003). *Exploring Mercury: The Iron Planet*. Dordrecht: Springer.

Strom, R. G., Malhotra, R., Ito, T., Yoshida, F. and Kring, D. A. (2005). The origin of planetary impactors in the inner Solar System. *Science*, **309**, 1847–50.

Sugden, D. E. and John, R. S. (1976). *Glaciers and Landscape*. New York, NY: Wiley.

Sullivan, R., Arvidson, R. E., Bell, J. F., Gellert, R., Golombeck, M. P., Greeley, R., Herkenhoff, K., Johnson, J., Thompson, S., Whelley, P. and Wray, J. (2008). Wind-driven particle mobility on Mars: Insights from Mars Exploration Rover observations at "El Dorado" and surroundings at Gusev Craters. *J. Geophys. Res.*, **113**, E06So7.

Sunshine, J. M., Farnham, T. L., Feaga, L., Groussin, O., Merlin, F., Milliken, R. E. and A'Hearn, M. A. (2009). Temporal and spatial variability of lunar hydration as observed by the Deep Impact spacecraft. *Science*, **265**, 565–8.

Suppe, J. (1983). Geometry and kinematics of fault-bend folding. *Am. J. Science*, **283**, 684–721.

(1985). *Principles of Structural Geology*. Englewood Cliffs, NJ: Prentice-Hall, Inc.

Taber, S. (1929). Frost heaving. *J. Geology*, **37**, 428–61.

Tennekes, H. and Lumley, J. L. (1972). *A First Course in Turbulence*. Cambridge, MA: MIT Press.

Terzaghi, K. (1943). *Theoretical Soil Mechanics*. New York, NY: John Wiley & Sons.

Theis, C. V. (1935). The relation between the lowering of the piezometric surface and the rate and duration of discharge of a well using ground water storage. *Trans. Am. Geophys. Union*, **16**, 519–24.

Thomas, P. (1988). Radii, shapes and topography of the satellites of Uranus from limb coordinates. *Icarus*, **73**, 427–41.

Thomas, P., Veverka, J. and Duxbury, T. (1978). Origin of the grooves on Phobos. *Nature*, **273**, 282–4.

Thomas, P., Burns, J. A., Rossier, L., Simonelli, D., Veverka, J., Chapman, C. R., Klaasen, K. P., V., J. T. and Belton, M. J. S. (1998). The small inner satellites of Jupiter. *Icarus*, **135**, 360–71.

Thomas, P. C. (1989). The shapes of small satellites. *Icarus*, **77**, 248–74.

Thomas, P. C., Burns, J. A., Helfenstein, P., Squyres, S. W., Veverka, J., Porco, C., Turtle, E. P., McEwen, A. S., Denk, T., Giese, B., Roatsch, T., Johnson, T. V. and Jacobson, R. A. (2007). Shapes of the Saturnian icy satellites and their significance. *Icarus*, **190**, 573–84.

Tilling and Dvorak, J. (1993). Anatomy of a basaltic volcano. *Nature*, **363**, 125–33.

Tjia, H. D. (1970). Lunar wrinkle ridges indicative of strike-slip faulting. *Geol. Soc. Amer. Bull.*, **81**, 3095–100.

Todhunter, I. (1962). *A History of the Mathematical Theories of Attraction and the Figure of the Earth.* New York, NY: Dover.

Tozer, D. C. (1965). Heat transfer and convection currents. *Philos. Trans. R. Soc. London, Ser. A*, **258**, 252–71.

  (1970). Factors determining the temperature evolution of thermally convecting earth models. *Phys. Earth Planet. Interiors*, **2**, 393–8.

Turcotte, D. L. (1979). Flexure. *Adv. in Geophys.*, **21**, 51–86.

Turcotte, D. L. and Schubert, G. (2002). *Geodynamics: Applications of Continuum Physics to Geological Problems*, 2nd edn. Cambridge, UK: Cambridge University Press.

Turcotte, D. L., Willemann, R. J., Haxby, W. F. and Norberry, J. (1981). Role of membrane stresses in the support of planetary topography. *J. Geophys. Res.*, **86**, 3951–9.

Turtle, E. P. and Melosh, H. J. (1997). Stress and flexural modeling of the Martian lithospheric response to Alba Patera. *Icarus*, **126**, 197–211.

Urai, J. L., Spiers, C. J., Zwart, H. J. and Lister, G. S. (1986). Weakening of rock salt by water during long-term creep. *Nature*, **324**, 554–7.

Urey, H. C. (1952). *The Planets: Their Origin and Development.* New Haven, CT: Yale University Press.

USGS (2003). *Color-Coded Contour Map of Mars*, M 25M RKT.

Van Schums, W. R. (1995). Natural radioactivity of the crust and mantle. In *Global Earth Physics*, ed. T. J. Ahrens. Washington, DC: American Geophysical Union, pp. 283–91.

Vening-Meinesz, F. A. (1947). Shear patterns of the Earth's crust. *Trans. Am. Geophys. Union*, **28**, 1–61.

Vervack, R. and Melosh, H. J. (1992). Wind interaction with falling ejecta: Origin of the parabolic features on Venus. *Geophys. Res. Lett.*, **19**, 525–8.

Wallace, M. H. and Melosh, H. J. (1994). Buckling of a pervasively faulted lithosphere. *PAGOPH*, **142**, 239–61.

Washburn, A. L. (1980). *Geocryology: A Study of Periglacial Processes and Environments.* New York, NY: Wiley.

Watson, E. B. (1982). Melt infiltration and magma evolution. *Geology*, **10**, 236–40.

Watters, T. R. and Schultz, R. A. (2010). *Planetary Tectonics.* Cambridge, UK: Cambridge University Press.

Watts, A. B. (2001). *Isostasy and Flexure of the Lithosphere.* Cambridge University Press.

Watts, A. B., Cochran, J. R. and Selzer, G. (1975). Gravity anomalies and flexure of the lithosphere: A three-dimensional study of the Great Meteor Seamount, NE Atlantic. *J. Geophys. Res.*, **80**, 1391–8.

Weast, R. C. (1972). *Chemical Rubber Handbook*. CRC Press.

Weertman, J. (1971). Theory of water-filled crevasses in glaciers applied to vertical magma transport beneath oceanic ridges. *J. Geophys. Res.*, **76**, 1171–83.

(1979). Height of mountains on Venus and the creep properties of rock. *Phys. Earth Planet. Int.*, **19**, 197–207.

Weertman, J. and Weertman, J. R. (1992). *Elementary Dislocation Theory*. New York, NY: Oxford University Press.

Weissel, J. K., Anderson, R. N. and Geller, C. A. (1980). Deformation of the Indo-Australian plate. *Nature*, **287**, 284–91.

Wenkert, D. D. (1979). The flow of salt glaciers. *Geophys. Res. Lett.*, **6**, 523–6.

Werner, B. T. (1995). Eolian dunes: Computer simulations and attractor interpretation. *Geology*, **23**, 1107–10.

Whalley, W. B. and Azizi, F. (2003). Rock glaciers and protalus landforms: Analogous forms and ice sources on Earth and Mars. *J. Geophys. Res.*, **108**, 10.1029/2002JE001864.

White, W. B. (1988). *Geomorphology and Hydrology of Karst Terrains*. New York, NY: Oxford.

Wieczorek, M. A. (2007). Gravity and topography of the terrestrial planets. In *Treatise on Geophysics*, ed. G. Schubert and T. Spohn. Amsterdam: Elsevier, pp. 165–206.

Wilhelms, D. E. (1987). *The Geologic History of the Moon*. Report *1348*.

(1989). *To a Rocky Moon*. Tucson, AZ: University of Arizona Press.

Willgoose, G. (2005). Mathematical modeling of whole landscape evolution. *Ann. Rev. Earth Planet. Sci.*, **33**, 443–59.

Williams, H. and McBirney, A. R. (1979). *Volcanology*. San Francisco, CA: Freeman.

Williams, P. J. and Smith, M. W. (1991). *The Frozen Earth*. Cambridge, UK: Cambridge University Press.

Willis, B. (1893). The mechanics of Appalachian structure. In *Thirteenth Annual Report of the United States Geologic Survey, 1891–1892*, ed. J. W. Powell. Washington, DC: Govt. Printing Office, pp. 211–82.

Winkler, E. M. and Singer, P. C. (1972). Crystallization pressure of salts in stone and concrete. *Geol. Soc. Amer. Bull.*, **83**, 3509–14.

Wise, D. U., Golombek, M. P. and McGill, G. E. (1979). Tharsis province of Mars: Geologic sequence, geometry and a deformation mechanism. *Icarus*, **38**, 456–72.

Wolfe, E. W. and Bailey, N. G. (1972). Lineaments of the Apennine Front – Apollo 15 landing site. In *Proceedings of the Third Lunar Science Conference*. Houston, TX: MIT Press, pp. 15–25.

Wood, J. A. (1997). Rock weathering on the surface of Venus. In *Venus II*, ed. S. W. Bougher, D. M. Hunten and R. J. Phillips. Tucson, AZ: University of Arizona Press, pp. 637–64.

Woodward, N. B., Boyer, S. E. and Suppe, J. (1989). *Balanced Geological Cross-Sections*. Washington, DC: American Geophysical Union.

Woronow, A. (1977). Crater saturation and equilibrium: A Monte-Carlo simulation. *J. Geophys. Res.*, **82**, 2447–56.

Wyllie, P. J. (1971). *The Dynamic Earth*. New York, NY: Wiley.

Yoder, C. F. (1995). Astrometric and geodetic properties of Earth and the Solar System. In *Global Earth Physics*, ed. T. J. Ahrens. Washington, DC: American Geophysical Union, pp. 1–31.

Zahnle, K., Schenk, P. M., Sobieszczyk, S., Dones, L. and Levison, H. F. (2001). Differential cratering of synchronously rotating satellites by ecliptic comets. *Icarus*, **153**, 111–29.

Zoback, M. D., Zoback, M. L., Mount, V. S., Suppe, J., Eaton, J. P., Healy, J. H., Oppenheimer, D., Reasenberg, P., Jones, L., Raleigh, C. B., Wong, I. G., Scotti, O. and Wentworth, W. (1987). New evidence for the state of stress on the San Andreas fault system. *Science*, **238**, 1105–11.

Index

of craters, 84, 85
on Venus, 86
Trans-Neptunian Objects, 3
Triton, 4
    geysers, 205
    seasons, 290
Tsilokovsky crater, 255
tuff
    airfall, 202, 203
    welded, 202, 203
tuff ring, *see* volcano: maar
Tunguska explosion, 254
turbidity currents, 339, 340, 428
Tycho crater, 14, 226, 285, 336
Tyndall, John, 434

Ubehebe crater, 204
undrained depressions, 395, 428
Uranus, 3
Urey, Harold, 299
Utopia basin, 17

Valhalla basin, 20, 227, 245
Valles Marineris, 17, 342
Varenius, 169
Variegated Glacier, 446
velocity
    escape, 6
    molecular, 6
    volcanic ejecta, 200–02
Venera 13, 313
Vening-Meinesz, Felix, 108, 152
Venus, 3, 15–16, 170
    "snow" on mountaintops, 299
    carbon dioxide, 299
    channels, 407, 415, 419
    cinder cones, 205
    coronas, 138, 157
        arachnid, 157
        nova, 157
    crater clusters, 254
    dark halos, 364
    fold belt, 129
    geoid anomalies, 99
    hypsometric curve, 39, 42
    isostatic compensation, 99
    pancake domes, 215, 216
    parabolas, 378
    shield volcano, 205, 206
    splotches, 255
    stone pavement, 313
    topographic power spectrum, 46
    topographic reference system, 38
    topography, 16
    transverse dunes, 374
    weathering
        oxidation of iron, 298

pyrite, 299
sulfur, 299
wind streaks, 377
yardangs, 377
vesicles, 199
Vesta, 5, 169, 180
    crust formation, 188
    magma percolation time, 188
Viking, 286, 313, 368, 417, 458
viscosity, 56, 57–58, 351
    basaltic melts, 183
    Bingham, 186
    dependence on crystal content, 185
    derived from topographic relaxation, 83
    effective, 75
    gas, 352
        pressure independence, 352
    granitic melts, 183
    Jeffreys equation, 213
    magma, 183
    Newtonian, 57
    of paint, 185
    of the Earth, 83, 84
    of the Moon, 84
    self-regulation, 125–26
    silicate melts, 184
volcanic ash, 198, 199, 206, 378
volcanic bombs, 199
    velocity, 201
volcanic eruptions
    deep-sea, 197
    explosive, 195–204
    fire-fountain, 197
    gas-rich, 196
    Impact-induced, *see* impact craters: volcanism
    mechanics, 194–204
    quiescent, 197
    Venusian, 197
    volume rate of, 170
volcano
    accordant summits, 189, 190
    caldera, 204, 207
    cinder cone, 205
        lunar, 205
    composite, 206–07
    ejection height, ballistic, 199
    ejection height, buoyant, 199
    ejection speed, 200–02
    eruption mechanics, 200, 203
    explosive eruption, 198
    fire-fountain, 205
    Hawaiian, 193
    laccolith, *see* laccolith
    maar, 204–05
    magma chamber, 193, 194, 205, 207
    plume, 203

volcano (*cont.*)
    buoyant, 203
    pyroclastic flow, 199, 339
    sector collapse, 344
    shield, 205–06
    spatter rampart, 206
volume
    expansion coefficient, 120
    expansion of water, 302
    sphere, 27
    triaxial ellipsoid, 27
Vostok station, 450
Vredefort crater, 267

Warrego Vallis, 400
Wasatch Mountains, 420
water
    expansion on freezing, 302
waves, 420–26
    amplitude, 422
    breaking, 423
    energy, 421
    energy flux, 421
    generation, 420
    gravity, 420
    group velocity, 421
    particle orbits, 422
    phase velocity, 420, 421
    refraction, 423–26
weathering, 293–311
    cavernous, 308–09
        honeycomb, 309
        tafoni, 308
    chemical, 295–300
        carbon dioxide, 299
        carbonation, 299
        clathrates, 297
        hydration, 297
        hydrolysis, 298
        iron oxidation, 295
        oxidation, 295, 299
        reactions, 296
        serpentinization, 298
        solution, 297

    spheroidal, 302
    sulfur, 299
    Urey reaction, 299
desert varnish, 309–10
duricrust, 308–09
physical, 300–06
    chemical potential, 305
    cracks, 300
    crystallization pressure, 303, 304
    frost shattering, 303, 304
    grus, 305
    insolation weathering, 306
    pressure of reaction, 305
    rock disintegration, 306
    salt weathering, 302, 305
    sheeting joints, 301
    spheroidal, 301
    stress corrosion cracking, 301
    supersaturation, 304
    thermal stress, 306
radiation, 287, 288
space weathering, 279, 285–87
sublimation, 307, 308
Weertman, Johannes, 188
Wessex Cliff, 277
Westerly granite, 140
Wiesbach, J., 403
Wild 2, 307, 333
Willis, Bailey, 128, 129
wind tunnel, 348
wollastonite, 299
wrinkle ridges, *see* tectonics: faults: mare ridges
wüstite, 298

Xanadu, 21, 413
Xanthe Terra, 405

Yakima Fold Belt, 152
Yellowstone, 195
Yukon Territory, 462
Yuma, AZ, 313
Yuty crater, 249

Zagros Mountains, 445